U. Meyer-Baese

Digital Signal Processing with Field Programmable Gate Arrays

Dear MyCopy Customer,

This Springer book is a monochrome print version of the eBook to which your library gives you access via SpringerLink. It is available to you at a subsidized price since your library subscribes to at least one Springer eBook subject collection.

Please note that MyCopy books are only offered to library patrons with access to at least one Springer eBook subject collection. MyCopy books are strictly for individual use only.

You may cite this book by referencing the bibliographic data and/or the DOI (Digital Object Identifier) found in the front matter. This book is an exact but monochrome copy of the print version of the eBook on SpringerLink.

Springer-Verlag Berlin Heidelberg GmbH

U. Meyer-Baese

Digital Signal Processing with Field Programmable Gate Arrays

With 213 Figures and 57 Tables

Springer

Dr. Uwe Meyer-Baese, Ph. D.
Florida State University
Dept. of Electrical and Computer Engineering
FAMU-FSU College Engineering
2525 Pottsdamer Street
Tallahassee, Fl 32310-6046, USA
e-mail: Uwe.Meyer-Baese@ieee.org

Library of Congress Cataloging-in-Publication-Data applied for

Die Deutsche Bibliothek - CIP-Einheitsaufnahme

Meyer-Bäse, Uwe:
Digital signal processing with field programmable gate arrays : with 57 tables / U. Meyer-Baese.
Dt. Ausg. u.d.T.: Meyer-Bäse, Uwe: Schnelle digitale Signalverarbeitung

DOI 10.1007/978-3-662-04613-5

http://www.springer.de

Originally published by Springer-Verlag Berlin Heidelberg New York in 2001
MyCopy version of the original edition 2001

Typesetting: Data delivered by author
Cover design: Design & Production, Heidelberg
Printed on acid free paper SPIN: 10789648 62/3020/M - 5 4 3 2 1 0
www.springer.com/mycopy

To my Parents
and
my wife Anke

Preface

Field-programmable gate arrays (FPGAs) are on the verge of revolutionizing digital signal processing in the manner that programmable digital signal processors (PDSPs) did nearly two decades ago. Many front-end digital signal processing (DSP) algorithms, such as FFTs, FIR or IIR filters, to name just a few, previously built with ASICs or PDSPs, are now most often replaced by FPGAs. Modern FPGA families provide DSP arithmetic support with fast-carry chains (Xilinx XC4000, Altera FLEX) which are used to implement multiply-accumulates (MACs) at high speed, with low overhead and low costs [1]. Previous FPGA families have most often targeted TTL "glue logic" and did not have the high gate count needed for DSP functions. The efficient implementation of these front-end algorithms is the main goal of this book.

At the beginning of the twenty-first century we find that the two programmable logic device (PLD) market leaders (Altera and Xilinx) both report revenues greater than US$1 billion. FPGAs have enjoyed steady growth of more than 20% in the last decade, outperforming ASIC and PDSPs by 10%. This comes from the fact that FPGAs have many features common with ASICs, such as reduction in size, weight, and power dissipation, higher throughput, better security against unauthorized copies, reduced device and inventory cost, and reduced board test costs, and claim advantages over ASICs, such as a reduction in development time (rapid prototyping), in-circuit reprogrammability, lower NRE costs, resulting in more economical designs for solutions requiring less than 1,000 units. Compared with PDSPs, FPGA design typically exploits parallelism, e.g., implementing multiple multiply-accumulate calls efficiency, e.g., zero product-terms are removed, and pipelining, i.e., each LE has a register, therefore pipelining requires no additional resources.

Another trend in the DSP hardware design world is the migration from graphical design entries to hardware description language (HDL). Although many DSP algorithms can be described with "signal flow graphs," it has been found that "code re-use" is much higher with HDL-based entries than with graphical design entries. There is a high demand for HDL design engineers and we already find undergraduate classes about logic design with HDLs [2]. Unfortunately *two* HDL languages are popular today. The US west coast and

Asia area prefer Verilog while US east coast and Europe more frequently use VHDL. For DSP with FPGAs both languages seem to be well suited, although are some VHDL examples are a little easier to read because of the supported signed arithmetic and multiply/divide operations in the IEEE VHDL 1076-1987 and 1076-1993 standards. The gap is expetected to disappear after approval of the new Verilog IEEE standard 1364-1999, as it also includes signed arithmetic. Other constraints may include personal preferences, EDA library and tool availability, data types, readability, capability, and language extensions using PLIs, as well as commercial, business, and marketing issues, to name just a few [3]. Tool providers acknowledge today that both languages have to be supported and this book covers examples in both design languages.

We are now also in the fortunate situation that "baseline" HDL compilers are available from different sources at essentially no cost for educational use. We take advantage of this fact in this book. It includes a CD-ROM with Altera's newest `MaxPlusII` software, which provides a complete set of design tools, from a content-sensitive editor, compiler, and simulator, to a bitstream generator. All examples presented are written in VHDL and Verilog and should be easily adapted to other propriety design-entry systems. Xilinx's "Foundation Series," ModelTech's ModelSim compiler, and Synopsys FC2 or FPGA Compiler should work without any changes in the VHDL or Verilog code.

The book is structured as follows. The first chapter starts with a snapshot of today's FPGA technology, and the devices and tools used to design state-of-the-art DSP systems. It also includes a detailed case study of a frequency synthesizer, including compilation steps, simulation, performance evaluation, power estimation, and floor planning. This case study is the basis for more than 30 other design examples in following chapters. The second chapter focuses on the computer arithmetic aspects, which include possible number representations for DSP FPGA algorithms as well as implementation of basic building blocks, such as adders, multipliers, or sum-of-product computations. At the end of the chapter we discuss two very useful computer arithmetic concepts for FPGAs: distributed arithmetic (DA) and the CORDIC algorithm. Chapters 3 and 4 deal with theory and implementation of FIR and IIR filters. We will review how to determine filter coefficients and discuss possible implementations optimized for size or speed. Chapter 5 covers many concepts used in multirate digital signal processing systems, such as decimation, interpolation, and filter banks. At the end of Chapter 5 we discuss the various possibilities for implementing wavelet processors with two-channel filter banks. In Chapter 6, implementation of the most important DFT and FFT algorithms is discussed. These include Rader, chirp-z, and Goertzel DFT algorithms, as well as Cooley-Tuckey, Good-Thomas, and Winograd FFT algorithms. In Chapter 7 we discuss more specialized algorithms, which seem to have great potential for improved FPGA implementation when compared with PDSPs.

These algorithms include number theoretic transforms, algorithms for cryptography and error-correction, and communication system implementations. The appendix includes an overview of the VHDL and Verilog languages, the examples in Verilog HDL, and a short introduction to the utility programs included on the CD-ROM.

Acknowledgements. This book is based on an FPGA communications system design class I taught four years at the Darmstadt University of Technology; my previous (German) books [4, 5]; and more than 60 Masters thesis projects I have supervised in the last 10 years at Darmstadt University of Technology and the University of Florida at Gainesville. I wish to thank all my colleagues who helped me with critical discussions in the lab and at conferences. Special thanks to: M. Acheroy, D. Achilles, F. Bock, C. Burrus, D. Chester, D. Childers, J. Conway, R. Crochiere, K. Damm, B. Delguette, A. Dempster, C. Dick, P. Duhamel, A. Drolshagen, W. Endres, H. Eveking, S. Foo, R. Games, A. Garcia, O. Ghitza, B. Harvey, W. Hilberg, W. Jenkins, A. Laine, R. Laur, J. Mangen, J. Massey, J. McClellan, F. Ohl, S. Orr, R. Perry, J. Ramirez, H. Scheich, H. Scheid, M. Schroeder, D. Schulz, F. Simons, M. Soderstrand, S. Stearns, P. Vaidyanathan, M. Vetterli, H. Walter, and J. Wietzke.

I would like to thank my students for the innumerable hours they have spent implementing my FPGA design ideas. Special thanks to: D. Abdolrahimi, E. Allmann, B. Annamaier, R. Bach, C. Brandt, M. Brauner, R. Bug, J. Burros, M. Burschel, H. Diehl, V. Dierkes, A. Dietrich, S. Dworak, W. Fieber, J. Guyot, T. Hattermann, T. Häuser, H. Hausmann, D. Herold, T. Heute, J. Hill, A. Hundt, R. Huthmann, T. Irmler, M. Katzenberger, S. Kenne, S. Kerkmann, V. Kleipa, M. Koch, T. Krüger, H. Leitel, J. Maier, A. Noll, T. Podzimek, W. Praefcke, R. Resch, M. Rösch, C. Scheerer, R. Schimpf, B. Schlanske, J. Schleichert, H. Schmitt, P. Schreiner, T. Schubert, D. Schulz, A. Schuppert, O. Six, O. Spiess, O. Tamm, W. Trautmann, S. Ullrich, R. Watzel, H. Wech, S. Wolf, T. Wolf, and F. Zahn.

For the English revision I wish to thank my wife Dr. Anke Meyer-Bäse, Dr. J. Harris, Dr. Fred Taylor from the University of Florida at Gainesville, and Paul DeGroot from Springer.

For financial support I would like to thank the DAAD, DFG, the European Space Agency, and the Max Kade Foundation.

If you find any errata or have any suggestions to improve this book, please contact me at `Uwe.Meyer-Baese@ieee.org` or through my publisher.

Tallahassee, May 2001 *Uwe Meyer-Bäse*

Contents

1. Introduction

This chapter gives an overview of the algorithms and technology we will discuss in the book. It starts with an introduction to digital signal processing and we will then discuss FPGA technology in particular. Finally, the Altera EPF10K20 and a larger design example, including chip synthesis, timing analysis, floorplan, and power consumption, will be studied.

1.1 Overview of Digital Signal Processing (DSP)

Signal processing has been used to transform or manipulate analog or digital signals for a long time. One of the most frequent applications is obviously the *filtering* of a signal, which will be discussed in Chapters 3 and 4. Digital signal processing has found many applications, ranging from data communications, speech, audio or biomedical signal processing, to instrumentation and robotics. Table 1.1 gives an overview of applications where DSP technology is used [6].

Digital signal processing (DSP) has become a mature technology and has replaced traditional analog signal processing systems in many applications. DSP systems enjoy several advantages, such as insensitivity to change in temperature, aging, or component tolerance. Historically, analog chip design yielded smaller die sizes, but now, with the noise associated with modern submicron designs, digital designs can often be much denser integrated than analog designs. This yields compact, low-power, and low-cost digital designs.

Two events have accelerated DSP development. One is the disclosure by Cooley and Tuckey (1965) of an efficient algorithm to compute the discrete Fourier Transform (DFT). This class of algorithms will be discussed in detail in Chapter 6. The other milestone was the introduction of the programmable digital signal processor (PDSP) in the late 1970s. This could compute a (fixed-point) "multiply-and-accumulate" in only one clock cycle, which was an essential improvement compared with the "Von Neuman" microprocessor-based systems in those days. Modern PDSPs may include more sophisticated functions, such as floating-point multipliers, barrelshifters, memory banks, or zero-overhead interfaces to A/D and D/A converters. EDN publishes every year a detailed overview of available PDSPs [7]. Fig. 1.1 shows a typical application used to implement an analog system by means of a PDSP. We

Table 1.1. Digital signal processing applications.

Area	DSP algorithm
General purpose	Filtering and convolution, adaptive filtering, detection and correlation, spectral estimation and Fourier transform
Speech processing	Coding and decoding, encryption and decryption, speech recognition and synthesis, speaker identification, echo cancellation, cochlea-implant signal processing
Audio processing	hi-fi encoding and decoding, noise cancellation, audio equalization, ambient acoustics emulation, audio mixing and editing, sound synthesis
Image processing	Compression and decompression, rotation, image transmission and decompositioning, image recognition, image enhancement, retina-implant signal processing
Information systems	Voice mail, facsimile (fax), modems, cellular telephones, modulators/demodulators, line equalizers, data encryption and decryption, digital communications and LANs, spread-spectrum technology, wireless LANs, radio and television, biomedical signal processing
Control	Servo control, disk control, printer control, engine control, guidance and navigation, vibration control, power system monitors, robots
Instrumentation	Beamforming, waveform generation, transient analysis, steady-state analysis, scientific instrumentation, radar and sonar

will return in Section 1.2.1 and Chapter 2 (p. 62) to PDSPs after we have studied FPGA architectures.

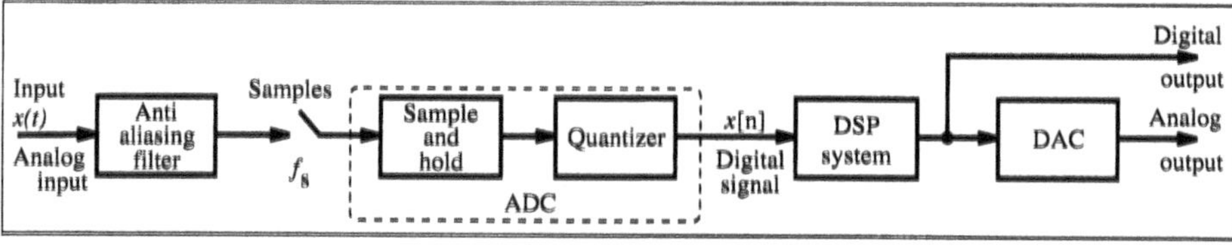

Fig. 1.1. A typical DSP application.

1.2 FPGA Technology

VLSI circuits can be classified as shown in Fig. 1.2. FPGAs are a member of a class of devices called field-programmable logic (FPL). FPLs are defined as programmable devices containing repeated fields of small logic blocks and elements[2]. It can be argued that an FPGA is an ASIC technology since FPGAs are application-specific ICs. It is, however, generally assumed that the design of a classic ASIC required additional semiconductor processing steps beyond those required for an FPL. The additional steps provide higher-order ASICs with their performance advantage, but also with high non-reoccurring engineering (NRE) costs. Gate arrays, on the other hand, typically consist of a "sea of NAND gates" whose functions are customer-provided in a "wire list." The wire list is used during the fabrication process to achieve the distinct definition of the final metal layer. The designer of a *programmable* gate array solution, however, has full control over the actual design implementation without the need (and delay) for any physical IC fabrication facility.

1.2.1 Classification by Granularity

Logic block size correlates to the *granularity* of a device which, in turn, relates to the effort required to complete the wiring between the blocks (routing channels). In general three different granularity classes can be found:

- Fine granularity (Pilkington or "sea of gates" architecture)
- Medium granularity (FPGA)
- Large granularity (CPLD)

Fine-Granularity Devices

Fine-grain devices were first licensed by Plessey and later by Motorola, being supplied by Pilkington Semiconductor. The basic logic cell consisted of a single NAND gate and a latch (see Fig. 1.3). Because it is possible to realize any binary logic function using NAND gates (see Exercise 1.1, p. 25), NAND gates are called *universal* functions. This technique is still in use for gate array designs along with approved logic synthesis tools, such as ESPRESSO. Wiring between gate-array NAND gates is accomplished by using additional metal layer(s). For programmable architectures, this becomes a bottleneck because the routing resources used are very high compared with the implemented logic functions. In addition, a high number of NAND gates is needed to build a simple DSP object. A fast 4-bit adder, for example, uses about 130 NAND gates. This makes fine-granularity technologies unattractive in implementing most DSP algorithms.

[2] Called configurable logic block (CLB) by Xilinx, logic cell (LC) or logic elements (LE) by Altera.

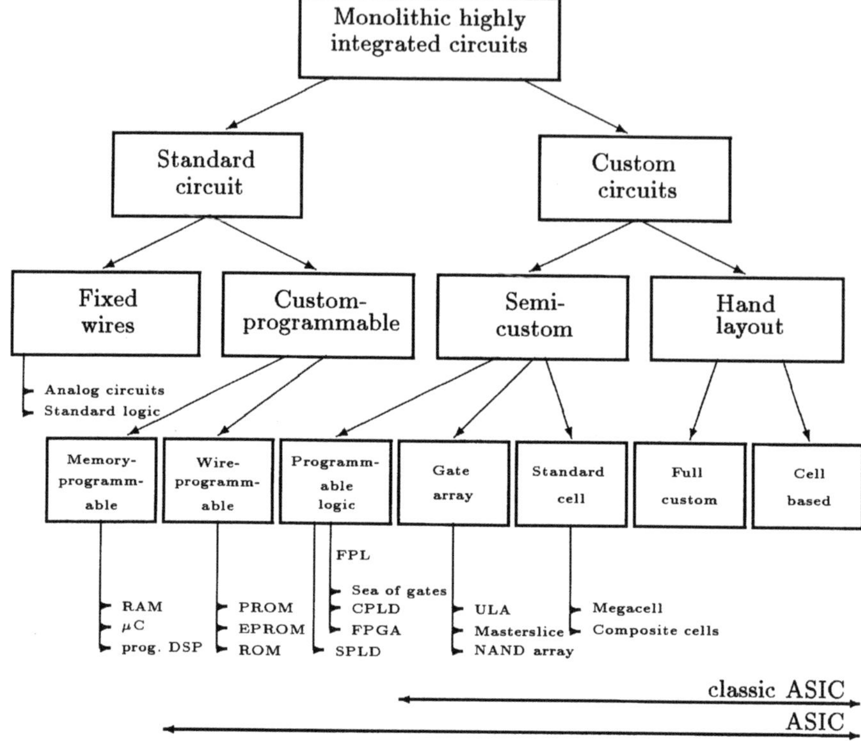

Fig. 1.2. Classification of VLSI circuits (©1995 VDI Press [4]).

Medium-Granularity Devices

The most common FPGA architecture is shown in Fig. 1.4(a). A concrete example of a contemporary medium-grain FPGA device is shown in Fig. 1.5. The elementary logic blocks are typically small tables (e.g., Xilinx XC2k-4k with 4- to 5-bit input tables, 1- or 2-bit output), or are realized with dedicated multiplexer (MPX) logic such as that used in Actel ACT-2 devices [9]. Routing channel choices range from short to long. A programmable I/O block with flip-flops is attached to the physical boundary of the device.

Large-Granularity Devices

Large granularity devices, such as complex programmable logic devices (CPLD), are characterized in Fig. 1.4(b). They are defined by combining so-called simple programmable logic devices (SPLDs), like the classic GAL16V8 shown in Fig. 1.6. This SPLD consists of a programmable logic array (PLA) implemented as an AND/OR array and a universal I/O logic block. The SPLDs used in CPLDs typically have 8 to 10 inputs, 3 to 4 outputs, and

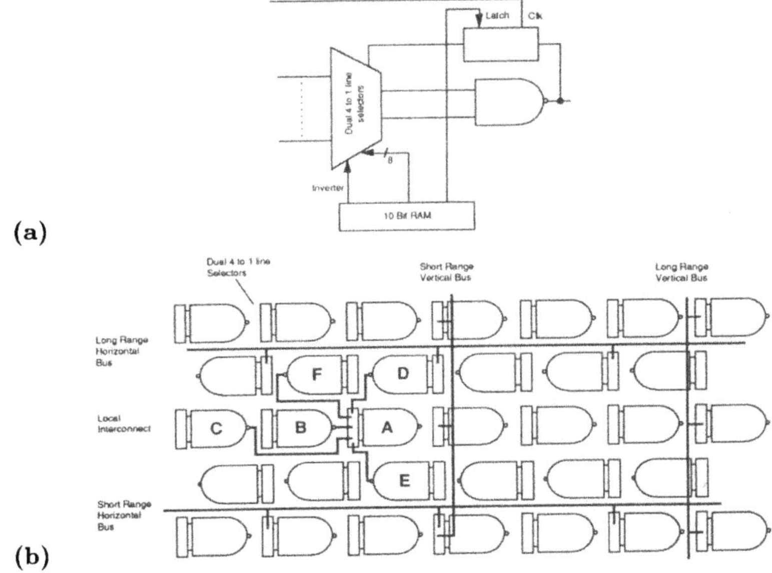

Fig. 1.3. Plessey ERA60100 architecture with 10K NAND logic blocks [8]. **(a)** Elementary logic block. **(b)** Routing architecture (©1990 Plessey).

support around 20 product terms. Between these SPLD blocks wide busses (called programmable interconnect arrays (PIAs) by Altera) with short delays are available. By combining the bus and the fixed SPLD timing, it is possible to provide predictable and short pin-to-pin delays with CPLDs.

1.2.2 Classification by Technology

FPLs are available in virtually all memory technologies: SRAM, EPROM, E^2PROM, and antifuse [10]. The specific technology defines whether the device is *reprogrammable* or *one-time programmable*. Most SRAM devices can be programmed by a single-bit stream that reduces the wiring requirements, but also increases programming time (typically in the ms range). SRAM devices, the dominate technology for FPGAs, are based on static CMOS memory technology, and are re- and in-system programmable. They require, however, an external "boot" device for configuration. Electrically programmable read-only memory (EPROM) devices are usually used in a one-time CMOS programmable mode because of the need to use ultraviolet light for erasure. CMOS electrically erasable programmable read-only memory (E^2PROM) can be used as re- and in-system programmable. EPROM and E^2PROM have the advantage of a short setup time. Because the programming information is not "downloaded" to the device, it is better protected against unauthorized use. A recent innovation, based on an EPROM technology, is called "flash"

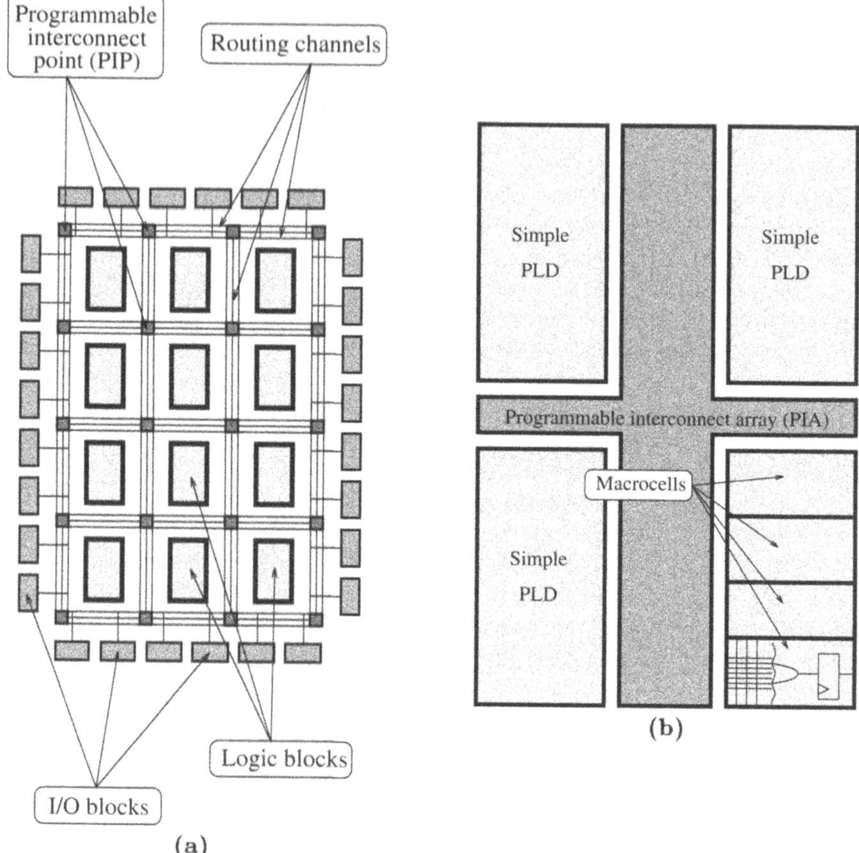

Fig. 1.4. **(a)** FPGA and **(b)** CPLD architecture (©1995 VDI Press [4]).

memory. These devices are usually viewed as "pagewise" in-system reprogrammable systems with physically smaller cells, equivalent to an E^{2}PROM device. Finally, the important advantages and disadvantages of different device technologies are summarized in Table 1.2.

1.2.3 Benchmark for FPLs

Providing objective benchmarks for FPL devices is a nontrivial task. Performance is often predicated on the experience and skills of the designer, along with design tool features. To establish valid benchmarks, the Programmable Electronic Performance Cooperative (PREP) was founded by Xilinx [11], Altera [12], and Actel [13], and has since expanded to more than 10 members. PREP has developed nine different benchmarks for FPLs that are summarized in Table 1.3. The central idea underlining the benchmarks is that each

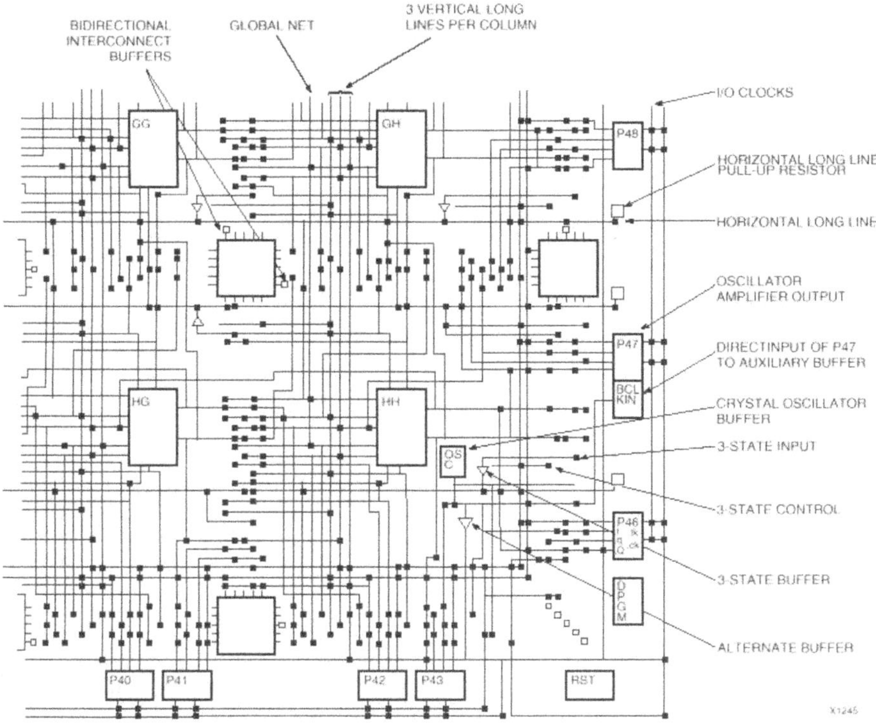

Fig. 1.5. Example of a medium-grain device (©1993 Xilinx).

vendor uses its own devices and software tools to implement the basic blocks

Table 1.2. FPL technology.

Technology	SRAM	EPROM	E^2PROM	Antifuse	Flash
Repro-grammable	✓	✓	✓	–	✓
In-system programmable	✓	–	✓	–	✓
Volatile	✓	–	–	–	–
Copy protected	–	✓	✓	✓	✓
Examples	Xilinx XC4K	Altera MAX5K	AMD MACH	Actel ACT	Xilinx XC9500
	Altera Flex	Xilinx XC7K	Altera MAX 9K		Cypress Ultra 37K

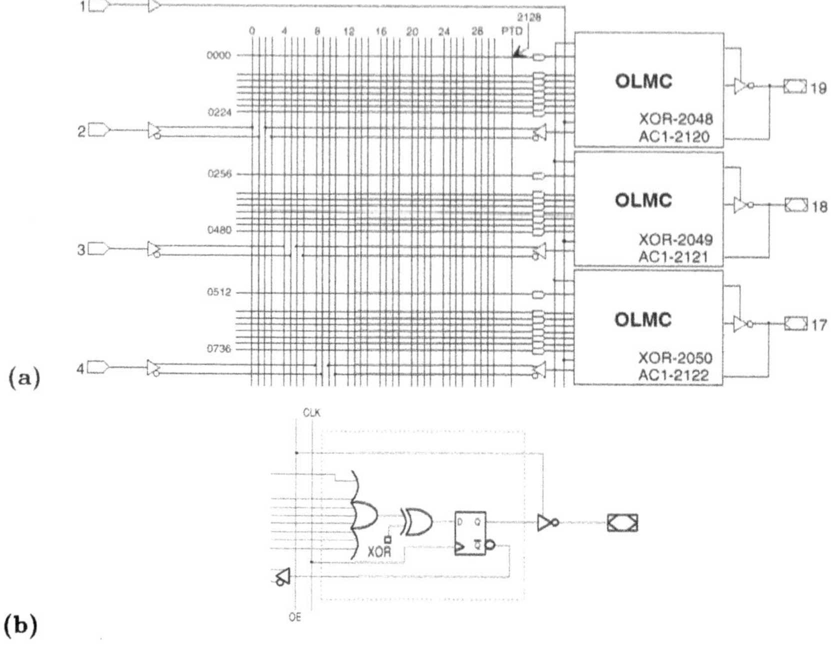

Fig. 1.6. The GAL16V8. **(a)** First three of eight macrocells. **(b)** The Output Logic macrocell (OLMC) (©1997 Lattice).

as many times as possible in the specified device, while attempting to maximize speed. The number of instantiations of the same logic block within one device is called the *repetition rate* and is the basis for all benchmarks. For DSP comparisons, benchmarks five and six of Table 1.3 are relevant. In Fig. 1.7, repetition rates are reported over frequency, for typical Actel (A_k), Altera (o_k), and Xilinx (x_k) devices. It can be concluded that modern FPGA families provide the best DSP complexity and maximum speed. This is attributed to the fact that modern devices provide fast carry logic (see Sect. 1.4.1, p. 15) with delays (less than 0.5 ns per bit) that allow fast adders with large bit width, without the need for expensive "carry look-ahead" decoders. Although PREP benchmarks are useful to compare equivalent gate counts and maximum speeds, for a concrete applications additional attributes are also important. They include:

- On-chip RAM or ROM
- Pin-to-pin delay
- Internal tri-state bus
- Readback- or boundary-scan decoder
- Programmable slew rate or voltage of I/O
- Power dissipation

Table 1.3. The PREP benchmarks for FPLs.

Number	Benchmark Name	Description
1	Data path	Eight 4-to-1 multiplixers drive a parallel-load 8-bit shift register
2	Timer counter	Two 8-bit values are clocked through 8-bit value registers and compared
3	Small state machine	An 8-state machine with 8 inputs and 8 outputs
4	Large state machine	A 16-state machine with 40 transitions, 8 inputs, and 8 outputs
5	Arithmetic	A 4-by-4 unsigned multiplier an 9-bit accumulator
6	Accumulator	A 16-bit accumulator
7	Up counter	A 16-bit loadable binary up counter
8	Down counter	A 16-bit loadable binary down counter
9	Memory map	The map decodes address spaces ranging in size from 4Kbyte to 1Kbyte

Fig. 1.8 summarizes the power dissipation of some typical FPL devices. It can be seen that CPLDs (Altera) usually have higher "standby" power consumption. For higher frequency applications, FPGAs (Xilinx and Actel) can be expected to have a higher power dissipation. A detailed power analysis example can be found in Sect. 1.4.2, p. 20.

1.3 DSP Technology Requirements

The PLD market share, by vendor, is presented in Fig. 1.9. PLDs, since their introduction in early eighties, have enjoyed steady growth of 20% per annum, outperforming ASIC growth by more than 10%. The reason seems to be related to the fact that FPLs can offer many of the advantages of ASICs such as:

- Reduction in size, weight, and power dissipation
- Higher throughput
- Better security again unauthorized copies
- Reduced device and inventory cost
- Reduced board test costs

without many of the disadvantages of ASICs such as:

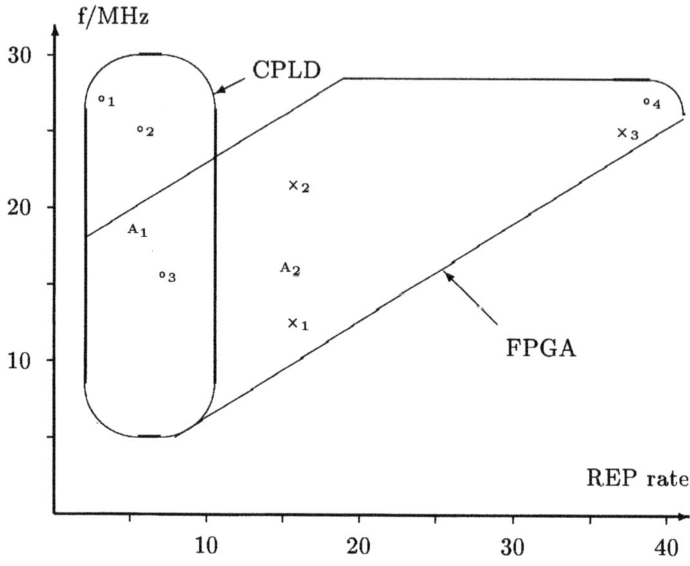

Fig. 1.7. Benchmarks for FPLs (©1995 VDI Press [4]).

- A reduction in development time (rapid prototyping) by three to four
- In-circuit reprogrammability
- Lower NRE costs resulting in more economical designs for solutions requiring less than 1,000 units

CBIC ASICs are used in high-end, high-volume applications (more than 1,000 copies). Compared to FPLs, CBIC ASICs typically have about ten times more gates for the same die size. An attempt to solve the second problem is the so-called hard wired FPGA, where a gate array is used to implement a verified FPGA design.

1.3.1 FPGA and Programmable Signal Processors

General purpose programmable digital signal processors (PDSPs) [14, 15, 6] have enjoyed tremendous success for the last two decades. They are based on a reduced instruction set computer (RISC) paradigm with an architecture consisting of at least one fast array multiplier (e.g., 16×16-bit to 24×24-bit fixed-point, or 32-bit floating-point), with an extended wordwidth accumulator. The PDSP advantage comes from the fact that most signal processing algorithms are multiply and accumulate (MAC) intensive. By using a multistage pipeline architecture, PDSPs can achieve MAC rates limited only by the speed of the array multiplier. It can be argued that an FPGA can also be used to implement MAC cells [16], but cost issues will most often give PDSPs an advantage, if the PDSP meets the desired MAC rate. On the other side we

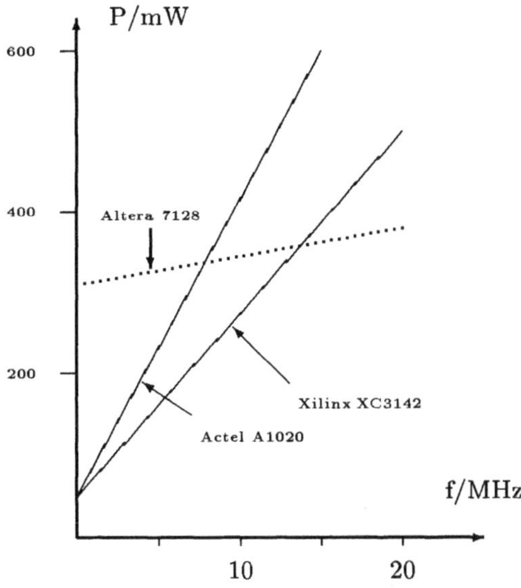

Fig. 1.8. Power dissipation for FPLs (©1995 VDI Press [4]).

now find many high-bandwidth signal-processing applications such as wireless, multimedia, or satellite transmission, and FPGA technology can provide more bandwidth through multiple MAC cells on one chip. In addition there are several algorithms such as CORDIC, NTT or error-correction algorithms, which will be discussed later, where FPL technology has been proven to be more efficient than a PDSP. It is assumed [17] that in the future PDSPs will dominate applications that require complicated algorithms (i.e., several `if-then-else` constructs), while FPGAs will dominate more front end (sensor) applications like FIR filters, CORDIC algorithms, or FFTs, which will be the focus of this book.

1.4 Design Implementation

The levels of detail commonly used in VLSI designs range from geometrical layout of full custom ASICs to system design using so-called set top boxes. Table 1.4 gives a survey. Layout and circuit-level activities are absent from FPGA design efforts because their physical structure is programmable but fixed. The best utilization of a device is typically achieved at the gate level using register transfer design languages. Time-to-market requirements, combined with the rapidly increasing complexity of FPGAs, are forcing a methodology shift towards the use of "Intellectual Property" (IP) macro cells or "mega core cells." Macro cells provide the designer with a collection of pre-

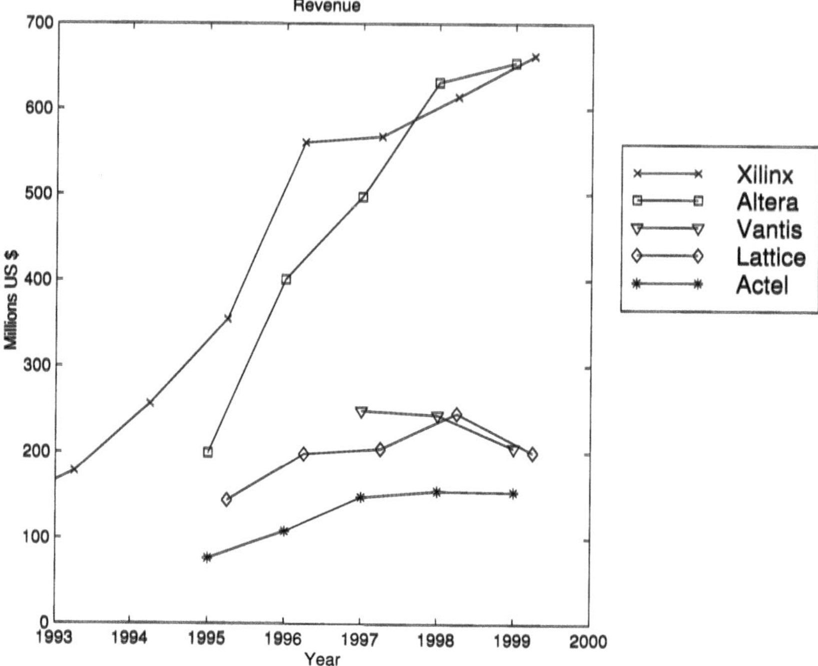

Fig. 1.9. Revenues of the top five vendors in the PLD/FPGA/CPLD market.

Table 1.4. VLSI design levels.

Object	Objectives	Example
System	Performance specifications	Computer, disk unit, radar
Chip	Algorithm	μp, RAM, ROM, UART, parallel port
Register	Data flow	Register, ALU, COUNTER, MUX
Gate	Boolean equations	AND, OR, XOR, FF
Circuit	Differential equations	Transistor, R, L, C
Layout	None	Geometrical shapes

defined functions, such as microprocessors or UARTs. The designer, therefore, need only to specify selected features and attributes (i.e., accuracy), and a "synthesizer" will generate a hardware description code or schematic for the resulting solution.

A key point in FPGA technology is, therefore, powerful design tools to

- Shorten the design cycle
- Provide good utilization of the device
- Provide synthesizer options, i.e., choose between optimization speed versus size of the design

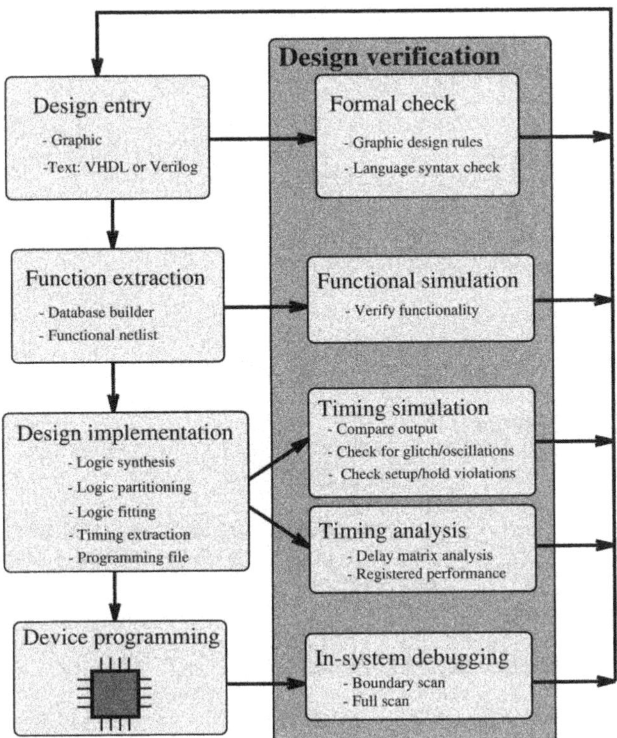

Fig. 1.10. CAD design circle.

A CAE tool taxonomy, as it applies to FPGA design flow, is presented in Fig. 1.10. In general, the decision whether to work within a graphical or a text design environment is a matter of personal taste and prior experience. A graphical presentation of a DSP solution can emphasize the highly regular dataflow associated with many DSP algorithms. The textual environment, however, is often preferred with regards to algorithm control design and allows a wider range of design styles as demonstrated in the following design example. Specifically, for Altera's MaxPlusII, it seemed that with text design more special attributes and more precise behavior can be assigned in the designs.

Example 1.1: Comparison of VHDL Design Styles

The following design example illustrates three design strategies in a VHDL context. Specifically, the techniques explored are:

- Component instantiation (stuctural style, i.e., graphical netlist design)
- Data flow
- Sequential design using **PROCESS** templates (i.e., behavioral style)

The VHDL design file **example.vhd**[4] follows (comments start with --):

```
PACKAGE eight_bit_int IS    -- User defined type
  SUBTYPE BYTE IS INTEGER RANGE -128 TO 127;
END eight_bit_int;

LIBRARY work;
USE work.eight_bit_int.ALL;

LIBRARY lpm;                      -- Using predefined packages
USE lpm.lpm_components.ALL;

LIBRARY ieee;
USE ieee.std_logic_1164.ALL;
USE ieee.std_logic_arith.ALL;

ENTITY example IS                               ------> Interface
  GENERIC (WIDTH : INTEGER := 8);   -- Bit width
  PORT (clk  :  IN STD_LOGIC;
        a, b :  IN BYTE;
        op1  :  IN STD_LOGIC_VECTOR(WIDTH-1 DOWNTO 0);
        sum  :  OUT STD_LOGIC_VECTOR(WIDTH-1 DOWNTO 0);
        d    :  OUT BYTE);
END example;

ARCHITECTURE flex OF example IS

  SIGNAL  c, s       :  BYTE;        -- Auxiliary variables
  SIGNAL  op2, op3   :  STD_LOGIC_VECTOR(WIDTH-1 DOWNTO 0);

BEGIN

  -- Conversion int -> logic vector
  op2 <= CONV_STD_LOGIC_VECTOR(b,8);

  add1: lpm_add_sub         ------> Component instantiation
    GENERIC MAP (LPM_WIDTH => WIDTH,
                 LPM_REPRESENTATION => "SIGNED",
                 LPM_DIRECTION => "ADD")
    PORT MAP (dataa => op1,
              datab => op2,
              result => op3);
  reg1: lpm_ff
    GENERIC MAP (LPM_WIDTH => WIDTH )
    PORT MAP (data => op3,
              q => sum,
              clock => clk);

  c <= a  + b ;                  ------> Data flow style

  p1: PROCESS                    ------> Behavioral style
  BEGIN
```

[4] The equivalent Verilog code **example.v** for this example can be found in Appendix A on page 343.

```
    WAIT UNTIL clk = '1';
    s <= c + s;             ----> Signal assignment statement
  END PROCESS p1;
  d <= s;

END flex;
```

1.1

After a successful functional (only) simulation of the design (for the MaxPlusII compiler mode select the option **Processing→Functional SNF Extractor**) we can proceed and start with the design implementation as reported in Fig. 1.10. To do this with the MaxPlusII compiler, we choose **Processing→Timing SNF Extractor**, and we will then notice that the compiler window now has three more entries, namely **Logic Synthesizer**, **Fitter**, and **Timing SNF Extractor**. After starting the compiler we can then conduct a simulation with timing, check for glitches, or measure the **Registered Performance** of the design, to name just a few options. After all these steps are successful, and if a hardware board (like the Altera University board) is available, we proceed with programming the device and may perform additional hardware tests using the "read back" methods, as reported in Fig. 1.10.

1.4.1 FPGA Structure

At the beginning of the twenty-first century two FPGA device families seemed to have the most attractive features for implementing DSP algorithms, due to the fact that these families provide fast carry logic, which allows implementations of 32-bit (nonpipelined) adders at speeds exceeding 50 MHz [1, 18, 19].

These two families are the Xilinx XC4000 family and the Altera FLEX 10K devices, which are Altera's 8K devices with additional 2Kbit RAM blocks called embedded array blocks (EAB). The Xilinx devices have the wide range of routing levels typical in FPGAs, while the Altera devices were based on the architecture with wide busses used in Altera's CPLDs. But the basic blocks of the FLEX 10K are no longer large PLAs as in CPLD. Instead the devices now have medium granularity, i.e., small look-up tables (LUTs), as is typical for FPGAs.

The basic logic elements of the Xilinx XC4000 family are called configurable logic blocks (CLB) and have two separate 4-input 1-output LUTs, fast carry, one additional 3-input 1-output LUT to combine the two separate LUTs, and two flip-flops, as shown in Fig. 1.11. The Xilinx device has five levels of routing, ranging from CLB to CLB, to long lines spanning the entire chip. Each CLB can be used as 16×2- or 32×1-bit RAM or ROM. Tables 1.5 shows some members of the Xilinx XC4000 family.

Table 1.5. The Xilinx XC4000 family.

Device	Total CLBs	Flip-flop bits	Max. RAM Kbits	Max. I/O
XC4003	100	360	3.2	80
XC4005	196	616	6.3	112
XC4010	400	1120	12.8	160
XC4025	1024	2560	32	256
XC4085	3136	7168	100	448
XC40150	5184	11520	165	448
XC40250	8464	18400	270	448

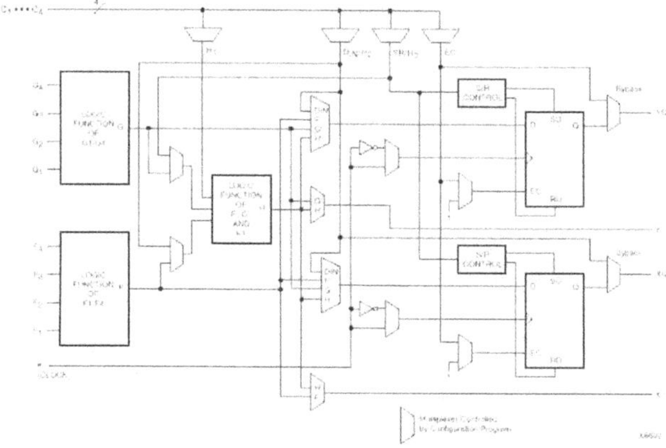

Fig. 1.11. XC4000 logic cell (©1993 Xilinx).

The basic block of the Altera FLEX 10K device achieves a medium granularity using small LUTs. The 10K device is an Altera 8K device with added 2Kbit RAM blocks, called embedded array blocks (EAB). The basic logic element in Altera FLEX 10K devices is called a logic element (LE)[5] and consists of a flip-flop, a 4-input 1-output LUT, or 3-input 1-output and a fast carry, logic or AND/OR product term expanders as shown in Fig. 1.12. Eight LCs are combined in a logic array block (LAB). Each row contains an embedded array block (EAB; i.e., a 2Kbit RAM or ROM) which can be configured as 256×8, 512×4, 1024×2, or 2048×1 memory devices. These EABs and LABs are connected through wide high-speed busses with 100 to 300 lines per column as shown in Fig. 1.13. Table 1.6 shows some members of the Altera FLEX 10K family.

If we compare the two routing strategies from Altera and Xilinx we find that both approaches have value: the Xilinx approach with more local and

[5] Sometimes also called logic cells (LCs) in a "design report file." See `example.rpt`.

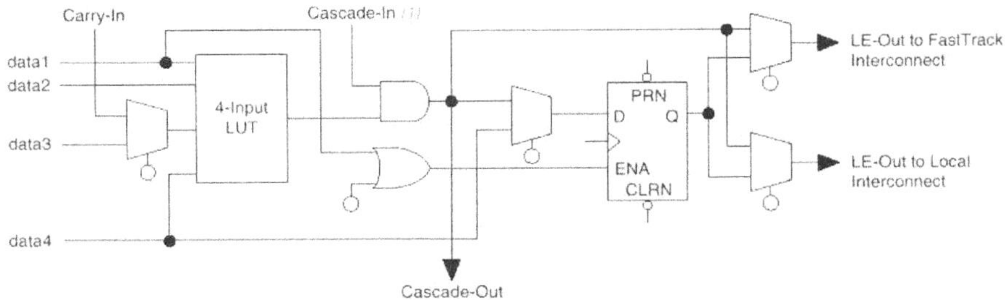

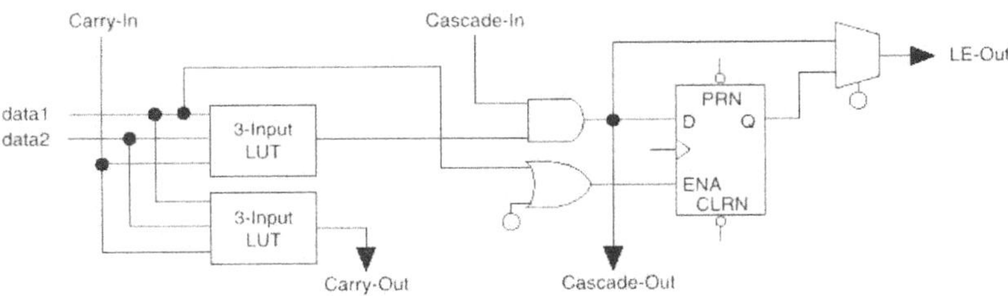

Fig. 1.12. FLEX logic cell (©1996 Altera).

less global routing resources is synergistic to DSP use because most digital signal processing algorithms process the data locally. The Altera approach, with wide busses, also has value, because typically not only are single bit processed in "bit slice" operations, but normally wide data vectors with 16 to 32 bits must be moved to the next DSP block.

Table 1.6. The FLEX 10K family.

Device	Total logic elements	Flip-flop bits	EAB Blocks	Max. RAM Kbits	Max. I/O
EPF10K10	576	720	3	6	134
EPF10K20	1152	1344	6	12	189
EPF10K30	1728	1968	6	12	246
EPF10K40	2304	2576	8	16	189
EPF10K50	2880	3184	10	20	310
EPF10K70	3744	4096	9	18	358
EPF10K100	4992	5392	12	24	406
EPF10K130	6656	7120	16	32	470
EPF10K250	12160	12624	20	40	470

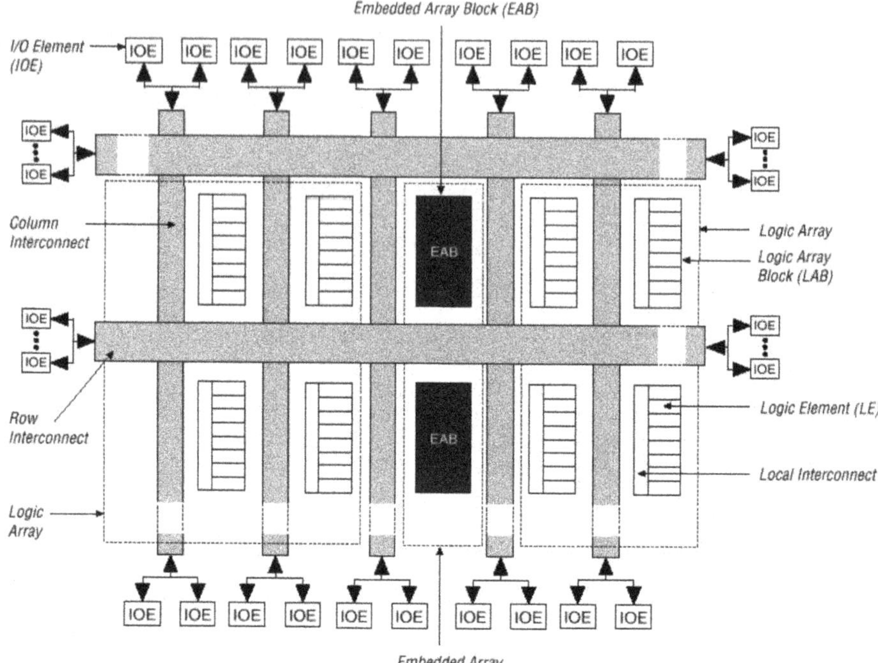

Fig. 1.13. Overall bus structure in FLEX 10K devices (©1996 Altera).

1.4.2 The Altera EPF10K20RC240-4

The Altera EPF10K20RC240-4 device, which is part of the demo board provided through Altera's University Program, is used throughout this book. The device nomenclature is interpreted as follows:

```
EPF10K20RC240-4
   |  |   |   |-> 4 ns device
   |  |   |-----> Package and pin number
   |  |---------> Equivalent gate count
   |------------> Device family
```

Specific design examples will, wherever possible, target Altera devices using Altera supplied software. The enclosed `MaxPlusII` software is a fully integrated system with VHDL and Verilog editor, synthesizer, simulator, and bitstream generator. Because all examples are available in VHDL and Verilog, any other simulator may also be used. For instance, the device-independent Synopsys FC2 or ModelTech compiler has successfully been used to compile the examples using the synthesizable code for `lpm` functions on the CD-ROM provided by EDIF.

Logic Resources

The EPF10K20 is a member of Altera 10K family and has a gate complexity equivaltent to about 20,000 two-input NAND gates. The maximum number of full adders which can be implemented may, however, be a more useful metric for DSP applications. From Table 1.6, it can be seen that the EPF10K20 device has 1,152 basic logic elements (LEs). This is also the maximum number of implementable full adders. Each LE can be used as a four-input LUT, or in the "arithmetic" mode, as a three-input LUT with an additional fast carry as shown in Fig. 1.12. Eight LEs are always combined into a logic array block (LAB). The number of LABs is therefore 1152/8=144. These 144 LABs are arranged in six rows and 24 columns. The device also includes one 2Kbit memory block (called an embedded array block, or EAB) in the center of each row. The EPF10K20 has therefore six EABs, or a total of 12Kbits of memory. Fig. 1.13 presents part of the device floorplan.

Routing Resources

Each LAB has 22 inputs from each row and eight signals coming from the logic elements. There are four additional LAB control signals (e.g., preset of registers) and two local carry and cascade interconnects. To connect the LABs, the EPF10K20 uses fast, wide row and column busses, called "Fast-Track Interconnects." Each row bus is 144 lines wide with 24 channels per column. For improved routability, Altera has divided the row interconnect into full-length (a total of 96 channels) and half-length channels ($2 \times 48 = 96$ channels). The half-length channels end toward the middle of the channel where the EABs are located. The EABs can access both half-length channels. It is also interesting to note that the long carry chains skip alternate rows, so that only each second EAB occupies the same carry chain (see Fig. 1.17, p. 24).

Timing Estimates

Altera's `MaxPlusII` software calculates various timing data, such as the `Delay Matrix`, `Registered Performance`, and `Setup/Hold Matrix`. For a full description of all timing parameters, refer to Altera's web-page [19]. To achieve optimal performance, it is necessary to understand how the software physically implements the design. It is useful, therefore, to produce a rough estimate of the solution and then determine how the design may be improved.

Example 1.2: Speed of an 16-bit Adder

Assume one is required to implement a 16-bit adder and estimate the design's maximum speed. The adder can be implemented in two LABs, each using the fast carry chain. The delay through the "same row" delay must be taken into account. The total delays are computed as follows: First, the two inputs must be stable t_{co}. Next, the first carry t_{cgen} must be generated, followed by seven more carries inside the first LAB. The signal then goes through the row interconnect $t_{samerow}$. Inside the second LAB, seven additional carries must be computed and the MSB then must run through an LUT to complete the sum. The results are then stored in the LE register. The following table summarizes these timing data:

LE register clock-to-output delay	t_{co} =	0.2 ns
Data-in to carry-out delay	t_{cgen} =	1.5 ns
Carry-in to carry-out delay	$7 \cdot t_{cico}$ =	$7 \cdot 0.3 = 2.1$ ns
Row routing delay	$t_{samerow}$ =	2.9 ns
Carry-in to carry-out delay	t_{cico} =	$7 \cdot 0.3 = 2.1$ ns
LE look-up table delay	t_{LUT} =	1.9 ns
LE register setup time	t_{su} =	2.7 ns
Total	=	13.4 ns

The estimated delay is 13.4 ns, or a rate of 74.6 MHz. The design is expected to use about 16 LEs (see also Exercise 1.7, p. 27). 1.2

If the two LABs used can not be placed in the same row then the same-column delay $t_{samecolumn} = 4.4$ ns applies (instead of $t_{samerow}$). The worst case occurs if the two LABs used are placed in different rows. The worst case delay becomes $t_{diffrow} = 10.1$ ns. It is therefore very important to check the floorplan and check for possible improvements "by hand" changes in the floorplan as described in the Altera "Getting Started" manual, pages 231-241 [20], or see `Ug/Maxiigs.pdf` on the CD-ROM.

Power Dissipation

The power consumption of an FPGA can be a critical design constraint, especially for mobile applications. Using 3.3V or 2.5V class devices is recommended in this case. To estimate the power dissipation of the Altera device EPF10K20RC240-4, three main sources must be considered, namely:

1) Standby power dissipation $I_{standby} \approx 0.5$ mA
2) I/O power dissipation $I_{I/O}$
3) Active power dissipation I_{active}

The first two are not design-dependent, and also the standby power in CMOS technology is generally small. The active current depends mainly on the clock frequency and the number of LEs in use. Altera provides the following empirical formula to estimate the active current:

$$I_{active} = 98 \cdot f_{max} \cdot N \cdot \tau_{LE} \cdot \frac{\mu A}{MHz \cdot LE} \tag{1.1}$$

where f_{max} is the maximum operating frequency in MHz, N is the total number of logic cells used in the device, and τ_{LE} the average percent of logic cells toggling at each clock (typically 12%). If, for instance, a design uses all LEs of the EPF10K20RC240-4 and the maximum frequency is 25 MHz, then the current will be estimated at 338 mA.

The following case study should be used as a detailed scheme for the examples and self-study problems in the next chapters.

1.4.3 Case Study: Frequency Synthesizer

The design objective in the following case study is to implement a classical frequency synthesizer based on the Philips PM5190 model (circa 1979, see Fig. 1.14). The synthesizer consists of a 32-bit accumulator, with the eight most significant bits (MSBs) wired to a SIN-ROM lookup table (LUT) to produce the desired output waveform. A *graphical* solution, using Altera's `MaxPlusII` software, is shown in Fig. 1.15, and can be found on the CD-ROM as **`book/vhdl/fun_graf.gdf`**. The following VHDL text file implements the design using "component instantiation," consisting of

1) Compilation of the design
2) Design results and floor plan
3) Simulation of the design, and
4) A performance evaluation

Design Compilation

To check and compile the file, start the MaxPlusII Software and select `File→Open` to load **`fun_text.vhd`**. Notice that the top and left menus have changed. The VHDL design[6] reads as follows:

[6] The equivalent Verilog code **`fun_text.v`** for this example can be found in Appendix A on page 344.

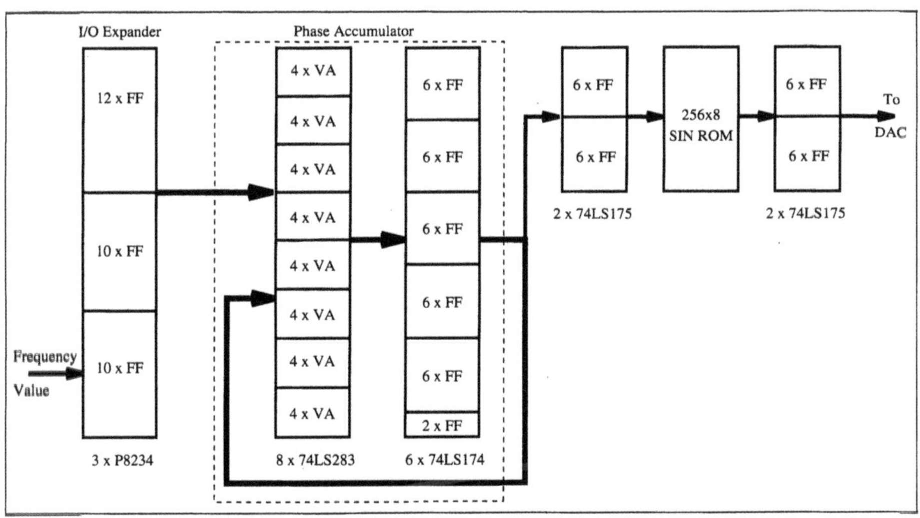

Fig. 1.14. PM5190 frequency synthesizer.

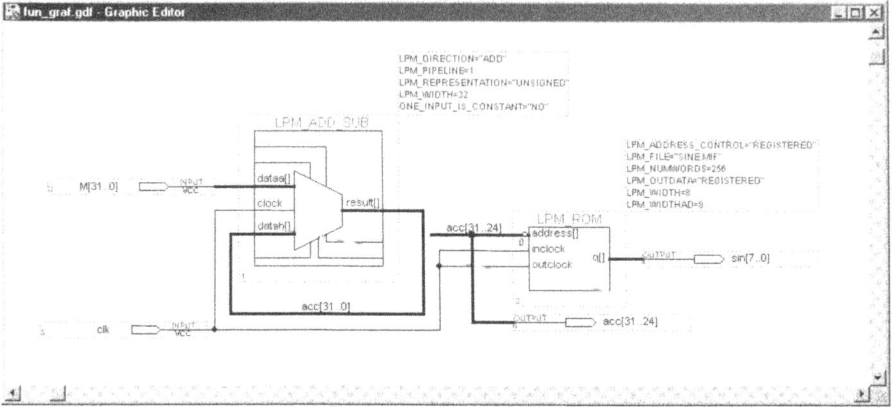

Fig. 1.15. Graphical design of frequency synthesizer.

```
--  A 32 bit function generator using accumulator and ROM

LIBRARY lpm;
USE lpm.lpm_components.ALL;

LIBRARY ieee;
USE ieee.std_logic_1164.ALL;
USE ieee.std_logic_arith.ALL;

ENTITY fun_text IS
  GENERIC ( WIDTH    : INTEGER := 32);     -- Bit width
  PORT ( M          : IN  STD_LOGIC_VECTOR(WIDTH-1 DOWNTO 0);
         sin, acc   : OUT STD_LOGIC_VECTOR(7 DOWNTO 0);
         clk        : IN  STD_LOGIC);
END fun_text;

ARCHITECTURE fun_gen OF fun_text IS

  SIGNAL s, acc32  : STD_LOGIC_VECTOR(WIDTH-1 DOWNTO 0);
  SIGNAL msbs      : STD_LOGIC_VECTOR(7 DOWNTO 0);
                                           -- Auxiliary vectors
BEGIN

  add1: lpm_add_sub                  -- Add M to acc32
    GENERIC MAP ( LPM_WIDTH => WIDTH,
                  LPM_REPRESENTATION => "SIGNED",
                  LPM_DIRECTION => "ADD",
                  LPM_PIPELINE => 0)
    PORT MAP ( dataa => M,
```

```
                datab => acc32,
                result => s );

    reg1: lpm_ff                        -- Save accu
      GENERIC MAP ( LPM_WIDTH => WIDTH)
      PORT MAP ( data => s,
                 q => acc32,
                 clock => clk);

    select1: PROCESS (acc32)
               VARIABLE i : INTEGER;
             BEGIN
               FOR i IN 7 DOWNTO 0 LOOP
                 msbs(i) <= acc32(31-7+i);
               END LOOP;
             END PROCESS select1;

    acc <= msbs;

    rom1: lpm_rom
      GENERIC MAP ( LPM_WIDTH => 8,
                    LPM_WIDTHAD => 8,
                    LPM_FILE => "sine.mif")
      PORT MAP ( address => msbs,
                 inclock => clk,
                 outclock => clk,
                 q => sin);

  END fun_gen;
```

The object `LIBRARY`, found early in the code, contains predefined modules and definitions. The `ENTITY` block specifies I/O ports of the device and generic variables. Using component instantiation, three blocks (see labels `add1, reg1, rom1`) are called like subroutines. The "`select1`" `PROCESS` construct is used to select the eight MSBs to address the ROM. To set the project to the current file, select `File` → `Project` → `Set Project to Current File`. To optimize the design for speed, choose the menu `Assign`→ `Global Project Logic Synthesis` option `Optimize 10` (Speed), and set `Global Project Synthesis Style` to `FAST`. Set the device type to FLEX10K20 by selecting in the menu `Assign`→ `Device` for `Device Family`, the option `FLEX10K`. For `Devices` we select `EPF10K20RC240-4`. Next, start the syntax checker with `<Ctrl+K>` or by selecting `File` → `Project` → `Save & Check`. The compiler checks for basic syntax errors and produces the netlist file `fun_text.cnf`. After the syntax check is successful, compilation can be

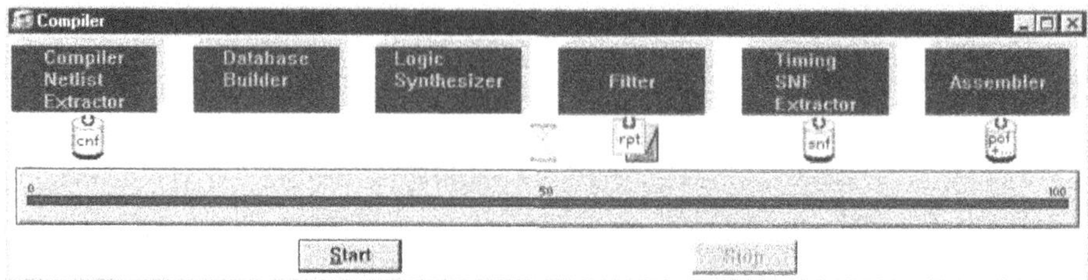

Fig. 1.16. Compilation steps in MaxPlusII.

started by pressing the **START** button in the compiler window or selecting **File → Project → Save & Compile**. If all compiler steps were successfully completed, the design is fully implemented. Fig. 1.16 summarizes all the processing steps of the compilation as shown in the MaxPlusII compiler window.

Floor Planing

The design results can be verified by opening **File→Open → fun_text.rpt**, or double click on the "rpt" button found in the compiler window (see Fig 1.16). Under **Utilities→ Find Text →LCs**, find in "device summary" the number of LCs and memory blocks used. In the report file, find the pin-out of the device and the result of the logic synthesis (i.e., the logic equations). Check the memory initialization file **sine.mif**, containing the sine table in offset binary form. This file was generated using the program **sine.exe** included on the CD-ROM under **book/util**. Select **MaxPlusII → Floorplan Editor** to view the physical implementation. Use the "reduce scale" button to produce the screen shown in Fig. 1.17. Notice that the accumulator uses

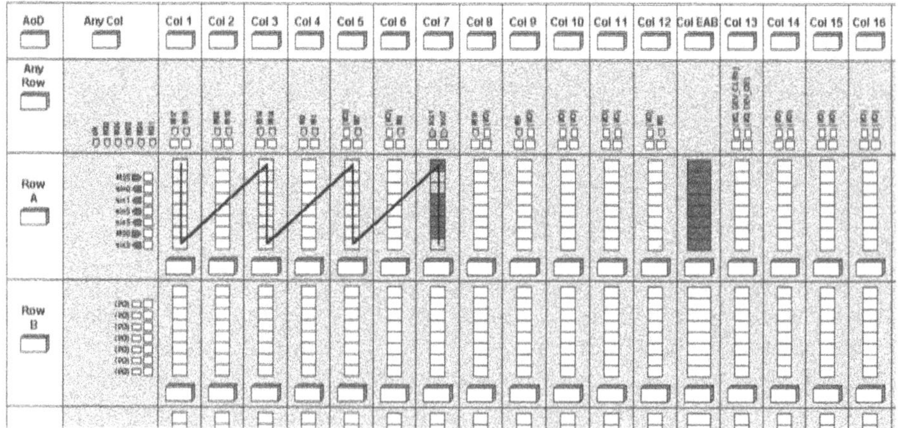

Fig. 1.17. Floorplan of frequency synthesizer design.

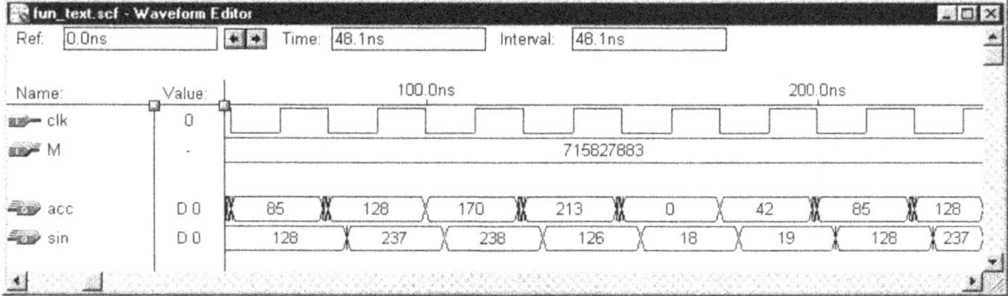

Fig. 1.18. VHDL simulation of frequency synthesizer design.

fast carry chains, and that only every second column has been used for the improved routing as explained in Sect. 1.4.2, p. 19.

Simulation

To simulate, open the prepared waveform `File→Open→fun_text.scf`. Notice that the top and left menu lines have changed. Set the time from the menu `File→End Time` to 1μs. In the `fun_text.scf` window, click on the `clk` symbol and set (left menu buttons) the `Clock Period` to 25 ns in the `Overwrite Clock` window. Set $M = 715827883$ ($M = 2^{32}/6$), so that the period of the synthesizer is 6 clock cycles long. Start the simulation by selecting `MaxPlusII→Simulator` and press the start button. The simulation should give an output similar to Fig. 1.18. Notice that the ROM has been coded in binary offset (i.e., zero=128). When complete, change the frequency so that a period of 8 cycles occurs, i.e., ($M = 2^{32}/8$), and repeat the simulation.

Performance Analysis

To initiate a performance analysis, enter the `MaxPlusII→Timing Analyzer`. Note that the menu line has again changed. Select `Analysis→Registered Performance` and the appropriate `Registered Performance` screen will appear. Click on the `Start` button to measure the register performance. The result should be similar to that shown in Fig. 1.19.

This concludes the case study of the frequency synthesizer.

Exercises

1.1: Use only two input NAND gates to implement a full adder:
(a) $s = a \oplus b \oplus c_{\text{in}}$
(Note: $\oplus$=XOR)
(b) $c_{\text{out}} = a \cdot b + c_{\text{in}} \cdot (a + b)$
(Note: +=OR; ·=AND)
(c) Show that the two-input NAND is *universal* by implementing NOT, AND, and OR with NAND gates.

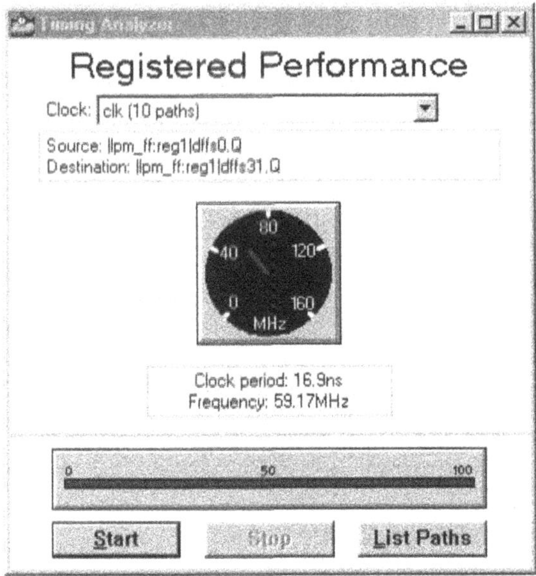

Fig. 1.19. Register performance of frequency synthesizer design.

Exercises Using MaxPlusII

1.2: **(a)** Compile the file **example.vhd** using the **MaxPlusII** compiler (see p. 13) in the functional mode. Select as compiler option **Processing→Functional SNF Extractor**.
(b) Simulate the design using the file **example.scf**.
Note: If you have no prior experience with the MaxPlusII software, refer to the case study found in Sect. 1.4.3, p. 21.
(c) Compile the file **example.vhd** using the **MaxPlusII** compiler with timing extraction. Select as compiler option **Processing→Timing SNF Extractor**.
(d) Simulate the design using the file **example.scf**.
(e) Turn on the option **Check Outputs** in the simulator window and compare the functional and implemented SNF.

1.3: **(a)** Generate a waveform file for **clk,a,b,op1** that approximates that shown in Fig. 1.20.
(b) Conduct a simulation using the VHDL code **example.vhd**.
(c) Explain the algebraic relation between **a,b,op1** and **sum,d**.

1.4: **(a)** Compile the file **fun_text.vhd** with the synthesis style (**Assign→Global Project Logic Synthesis**) **Fast** and **Normal**.
(b) Evaluate **Registered Performance** and the LC's utilization of the two designs from (a). Explain the results.

1.5: **(a)** Compile the file **fun_text.vhd** with the synthesis style (**Assign → Global Project Logic Synthesis**) **Fast** and compiler option **Processing→ Timing SNF Extractor**.
Use the waveform file **fun_text.snf** and
(b1) Set the period of the clock signal to 20 ns and use the simulator to check

Setup/Hold, **Check Ouputs**, **Oscillation**, and **Glitch**.
(b2) Set the period of the clock signal to 15 ns and use the simulator to check **Setup/Hold**, **Check Ouputs**, **Oscillation**, and **Glitch**.

1.6: **(a)** Open the file **fun_text.scf** and start the simulation.
(b) Select the simulator window with the top menu line labelled **Initialize**. Select **Initialize Memory** and export the ROM table in Intel HEX format as **sine.hex**.
(c) Change the **fun_text.vhd** file so that it uses the Intel HEX file **sine.hex** for the ROM table, and verify the correct results through a simulation.

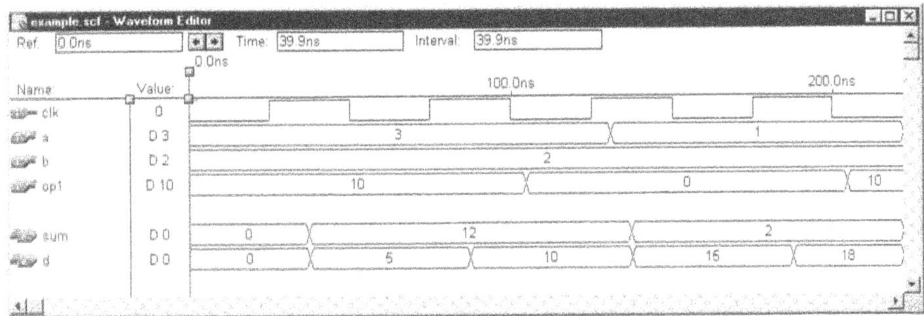

Fig. 1.20. Waveform file for example 1.1 on p. 13.

1.7: **(a)** Design a 16-bit adder using the **LPM_ADD_SUB** macro with the MaxPlusII software.
(b) Measure the **Registered Performance** and compare the result with the data from Example 1.2 (p. 19).

2. Computer Arithmetic

2.1 Introduction

In computer arithmetic two fundamental design principles are of great importance: number representation and the implementation of algebraic operations [21, 22, 23, 24, 25]. We will first discuss possible number representations, (e.g., fixed-point or floating-point), then basic operations like adder and multiplier, and finally efficient implementation of more difficult operations such as square-roots, and the computation of trigonometric functions using the CORDIC algorithm.

FPGAs allow a wide variety of computer arithmetic implementations for the desired digital signal processing algorithms, because of the physical bit-level programming architecture. This contrasts with to the programmable digital signal processors (PDSPs), with the fixed multiply accumulator core. Careful choice of the bit width in FPGA design can result in substantial savings.

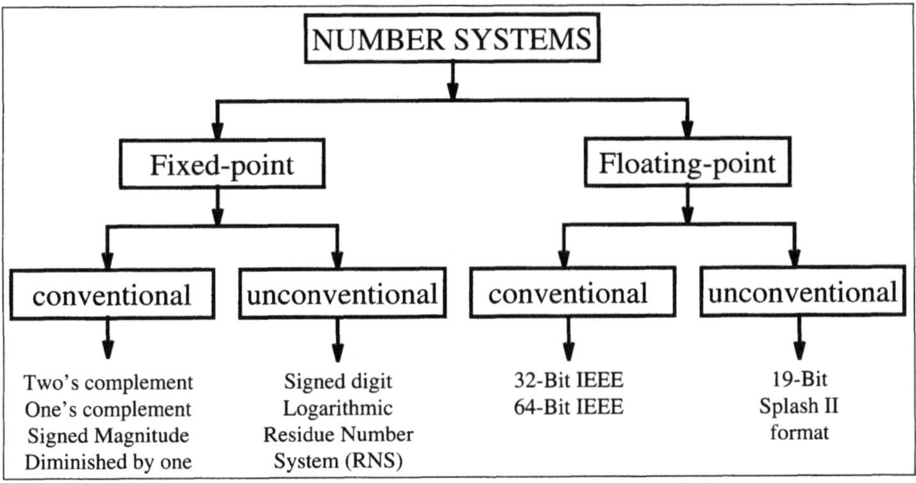

Fig. 2.1. Survey of number representations.

2.2 Number Representation

Deciding whether fixed- or floating-point is more appropriate for the problem must be done carefully, preferablly at an early phase in the project. In general it can be assumed that fixed-point implementations have higher speed and lower cost, while floating-point has higher dynamic range and no need for scaling, which may be attractive for more complicated algorithms. Fig. 2.1 is a survey of conventional and less conventional fixed- and floating-point number representations. Both systems are covered by a number of standards but may, if desired, be implemented in a proprietary form.

2.2.1 Fixed-Point Numbers

We will first review the fixed-point number systems shown in Fig. 2.1 and Table 2.1.

Unsigned Integer

Let X be an N-bit unsigned binary number. Then the range is $[0, 2^N - 1]$ and the representation is given by:

$$X = \sum_{n=0}^{N-1} x_n 2^n. \tag{2.1}$$

where x_n is the nth binary digit of X (i.e., $x_n \in [0, 1]$). The digit x_0 is called the least significant bit (LSB) and has a relative weight of unity. The digit x_{N-1} is the most significant bit (MSB) and has a relative weight of 2^{N-1}.

Signed Magnitude (SM)

In signed magnitude systems the magnitude and the sign are represented separately. The first bit represents the sign and the remaining $N-1$ bits the magnitude. The representation becomes:

$$X = \begin{cases} \sum_{n=0}^{N-1} x_n 2^n & X \geq 0 \\ -\sum_{n=0}^{N-1} x_n 2^n & X < 0 \end{cases} \tag{2.2}$$

The range of this representation is $[-2^{N-1}, 2^{N-1}]$. The advantage of the signed magnitude representation is simplified prevention of overflows, but the disadvantage is that addition must be split depending on which operand is larger.

Two's Complement (2C)

An N-bit two's complement representation of a signed integer, over the range $[-2^{N-1}, 2^{N-1}-1]$, is given by:

$$X = \begin{cases} \sum_{n=0}^{N-1} x_n 2^n & X \geq 0 \\ 2^N - \sum_{n=0}^{N-1} x_n 2^n & X < 0 \end{cases} \quad (2.3)$$

The two's complement (2C) system is by far the most popular signed numbering system in DSP use today. That's because it's possible to add several signed numbers, and as long as the final sum is in the N-bit range, we can ignore *any* overflow in the arithmetic. For instance if we add two 3-bit numbers as follows

$$\begin{array}{rcr} 3_{10} & \longleftrightarrow & 0\,1\,1_{2C} \\ -2_{10} & \longleftrightarrow & 1\,1\,0_{2C} \\ 1_{10} & \longleftrightarrow & 1.\,0\,0\,1_{2C} \end{array}$$

the overflow can be ignored. All computations are modulo 2^N. It follows that it is possible to have intermediate values that can not be correctly represented, but if the final value is valid then the result is correct. For instance, if we add the 3-bit numbers $2+2-3$, we would have an intermediate value of $010+010=100_{2C}$, i.e., -4_{10}, but the result $100-011=100+101=001_{2C}$ is correct.

Two's complement numbers can also be used to implement modulo 2^N arithmetic without any change in the arithmetic. This is what we will use in Chapter 5 to design CIC filters.

One's Complement (1C)

An N-bit one's complement system (1C) can represent integers over the range $[-2^{N-1}-1, 2^{N-1}-1]$. In a one's complement code, positive and negative numbers have the same representation except for the sign bit. There is, in fact a redundant representation of zero (see Table 2.1). The representation of signed numbers in a 1C system is formally given by:

$$X = \begin{cases} \sum_{n=0}^{N-1} x_n 2^n & X \geq 0 \\ 2^N - 1 - \sum_{n=0}^{N-1} x_n 2^n & X < 0 \end{cases} \quad (2.4)$$

For example, the three-bit 1C representation of the numbers -3 to 3 is shown in the third column of Table 2.1.

From the following simple example

Table 2.1. Conventional coding of signed binary numbers.

Binary	2C	1C	D1	SM
011	3	3	4	3
010	2	2	3	2
001	1	1	2	1
000	0	0	1	0
111	−1	−0	−1	−3
110	−2	−1	−2	−2
101	−3	−2	−3	−1
100	−4	−3	−4	−0

$$
\begin{array}{lcll}
3_{10} & \longleftrightarrow & & 0\ 1\ 1_{1C} \\
-2_{10} & \longleftrightarrow & & 1\ 0\ 1_{1C} \\
1_{10} & \longleftrightarrow & 1. & 0\ 0\ 0_{1C} \\
\text{Carry} & & \hookrightarrow & \rightarrow\ \rightarrow\ 1_{1C} \\
1_{10} & \longleftrightarrow & & 0\ 0\ 1_{1C}
\end{array}
$$

we remember that in one's complement a "carry wrap-around" addition is needed. A carry occurring at the MSB must be added to the LSB to get the correct final result.

The system can, however, efficiently be used to implement modulo $2^N - 1$ arithmetic without correction. As a result, one's complement has specialized value in implementing selected DSP algorithms (e.g., Mersenne transforms over the integer ring $2^N - 1$; see Chapter 7).

Diminished one System (D1)

A diminished ones (D1) system is a biased system. The positive numbers are, compared with the 2C, diminished by 1. The range for N-bit D1 numbers is $[-2^{N-1}, 2^{N-1}]$, excluding 0. The coding rule for a D1 system is defined as follows:

$$
X = \begin{cases} \sum_{n=0}^{N-1} x_n 2^n - 1 & X \geq 0 \\ 2^N - \sum_{n=0}^{N-1} x_n 2^n & X < 0 \\ 2^N & X = 0 \end{cases} \tag{2.5}
$$

From adding two D1 numbers

$$
\begin{array}{lcll}
3_{10} & \longleftrightarrow & & 0\ 1\ 0_{D1} \\
-2_{10} & \longleftrightarrow & & 1\ 1\ 0_{D1} \\
1_{10} & \longleftrightarrow & 1. & 0\ 0\ 0_{D1} \\
\text{Carry} & & \hookrightarrow & \boxed{\cdot - 1} \rightarrow 0_{D1} \\
1_{10} & \longleftrightarrow & & 0\ 0\ 0_{D1}
\end{array}
$$

we see that, for D1 a complement and add of the *inverted* carry must be computed.

D1 numbers can efficiently be used to implement modulo 2^N+1 arithmetic without any change in the arithmetic. This fact will be used in Chapter 7 to implement Fermat NTTs in the ring 2^N+1.

2.2.2 Unconventional Fixed-Point Numbers

In the following we continue the review of number systems according to Fig. 2.1 (p. 29). The unconventional fixed-point number systems discussed in the following are not as often used, as for instance the 2C system, but can yield significant improvements for particular applications or problems.

Signed Digit Numbers (SD)

The signed digit (SD) system differs from the traditional binary systems presented in the previous section in the fact that it is ternary valued (i.e., digits have the value $\{0,1,-1\}$, where -1 is sometimes denoted as $\bar{1}$).

SD numbers have proven to be useful in carry-free adders or multipliers with less complexity, because the effort in multiplication can be typically estimated through the number of nonzero elements, which can be reduced by using SD numbers. Statistically, half the digits in the two's complement coding of a number are zero. For an SD code, the density of zeros increases to two thirds as the following example shows:

Example 2.1: SD coding

Consider coding the decimal number $15 = 1111_2$ using a 5-bit binary and an SD code. Their representations are as follows:

1) $15_{10} = 16_{10} - 1_{10} = 1000\bar{1}_{SD}$.
2) $15_{10} = 16_{10} - 2_{10} + 1_{10} = 100\bar{1}1_{SD}$.
3) $15_{10} = 16_{10} - 4_{10} + 3_{10} = 10\bar{1}10_{SD}$.
4) etc.

2.1

The SD representation, unlike a 2C code, is nonunique. We call a *canonic signed digit* system, or CSD, the system with the minimum number of nonezero elements. The following algorithms can be used to produce a "classical" CSD code.

Algorithm 2.2: **Classical CSD Coding**
Starting with the LSB substitute all 1 sequences equal or larger two, with $10\ldots0\bar{1}$.

This CSD coding is the basis for the C utility program `csd.exe` on the CD-ROM. This classical CSD code is also unique and an additional property is

Table 2.2. Adding carry-free binaries using SD representation.

$x_k y_k$ $x_{k-1} y_{k-1}$	00 –	01 neither is $\bar{1}$	01 at least one is $\bar{1}$	$0\bar{1}$ neither is $\bar{1}$	$0\bar{1}$ at least one is $\bar{1}$	11 –	$\bar{1}\bar{1}$ –
c_k	0	1	0	0	$\bar{1}$	1	$\bar{1}$
u_k	0	$\bar{1}$	1	$\bar{1}$	1	0	0

that the resulting representation has at least one zero between two digits, which may have values 1, $\bar{1}$, or 0.

Example 2.3: Classical CSD Code

Consider again coding the decimal number 15 using a 5-bit binary and a CSD code. Their representations are: $1111_2 = 1000\bar{1}_{CSD}$. We notice from a comparison with the SD coding from Example 2.1 that only the first representation is a CSD code.

As another example consider the coding of

$$27_{10} = 11011_2 = 1110\bar{1}_{SD} = 100\bar{1}0\bar{1}_{CSD} \quad (2.6)$$

We notice that although the first substitution of $011 \rightarrow 10\bar{1}$ does not reduce the complexity, this produces a length three strike, and the complexity reduces from 3 additions to two subtractions. 2.3

On the other side, the classical CSD coding does not always produce the "optimal" CSD coding in terms of hardware complexity, because in Algorithm 2.2 additions are also substituted by subtractions, when there should be no such substitution. For instance 011_2 is coded as $10\bar{1}_{CSD}$, and if this coding is used to produce a constant multiplier the subtraction will need a full adder instead of a halfadder for the LSB. The CSD coding given in the following will produce a CSD with the minimum number of none-zero terms, but also with the minimum number of subtractions.

Algorithm 2.4: Optimal CSD Coding

1) Starting with the LSB substitute all 1 sequences larger than two with $10\ldots0\bar{1}$. Also substitute 1011 with $110\bar{1}$.
2) Starting with the MSB, substitute $10\bar{1}$ with 011.

Carry-free Adder

The SD number representation can be used to implement a carry-free adder. Tagaki et al. [26] introduced the scheme presented in Table 2.2. Here, u_k is the interim sum and c_k is the carry of the kth bit (i.e., to be added to u_{k+1}).

Example 2.5: Carry-Free Addition

The addition of 29 to -9 in the SD system is performed below.

$$\begin{array}{rl} 1\,0\,0\,\bar{1}\,0\,1 & x_k \\ +\quad 0\,\bar{1}\,1\,\bar{1}\,1\,1 & y_k \\ \hline 0\,0\,0\,1\,1\,1 & c_k \\ 1\,\bar{1}\,1\,0\,\bar{1}\,0 & u_k \\ \hline 1\,1\,0\,1\,0\,0 & s_k \end{array}$$

2.5

However, due to the ternary logic burden, implementing Table 2.2 with FPGAs requires 4-input operands for the c_k and u_k. This translates into a $2^8 \times 4$-bit LUT when implementing Table 2.2.

Multiplier Adder Graph (MAG)

We have seen that the cost of multiplication is a direct function of the number of nonzero elements a_k in A. The CSD system minimizes this cost. The CSD is also the basis for the Booth multiplier [21] discussed in Exercise 2.2 (p. 76).

It can, however, sometimes be more efficient first to factor the coefficient into several factors, and realize the individual factors in an optimal CSD sense [27, 28, 29, 30]. Fig. 2.2 illustrates this option for the coefficient 93. The direct binary and CSD codes are given by $93_{10} = 1011101_2 = 1100\bar{1}01$, with the 2C requiring four adders, and the CSD requiring three adders. The coefficient 93 can also be represented as $93 = 3 \cdot 31$ which require one adder for each factor (see Fig. 2.2). The complexity for the factor number is reduced to two. There are several ways to combine these different factors. The number of adders required is often referred to as the cost of the constant coefficient multiplier. Fig. 2.3, suggested by Dempster et al. [29], shows all possible configurations for one to four adders. Using this graph, all coefficients with a cost ranging from one to four can be synthesized with $k_i \in \mathbb{N}_0$, according to:

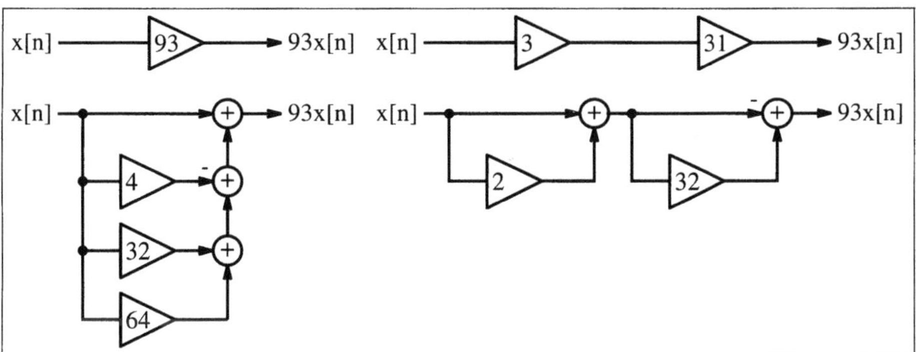

Fig. 2.2. Two realizations for the constant factor 93.

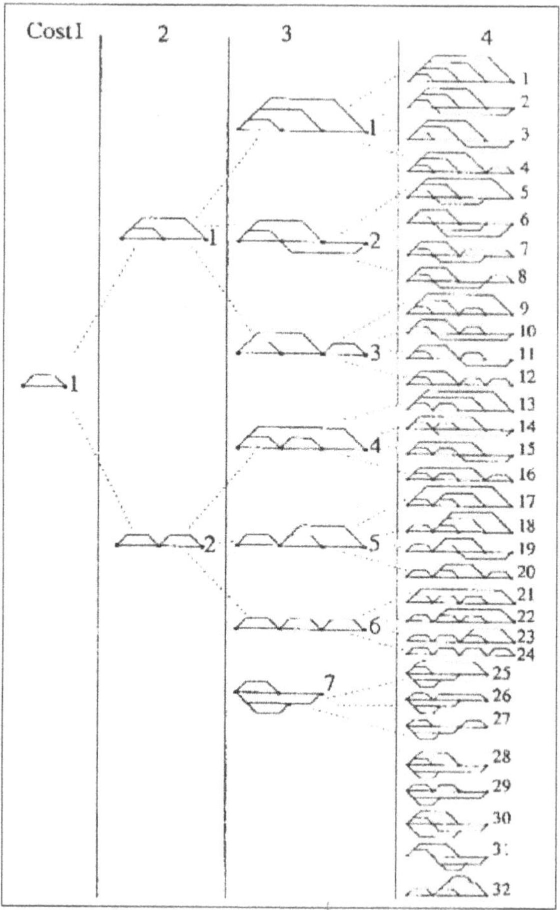

Fig. 2.3. Possible cost one to four graphs. Each node is either an adder or subtractor and each edge is associated with a power-of-two factor (©1995 IEEE [29]).

Cost 1: 1) $A = 2^{k_0}(2^{k_1} \pm 2^{k_2})$
Cost 2: 1) $A = 2^{k_0}(2^{k_1} \pm 2^{k_2} \pm 2^{k_3})$
2) $A = 2^{k_0}(2^{k_1} \pm 2^{k_2})(2^{k_3} \pm 2^{k_4})$
Cost 3: 1) $A = 2^{k_0}(2^{k_1} \pm 2^{k_2} \pm 2^{k_3} \pm 2^{k_4})$
$\vdots$

Using this technique, Table 2.3 shows the optimal coding for all eight-bit integers having a cost between zero and three [5].

Table 2.3. Cost C (i.e., number of adder) for all eight-bit numbers using the multiplier adder graph (MAG) technique.

C	Coefficient
0	4, 8, 16, 32, 64, 128, 256
1	3, 5, 6, 7, 9, 10, 12, 14, 15, 17, 18, 20, 24, 28, 30, 31, 33, 34, 36, 40, 48, 56, 60, 62, 63, 65, 66, 68, 72, 80, 96, 112, 120, 124, 126, 127, 129, 130, 132, 136, 144, 160, 192, 224, 240, 248, 252, 254, 255
2	11, 13, 19, 21, 22, 23, 25, 26, 27, 29, 35, 37, 38, 39, 41, 42, 44, 46, 47, 49, 50, 52, 54, 55, 57, 58, 59, 61, 67, 69, 70, 71, 73, 74, 76, 78, 79, 81, 82, 84, 88, 92, 94, 95, 97, 98, 100, 104, 108, 110, 111, 113, 114, 116, 118, 119, 121, 122, 123, 125, 131, 133, 134, 135, 137, 138, 140, 142, 143, 145, 146, 148, 152, 156, 158, 159, 161, 162, 164, 168, 176, 184, 188, 190, 191, 193, 194, 196, 200, 208, 216, 220, 222, 223, 225, 226, 228, 232, 236, 238, 239, 241, 242, 244, 246, 247, 249, 250, 251, 253
3	43, 45, 51, 53, 75, 77, 83, 85, 86, 87, 89, 90, 91, 93, 99, 101, 102, 103, 105, 106, 107, 109, 115, 117, 139, 141, 147, 149, 150, 151, 153, 154, 155, 157, 163, 165, 166, 167, 169, 170, 172, 174, 175, 177, 178, 180, 182, 183, 185, 186, 187, 189, 195, 197, 198, 199, 201, 202, 204, 206, 207, 209, 210, 212, 214, 215, 217, 218, 219, 221, 227, 229, 230, 231, 233, 234, 235, 237, 243, 245
4	171, 173, 179, 181, 203, 205, 211, 213
	Minimum costs through factorization
2	$45 = 5 \cdot 9, 51 = 3 \cdot 17, 75 = 5 \cdot 15, 85 = 5 \cdot 17, 90 = 2 \cdot 9 \cdot 5, 93 = 3 \cdot 31, 99 = 3 \cdot 33, 102 = 2 \cdot 3 \cdot 17, 105 = 7 \cdot 15, 150 = 2 \cdot 5 \cdot 15, 153 = 9 \cdot 17, 155 = 5 \cdot 31, 165 = 5 \cdot 33, 170 = 2 \cdot 5 \cdot 17, 180 = 4 \cdot 5 \cdot 9, 186 = 2 \cdot 3 \cdot 31, 189 = 7 \cdot 9, 195 = 3 \cdot 65, 198 = 2 \cdot 3 \cdot 33, 204 = 4 \cdot 3 \cdot 17, 210 = 2 \cdot 7 \cdot 15, 217 = 7 \cdot 31, 231 = 7 \cdot 33$
3	$171 = 3 \cdot 57, 173 = 8 + 165, 179 = 51 + 128, 181 = 1 + 180, 211 = 1 + 210, 213 = 3 \cdot 71, 205 = 5 \cdot 41, 203 = 7 \cdot 29$

Logarithmic Number System (LNS)

The logarithmic number system (LNS) [31, 32] is anologous to the floating-point system with a fixed mantissa and a fractional exponent. In the LNS, a number x is represented as:

$$x = \pm r^{\pm e_x} \tag{2.7}$$

where r is the system's radix, and e_x is the LNS exponent. The LNS format consists of a sign-bit for the number and exponent, and an exponent assigned I integer bits and F fractional bits of precision. The format in graphical form is shown below:

sign S_x	exponent sign S_e	exponent integer bits I	exponent fractional bits F

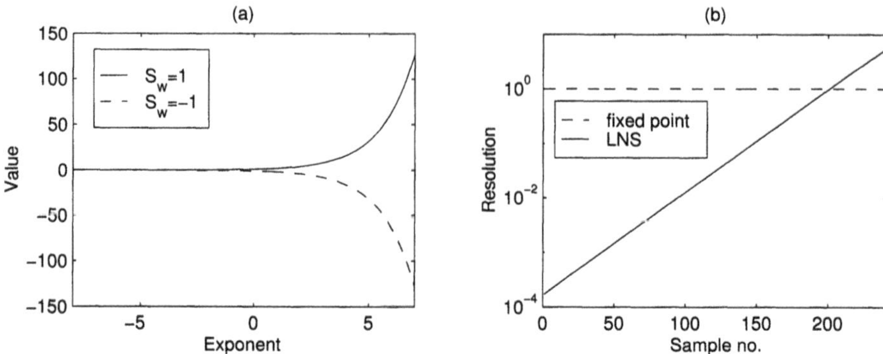

Fig. 2.4. LNS processing. **(a)** Values. **(b)** Resolution.

The LNS, like floating-point, carries a nonuniform precision. Small values of x are highly resolved while large values of x are more coarsely resolved as the following example shows.

Example 2.6: LNS Coding

Consider a radix-2 9-bit LNS word with two sign-bits, three bits for integer precision and four-bit fractional precision. How can, for instance, the LNS coding 00 011.0010 be translated into the real number system? The two sign bits indicate that the whole number and the exponent are positive. The integer part is 3 and the fractional part $2^{-3} = 1/8$. The real number representation is therefore $2^{3+1/8} = 2^{3.125} = 8.724$. We find also that $-2^{3.125} = 10\,011.0010$ and $2^{-3.125} = 01\,100.1110$. The largest number which can be represented with this 9-bit LNS format is $2^{8-1/16} \approx 256$ and the smallest is $2^{-8} = 0.0039$, as graphically interpreted in Fig. 2.4(a). In contrast, an 8-bit plus sign fixed-point number has a maximal positive value of $2^8 - 1 = 255$, and the smallest nonzero positive value is one. A comparison of the two 9-bit systems is shown in Fig. 2.4(b). 2.6

The historical attraction of the LNS lies in its ability to efficiently implement multiplication, division, square-rooting, or squaring. For example, the product $C = A \times B$, where A, B, and C are LNS words, is given by:

$$C = r^{e_a} \times r^{e_b} = r^{e_a + e_b} = r^{e_c} \tag{2.8}$$

That is, the exponent of the LNS product is simply the sum of the two exponents. Division and high-order operations immediately follow. Unfortunately, addition or subtraction are by comparison far more complex. Addition and subtraction operations are based on the following procedure, where it is assumed that $A > B$.

$$C = A + B = 2^{e_a} + 2^{e_b} = 2^{e_a} \underbrace{\left(1 + 2^{e_b - e_a}\right)}_{\Phi^+(\Delta)} = 2^{e_c}. \tag{2.9}$$

Solving for the exponent e_c, one obtains $e_c = e_a + \phi^+(\Delta)$ where $\Delta = e_b - e_a$ and $\phi^+(u) = \log_2(\Phi^+(\Delta))$. For subtraction a similar table, $\phi^-(u) = \log_2(\Phi^-(\Delta))$, $\Phi^-(\Delta) = (1 - 2^{e_b - e_a})$, can be used. Such tables have been historically used for rational numbers as described in "Logarithmorm Completus," Jurij Vega (1754-1802), containing tables computed by Zech. As a result, the term $\log_2(1 - 2^u)$ is usually referred as Zech logarithms.

LNS arithmetic is performed in the following manner [31]. Let $A = 2^{e_a}, B = 2^{e_b}, C = r^{e_c}$, with S_A, S_B, S_C denoting the sign-bit for each word:

Operation		Action
Multiply	$C = AB$	$e_c = e_a + e_b; S_c = S_a \text{ XOR } S_b$
Divide	$C = A/B$	$e_c = e_a - e_b; S_c = S_a \text{ XOR } S_b$
Add	$C = A + B$	$e_c = \begin{cases} e_a + \phi^+(e_b - e_a) & A \geq B \\ e_b + \phi^+(e_a - e_b) & B > A \end{cases}$
Subtract	$C = A - B$	$e_c = \begin{cases} e_a + \phi^-(e_b - e_a) & A \geq B \\ e_b + \phi^-(e_a - e_b) & B > A \end{cases}$
Square root	$C = \sqrt{A}$	$e_c = e_a/2$
Square	$C = A^2$	$e_c = 2e_a$

Methods have been developed to reduce the necessary table size for the Zech logarithm by using partial tables [31] or using linear interpolation techniques [33]. These techniques are beyond the scope of the discussion presented here.

Residue Number System (RNS)

The RNS is actually an ancient algebraic system whose history can be traced back 2,000 years. The RNS is an integer arithmetic system in which the primitive operations of addition, subtraction, and multiplication are defined. The primitive operations are performed concurrently within noncommunicating small wordlength channels [34, 35]. An RNS system is defined with respect to a positive integer basis set $\{m_1, m_2, \ldots, m_L\}$, where the $m_l's$ are all relatively (pairwise) prime. The dynamic range of the resulting system is M where $M = \prod_{l=1}^{L} m_l$. For signed-number applications, the integer value of X is assumed to be constrained to $X \in [-M/2, M/2)$. RNS arithmetic is defined within a ring isomorphism:

$$\mathbb{Z}_M \cong \mathbb{Z}_{m_1} \times \mathbb{Z}_{m_2} \times \cdots \times \mathbb{Z}_{m_L} \tag{2.10}$$

where $\mathbb{Z}_M = \mathbb{Z}/(M)$ corresponds to the ring of integers modulo M, called the residue class mod$M$. The mapping of an integer X into a RNS L-tuple $X \leftrightarrow (x_1, x_2, \ldots, x_L)$ is defined by $x_l = X \bmod m_l$, for $l = 1, 2, \ldots L$. Defining $\square$ to be the algebraic operations $+, -$ or $*$, it follows that if $Z, X, Y \in \mathbb{Z}_M$, then:

$$Z = X \square Y \mod M \tag{2.11}$$

is isomorphic to $Z \leftrightarrow (z_1, z_2, \ldots, z_L)$,. Specifically:

$$\begin{array}{ccc} X & \overset{(m_1,m_2,\ldots,m_L)}{\longleftrightarrow} & (\ \langle X\rangle_{m_1}\ ,\ \langle X\rangle_{m_2}\ ,\ldots,\ \langle X\rangle_{m_L}\ ) \\ Y & \overset{(m_1,m_2,\ldots,m_L)}{\longleftrightarrow} & (\ \langle Y\rangle_{m_1}\ ,\ \langle Y\rangle_{m_2}\ ,\ldots,\ \langle Y\rangle_{m_L}\ ) \\ \hline Z = X \square Y & \overset{(m_1,m_2,\ldots,m_L)}{\longleftrightarrow} & (\langle X\square Y\rangle_{m_1}, \langle X\square Y\rangle_{m_2}, \ldots, \langle X\square Y\rangle_{m_L},) \end{array}$$

As a result, RNS arithmetic is "pairwise" defined. The L elements of $Z = (X \square Y) \mod M$ are computed concurrently within L small wordlength mod (m_l) channels whose width is bounded by $w_l = \lceil \log_2(m_l) \rceil$ bits (typical 4- to 8-bits). In practice, most RNS arithmetic systems use small RAM or ROM tables to implement the modular mappings $z_l = x_l \square y_l \mod m_l$.

Example 2.7: RNS Arithmetic

Consider an RNS system based on the relatively prime moduli set $\{2, 3, 5\}$ having a dynamic range of $M = 2 \cdot 3 \cdot 5 = 30$. Two integers in $\mathbb{Z}_{30}$, say 7_{10} and 4_{10}, have RNS representations $7 = (1, 1, 2)_{\mathrm{RNS}}$ and $4 = (0, 1, 4)_{\mathrm{RNS}}$ respectively. Their sum, difference, and products are 11, 3, and 28 respectively, which are all within $\mathbb{Z}_{30}$. Their computation is shown below.

$$\begin{array}{rcl} 7 & \overset{(2,3,5)}{\longleftrightarrow} & (1,1,2) \\ +4 & \overset{(2,3,5)}{\longleftrightarrow} & +(0,1,4) \\ \hline 11 & \overset{(2,3,5)}{\longleftrightarrow} & (1,2,1) \end{array} \qquad \begin{array}{rcl} 7 & \overset{(2,3,5)}{\longleftrightarrow} & (1,1,2) \\ -4 & \overset{(2,3,5)}{\longleftrightarrow} & -(0,1,4) \\ \hline 3 & \overset{(2,3,5)}{\longleftrightarrow} & (1,0,3) \end{array} \qquad \begin{array}{rcl} 7 & \overset{(2,3,5)}{\longleftrightarrow} & (1,1,2) \\ \times 4 & \overset{(2,3,5)}{\longleftrightarrow} & \times(0,1,4) \\ \hline 28 & \overset{(2,3,5)}{\longleftrightarrow} & (0,1,3) \end{array}$$

2.7

RNS systems have been built as custom VLSI devices [36], GaAs, and LSI [35]. It has been shown that for small wordlengths, the RNS can provide a significant speed-up using the $2^4 \times 2$-bit tables found in Xilinx XC4000 FPGAs [37]. For larger moduli, the $2^8 \times 8$-bit tables belonging to the Altera FLEX are beneficial in designing RNS arithmetic and RNS-to-integer converters. With the ability to support larger moduli, the design of high-precision high-speed FPGA systems becomes a practical reality.

A historical barrier to implementing practical RNS systems, until recently, has been decoding [38]. Implementing RNS-to-integer decoder, division, or magnitude scaling, requires that data first be converted from an RNS format to an integer. The commonly referenced RNS-to-integer conversion methods are called the Chinese Remainder Theorem (CRT) and mixed-radix-conversion (MRC) algorithm [34]. The MRC actually produced the digits of a weighted number system representation of an integer while the CRT maps an RNS L-tuple directly to an integer. The CRT is defined below.

$$X \mod M \equiv \sum_{l=0}^{L-1} \hat{m}_l \langle \hat{m}_l^{-1} x_l \rangle_{m_l} \mod M \tag{2.12}$$

where $\hat{m}_l = M/m_l$ is an integer, and $\hat{m}_l^{-1}$ is the multiplicative inverse of $\hat{m}_l \bmod m_l$, i.e., $\hat{m}_l\hat{m}_l^{-1} \equiv 1 \bmod m_l$. Typically, the desired output of an RNS computation is much less than the maximum dynamic range M. In such cases, a highly efficient algorithm, called the $\varepsilon-$CRT [39], can be used to implement a time- and area-efficient RNS to (scaled) integer conversion.

Index Multiplier

There are, in fact, several variations of the RNS. One in common use is based on the use of "index" arithmetic [34]. It is similar in some respects to logarithmic arithmetic. Computation in the index domain is based on the fact that if all the moduli are primes, it is known from number theory that there exists a primitive element, a *generator* g, such that:

$$a \equiv g^{\alpha} \bmod p \tag{2.13}$$

that generates all elements in the field $\mathbb{Z}_p$, excluding zero (denoted $\mathbb{Z}_p/\{0\}$). There is, in fact, a one-to-one correspondence between the integers a in $\mathbb{Z}_p/\{0\}$ and the exponents α in $\mathbb{Z}_{p-1}$. As a point of terminology, the index α, with respect to the generator g and integer a, is denoted $\alpha = \mathrm{ind}_g(a)$.

Example 2.8: Index Coding

Consider a prime moduli $p = 17$, a generator $g = 3$ will generates the elements of $\mathbb{Z}_p/\{0\}$. The encoding table is shown below. For notational purposes, the case $a = 0$ is denoted by $g^{-\infty} = 0$.

a	0	1	2	3	4	5	6	7	8	9	10	11	12	13	14	15	16
$\mathrm{ind}_3(a)$	$-\infty$	0	14	1	12	5	15	11	10	2	3	7	13	4	9	6	8

2.8

Multiplication of RNS numbers can be performed as follows:

1) Map a and b into the index domain, i.e., $a = g^{\alpha}$ and $b = g^{\beta}$
2) Add the index values modulo $p - 1$, i.e., $\nu = (\alpha + \beta) \bmod (p - 1)$
3) Map the sum back to the original domain, i.e., $n = g^{\nu}$

If the data being processed is in index form, then only exponent addition $\bmod(p - 1)$ is required. This can be illustrated by the following example.

Example 2.9: Index Multiplication

Consider the prime moduli $p = 17$, generator $g = 3$, and the results shown in Example 2.8. The multiplication of $a = 2$ and $b = 4$ proceeds as follows:

$$(\mathrm{ind}_g(2) + \mathrm{ind}_g(4)) \bmod 16 = (14 + 12) \bmod 16 = 10$$

From the table in Example 2.8 it is seen that $\mathrm{ind}_3(8) = 10$, which corresponds to the integer 8, which is the expected result. 2.9

Addition in the Index Domain

Most often, DSP algorithms require both multiplication *and* addition. Index arithmetic is well suited to multiplication, but addition is no longer trivial. Technically addition can be performed by converting index RNS data back into the RNS where addition is simple to implement. Once the sum is computed the result is mapped back into the index domain. Another approach is based on Zech logarithms. The sum of index-coded numbers a and b is expressed as:

$$d = a + b = g^\delta = g^\alpha + g^\beta = g^\alpha\left(1 + g^{\beta-\alpha}\right) = g^\beta\left(1 + g^{\alpha-\beta}\right) \qquad (2.14)$$

If we now define the Zech logarithms as

Definition 2.10: Zech Logarithm

$$Z(n) = \mathrm{ind}_g(1 + g^n) \quad \longleftrightarrow \quad g^{Z(n)} = 1 + g^n \qquad (2.15)$$

Then we can rewrite (2.14) in the following way:

$$g^\delta = g^\beta \cdot g^{Z(\alpha-\beta)} \quad \longleftrightarrow \quad \delta = \beta + Z(\alpha - \beta). \qquad (2.16)$$

Adding numbers in the index domain, therefore, requires one addition, one subtraction, and a Zech LUT. The following small example illustrates the principle of adding $2 + 5$ in the index domain.

Example 2.11: Zech Logarithms

A table of Zech logarithms, for a prime moduli 17 and $g = 3$, is shown below.

n	$-\infty$	0	1	2	3	4	5	6	7	8	9	10	11	12	13	14	15
$Z(n)$	0	14	12	3	7	9	15	8	13	$-\infty$	6	2	10	5	4	1	11

The index values for 2 and 5 are defined in the tables found in Example 2.8 (p. 41). It therefore follows that:

$$2 + 5 = 3^{14} + 3^5 = 3^5(1 + 3^9) = 3^{5+Z(9)} = 3^{11} \equiv 7 \bmod 17.$$

2.11

The case where $a + b \equiv 0$ needs special attention, corresponding to the case where [40]:

$$-X \equiv Y \bmod p \quad \longleftrightarrow \quad g^{\alpha+(p-1)/2} \equiv g^\beta \bmod p.$$

That is, the sum is zero if, in the index domain, $\beta = \alpha + (p-1)/2 \bmod (p-1)$. An example follows.

Example 2.12: The addition of 5 and 12 in the original domain is given by $5 + 12 = 3^5 + 3^{13} = 3^5(1 + 3^8) = 3^{5+Z(8)} \equiv 3^{-\infty} \equiv 0 \bmod 17.$ 2.12

Complex Multiplication using QRNS

Another interesting property of the RNS arises if we process complex data. This special representation called QRNS allows very efficient multiplication, which we wish to discuss next.

When the real and imaginary components are coded as RNS digits, the resulting system is called the complex RNS or CRNS. Complex addition in the CRNS requires that two real adds be performed. Complex RNS (CRNS) multiplication is defined in terms of four real products, an addition, and a subtraction. This condition is radically changed when using a variant of the RNS, called the quadratic RNS, or QRNS. The QRNS is based on known properties of Gaussian primes of the form $p = 4k + 1$, where k is a positive integer. The importance of this choice of moduli is found in the factorization of the polynomial $x^2 + 1$ in $\mathbb{Z}_p$. The polynomial has two roots, $\hat{\jmath}$ and $-\hat{\jmath}$, where $\hat{\jmath}$ and $-\hat{\jmath}$ are real integers belonging to the residue class $\mathbb{Z}_p$. This is in sharp contrast with the factoring of $x^2 + 1$ over the complex field. Here, the roots are complex and have the form $x_{1,2} = \alpha \pm j\beta$ where $j = \sqrt{-1}$ is the imaginary operator. Converting a CRNS number into the QRNS is accomplished by the transform $f : \mathbb{Z}_p^2 \to \mathbb{Z}_p^2$, defined as follows:

$$f(a + jb) = ((a + \hat{\jmath}b) \bmod p, (a - \hat{\jmath}b) \bmod p) = (A, B) \qquad (2.17)$$

In the QRNS, addition and multiplication is realized componentwise, and is defined as

$$(a + ja) + (c + jd) \leftrightarrow (A + C, B + D) \quad \bmod p \qquad (2.18)$$

$$(a + jb)(c + jd) \leftrightarrow (AC, BD) \quad \bmod p \qquad (2.19)$$

and the square of the absolute value can be computed with

$$|a + jb|^2 \leftrightarrow (A \cdot B) \quad \bmod p \qquad (2.20)$$

The inverse mapping from QRNS digit back to the CRNS is defined by:

$$f^{-1}(A, B) = 2^{-1}(A + B) + j\ (2\hat{\jmath})^{-1}(A - B) \quad \bmod p, \qquad (2.21)$$

Consider the Gaussian prime $p = 13$ and the complex product of $(a + jb) = (2 + j1)$, $(c + jd) = (3 + j2)$, is $(2 + j1) \cdot (3 + j2) = (4 + j7) \bmod 13$. In this case four real multiplies, a real add, and real subtraction are required to complete the product.

Example 2.13: QRNS Multiplication

The quadratic equation $x^2 \equiv (-1) \bmod 13$ has two roots: $\hat{\jmath} = 5$ and $-\hat{\jmath} = -5 \equiv 8 \bmod 13$. The QRNS coded data become:

$(a + jb) = 2 + j \leftrightarrow (2 + 5 \cdot 1, 2 + 8 \cdot 1) = (A, B) = (7, 10) \bmod 13$
$(c + jd) = 3 + j2 \leftrightarrow (3 + 5 \cdot 2, 3 + 8 \cdot 2) = (C, D) = (0, 6) \bmod 13.$

Componentwise multiplication yields $(A, B)(C, D) = (7, 10)(0, 6) \equiv (0, 8) \bmod 13$ requiring only two real multiplies. The inverse mapping to the CRNS is defined in terms of (2.21), where $2^{-1} \equiv 7$ and $(2\hat{\jmath})^{-1} = 10^{-1} \equiv 4$. Solving the equations for $2 \cdot x \equiv 1 \bmod 13$ and $10 \cdot x \equiv 1 \bmod 13$, produces 7 and 4, respectively. It then follows that

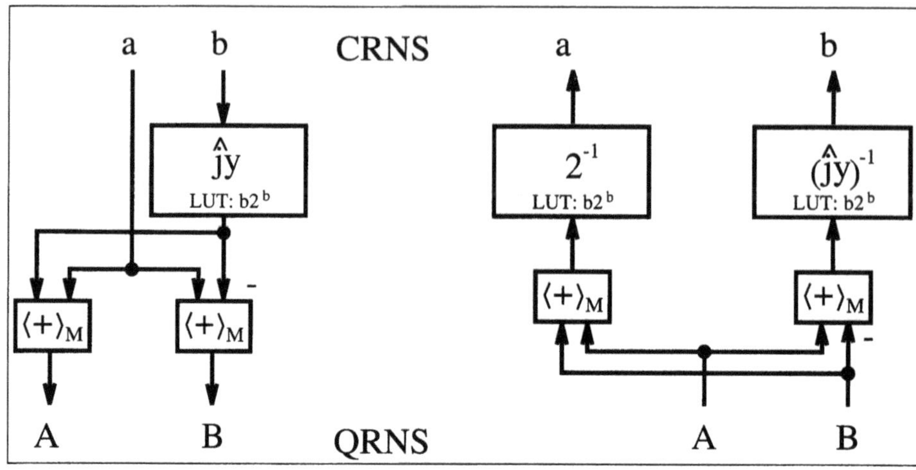

Fig. 2.5. CRNS ↔ QRNS conversion.

$$f^{-1}(0,8) = 7(0+8) + \jmath\, 4(0-8) \bmod 13 \equiv 4 + \jmath 7 \bmod 13. \checkmark$$

2.13

Fig. 2.5 shows a graphical interpretation of the mapping between CRNS and QRNS.

2.2.3 Floating-Point Numbers

Floating-point systems were developed to provide high resolution over a large dynamic range. Floating-point systems often can provide a solution when fixed-point systems, with their limited precision and dynamic range, fail. Floating-point systems, however, bring a speed and complexity penalty. Most floating-point systems comply with the published single- or double-precision IEEE floating-point standard [41]. A standard floating-point word consists of a sign-bit S, exponent e, and an unsigned (fractional) normalized mantissa m, arranged as follows:

S	Exponent e	unsigned mantissa m

Algebraically, a floating-point word is represented by:

$$X = (-1)^S \cdot 1.m \cdot 2^{e-\text{bias}} \tag{2.22}$$

The exponent $e = 1 \ldots 1_2$ is reserved for ∞, and $e = 0$ for zero. The other parameters of the IEEE single and double format can be seen from Table 2.4.

In floating-point multiplication the mantissas have to be multiplied like fixed-point numbers, and the exponent must be added. Floating-point addition is in general more complicated, because the mantissas must first be

Table 2.4. IEEE floating-point standard.

	Single	Double
Word length	32	64
Mantissa	23	52
Exponent	8	11
Bias	127	1023
Range	$2^{138} \approx 3.8 \cdot 10^{38}$	$2^{1024} \approx 9 \cdot 10^{307}$

normalized so that both numbers are adjusted to the larger exponent. Then the two mantissas can be added. For both multiplication and addition a final normalization is necessary to achieve again the fractional $1.m$ representation for the mantissa.

The single IEEE floating-point format uses FPGA resources intensively due to the 23×23-bit fast integer multiplier. Therefore Shirazi et al. [42] have developed a modified format to implement various algorithms on their custom computing machine called SPLASH-2, a multiple-FPGA board based on Xilinx XC4010 devices (see Table 1.5, p. 16). They used an 18-bit format so that they can transport two operands over the 36-bit wide system bus of the multiple-FPGA board. The 18-bit format has a 10-bit mantissa, 7-bit exponent and a sign bit, and can represents a range of $3.7 \cdot 10^{19}$. They implemented an adder and a multiplier with three pipeline stages. The data for a single floating-point multiplier and adder [42] can be seen from Table 2.5.

Table 2.5. Custom 18-bit floating-point design using Xilinx XC4010.

	Adder	Multiplier
Function Generator (i.e., LE)	224	352
Flip-flops	112	112
Speed	8.6 MHz	4.9 MHz

2.3 Binary Adders

A basic binary N-bit adder/subtractor consists of N full adders (FA). A full adder implements the following Boolean equations

$$s_k = x_k \quad \text{XOR} \quad y_k \quad \text{XOR} \quad c_k \tag{2.23}$$

$$= x_k \oplus y_k \oplus c_k \tag{2.24}$$

that define the sum-bit. The carry (out) bit is computed with:

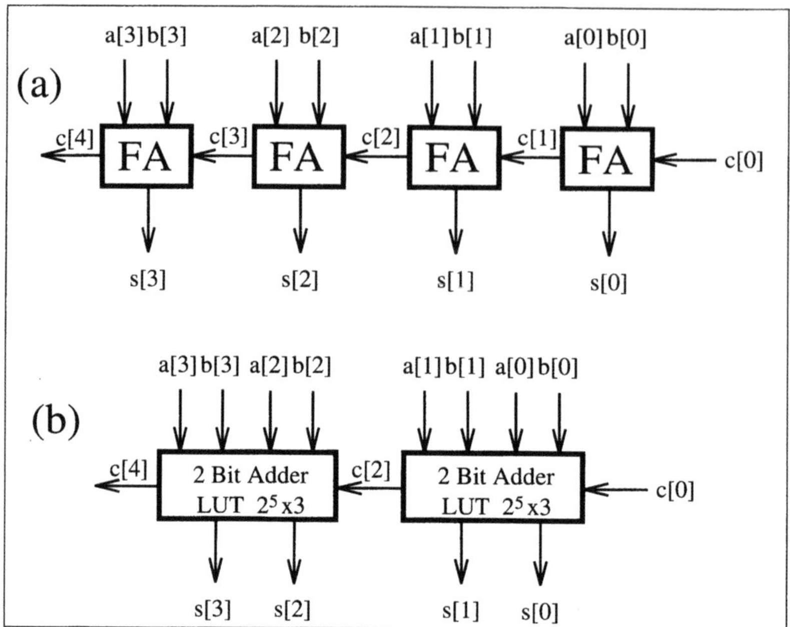

Fig. 2.6. Two's complement adders.

$$c_{k+1} = (x_k \text{ AND } y_k) \text{ OR } (x_k \text{ AND } c_k) \text{ OR } (y_k \text{ AND } c_k) \tag{2.25}$$
$$= (x_k \cdot y_k) + (x_k \cdot c_k) + (y_k \cdot c_k) \tag{2.26}$$

In the case of a 2C adder, the LSB can be reduced to a half adder because the carry input is zero.

The simplest adder structure is called the "ripple carry adder" as shown in Fig. 2.6(a) in a bit-serial form. If larger tables are available in the FPGA, several bits can be grouped together into one LUT, as shown in Fig. 2.6(b). For this "two bit at a time" adder the longest delay comes from the ripple of the carry through all stages. Attempts have been made to reduce the carry delays using techniques such as the carry-skip, carry lookahead, conditional sum, or carry-select adders. These techniques can speed up addition and can be used with older generation FPGA families (e.g., XC 3000 from Xilinx) since these devices do not provide internal fast carry logic. Modern families, such as the Xilinx XC4K or Altera Flex, posses very fast "ripple carry logic" that is about a magnitude faster than the delay through a regular logic LUT [1]. Altera uses fast tables (see Fig. 1.12, p. 17), while the Xilinx XC4K uses hardwired decoders for implementing carry logic based on the multiplexer structure shown in Fig. 2.7. The presence of the fast carry logic in modern FPGA families removes the need to develop hardware intensive carry look-ahead schemes.

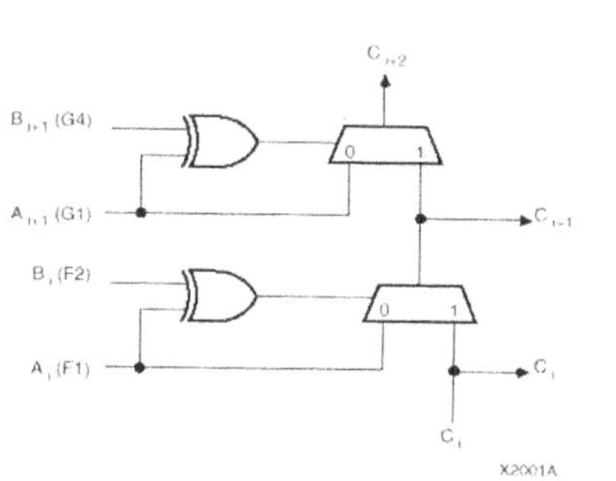

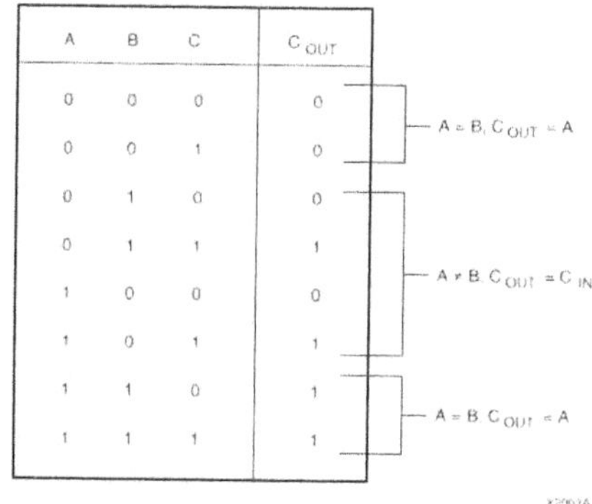

A	B	C	C OUT
0	0	0	0
0	0	1	0
0	1	0	0
0	1	1	1
1	0	0	0
1	0	1	1
1	1	0	1
1	1	1	1

Fig. 2.7. XC4000 fast carry logic (©1993 Xilinx).

Fig. 2.8 summarizes the size and **Registered Performance** of N-bit binary adders, if implemented with the **lpm_add_sub** megafunction component. If the operands are applied through I/O cells, the delays through the busses of a FLEX device are dominant and performance decreases if the registers of the I/O cells are used (Option: **Assign→ Global Project Logic Synthesis→Automatic Fast I/O**). If the data are routed from local registers, performance improves. This type of additional LC register allocation will appear (in the project report file) as increased LC use by a factor of three. A synchronous registered design would not consume any additional resources. A typical design will achieve a speed between these two cases.

2.3.1 Pipelined Adders

Pipelining is extensively used in DSP solutions due the intrinsic data-flow regularity of DSP algorithms. Programmable digital signal processor MACs [14, 15, 6] typically carry at least four pipelined stages. The processor:

1) Decodes the command
2) Loads the operands in registers
3) Performs multiplication and stores the product, and
4) Accumulates the products, all concurrently.

The pipelining principle can be applied to FPGA designs as well, at little or no additional cost since each logic element contains a flip-flop, which is otherwise unused, to save routing resources. With pipelining it is possible to break an arithmetic operation into small primitive operations, save the carry and the intermediate values in registers, and continue the calculation in the next clock cycle. Such adders are sometimes called carry save adders (CSAs) [1] in the literature. Then the question arises: In how many pieces should we

[1] The name carry save adder is also used in the context of Wallace-multiplier, see Exercise 2.1, p. 75.

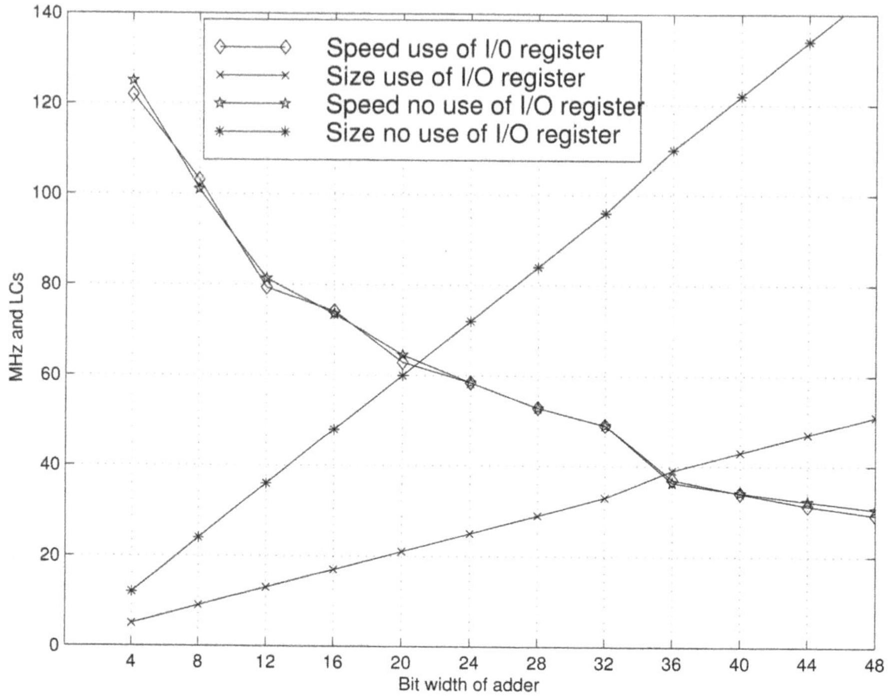

Fig. 2.8. Adder speed and size for Flex 10K.

divide the adder? Should we use bit level? For Altera's Flex 10K devices a reasonable choice will be always using an LAB with 8 LCs and 8 FFs for one pipeline element. In fact it can be shown that if we try to pipeline (for instance) a 5-bit adder, the performance drops, as reported in Table 2.6, because the pipelined 5-bit adder does not fit in one LAB.

Table 2.6. Performance of a 5-bit pipelined adder using synthesis of predefined LPM modules with pipeline option.

Pipeline-stages	Use I/O register		No use of I/O register	
	MHz	LCs	MHz	LCs
0	123.45	6	120.48	15
1	121.95	10	123.45	18
2	112.35	14	123.45	24
3	112.35	20	120.48	32
4	112.35	28	123.45	41
5	103.91	31	121.95	45

Table 2.7. Performance of pipelined adders. Size and speed are for the maximum bit width, for 15-, 22-, and 29-bit adders.

Bit width	Use I/O register		No use of I/O register		Pipeline stages	Design file name
	MHz	LCs	MHz	LCs		
9-15	97.08	26	94.33	61	1	`add_1p.vhd`
16-22	94.33	58	91.74	113	2	`add_2p.vhd`
23-29	91.74	105	90.90	180	3	`add_3p.vhd`

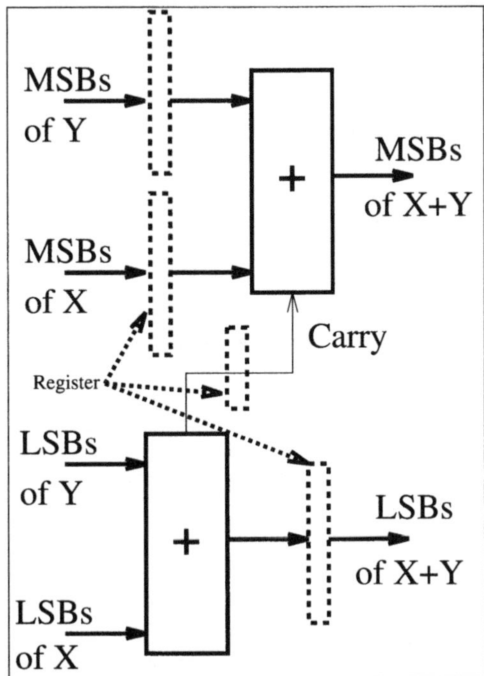

Fig. 2.9. Pipelined adder.

Because the number of flip-flops in one LAB is 8 and we need an extra flip-flop for the carry-out, we should use a maximum block size of 7 bits for maximum `Registered Performance`. Only the blocks with the MSBs can be 8 bits wide, because we do not need the extra flip-flop for the carry. This observation leads to the following conclusions:

1) With one additional pipeline stage we can build adders up to a length $7 + 8 = 15$.

2) With two pipeline stages we can build adders with up to $7+7+8 = 22$-bit length.

3) With three pipeline stages we can build adders with up to $7+7+7+8 =$ 29-bit length.

Table 2.7 shows the `Registered Performance` and LC utilization of this kind of pipelined adder. From Table 2.7 it can be concluded that although the bit width increases the `Registered Performance` remains almost the same if we add the appropriate number of pipeline stages.

The following example shows the code of a 15-bit pipelined adder which is graphically interpreted by Fig. 2.9.

Example 2.14: VHDL Design of 15-bit Pipelined Adder

Consider the VHDL code[2] of a 15-bit pipelined adder that is graphically interpreted in Fig. 2.9. Depending on the synthesis style, the design runs at 90 to 104 MHz.

```
LIBRARY lpm;
USE lpm.lpm_components.ALL;

LIBRARY ieee;
USE ieee.std_logic_1164.ALL;
USE ieee.std_logic_arith.ALL;

ENTITY add_1p IS
  GENERIC (WIDTH  : INTEGER := 15; -- Total bit width
           WIDTH1 : INTEGER := 7;  -- Bit width of LSBs
           WIDTH2 : INTEGER := 8;  -- Bit width of MSBs
           ONE    : INTEGER := 1); -- 1 bit for carry reg.
  PORT (x,y : IN  STD_LOGIC_VECTOR(WIDTH-1 DOWNTO 0);
                                                 -- Inputs
        sum : OUT STD_LOGIC_VECTOR(WIDTH-1 DOWNTO 0);
                                                 -- Result
        clk : IN  STD_LOGIC);
END add_1p;

ARCHITECTURE flex OF add_1p IS
  SIGNAL l1, l2, r1, q1                   -- LSBs of inputs
                     : STD_LOGIC_VECTOR(WIDTH1-1 DOWNTO 0);
  SIGNAL l3, l4, r2, q2, u2, h2           -- MSBs of inputs
                     : STD_LOGIC_VECTOR(WIDTH2-1 DOWNTO 0);
  SIGNAL s           : STD_LOGIC_VECTOR(WIDTH-1 DOWNTO 0);
                                          -- Output register
  SIGNAL cr1, cq1    : STD_LOGIC_VECTOR(ONE-1 DOWNTO 0);
                                          -- LSBs carry signal


BEGIN
  PROCESS  -- Split in MSBs and LSBs and store in registers
  BEGIN
    WAIT UNTIL clk = '1';
    -- Split LSBs from input x,y
    FOR k IN WIDTH1-1 DOWNTO 0 LOOP
```

[2] The equivalent Verilog code **add_1p.v** for this example can be found in Appendix A on page 345.

```
      l1(k) <= x(k);
      l2(k) <= y(k);
    END LOOP;
    -- Split MSBs from input x,y
    FOR k IN WIDTH2-1 DOWNTO 0 LOOP
      l3(k) <= x(k+WIDTH1);
      l4(k) <= y(k+WIDTH1);
    END LOOP;
  END PROCESS;
-------------- First stage of the adder -----------------
  add_1: lpm_add_sub                      -- Add LSBs of x and y
         GENERIC MAP ( LPM_WIDTH => WIDTH1,
                       LPM_REPRESENTATION => "UNSIGNED",
                       LPM_DIRECTION => "ADD")
         PORT MAP ( dataa => l1, datab => l2,
                    result => r1,  cout => cr1(0));
  reg_1: lpm_ff           -- Save LSBs of x+y and carry
         GENERIC MAP ( LPM_WIDTH => WIDTH1 )
         PORT MAP ( data => r1, q => q1,clock => clk );
  reg_2: lpm_ff
         GENERIC MAP ( LPM_WIDTH => ONE )
         PORT MAP ( data => cr1, q => cq1, clock => clk );

  add_2: lpm_add_sub                      -- Add MSBs of x and y
         GENERIC MAP ( LPM_WIDTH => WIDTH2,
                       LPM_REPRESENTATION => "UNSIGNED",
                       LPM_DIRECTION => "ADD")
         PORT MAP (dataa => l3, datab => l4, result => r2);
  reg_3: lpm_ff                       -- Save MSBs of x+y
         GENERIC MAP ( LPM_WIDTH => WIDTH2 )
         PORT MAP ( data => r2, q => q2, clock => clk );
------------ Second stage of the adder -------------------
  -- One operand is zero
  h2 <= (OTHERS => '0');

  -- Add result from MSBs (x+y) and carry from LSBs
  add_3: lpm_add_sub
         GENERIC MAP ( LPM_WIDTH => WIDTH2,
                       LPM_REPRESENTATION => "UNSIGNED",
                       LPM_DIRECTION => "ADD")
         PORT MAP ( cin => cq1(0), dataa => q2,
                    datab => h2, result => u2 );

  PROCESS                 -- Build a single registered output
  BEGIN                   -- word of WIDTH=WIDTH1+WIDHT2
    WAIT UNTIL clk = '1';
    FOR k IN WIDTH1-1 DOWNTO 0 LOOP
      s(k) <= q1(k);
    END LOOP;
    FOR k IN WIDTH2-1 DOWNTO 0 LOOP
      s(k+WIDTH1) <= u2(k);
    END LOOP;
  END PROCESS;
```

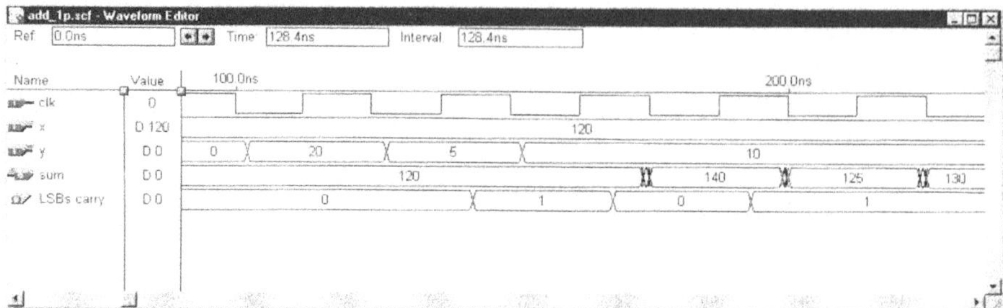

Fig. 2.10. Simulation results for a pipelined adder.

```
    sum <= s ;     -- Connect s to output pins
  END flex;
```

The simulated performance of the 15-bit pipelined adder is shown in Fig. 2.10. Note that the addition of 140 and 130 produces a carry from the lower 7-bit adder, but there is no carry for $120 + 5 = 125 < 127$. 2.14

2.3.2 Modulo Adders

Modulo adders are the most important building blocks in RNS-DSP designs. They are used for both additions and, via index arithmetic, for multiplications. We wish to describe some design options for FPGAs in the following discussion.

A wide variety of *modular* addition designs exists [43]. Using LEs only, the design of Fig. 2.11(a) is viable for FPGAs. The Altera FLEX devices contain a small number of 2Kbit ROMs or RAMs (EABs) which can be configured as $2^8 \times 8, 2^9 \times 4, 2^{10} \times 2$ or $2^{11} \times 1$ tables and can be used for modulo m_l correction. The next table shows size and `Registered Performance` 6, 7, and 8-bit modulo adder [44].

	Pipeline stages	Bits 6	7	8
MPX	0	41.3 MSPS 27 LE	46.5 MSPS 31 LE	33.7 MSPS 35 LE
MPX	2	76.3 MSPS 16 LE	62.5 MSPS 18 LE	60.9 MSPS 20 LE
MPX	3	151.5 MSPS 27 LE	138.9 MSPS 31 LE	123.5 MSPS 35 LE
ROM	3	86.2 MSPS 7 LE 1 EAB	86.2 MSPS 8 LE 1 EAB	86.2 MSPS 9 LE 2 EAB

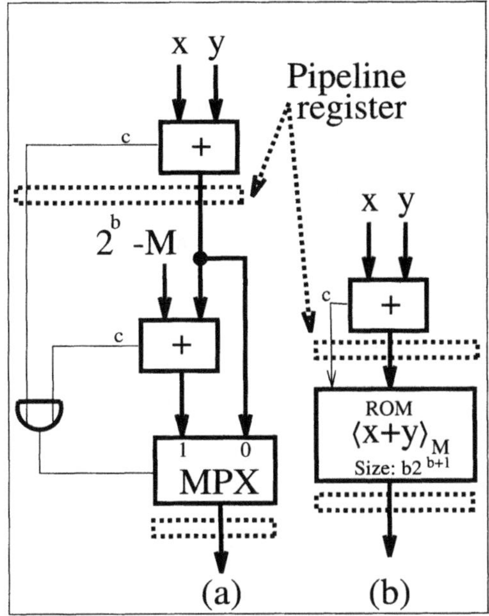

Fig. 2.11. Modular additions. **(a)** MPX-Add and MPX-Add-Pipe. **(b)** ROM-Pipe.

Although the ROM shown in Fig 2.11 provides high speed, the ROM itself produces a four-cycle pipeline delay and the number of ROMs is limited. ROMs, however, are mandatory for the scaling schemes discussed before. The multiplexed-adder (MPX-Add) has a comparatively reduced speed even if a carry chain is added to each column. The pipelined version usually needs the same number of LEs as the unpipelined version but runs about twice as fast. Maximum throughput occurs when the adders are implemented in two blocks within 6-bit pipelined channels.

2.4 Binary Multipliers

The product of two N-bit binary numbers, say X and $A = \sum_{k=0}^{N-1} a_k 2^k$, is given by the "pencil and paper" method as:

$$P = A \cdot X = \sum_{k=0}^{N-1} a_k 2^k X \tag{2.27}$$

It can be seen that the input X is successively shifted by k positions and whenever $a_k \neq 0$, then $X2^k$ is accumulated. If $a_k = 0$, then the corresponding shift-add can be ignored (i.e., nop). The following VHDL example uses this "pencil and paper" scheme to multiply two 8-bit integers.

Example 2.15: 8-bit Multiplier

The VHDL description[3] of an 8-bit multiplier is developed below. Multiplication is performed in three stages. First, the 8-bit operands are "loaded" and the product register reset. In the second stage, **s1**, the actual serial-parallel multiplication takes place. In the third step, **s2**, the product is transferred to the output register **y**.

```
PACKAGE eight_bit_int IS              -- User defined types
  SUBTYPE BYTE IS integer RANGE -128 TO 127;
  SUBTYPE TWOBYTES IS integer RANGE -32768 TO 32767;
END eight_bit_int;

LIBRARY work;
USE work.eight_bit_int.ALL;

LIBRARY ieee;                       -- Using predefined packages
USE ieee.std_logic_1164.ALL;
USE ieee.std_logic_arith.ALL;

ENTITY mul_ser IS                               ------> Interface
  PORT ( clk   : IN  STD_LOGIC;
         x     : IN  BYTE;
         a     : IN  STD_LOGIC_VECTOR(7 DOWNTO 0);
         y     : OUT TWOBYTES);
END mul_ser;

ARCHITECTURE flex OF mul_ser IS

  TYPE STATE_TYPE IS (s0, s1, s2);
  SIGNAL state    : STATE_TYPE;

BEGIN

  States: PROCESS     ------> Multiplier in behavioral style
    VARIABLE  p, t  : TWOBYTES;          -- Double bit width
    VARIABLE count  : integer RANGE 0 TO 7;
  BEGIN
    WAIT UNTIL clk = '1';
    CASE state IS
      WHEN s0 =>          -- Initialization step
        state <= s1;
        count := 0;
        p := 0;          -- Product register reset
        t := x;          -- Set temporary shift register to x
      WHEN s1 =>            -- Processing step
        IF count = 7 THEN -- Multiplication ready
          state <= s2;
          ELSE
          IF a(count) = '1' THEN
            p := p + t;      -- Add 2^k
          END IF;
```

[3] The equivalent Verilog code `mul_ser.v` for this example can be found in Appendix A on page 353.

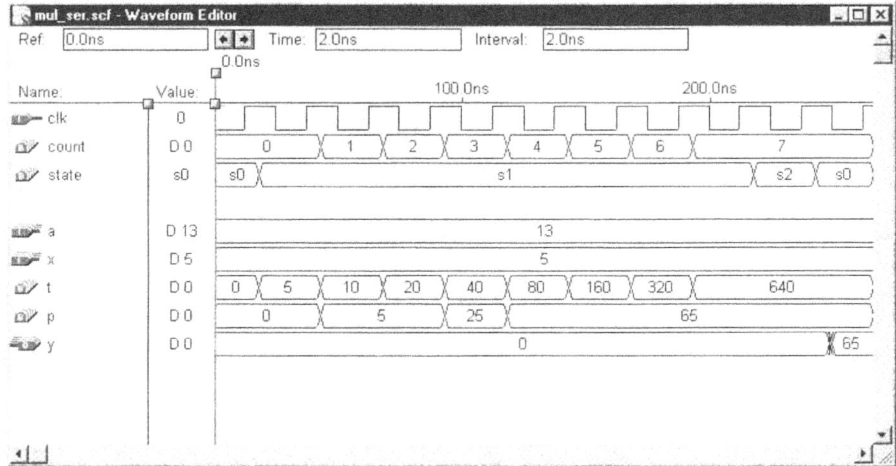

Fig. 2.12. Simulation results for a shift add multiplier.

```
          t := t * 2;
          count := count + 1;
          state <= s1;
        END IF;
      WHEN s2 =>            -- Output of result to y and
        y <= p;             -- start next multiplication
        state <= s0;
    END CASE;
  END PROCESS States;

END flex;
```

Fig. 2.12 shows the simulation result of a multiplication of 13 and 5. The register **t** shows the partial product sequence of $5, 10, 20, \ldots$. Since $13_{10} = 00001101_{2C}$, the product register **p** is updated only three times in the production of the final result, 65. In state **s2** the result 65 is transferred to the output **y** of the multiplier. 2.15

Because one operand is used in parallel (i.e., X) and the second operand A is used bitwise, the multipliers we just described are called serial/parallel multipliers. If both operands are used serial, the scheme is called a serial/serial multiplier [45], and such a multiplier only needs one full adder, but the latency of serial/serial multipliers is high $\mathcal{O}(N^2)$, because the state machine needs about N^2 cycles.

Another approach, which trades speed for increased complexity, is called an "array," or parallel/parallel multiplier. A 4×4-bit array multiplier is shown in Fig. 2.13. Notice that both operands are presented in parallel to an adder array of N^2 adder cells.

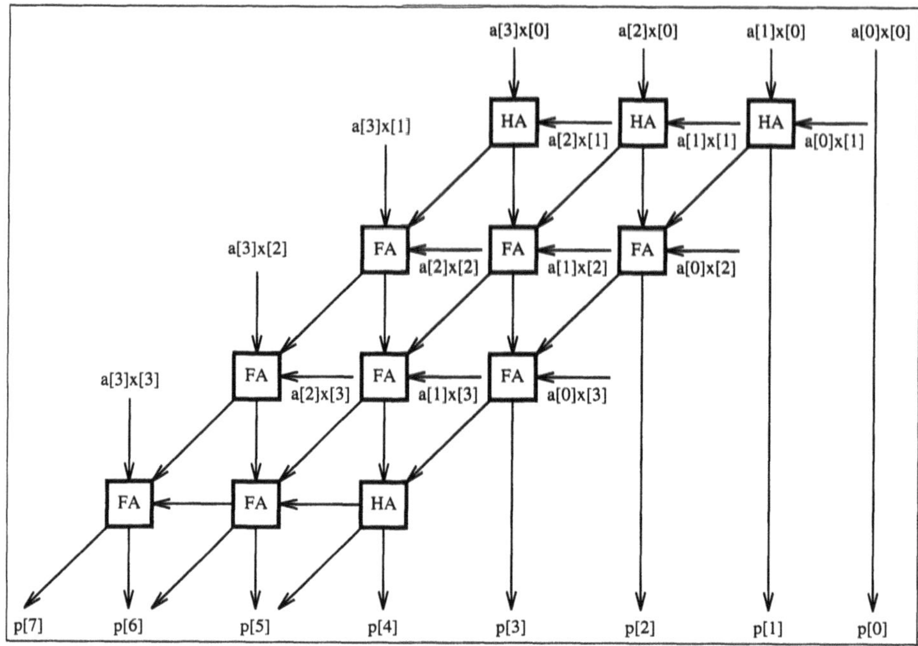

Fig. 2.13. A 4-bit array multiplier.

This arrangement is viable if the time required to complete the carry and sum calculations are the same. For a modern FPGA, however, the carry computation is performed faster than the sum calculation and a different architecture is more efficient for FPGAs. The approach for this array multiplier is shown in Fig. 2.14, for an 8×8-bit multiplier. This scheme combines in the first stage two neighboring partial products $a_n X 2^n$ and $a_{n+1} X 2^{n+1}$ and the results are added to arrrive at the final output product. This is a direct array form of the "pencil and paper" method and must therefore produce a valid product.

We recognize from Fig. 2.14 that this type of array multiplier gives the opportunity to realize a (parallel) *binary tree* of the multiplier with a total:

$$\text{number of stages in the binary tree multiplier} = \log_2(N). \qquad (2.28)$$

This alternative architecture also makes it easier to introduce pipeline stages after each tree level. The necessary number of pipeline stages, according to (2.28), to achieve maximum throughput is:

Bit width	2	$3-4$	$5-8$	$9-16$	$17-32$
Optimal number of pipeline stages	1	2	3	4	5

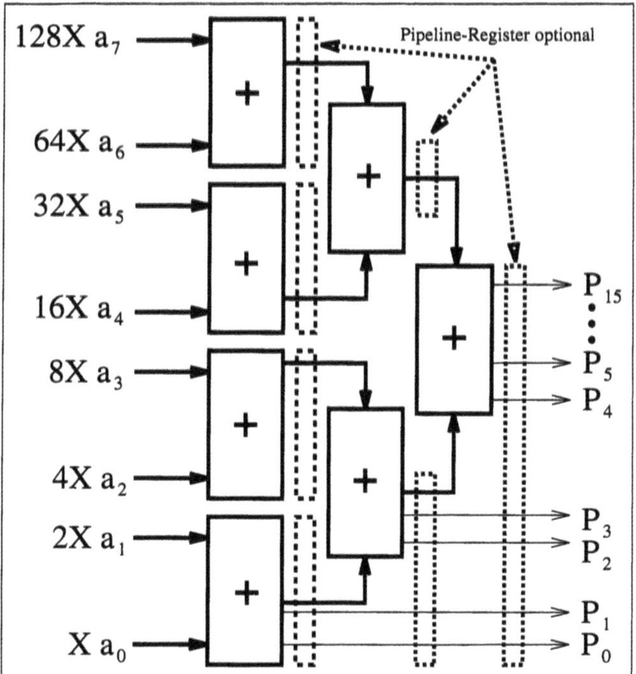

Fig. 2.14. Fast array multiplier for FPGAs.

Fig. 2.15 reports the `Registered Performance` of three pipelined $N \times N$-bit multipliers, using the MaxPlusII `lpm_mult` function, in a range from 4×4 to 16×16 bits without (dotted line) and with (solid line) the use of I/O cell registers. Fig. 2.16 shows the effort for the multiplier, with (solid line) and without (dotted line) using the I/O cell registers. Placing the input register close to the multiplier (turn off option: `Assign`→`Global Project Logic Synthesis`→`Automatic Fast I/O`), produces a slight gain in performance. The maximum size of multiplier that fits in the FLEX10K20 is a 20×20-bit unit that uses 872 LCs and runs at 37.03 MHz `Registered Performance` with four pipeline stages.

Other multiplier architectures typically used in the ASIC world include Booth multipliers and Wallace-tree multipliers. They are discussed in Exercises 2.2 (p. 76) and 2.1 (p. 75) but are rarely used in connection with FPGAs.

2.4.1 Multiplier Blocks

A $2N \times 2N$ multiplier can be defined in terms of an $N \times N$ multiplier block. The resulting multiplication is defined as:

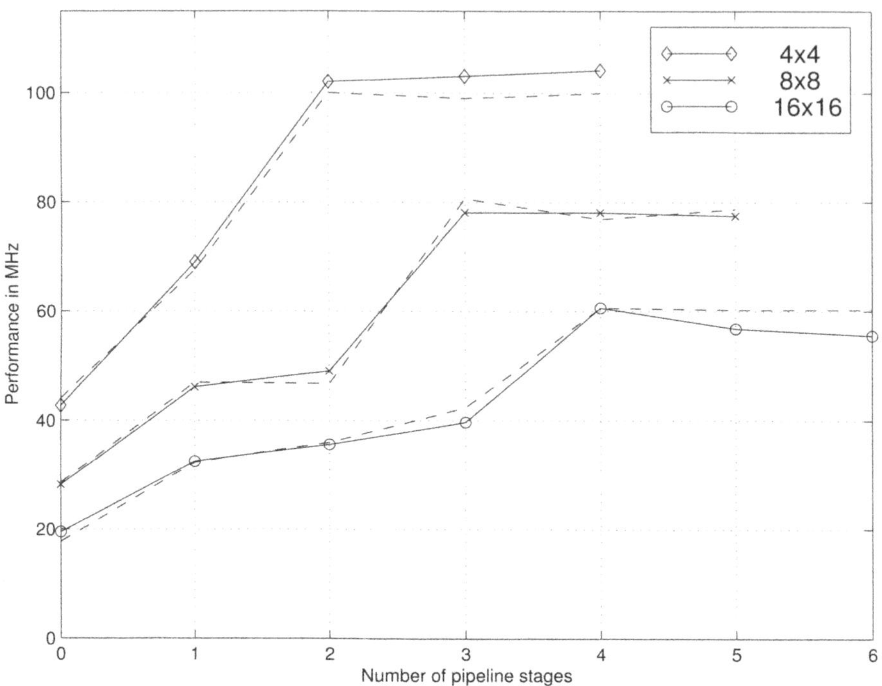

Fig. 2.15. Performance of array multiplier for FPGAs.

$$
\begin{aligned}
P = Y \cdot X &= (Y_2 2^N + Y_1)(X_2 2^N + Y_1) \\
&= Y_2 Y_2 2^{2N} + (Y_2 X_1 + Y_1 X_2) 2^N + Y_1 X_1
\end{aligned}
\tag{2.29}
$$

where the indices 2 and 1 indicate the most significant half and least significant N-bit halves respectively. This partitioning scheme can be used if the capacity of the FPGA is insufficient to implement a multiplier of desired size, or used to implement a multiplier using 2Kbit EAB blocks. The number of EAB blocks in the FLEX10K20 are limited to six, and therefore the maximum symmetric multiplier is 8×8. Table 2.8 shows the data for an EAB-based multiplier. Comparing the data of Table 2.8 with the data from Fig. 2.15 (p. 58) and 2.16 (p. 59), it can be seen that the EAB-based multiplier reduces the number of LCs but does not improve the `Registered Performance`.

2.5 Multiply-Accumulator (MAC) and Sum of Product (SOP)

DSP algorithms are known to be multiply-accumulate (MAC) intensive. To illustrate, consider the linear convolution sum given by

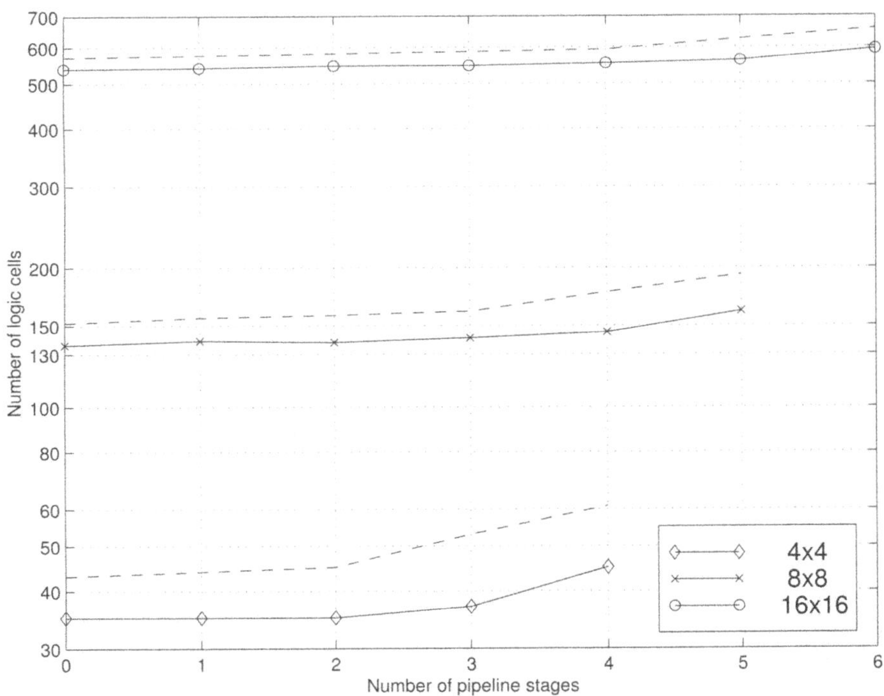

Fig. 2.16. Effort in LCs for array multipliers.

Table 2.8. Data for EAB-based multipliers.

Size	Use I/O register	LCs	EABs	Pipeline stages	Registered Performance
4×4	✓	0	1	0	35.58 MHz
4×4	✓	0	1	1	51.54 MHz
4×4	✓	0	1	2	76.33 MHz
4×4	–	16	1	0	37.87 MHz
4×4	–	16	1	1	51.81 MHz
4×4	–	16	1	2	76.33 MHz
8×8	✓	29	4	0	23.80 MHz
8×8	✓	29	4	1	40.81 MHz
8×8	✓	29	4	2	40.81 MHz
8×8	–	49	4	0	24.15 MHz
8×8	–	49	4	1	40.81 MHz
8×8	–	49	4	2	40.81 MHz

$$y[n] = f[n] * x[n] = \sum_{k=0}^{L-1} f[k]x[n-k] \tag{2.30}$$

requiring L consecutive multiplications and $L-1$ addition operations per sample $y[n]$ to compute the sum of products (SOPs). This suggests that $N \times N$-bit multipliers need to be fused together with an accumulator. A full-precision $N \times N$-bit product is $2N$ bits wide. If both operands are (symmetric) signed numbers, the product will only have $2N-1$ significant bits, i.e., two sign bits. The accumulator, in order to maintain sufficient dynamic range, is often designed to have an extra K-bits in width, as demonstrated in the following example.

Example 2.16: The Analog Devices PDSP family ADSP21xx contains a 16×16 array multiplier and an accumulator with an extra 8-bit width (for a total accumulator width of $32+8=40$ bits). With this 8 extra bits, at least 2^8 accumulations are possible without sacrificing the output. If both operands are signed 2^9 accumulation can be performed. In order to produce the desired output format, such modern PDSPs include also a barrelshifter, which allows the desired adjustment within one clock cycle. 2.16

This overflow consideration in fixed-point PDSP is important to mainstream digital signal processing, which requires that DSP objects be computed in real-time without unexpected interruptions. Recall that checking and servicing accumulator overflow interrupts the data flow and carries a significant temporal liability. By choosing the number of guard bits correctly, the liability can be eliminated.

An alternative approach to the MAC of a conventional PDSP for computing a sum of product will be discussed in the next section.

2.5.1 Distributed Arithmetic Fundamentals

Distributed arithmetic (DA) is an important FPGA technology. It is extensively used in computing the sum of products

$$y = \langle \boldsymbol{c}, \boldsymbol{x} \rangle = \sum_{n=0}^{N-1} c[n]x[n]. \tag{2.31}$$

Besides convolution, correlation, DFT computation and the RNS inverse mapping discussed earlier can also be formulated as such a "sum of products" (SOPs). Completing a filter cycle, when using a conventional arithmetic unit, would take approximately N MAC cycles. This amount can be shortened with pipelining but can, nevertheless, be prohibitively long. This is a fundamental problem when general purpose multipliers are used.

In many DSP applications, a general purpose multiplication is technically not required. If the filter coefficients $c[n]$ are known apriori, then technically the partial product term $c[n]x[n]$ becomes a multiplication with a constant (i.e., scaling). This is an important difference and is a prerequisite for a DA design.

The first discussion of DA can be traced to a 1973 paper by Croisier [46] and DA was popularized by Peled and Liu [47]. Yiu [48] extended DA to signed numbers, and Kammeyer [49] and Taylor [50] studied quantization effects in DA systems. DA tutorials are available from White [51] and Kammeyer [52]. DA also is addressed in textbooks [53, 54]. To understand the DA design paradigm, consider the "sum of products" inner product shown below:

$$
\begin{aligned}
y = \langle \boldsymbol{c}, \boldsymbol{x} \rangle &= \sum_{n=0}^{N-1} c[n] \cdot x[n] \\
&= c[0]x[0] + c[1]x[1] + \ldots + c[N-1]x[N-1],
\end{aligned} \tag{2.32}
$$

Assume further that the coefficients $c[n]$ are known constants and $x[n]$ is a variable. An unsigned DA system assumes that the variable $x[n]$ is represented by:

$$
x[n] = \sum_{b=0}^{B-1} x_b[n] \cdot 2^b \quad \text{with } x_b[n] \in [0, 1], \tag{2.33}
$$

where $x_b[n]$ denotes the bth bit of $x[n]$, i.e., the nth sample of $\boldsymbol{x}$. The inner product y can, therefore, be represented as:

$$
y = \sum_{n=0}^{N-1} c[n] \cdot \sum_{b=0}^{B-1} x_b[k] \cdot 2^b \tag{2.34}
$$

Re-distributing the order of summation (thus the name "distributed arithmetic") results in:

$$
\begin{aligned}
y = \; & c[0] \left(x_{B-1}[0]2^{B-1} + x_{B-2}[0]2^{B-2} + \ldots + x_0[0]2^0 \right) \\
& + c[1] \left(x_{B-1}[1]2^{B-1} + x_{B-2}[1]2^{B-2} + \ldots + x_0[1]2^0 \right) \\
& \vdots \\
& + c[N-1] \left(x_{B-1}[N-1]2^{B-1} + \ldots + x_0[N-1]2^0 \right) \\
= \; & \left(c[0]x_{B-1}[0] + c[1]x_{B-1}[1] + \ldots + c[N-1]x_{B-1}[N-1] \right) 2^{B-1} \\
& + \left(c[0]x_{B-2}[0] + c[1]x_{B-2}[1] + \ldots + c[N-1]x_{B-2}[N-1] \right) 2^{B-2} \\
& \vdots \\
& + \left(c[0]x_0[0] + c[1]x_0[1] + \ldots + c[N-1]x_0[N-1] \right) 2^0
\end{aligned}
$$

or in more compact form

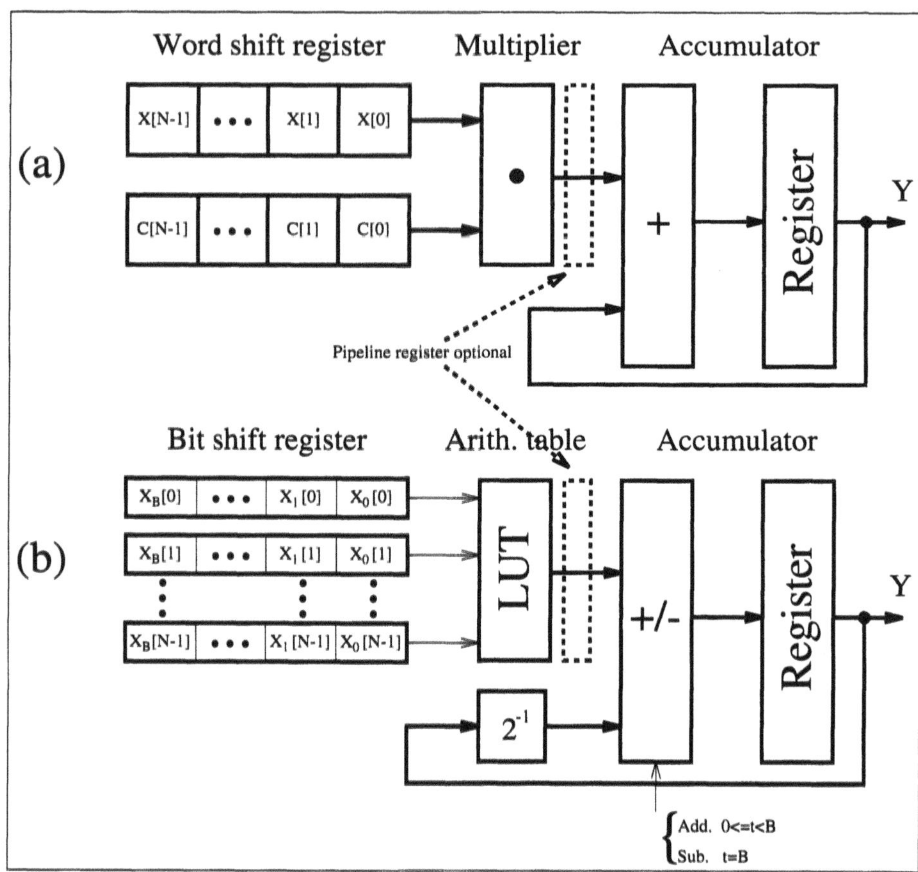

Fig. 2.17. Conventional PDSP and Shift-Adder DA Architecture.

$$y = \sum_{b=0}^{B-1} 2^b \cdot \sum_{n=0}^{N-1} \underbrace{c[n] \cdot x_b[n]}_{f(c[n],x_b[n])} = \sum_{b=0}^{B-1} 2^b \cdot \sum_{n=0}^{N-1} f\left(c[n], x_b[n]\right) \qquad (2.35)$$

Implementation of the function $f(c[n], x_b[n])$ requires special attention. The preferred implementation method is to realize the mapping $f(c[n], x_b[n])$ using one LUT. That is, a 2^N-word LUT is preprogrammed to accept an N-bit input vector $\boldsymbol{x}_b = [x_b[0], x_b[1], \cdots, x_b[N-1]]$, and output $f(c[n], x_b[n])$. The individual mappings $f(c[n], x_b[n])$ are weighted by the appropriate power-of-two factor and accumulated. The accumulation can be efficiently implemented using a shift-adder as shown in Fig. 2.17(b). After N look-up cycles, the inner product y is computed.

Example 2.17: Unsigned DA Convolution

A third-order innerproduct is defined by the inner product equation $y = \langle c, x \rangle = \sum_{n=0}^{2} c[n]x[n]$. Assume that the 3-bit coefficients have the values $c[0] = 2$, $c[1] = 3$, and $c[2] = 1$. The resulting LUT, which implements $f(c[n], x_b[n])$, is defined below:

$x_b[2]$	$x_b[1]$	$x_b[0]$	$f(c[n], x[n])$
0	0	0	$1 \cdot 0{+}3 \cdot 0{+}2 \cdot 0{=}0_{10}{=}000_2$
0	0	1	$1 \cdot 0{+}3 \cdot 0{+}2 \cdot 1{=}2_{10}{=}001_2$
0	1	0	$1 \cdot 0{+}3 \cdot 1{+}2 \cdot 0{=}3_{10}{=}011_2$
0	1	1	$1 \cdot 0{+}3 \cdot 1{+}2 \cdot 1{=}5_{10}{=}101_2$
1	0	0	$1 \cdot 1{+}3 \cdot 0{+}2 \cdot 0{=}1_{10}{=}001_2$
1	0	1	$1 \cdot 1{+}3 \cdot 0{+}2 \cdot 1{=}3_{10}{=}011_2$
1	1	0	$1 \cdot 1{+}3 \cdot 1{+}2 \cdot 0{=}4_{10}{=}100_2$
1	1	1	$1 \cdot 1{+}3 \cdot 1{+}2 \cdot 1{=}6_{10}{=}110_2$

The inner-product, with respect to $x[n] = \{x[0] = 1_{10} = 001_2, x[1] = 3_{10} = 011_2, x[2] = 7_{10} = 111_2\}$, is obtained as follows:

Step t	$x_t[2]$	$x_t[1]$	$x_t[0]$	$f[t]$	$+ACC[t-1]$	=	$ACC[t]$
0	1	1	1	$6 \cdot 2^0 +$	0	=	6
1	1	1	0	$4 \cdot 2^1 +$	6	=	14
2	1	0	0	$1 \cdot 2^2 +$	14	=	18

As a numerical check, note that

$$y = \langle c, x \rangle = c[0]x[0] + c[1]x[1] + c[2]x[2]$$
$$= 2 \cdot 1 + 3 \cdot 3 + 1 \cdot 7 = 18.$$

2.17

For a hardware implementation, instead of shifting each intermediate value by b (which will demand an expensive barrelshifter) it is more appropriate to shift the accumulator content itself in each iteration one bit to the right. It is easy to verify that this will give the same results.

The bandwidth of an Nth-order B-bit linear convolution, using general purpose MACs and DA hardware, can be compared. Fig. 2.17 shows the architectures of a conventional PDSP and the same realization using distributed arithmetic.

Assume that a LUT and a general purpose multiplier have the same delay $\tau = \tau(\text{LUT}) = \tau(\text{MUL})$. The computational latencies are then $B\tau(\text{LUT})$ for DA and $N\tau(\text{MUL})$ for the PDSP. In the case of small-bit-width B, the speed of the DA design can therefore be significantly faster than a MAC-based design. In Chapter 3, comparisons will be made for specific filter design examples.

2.5.2 Signed DA Systems

In the following we wish to discuss how (2.32) should be modified, in order to process a signed two's complement number. In two's complement, the MSB is used to distinguish between positive and negative numbers. For instance, from Table 2.1 (p. 32) we see that decimal -3 is coded as $101_2 = -4+0+1 = -3_{10}$. We use, therefore, the following $(B+1)$ bit representation

$$x[n] = -2^b \cdot x_B[n] + \sum_{b=0}^{B-1} x_b[n] \cdot 2^b. \tag{2.36}$$

Combining this with (2.34), the outcome y is defined by:

$$y = -2^b \cdot f(c[n], x_B[n]) + \sum_{b=0}^{B-1} 2^b \cdot \sum_{n=0}^{N-1} f\left(c[n], x_b[n]\right). \tag{2.37}$$

To achieve the signed DA system we therefore have two choices to modify the unsigned DA system. They are

- An accumulator with add/subtract control
- Using a ROM with one additional input

Most often the switchable accumulator is preferred, because the additional input bit in the table requires a table with twice as many words. The following example demonstrates the processing steps for the add/sub switch design.

Example 2.18: Signed DA Inner Product

Consider again a third-order inner product defined by the convolution sum $y = \langle \boldsymbol{c}, \boldsymbol{x} \rangle = \sum_{n=0}^{2} c[n]x[n]$. Assume that the data is given a $N = 4$-bit two's complement encoding and that the coefficients are $c[0] = -2$, $c[1] = 3$, and $c[2] = 1$. The corresponding LUT table is given below:

$x_b[2]$	$x_b[1]$	$x_b[0]$	$f(c[k], x[n])$
0	0	0	$1 \cdot 0 + 3 \cdot 0 - 2 \cdot 0 = 0_{10}$
0	0	1	$1 \cdot 0 + 3 \cdot 0 - 2 \cdot 1 = -2_{10}$
0	1	0	$1 \cdot 0 + 3 \cdot 1 - 2 \cdot 0 = 3_{10}$
0	1	1	$1 \cdot 0 + 3 \cdot 1 - 2 \cdot 1 = 1_{10}$
1	0	0	$1 \cdot 1 + 3 \cdot 0 - 2 \cdot 0 = 1_{10}$
1	0	1	$1 \cdot 1 + 3 \cdot 0 - 2 \cdot 1 = -1_{10}$
1	1	0	$1 \cdot 1 + 3 \cdot 1 - 2 \cdot 0 = 4_{10}$
1	1	1	$1 \cdot 1 + 3 \cdot 1 - 2 \cdot 1 = 2_{10}$

The values of $x[k]$ are $x[0] = 1_{10} = 0001_{2C}$, $x[1] = -3_{10} = 1101_{2C}$, and $x[2] = 7_{10} = 0111_{2C}$. The output at sample index k, namely y, is defined as follows:

Step t	$x_t[2]$	$x_t[1]$	$x_t[0]$	$f[t] \cdot 2^t$	$+Y[t-1]$	$=Y[t]$
0	1	1	1	$2 \cdot 2^0$	+ 0	= 2
1	1	0	0	$1 \cdot 2^1$	+ 2	= 4
2	1	1	0	$4 \cdot 2^2$	+ 4	= 20
	$x_t[2]$	$x_t[1]$	$x_t[0]$	$-f[t] \cdot 2^t$	$+Y[t-1]$	$=Y[t]$
3	0	1	0	$-3 \cdot 2^3$	+ 20	= −4

A numerical check results in $c[0]x[0]+c[1]x[1]+c[2]x[2] = -2\cdot1+3\cdot(-3)+1\cdot7 = -4$ ✓ 2.18

2.5.3 Modified DA Solutions

In the following we wish to discuss two interesting modifications to the basic DA concept, where the first variation reduces the size, and the second increases the speed.

If the number of coefficients N is too large to implement the full word with a single LUT (recall that input LUT bit width = number of coefficients), then we can use partial tables and add the results. If we also add pipeline registers, this modification will not reduce the speed, but can dramatically reduce the size of the design, because the size of a LUT grows exponentially with the address space, i.e., the number of input coefficients N. Suppose the length LN inner product

$$y = \langle \boldsymbol{c}, \boldsymbol{x} \rangle = \sum_{n=0}^{LN-1} c[n]x[n] \tag{2.38}$$

is to be implemented using a DA architecture. The sum can be partitioned into L independent N^{th} parallel DA LUTs resulting in

$$y = \langle \boldsymbol{c}, \boldsymbol{x} \rangle = \sum_{l=0}^{L-1} \sum_{n=0}^{N-1} c[Ll+n]x[Ll+n] \tag{2.39}$$

This is shown in Fig. 2.18 for a realization of a $4N$ DA design requiring three post-additional adders. The size of the table is reduced from one $4N \times 2^B$ LUT to four $N \times 2^B$ tables.

Another variation of the DA architecture increases speed at the expense of additional LUTs, registers, and adders. A basic DA architecture, for a length N^{th} sum-of-product computation, accepts one bit from each of N words. If two bits per word are accepted, then the computational speed can be essentially doubled. The maximum speed can be achieved with the fully pipelined word-parallel architecture shown in Fig. 2.19. Here, a new result of a length four sum-of-product is computed for 4-bit signed coefficients at each LUT cycle. For maximum speed, we have to provide a separate ROM (with identical content) for each bit vector $x_b[n]$. But the maximum speed can become expensive: If we double the input bit width, we need twice as many LUTs, adders and registers. If the number of coefficients N is limited to four or eight this modification gives attractive performance, essentially outperforming all commercially available programmable signal processors, as we will see in Chapter 3.

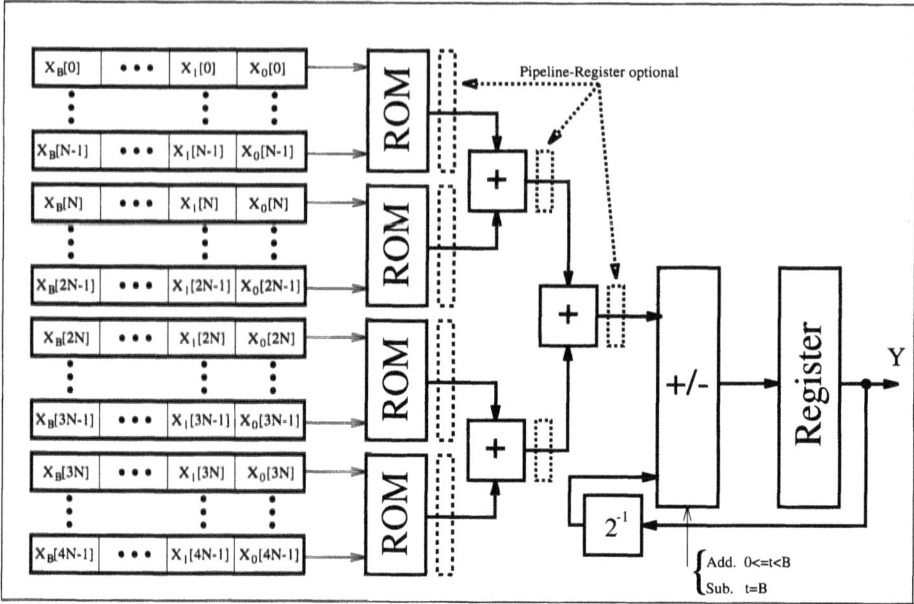

Fig. 2.18. Distributed arithmetic with table partitioning to yield a reduced size.

2.6 Computation of Special Functions Using CORDIC

If a digital signal processing algorithm is implemented with FPGAs and the algorithm uses a nontrivial (transcendental) algebraic function, like $\sqrt{x}$ or $\arctan y/x$, we can always use the Taylor series to approximate this function, i.e.,

$$f(x_0) = \sum_{k=0}^{K} \frac{\mathrm{d}f^{(k)}(x - x_0)}{\mathrm{d}x} x^k \bigg|_{x=x_0} \tag{2.40}$$

and the problem is reduced to a sequence of multiply and add operations. A more efficient, alternative approach, based on the *Coordinate Rotation Digital Computer* (CORDIC) algorithm can also be considered. The CORDIC algorithm is found in numerous applications, such as pocket calculators [55], and in mainstream DSP objects, such as adaptive filters, FFTs, DCTs [56], demodulators [57], and neural networks [36]. The basic CORDIC algorithm can be found in two classic papers by Volder [58] and Walther [59]. Some theoretical extensions have been made, such as the extension of range in the hyperbolic mode, or the quantization error analysis by Hu et al. [60], and Meyer-Bäse et al. [57]. VLSI implementations have been discussed in Ph.D. theses, such as those by Timmermann [61] or Hahn [62]. The first FPGA implementations were investigated by Meyer-Bäse et al. [4, 57]. The realization of the CORDIC algorithm in distributed arithmetic was investigated by Ma

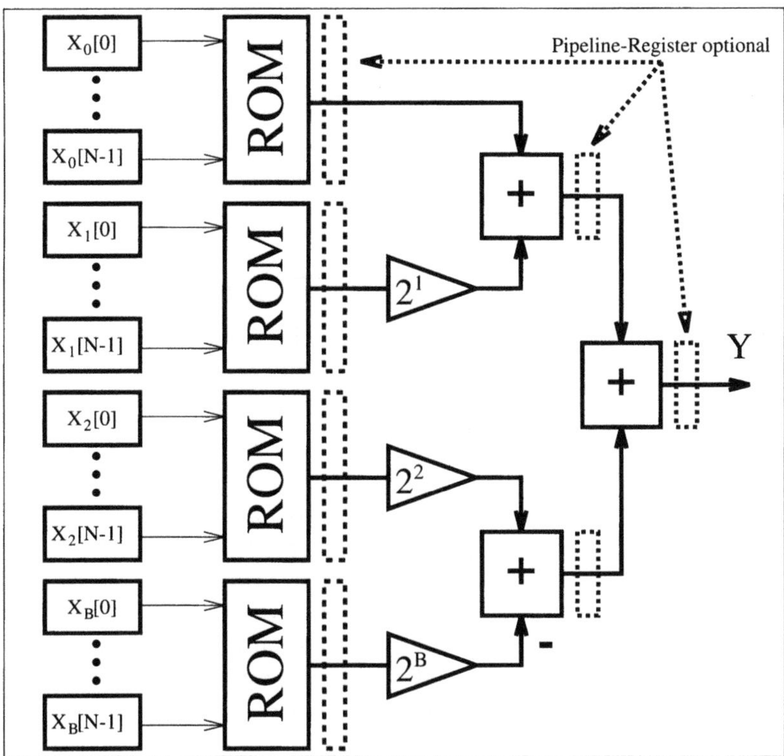

Fig. 2.19. Higher-order distributed arithmetic optimized for speed.

[63]. A very detailed overview including details of several applications, was provided by Y. Hu [56] in a 1992 IEEE Signal Processing Magazine review paper.

The original CORDIC algorithm by Volder [58] computes a multiplier-free coordinate conversion between rectangular (x, y) and polar (R, θ) coordinates. Walther [59] generalized the CORDIC algorithm to include circular- $(m = 1)$, linear- $(m = 0)$, and hyperbolic- $(m = -1)$ transforms. For each mode, two rotation directions are identified. For *vectoring*, a vector with starting coordinates (X_0, Y_0) is rotated in such a way that the vector finally lies on the abscissa (i.e., x axis) by iteratively converging Y_K to zero. For *rotation*, a vector with a starting coordinate (X_0, Y_0) is rotated by an angle θ_0 in such a way that the final value of the angle register, denoted Z, converges to zero. The angle θ_k is chosen so that each iteration can be performed with an addition and a binary shift. Table 2.9 shows in the second column the choice for the rotation angle for the three modes $m = 1, 0$, and -1.

Now we can formally define the CORDIC algorithm as follows:

Table 2.9. CORDIC algorithm modes.

Modus	Angle θ_k	Shift-Sequence	Radius-Factor
circular $m = 1$	$\tan^{-1}(2^{-k})$	$0, 1, 2, \ldots$	$K_1 = 1.65$
linear $m = 0$	2^{-k}	$1, 2, \ldots$	$K_0 = 1.0$
hyperbolic $m = -1$	$\tanh^{-1}(2^{-k})$	$1, 2, 3, 4, 4, \ldots$	$K_{-1} = 0.80$

Algorithm 2.19: CORDIC Algorithm

At each iteration, the CORDIC algorithm implements the mapping:

$$\begin{bmatrix} X_{k+1} \\ Y_{k+1} \end{bmatrix} = \begin{bmatrix} 1 & m\delta_k 2^{-k} \\ m\delta_k 2^{-k} & 1 \end{bmatrix} \begin{bmatrix} X_k \\ Y_k \end{bmatrix} \tag{2.41}$$

$$Z_{k+1} = Z_k + \delta_k \theta_k$$

where the two rotation directions are $Z_K \to 0$ and $Y_K \to 0$.

This means that six operational modes exist, and they are summarized in Table 2.10. A consequence is that nearly all transcendental functions can be computed with the CORDIC algorithm. With a proper choice of the initial values, the function $X \cdot Y, Y/X, \sin(Z), \cos(Z), \tan^{-1}(Y)$, $\sinh(Z), \cosh(Z)$, and $\tanh(Z)$ can directly be computed. Additional functions may be generated by choosing appropriate initialization, sometimes combined with multiple modes of operation, as shown in the following listing:

$\tan(Z)=\sin(Z)/\cos(Z)$	Modes: $m = 1, 0$
$\tanh(Z)=\sinh(Z)/\cosh(Z)$	Modes: $m = -1, 0$
$\exp(Z)=\sinh(Z)+\cosh(Z)$	Modes: $m = -1$; $x = y = 1$
$\log_e(W)=2\tanh^{-1}(Y/X)$	Modes: $m = -1$ with $X = W + 1, Y = W - 1$
$\sqrt{W}=\sqrt{X^2 - Y^2}$	Modes: $m = 1$ with $X = W + \frac{1}{4}, Y = W - \frac{1}{4}$,

A careful analysis of (2.41) reveals that the iteration vectors only approach the curves shown in Fig. 2.20(a). The length of the vectors changes with each

Table 2.10. Modes m of operation for the CORDIC algorithm.

m	$Z_K \to 0$	$Y_K \to 0$
1	$X_K = K_1(X_0 \cos(Z_0) - Y_0 \sin(Z_0))$ $Y_K = K_1(X_0 \cos(Z_0) + Y_0 \sin(Z_0))$	$X_K = K_1\sqrt{X_0^2 + Y_0^2}$ $Z_K = Z_0 + \arctan(Y_0/X_0)$
0	$X_K = X_0$ $Y_K = Y_0 + X_0 \cdot Z_0$	$X_K = X_0$ $Z_K = Z_0 + Y_0/X_0$
−1	$X_K = K_{-1}(X_0 \cosh(Z_0) - Y_0 \sinh(Z_0))$ $Y_K = K_{-1}(X_0 \cosh(Z_0) + Y_0 \sinh(Z_0))$	$X_K = K_{-1}\sqrt{X_0^2 + Y_0^2}$ $Z_K = Z_0 + \tanh^{-1}(Y_0/X_0)$

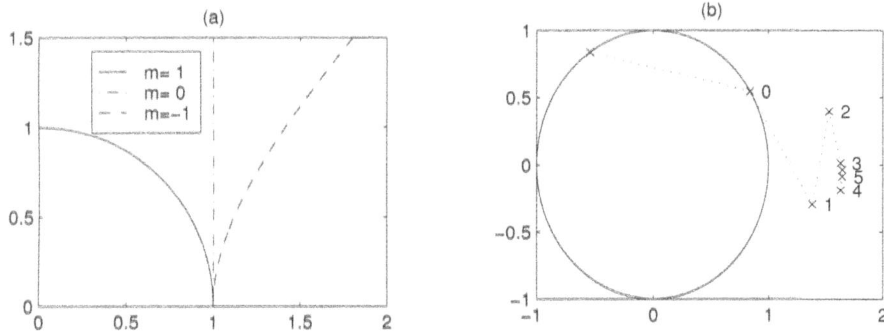

Fig. 2.20. CORDIC. **(a)** Modes. **(b)** Example for circular vectoring.

iteration, as shown in Fig. 2.20(b). This change in length does *not* depend on the starting angle and after K iterations the same change always occurs. In the last column of Table 2.9 the radius factors are shown. To ensure that the CORDIC algorithm converges, the sum of all remaining rotation angles must be larger than the actual rotation angle. This is the case for linear and circular transforms. For the hyperbolic mode, all iterations of the form $n_{k+1} = 3n_k + 1$ have to be repeated. These are the iterations $4, 13, 40, 121 \ldots$.

Output precision can be estimated using a procedure developed by Hu [64] and illustrated in Fig. 2.21. The graph shows the effective bit precision for the circular mode, depending on the X, Y path width, and the number of iterations. If b bits is the desired output precision, the "rule of thumb"

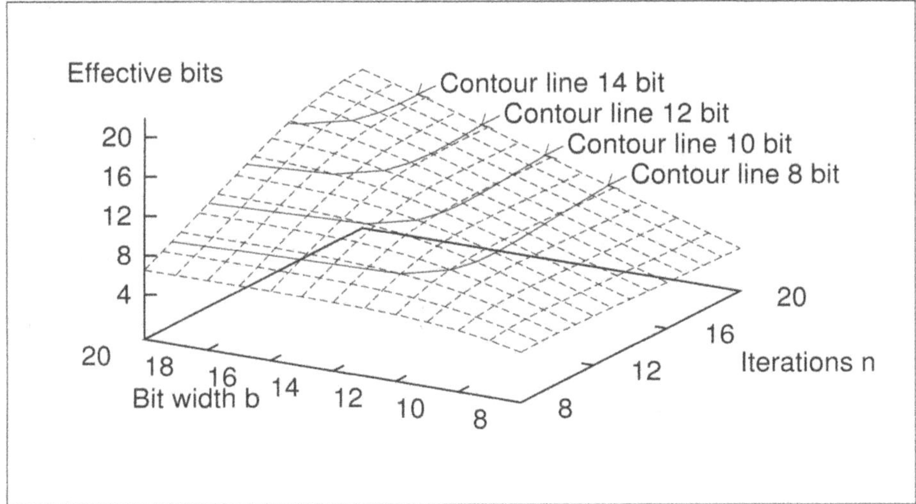

Fig. 2.21. Effective bits in circular mode.

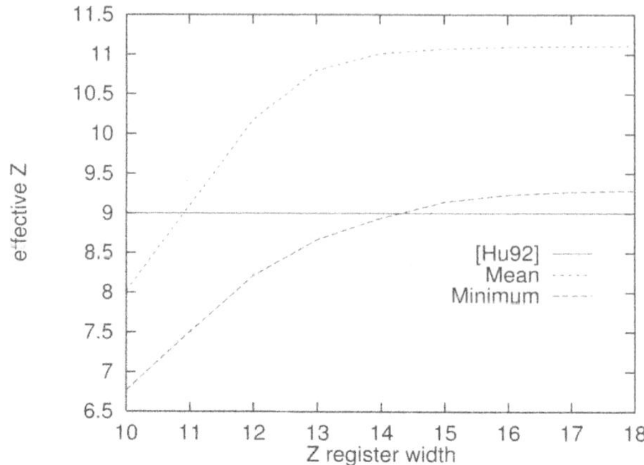

Fig. 2.22. Resolution of phase for circular mode.

suggests that the X, Y path should have $\log_2(b)$ additional guard bits. From Fig. 2.22, it can also be seen that the bit width of the Z path should have the same precision as that for X and Y.

In contrast to the circular CORDIC algorithm, the effective resolution of a hyperbolic CORDIC cannot be computed analytically because the precision depends on the angular values of $z(k)$ at iteration k. Hyperbolic precision can, however, be estimated using simulation. Fig. 2.23 shows the minimum accuracy estimate computed over 1,000 test values for each bit-width/number combination of the possible iterations. The 3D representation shows the number of iterations, the bit width of the X/Y path, and the resulting minimum precision of the result in terms of effective bits. The contour lines allow an exchange between the number of iterations and the bit width. For example, to achieve 10-bit precision, one can use a 21-bit X/Y path and 18 iterations, or 14 iterations at 24 bits.

2.6.1 CORDIC Architectures

Two basic structures are used to implement a CORDIC architecture: the more compact state machine or the high-speed, fully pipelined processor.

If computation time is not critical, then a state machine as shown in Fig. 2.24 is applicable. In each cycle, exactly one iteration of (2.41) will be computed. The most complex part of this design is the two barrelshifters. The two barrelshifters can be replaced by a single barrelshifter, using a multiplexer as shown in Fig. 2.25, or a serial (right, or right/left) shifter. Table 2.11 compares different design options for a 13-bit implementation using Xilinx XC3K FPGAs.

If high speed is needed, a fully pipelined version of the design shown in Fig. 2.26 can be used. Fig. 2.26 shows eight iterations of a circular CORDIC.

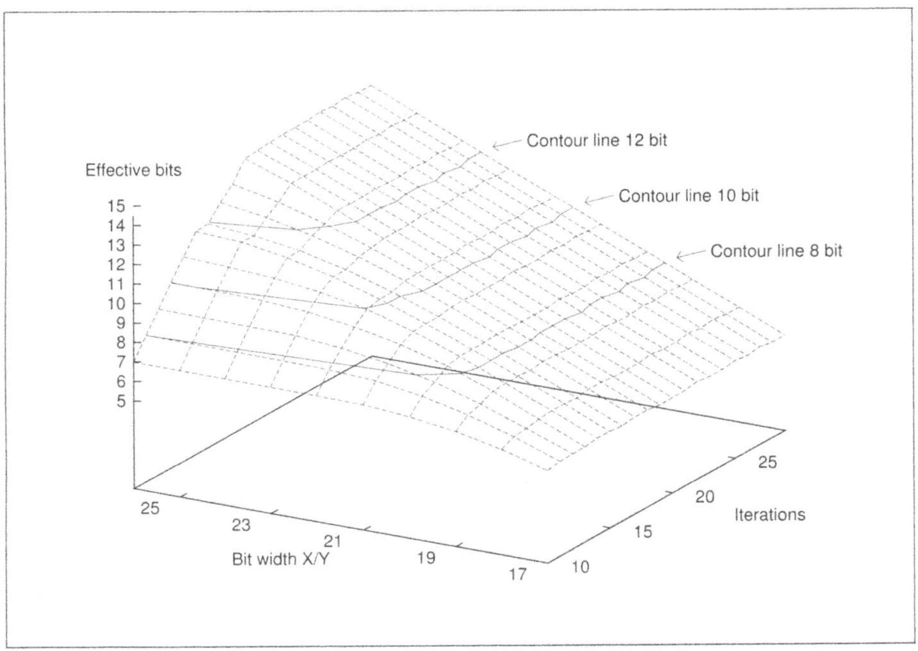

Fig. 2.23. Effective bits in hyperbolic mode.

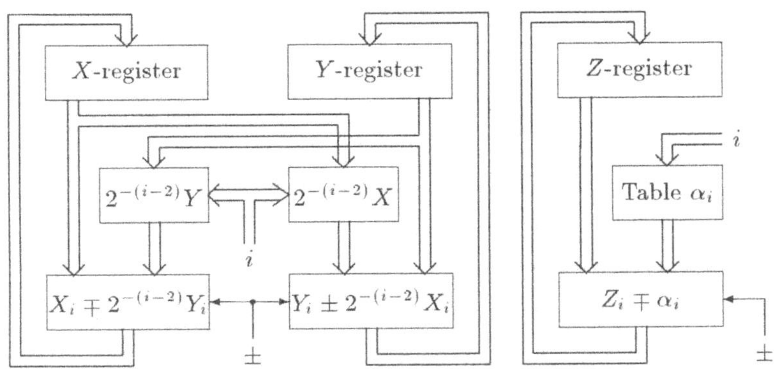

Fig. 2.24. CORDIC state machine.

After an initial delay of K cycles, a new output value becomes available after each cycle. As with array multipliers, CORDIC implementations have a quadratic growth in LE complexity as the bit width increases (see Table 2.9).

The following example shows the first four steps of a circular-vectoring fully pipelined design.

Table 2.11. Effort estimation (Xilinx XC3K) for a CORDIC a machine with 13-bits plus sign for X/Y path. (Abbreviations: Ac=accumulator; BS=barrelshifter; RS=serial right shifter; LRS=serial left right shifter)

Structure	Registers	Multiplexer	Adder	Shifter	$\sum$LE	Cycle
2BS+2Ac	2·7	0	2·14	2·19.5	81	12
2RS+2Ac	2·7	0	2·14	2·6.5	55	46
2LRS+2Ac	2·7	0	2·14	2·8	58	39
1BS+2Ac	7	3·7	2·14	19,5	75.5	20
1RS+2Ac	7	3·7	2·14	6,5	62.5	56
1LRS+2Ac	7	3·7	2·14	8	64	74
1BS+1Ac	3·7	2·7	14	19,5	68.5	20
1RS+1Ac	3·7	2·7	14	6,5	55.5	92
1LRS+1Ac	3·7	2·7	14	8	57	74

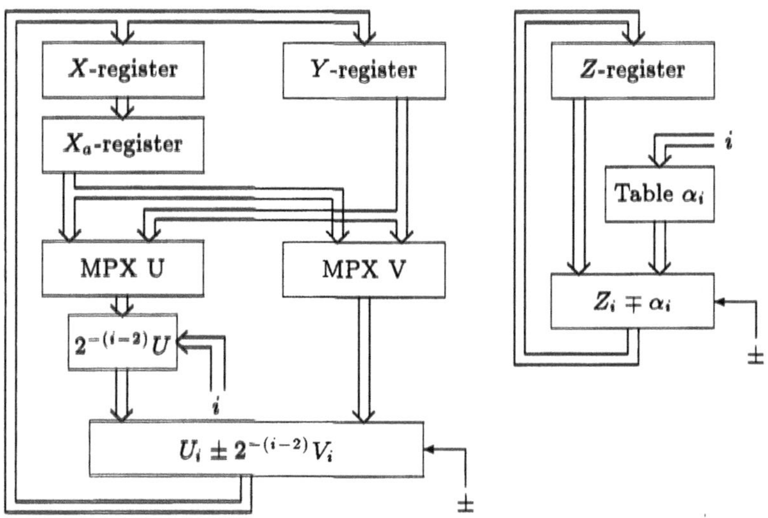

Fig. 2.25. CORDIC machine with reduced complexity.

Example 2.20: Circular CORDIC in Vectoring Mode

The first iteration rotates the vectors from the second or third quadrant to the first or fourth, respectively. The shift sequence is 0,0,1, and 2. The rotation angle of the first four steps becomes: $\arctan(\infty) = 90°, \arctan(2^0) = 45°, \arctan(2^{-1}) = 26.5°$, and $\arctan(2^{-2}) = 14°$. The VHDL code[4] for 8-bit data can be implemented as follows:

```
PACKAGE eight_bit_int IS    -- User defined types
```

[4] The equivalent Verilog code `cordic.v` for this example can be found in Appendix A on page 354.

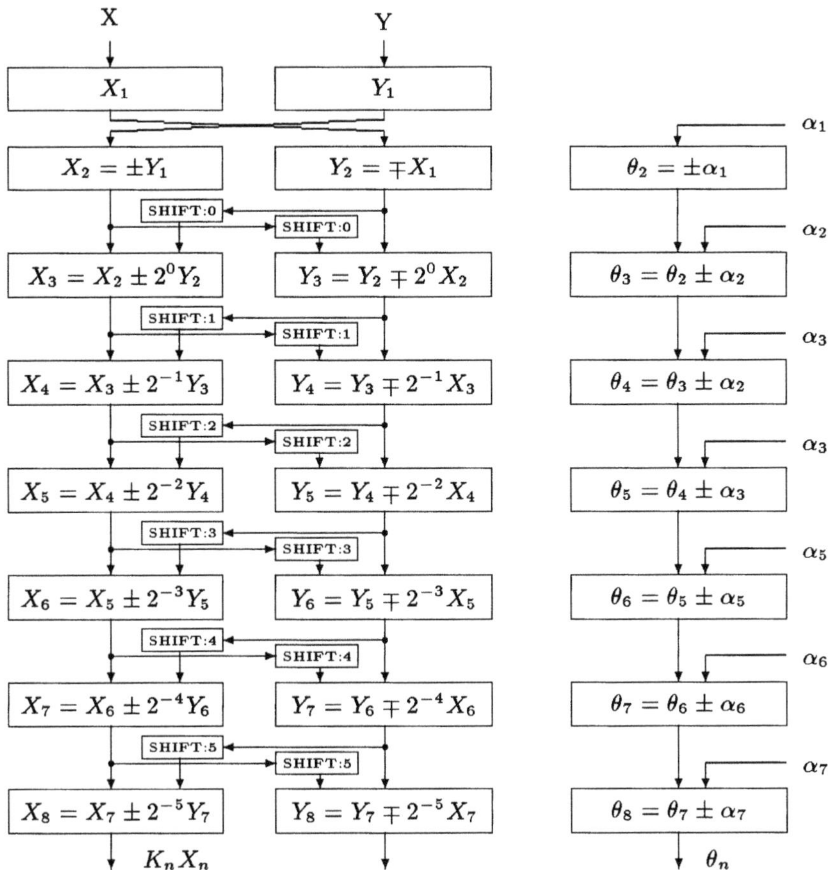

Fig. 2.26. Fast CORDIC pipeline.

```
  SUBTYPE BYTE IS INTEGER RANGE -128 TO 127;
  TYPE ARRAY_BYTE IS ARRAY (0 TO 3) OF BYTE;
END eight_bit_int;

LIBRARY work;
USE work.eight_bit_int.ALL;

LIBRARY ieee;
USE ieee.std_logic_1164.ALL;
USE ieee.std_logic_arith.ALL;

ENTITY cordic IS                          ------> Interface
       PORT (clk          : IN  STD_LOGIC;
             x_in , y_in : IN  BYTE;
             r, phi, eps : OUT BYTE);
END cordic;
```

```
ARCHITECTURE flex OF cordic IS
  SIGNAL  x, y, z  :   ARRAY_BYTE; -- Array of Bytes
BEGIN

  PROCESS                  ------> Behavioral Style
  BEGIN
    WAIT UNTIL clk = '1'; -- Compute last value first in
    r <= x(3);            -- sequential VHDL statements !!
    phi <= z(3);
    eps <= y(3);

    IF y(2) > 0 THEN                      -- Rotate 14 degrees
      x(3) <= x(2) + y(2) /4;
      y(3) <= y(2) - x(2) /4;
      z(3) <= z(2) + 14;
    ELSE
      x(3) <= x(2) - y(2) /4;
      y(3) <= y(2) + x(2) /4;
      z(3) <= z(2) - 14;
    END IF;

    IF y(1) > 0 THEN                      -- Rotate 26 degrees
      x(2) <= x(1) + y(1) /2;
      y(2) <= y(1) - x(1) /2;
      z(2) <= z(1) + 26;
    ELSE
      x(2) <= x(1) - y(1) /2;
      y(2) <= y(1) + x(1) /2;
      z(2) <= z(1) - 26;
    END IF;

    IF y(0) > 0 THEN                     -- Rotate  45 degrees
      x(1) <= x(0) + y(0);
      y(1) <= y(0) - x(0);
      z(1) <= z(0) + 45;
    ELSE
      x(1) <= x(0) - y(0);
      y(1) <= y(0) + x(0);
      z(1) <= z(0) - 45;
    END IF;

-- Test for x_in < 0 rotate 0,+90, or -90 degrees
    IF x_in > 0 THEN
      x(0) <= x_in;       -- Input in register 0
      y(0) <= y_in;
      z(0) <= 0;
    ELSIF y_in > 0 THEN
      x(0) <= y_in;
      y(0) <= - x_in;
      z(0) <= 90;
    ELSE
      x(0) <= - y_in;
      y(0) <= x_in;
```

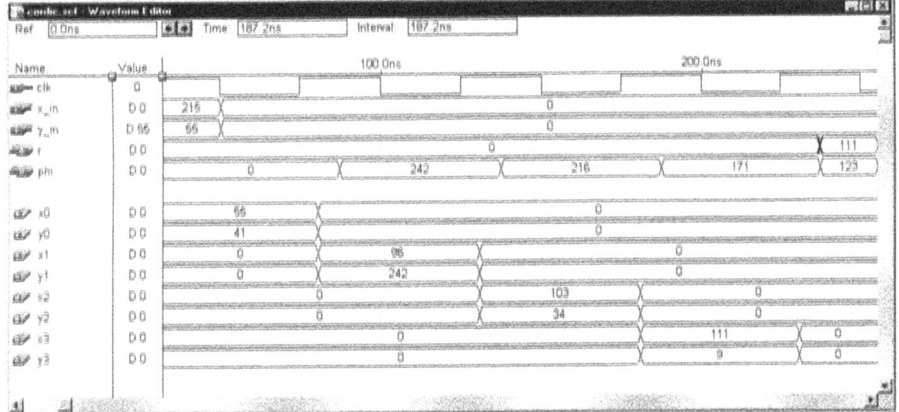

Fig. 2.27. CORDIC simulation results.

```
      z(0) <= -90;
    END IF;
  END PROCESS;

END flex;
```

Fig. 2.27 shows the simulation of the conversion of $X_0 = 215 = -41 \bmod 256$, and $Y_0 = 55$. Note that the radius is enlarged to $R = X_K = 111 = 1.618\sqrt{X_0^2 + Y_0^2}$ and the accumulated angle in degrees is $\arctan(Y_0/X_0) = 123°$. The design requires 256 LCs and runs with a **fast** synthesis selection at 43.47 MHz using I/O cell registers. Without using the I/O registers, the design consumes 280 LCs and runs at 48.54 MHz. 2.20

The actual LC count in the previous example is larger than that expected for a four-stage 8-bit pipeline design that is $5 \cdot 8 \cdot 3 = 120$ LCs. The increase by a factor of two comes from the fact that a FLEX device uses an N-bit switchable `LPM_ADD_SUB` megafunction which needs $2N$ LCs. It needs $2N$ LCs because the LC has only three inputs in the fast arithmetic mode, and the switch mode needs four input LUTs. A Xilinx XC4K series device would be needed, with four inputs per LC, to reduce the count by a factor of two.

Exercises

2.1: Wallace has introduced an alternative scheme for a fast multiplier. The basic building block of this type of multiplier is a carry-save adder (CSA). A CSA takes three n-bit operands and produces two n-bit outputs. Because there is no propagation of the carry, this type of adder is sometimes called a 3:2 compress or counter. For an $n \times n$-bit multiplier we need a total total of $n-2$ CSAs to reduce the output to two operands. These operands then have to be added by a (fast) $2n$-bit ripple-carry adder to compute the final result of the multiplier.

(a) The CSA computation can be done in parallel. Determine the minimum number of levels for an $n \times n$-bit multiplier with $n \in [0, 16]$.
(b) Explain why, for FPGAs with fast two's complement adders, these multipliers are not more attractive than the usual array multiplier.
(c) Explain how a pipelined adder in the final adder stage can be used to implement a faster multiplier. Use the data from Table 2.7 (p. 49) to estimate the necessary LC usage and possible speed for
(c1) 8 × 8-bit multiplier.
(c2) 16 × 16-bit multiplier.

2.2: The Booth multiplier used the classical CSD code to reduce the number of necessary add/subtract operations. Starting with the LSB, typically two or three bits (called radix-4 and radix-8 algorithms) are processed in one step. The following table demonstrates possible radix-4 patterns and actions:

x_{k+1}	x_k	x_{k-1}	Accumulator activity	Comment
0	0	0	ACC→ACC +R* (0)	within a string of "0's"
0	0	1	ACC→ACC +R* (X)	end of a string of "1's"
0	1	0	ACC→ACC +R* (X)	
0	1	1	ACC→ACC +R* ($2X$)	end of a string of "1's"
1	0	0	ACC→ACC +R* (-2X)	beginning of a string of "1's"
1	0	1	ACC→ACC +R* ($-X$)	
1	1	0	ACC→ACC +R* (-X)	beginning of a string of "1's"
1	1	1	ACC→ACC +R* (0)	within a string of "1's"

The hardware requirements for a state machine implementation are an accumulator and a two's complement shifter.
(a) Let X be a signed 6-bit two's complement representation of $-10 = 110110_{2C}$. Complete the following table for the Booth product $P = XY = -10Y$ and indicate the accumulator activity in each step.

Step	x_5	x_4	x_3	x_2	x_1	x_0	x_{-1}	ACC	ACC + Booth rule
Start	1	1	0	1	1	0	0		
0									
1									
2									

(2.42)

(b) Compare the latency of the Booth multiplier, with the serial/parallel multiplier from Example 2.15 (p. 54), for the radix-4 and radix-8 algorithms.

Exercises Using MaxPlusII

2.3: **(a)** Compile the file `add_2p.vhd` with the `MaxPlusII` compiler with optimization for speed and size. How many LCs are needed? Explain the results.
(b) Conduct a simulation with $15 + 102$.

2.4: Explain how to modify the design `add_2p.vhd` for subtraction. Modify the design and simulate as an example $3 - 2$.

2.5: **(a)** Compile the file `mul_ser.vhd` with the MaxPlusII compiler.
(b) Determine the `Registered Performance` and size of the 8-bit design. What is the total multiplication latency?

2.6: Modify the design file `mul_ser.vhd` to multiply 12×12-bit numbers.
(a) Simulate the new design with the values $1000 \cdot 2000$.
(b) Measure the `Registered Performance` and the size.
(c) What is the total multiplication latency of the 12×12-bit multiplier?

2.7: **(a)** Design a stage machine in MaxPlusII to implement the Booth multiplier (see Exercise 2.2) for 6×6 bit signed inputs.
(b) Simulate the four data $\pm 5 \cdot (\pm 9)$.
(c) Determine the `Registered Performance.`
(d) Determine LC utilization for maximum speed.

2.8: **(a)** Design a `generic` CSA component which is used to build a Wallace-tree multiplier for an 8×8-bit multiplier.
(b) Implement the 8×8 Wallace tree using MaxPlusII.
(c) Use a final adder to compute the product, and test your multiplier with a multiplication of 100×63.
(d) Pipeline the Wallace tree. What is the maximum throughput of the pipelined design?
(e) Substitute the 16-bit output adder with the pipeline adder from the CD-ROM. What is the `Registered Performance` of this design?

2.9: **(a)** Use the principle of component instantiation, using the predefined macros `LPM_ADD_SUB` and `LPM_MULT`, to write the VHDL code for a pipelined complex 8-bit multiplier, (i.e., $(a + jb)(c + jd) = ac - bd + j(ad + bd)$), with all operands $a, b, c,$ and d in 8-bit.
(b) Determine the `Registered Performance.`
(c) Determine LC utilization for maximum speed synthesis.
(d) How many pipeline stages does the optimal single `LPM_MULT` multiplier have?
(e) How many pipeline stages does the optimal complex multiplier have in total?

2.10: An alternative algorithm for a complex multiplier is:

$$\begin{array}{ccc} s[1] = a - b & s[2] = c - d & s[3] = c + d \\ m[1] = s[1]d & m[2] = s[2]a & m[3] = s[3]b \\ s[4] = m[1] + m[2] & s[5] = m[1] + m[3] & \\ & (a + jb)(c + jd) = s[4] + js[5] & \end{array} \qquad (2.43)$$

which, in general, needs five adders and three multipliers. Verify that if one coefficient, say $c + jd$ is known, then $s[2], s[3]$, and d can be prestored and the algorithm reduces to three adds and three multiplications. Also
(a) Design a pipelined 5/3 complex multiplier using the above algorithm for 8-bit signed inputs. Use the predefined macros `LPM_ADD_SUB` and `LPM_MULT`.
(b) Measure register performance and size for maximum speed synthesis.
(c) How many pipeline stages does the single `LPM_MULT` multiplier have?
(d) How many pipeline stages does the complex multiplier have in total?

2.11: Compile the file `cordic.vhd` with the MaxPlusII compiler, and
(a) Conduct a simulation (using the waveform file `cordic.scf`) with `x_in=`± 50 and `y_in=`± 70. Determine the radius factor for.all four simulations.

(b) Determine the maximum errors for radius and phase, compared with an unquantized computation.

2.12: Modify the design `cordic.vhd` to implement stage 4 and 5 of the CORDIC pipeline.
(a) Compute the rotation angle, and compile the VHDL code.
(b) Conduct a simulation with values `x_in`$=\pm 50$ and `y_in`$=\pm 70$.
(c) What are the maximum errors for radius and phase, compared with the unquantized computation?

3. Finite Impulse Response (FIR) Digital Filters

3.1 Digital Filters

Digital filters are typically used to modify or alter the attributes of a signal in the time or frequency domain. The most common digital filter is the linear time-invariant (LTI) filter. An LTI interacts with its input signal through a process called linear convolution, denoted by $y = f * x$ where f is the filter's impulse response, x is the input signal, and y is the convolved output. The linear convolution process is formally defined by:

$$y[n] = x[n] * f[n] = \sum_k x[k]f[n-k] = \sum_k f[k]x[n-k]. \tag{3.1}$$

LTI digital filters are generally classified as being *finite impulse response* (i.e., FIR), or *infinite impulse response* (i.e., IIR). As the name implies, an FIR filter consists of a finite number of sample values, reducing the above convolution sum to a finite sum per output sample instant. An IIR filter, however, requires that an infinite sum be performed. An FIR design and implementation methodology is discussed in this chapter, while IIR filter issues are addressed in Chapter 4.

The motivation for studying digital filters is found in their growing popularity as a primary DSP operation. Digital filters are rapidly replacing classic analog filters, which were implemented using RLC components and operational amplifiers. Analog filters were mathematically modeled using ordinary differential equations of Laplace transforms. They were analyzed in the time or s (also known as Laplace) domain. Analog prototypes are now only used in IIR design, while FIR are typically designed using direct computer specifications and algorithms.

In this chapter it is assumed that a digital filter, an FIR in particular, has been designed and selected for implementation. The FIR design process will be briefly reviewed, followed by a discussion of FPGA implementation variations.

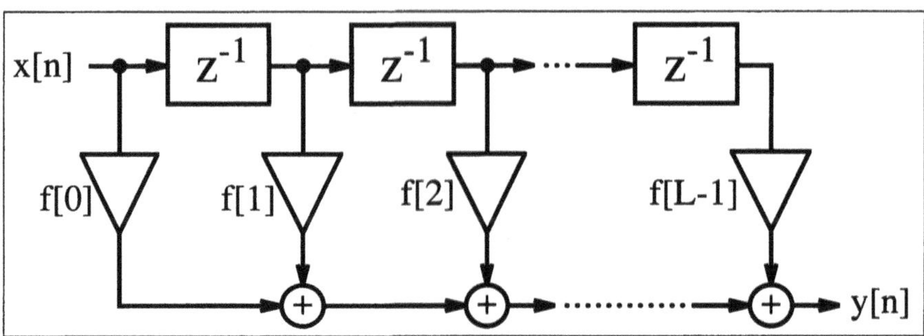

Fig. 3.1. Direct form FIR filter.

3.2 FIR Theory

An FIR with constant coefficients is an LTI digital filter. The output of an FIR of order or length L, to an input time-series $x[n]$, is given by a *finite* version of the convolution sum given in (3.1), namely:

$$y[n] = x[n] * f[n] = \sum_{k=0}^{L-1} x[k]f[n-k] \tag{3.2}$$

where $f[0] \neq 0$ through $f[L-1] \neq 0$ are the filter's L coefficients. They also correspond to the FIR's impulse response. For LTI systems it is sometimes more convenient to express (3.2) in the z-domain with

$$Y(z) = F(z)X(z) \tag{3.3}$$

where $F(z)$ is the FIR's *transfer function* defined in the z-domain by

$$F(z) = \sum_{k=0}^{L-1} f[k]z^{-k}. \tag{3.4}$$

The Lth order LTI FIR filter is graphically interpreted in Fig. 3.1. It can be seen to consist of a collection of a "tapped delay line," adders, and multipliers. One of the operands presented to each multiplier is an FIR coefficient, often referred to as a "tap weight" for obvious reasons. Historically, the FIR filter is also known by the name "transversal filter," suggesting its "tapped delay line" structure.

The *roots* of polynomial $F(z)$ in (3.4) define the zeros of the filter. The presence of only zeros is the reason that FIRs are sometimes called *all zero filters*. In Chapter 5 we will discuss an important class of FIR filters (called CIC filters) which are *recursive* but also FIR. This is possible because the poles produced by the recursive part are canceled by the nonrecursive part of the filter. The effective pole/zero plot also has then *only* zeros, i.e., is an *all zero filter* or FIR. We note that nonrecursive filter are always FIR, but recursive filter can be either FIR or IIR. Fig. 3.2 illustrates this dependence.

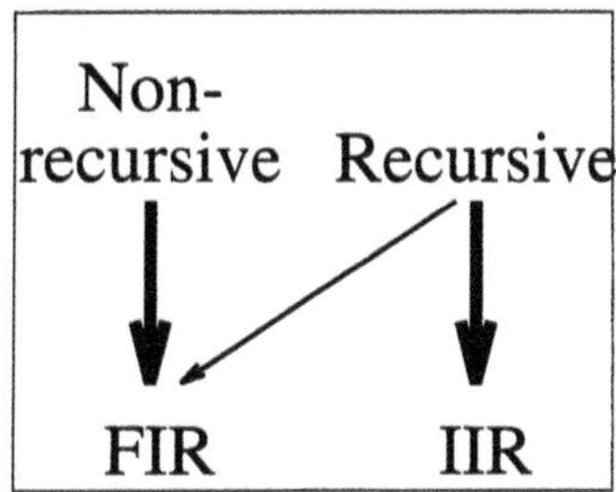

Fig. 3.2. Relation between structure and impulse length.

3.2.1 FIR Filter with Transposed Structure

A variation of the direct FIR model is called the *transposed FIR filter*. It can be constructed from the FIR filter in Fig. 3.1 by:

- Exchanging the input and output
- Inverting the direction of signal flow
- Substituting an adder by a fork, and vice versa

A transposed FIR filter is shown in Fig. 3.3 and is, in general, the preferred implementation of an FIR filter. The benefit of this filter is that we do not need an extra shift register for $x[n]$, and there is no need for an extra pipeline stage for the adder (tree) of the products to achieve high throughput.

The following examples show a direct implementation of the transposed filter.

Example 3.1: Programmable FIR Filter

We recall from the discussion of sum-of-product (SOP) computations using a PDSP (see Sect. 2.5, p. 58) that, for B_x data/coefficient bit width and filter length L, additional $\log_2(L)$ bits for unsigned SOP and $\log_2(L)-1$ guard bits for signed arithmetic must be provided. For a 9-bit signed data/coefficient and $L=4$, the adder width must be $9+9+\log_2(4)-1=19$.

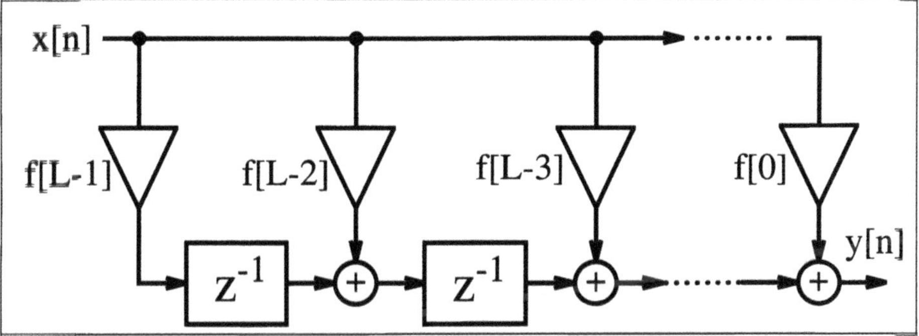

Fig. 3.3. FIR filter in the transposed structure.

The following VHDL code[2] shows the generic specification for an implementation for a length-4 filter.

```
-- This is a generic FIR filter generator
-- It uses W1 bit data/coefficients bits
LIBRARY lpm;                        -- Using predefined packages
USE lpm.lpm_components.ALL;

LIBRARY ieee;
USE ieee.std_logic_1164.ALL;
USE ieee.std_logic_arith.ALL;
USE ieee.std_logic_unsigned.ALL;

ENTITY fir_gen IS                              ------> Interface
  GENERIC (W1 : integer := 9; -- Input bit width
           W2 : integer := 18;-- Multiplier bit width 2*W1
           W3 : integer := 19;-- Adder width = W2+log2(L)-1
           W4 : integer := 11;-- Output bit width
           L  : integer := 4; -- Filter length
        Mpipe : integer := 3-- Pipeline steps of multiplier
           );
  PORT ( clk    : IN STD_LOGIC;
         Load_x : IN  STD_LOGIC;
         x_in   : IN  STD_LOGIC_VECTOR(W1-1 DOWNTO 0);
         c_in   : IN  STD_LOGIC_VECTOR(W1-1 DOWNTO 0);
         y_out  : OUT STD_LOGIC_VECTOR(W4-1 DOWNTO 0));
END fir_gen;

ARCHITECTURE flex OF fir_gen IS

  SUBTYPE N1BIT IS STD_LOGIC_VECTOR(W1-1 DOWNTO 0);
  SUBTYPE N2BIT IS STD_LOGIC_VECTOR(W2-1 DOWNTO 0);
  SUBTYPE N3BIT IS STD_LOGIC_VECTOR(W3-1 DOWNTO 0);
  TYPE ARRAY_N1BIT IS ARRAY (0 TO L-1) OF N1BIT;
  TYPE ARRAY_N2BIT IS ARRAY (0 TO L-1) OF N2BIT;
  TYPE ARRAY_N3BIT IS ARRAY (0 TO L-1) OF N3BIT;

  SIGNAL  x  :  N1BIT;
  SIGNAL  y  :  N3BIT;
  SIGNAL  c  :  ARRAY_N1BIT; -- Coefficient array
  SIGNAL  p  :  ARRAY_N2BIT; -- Product array
  SIGNAL  a  :  ARRAY_N3BIT; -- Adder array

BEGIN

  Load: PROCESS              ------> Load data or coefficient
  BEGIN
    WAIT UNTIL clk = '1';
    IF (Load_x = '0') THEN
      c(L-1) <= c_in;       -- Store coefficient in register
      FOR I IN L-2 DOWNTO 0 LOOP  -- Coefficients shift one
        c(I) <= c(I+1);
```

[2] The equivalent Verilog code fir_gen.v for this example can be found in Appendix A on page 356.

```
      END LOOP;
    ELSE
      x <= x_in;             -- Get one data sample at a time
    END IF;
  END PROCESS Load;


  SOP: PROCESS (clk)         ------> Compute sum-of-products
  BEGIN
    IF clk'event and (clk = '1') THEN
    FOR I IN 0 TO L-2  LOOP        -- Compute the transposed
      a(I) <= (p(I)(W2-1) & p(I)) + a(I+1); -- filter adds
    END LOOP;
    a(L-1) <= p(L-1)(W2-1) & p(L-1);        -- First TAP has
    END IF;                                 -- only a register
    y <= a(0);
  END PROCESS SOP;

  -- Instantiate L pipelined multiplier
  MulGen: FOR I IN 0 TO L-1 GENERATE
  Muls: lpm_mult                   -- Multiply p(i) = c(i) * x;
        GENERIC MAP ( LPM_WIDTHA => W1, LPM_WIDTHB => W1,
                      LPM_PIPELINE => Mpipe,
                      LPM_REPRESENTATION => "SIGNED",
                      LPM_WIDTHP => W2,
                      LPM_WIDTHS => W2)
        PORT MAP ( clock => clk, dataa => x,
                   datab => c(I), result => p(I));
        END GENERATE;

  y_out <= y(W3-1 DOWNTO W3-W4);

END flex;
```

The first process, **Load**, is used to load the coefficient in a tapped delay line if `Load_x=0`. Otherwise, a data word is loaded in the **x** register. The second process called **SOP**, implements the sum-of-product computation. The products `p(I)` are sign-extended by one bit and added to the previous partial SOP. Note also that all multipliers are instantiated by a **generate** statement, which allows the assignment of extra pipeline stages. Finally, the output **y_out** is assigned the value of the SOP, divided by 256, because the coefficients are assumed to be all fractional (i.e., $|f[k]| \leq 1.0$) The design uses 890 LCs and runs with 46.72 MHz **Registered Performance**.
To simulate this length-4 filter consider a Daubechies DB4 filter coefficient with

$$G(z) = \left((1+\sqrt{3}) + (3+\sqrt{3})z^{-1} + (3-\sqrt{3})z^{-2} + (1-\sqrt{3})z^{-3}\right)\frac{1}{4\sqrt{2}}$$

$$G(z) = 0.48301 + 0.8365z^{-1} + 0.2241z^{-2} - 0.1294z^{-3}$$

Quantizing the coefficients to 8 bits (plus sign bit) of precision results in the following model:

$$G(z) = \left(124 + 214z^{-1} + 57z^{-2} - 33z^{-3}\right)/256$$

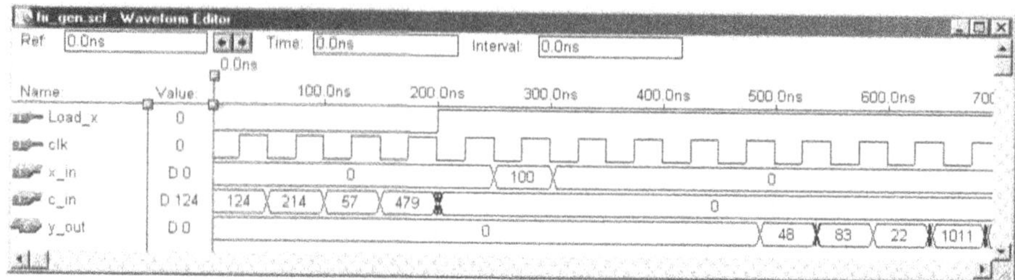

Fig. 3.4. Simulation of the 4 tap programmable FIR filter with Daubechies filter coefficient loaded.

$$= \frac{124}{256} + \frac{214}{256}z^{-1} + \frac{57}{256}z^{-2} + -\frac{33}{256}z^{-3}$$

As can be seen from Fig. 3.4, in the first four steps we load the coefficients $\{124, 214, 57, -33\}$ into the tapped delay line. Note that MaxPlusII diplays -33 as unsigned number, i.e., $512 - 33 = 479$. Then we check the impulse response of the filter by loading 100 into the **x** register. The first valid output is then available after 450 ns, as can be seen from Fig. 3.4. 3.1

3.2.2 Symmetry in FIR Filters

The center of an FIR's impulse response is an important point of symmetry. It is sometimes convenient to define this point as the 0th sample instant. Such filter descriptions are *a-causal* (centered notation). For an odd-length FIR, the a-causal filter model is given by:

$$F(z) = \sum_{k=-(L-1)/2}^{(L-1)/2} f[k]z^{-k}. \tag{3.5}$$

The FIR's frequency response can be computed by evaluating the filter's transfer function about the periphery of the unity circle, by setting $z = e^{j\omega T}$. It then follows that:

$$F(\omega) = F(e^{j\omega T}) = \sum_{k} f[k]e^{-j\omega kT}. \tag{3.6}$$

We then denote with $|F(\omega)|$ the filter's *magnitude frequency response* and $\phi(\omega)$ denotes the *phase response*, and satisfies:

$$\phi(\omega) = \arctan\left(\frac{\Im(F(\omega))}{\Re(F(\omega))}\right). \tag{3.7}$$

Digital filters are more often characterized by phase and magnitude than by the z-domain transfer function or the complex frequency transform.

Table 3.1. Four possible linear-phase FIR filters $F(z) = \sum_k f[k]z^{-k}$.

Symmetry	$f[n] = f[-n]$	$f[n] = f[-n]$	$f[n] = -f[-n]$	$f[n] = -f[-n]$
L	odd	even	odd	even
Zeros at		π	0 and π	0
Example	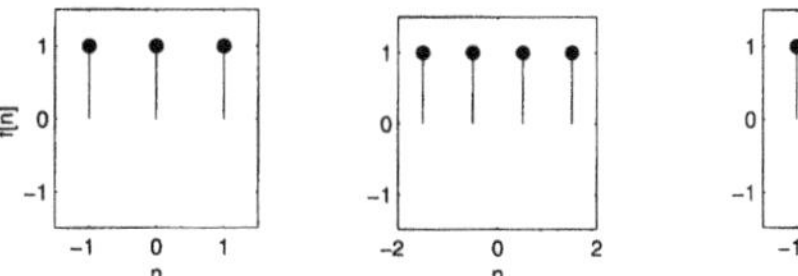		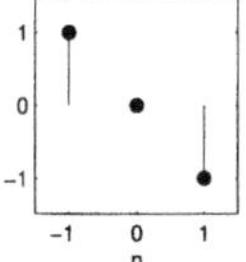	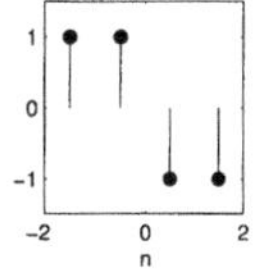

3.2.3 Linear-Phase FIR Filters

Maintaining phase integrity across a range of frequencies is a desired system attribute in many applications such as communications and image processing. As a result, designing filters that establish linear-phase versus frequency is often mandatory. The standard measure of the phase linearity of a system is the "group delay" defined by:

$$\tau(\omega) = \frac{\mathrm{d}\phi(\omega)}{\mathrm{d}\omega} \tag{3.8}$$

A perfectly linear-phase filter has a group delay that is constant over a range of frequencies. It can be shown that linear-phase is achieved if the filter is symmetric or anti-symmetric, and it is therefore preferable to use the a-causal framework of (3.5). From (3.7) it can be seen that a constant group delay can only be achieved if the frequency response $F(\omega)$ is a purely real or imaginary function. This implies that the filter's impulse response possesses even or odd symmetry. That is:

$$f[n] = f[-n] \quad \text{or} \quad f[n] = -f[-n]. \tag{3.9}$$

An odd-order even-symmetry FIR filter would, for example, have a frequency response given by:

$$F(\omega) = f[0] + \sum_k f[k]e^{-jk\omega T} + f[-k]e^{jk\omega T} \tag{3.10}$$

$$= f[0] + 2\sum_{k>0} f[k]\cos(k\omega T). \tag{3.11}$$

which is seen to be a purely real function of frequency. Table 3.1 summarizes the four possible choices of symmetry, anti-symmetry, even order and odd order. In addition Table 3.1 graphically displays an example of each class of linear-phase FIR.

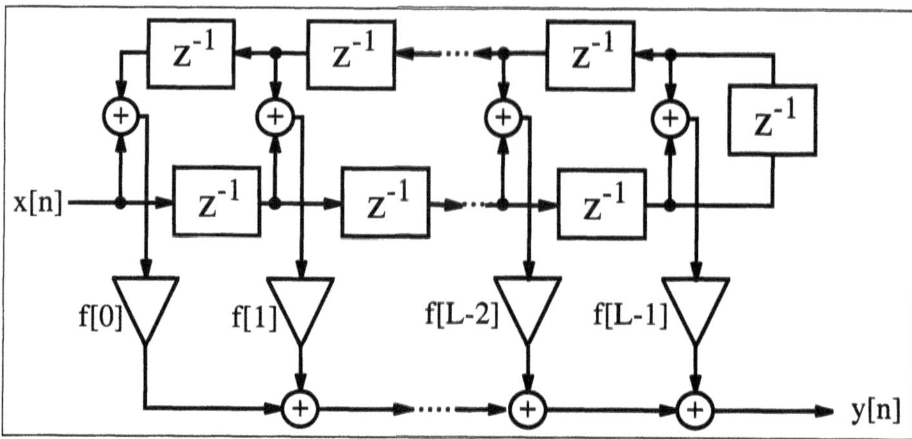

Fig. 3.5. Linear phase filter with reduced number of multipliers.

The symmetry properties intrinsic to a linear-phase FIR can also be used to reduce the necessary number of multipliers L, as shown in Fig. 3.1. Consider the linear-phase FIR shown in Fig. 3.5 (even symmetry assumed), which fully exploits coefficient symmetry. Observe that the "symmetric" architecture has a multiplier budget per filter cycle exactly half of that found in the direct architecture shown in Fig. 3.1 (L versus $L/2$) while the number of adders remains constant at $L-1$.

3.3 Designing FIR Filters

Modern digital FIR filters are designed using computer-aided engineering (CAE) tools. The filters used in this chapter are designed using the MATLAB Signal Processing toolbox. The toolbox includes an "Interactive Lowpass Filter Design" demo example that covers many typical digital filter designs, including:

- Equirriple (also known as minimax) FIR design, which uses the Parks-McClellan and Remez exchange methods for designing a linear-phase (symmetric) equirriple FIR. This equirriple design may also be used to design a differentiator or Hilbert transformer.
- Kaiser window design using the inverse DFT method weighted by a Kaiser window.
- Least square FIR method. This filter design also has ripple in the passband and stopband, but the mean least square error is minimized.
- Four IIR filter design methods (Butterworth, Chebyshev I and II, and Elliptic) which will be discussed in Chapter 4.

The FIR methods are individually developed in this section. Most often we already know the transfer function (i.e., magnitude of the frequency response) of the desired filter. Such a lowpass specification typically consists of the passband $[0 \ldots \omega_p]$, the transition band $[\omega_p \ldots \omega_s]$, and the stopband $[\omega_s \ldots \pi]$ specification, where the sampling frequency is assumed to be 2π. To compute the filter coefficients we may therefore apply the *direct frequency* method discussed next.

3.3.1 Direct Window Design Method

The discrete Fourier Transform (DFT) establishes a direct connection between the frequency and time domains. Since the frequency domain is the domain of filter definition, the DFT can be used to calculate a set of FIR filter coefficients which produce a filter that approximates the frequency response of the target filter. A filter designed in this manner is called a *direct FIR filter*. A direct FIR filter is defined by:

$$f[n] = \mathrm{IDFT}(F[k]) = \sum_k F[k] e^{j2\pi kn/L} \tag{3.12}$$

From basic signals and systems theory, it is known that the spectrum of a real signal is Hermitian. That is, the real spectrum has even symmetry and the imaginary spectrum has odd symmetry. If the synthesized filter should have only real coefficients, the target DFT design spectrum must therefore be Hermitian or $F[k] = F^*[-k]$, where the $*$ denotes conjugate complex.

Consider a length-16 direct FIR filter design with a rectangular window, shown in Fig. 3.6(a), with the passband ripple shown in 3.6(b). Notice that the filter provides a reasonable approximation to the ideal lowpass filter with the greatest mismatch occurring at the edges of the transition band. The observed "ringing" is due to Gibbs phenomenon, which relates to the inability of a finite Fourier spectrum to reproduce sharp edges. The Gibbs ringing is implicit in the direct inverse DFT method and can be expected to be about $\pm 7\%$ over a wide range of filter orders. To illustrate this, consider the example filter with length 128, shown in Fig. 3.6(c), with the passband ripple shown in (d). Although the filter length is essentially increased (from 16 to 128) the ringing at the edge still has about the same quantity. The effects of ringing can only be suppressed with the use of a data "window" which tapers smoothly to zero on both sides. Data windows overlay the FIR's impulse response, resulting in a "smoother" magnitude frequency response with an attendant widening of the transition band. If, for instance, a Kaiser window is applied to the FIR, the Gibbs ringing can be reduced as shown in Fig. 3.7(upper). The deleterious effect on the transition band can also be seen. Other classic window functions are summarized below. They differ in terms of their ability to make tradeoffs between "ringing" and transition bandwidth extension. The number of recognized and published window functions is large. The most common windows, denoted $w[n]$, are:

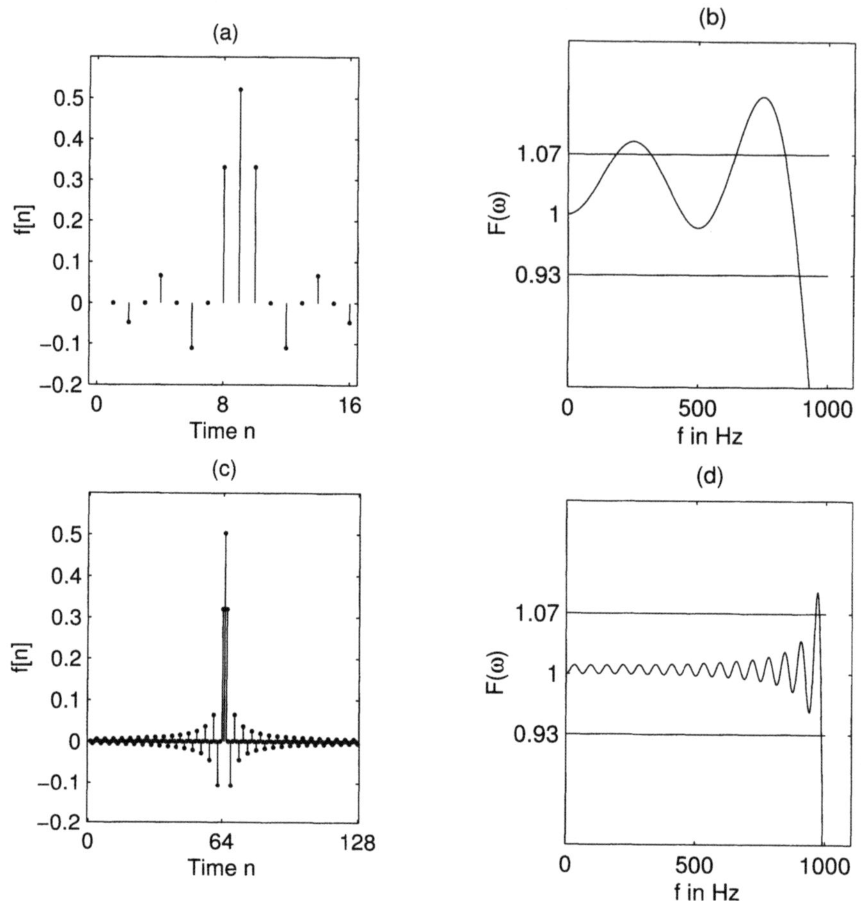

Fig. 3.6. Gibbs phenomenon. **(a)** Impulse response of FIR lowpass with $L = 16$. **(b)** Passband of transfer function $L = 16$. **(c)** Impulse response of FIR lowpass with $L = 128$. **(d)** Passband of transfer function $L = 128$.

- Rectangular: $w[n] = 1$
- Bartlett (triangular) : $w[n] = 2n/N$
- Hanning: $w[n] = 0.5\,(1 - \cos(2\pi n/L)$
- Hamming: $w[n] = 0.54 - 0.46\cos(2\pi n/L)$
- Blackman: $w[n] = 0.42 - 0.5\cos(2\pi n/L) + 0.08\cos(4\pi n/L)$
- Kaiser: $w[n] = \mathrm{I}_0\left(\beta\sqrt{1 - (n - L/2)^2/(L/2)^2}\right)$

Table 3.2 shows the most important parameters of these windows.

The 3 dB bandwidth shown in Table 3.2 is the bandwidth where the transfer function is decreased from DC by 3 dB or $\approx 1/\sqrt{2}$. Data windows also generates sidelobes, to various degrees, away from the 0th harmonic.

Table 3.2. Parameter of common used window functions.

Name	3 dB bandwidth	First zero	Maximum sidelobe	Sidelobe decrease per octave	Equivalent Kaiser β
Rectangular	$0.89/T$	$1/T$	−13 dB	−6 dB	0
Bartlett	$1.28/T$	$2/T$	−27 dB	−12 dB	1.33
Hanning	$1.44/T$	$2/T$	−32 dB	−18 dB	3.86
Hamming	$1.33/T$	$2/T$	−42 dB	−6 dB	4.86
Blackman	$1.79\ /T$	$3/T$	−74 dB	−6 dB	7.04
Kaiser	$1.44/T$	$2/T$	−38 dB	−18 dB	3

Depending on the smoothness of the window, the third column in Table 3.2 shows that some windows do not have a zero at the first or second zero DFT frequency $1/T$. The maximum sidelobe gain is measured relative to the 0th harmonic value. The fifth column describes the asymptotic decrease of the window per octave. Finally, the last column describes the value β for a Kaiser window that emulates the corresponding window properties. The Kaiser window, based on the first-order Bessel function I_0, is special in two respects. It is nearly optimal in terms of the relationship between "ringing" suppression and transition width, and second, it can be tuned by β, which determines the ringing of the filter. This can be seen from the following equation credited to Kaiser.

$$\beta = \begin{cases} 0.1102(A-8.7) & A > 50, \\ 0.5842(A-21)^{0.4} + 0.07886(A-21) & 21 \leq A \leq 50, \\ 0 & A < 21. \end{cases} \tag{3.13}$$

where $A = 20\log_{10}\varepsilon_r$ is both stopband attenuation and the passband ripple in dBs. The Kaiser window length to achieve a desired level of suppression can be estimated:

$$L = \frac{A-8}{2.285(\omega_s - \omega_p)} + 1. \tag{3.14}$$

The length is generally correct within an error of ± 2 taps.

3.3.2 Equiripple Design Method

A typical filter specification not only includes the specification of passband ω_p and stopband ω_s frequencies and ideal gains, but also the allowed deviation (or ripple) from the desired transfer function. The transition band is most often assumed to be arbitrary in terms of ripples. A special class of FIR filter that is particularly effective in meeting such specifications is called the equiripple FIR. An equiripple design protocol minimizes the maximal deviations (ripple error) from the ideal transfer function. The equiripple algorithm applies to a number of FIR design instances. The most popular are:

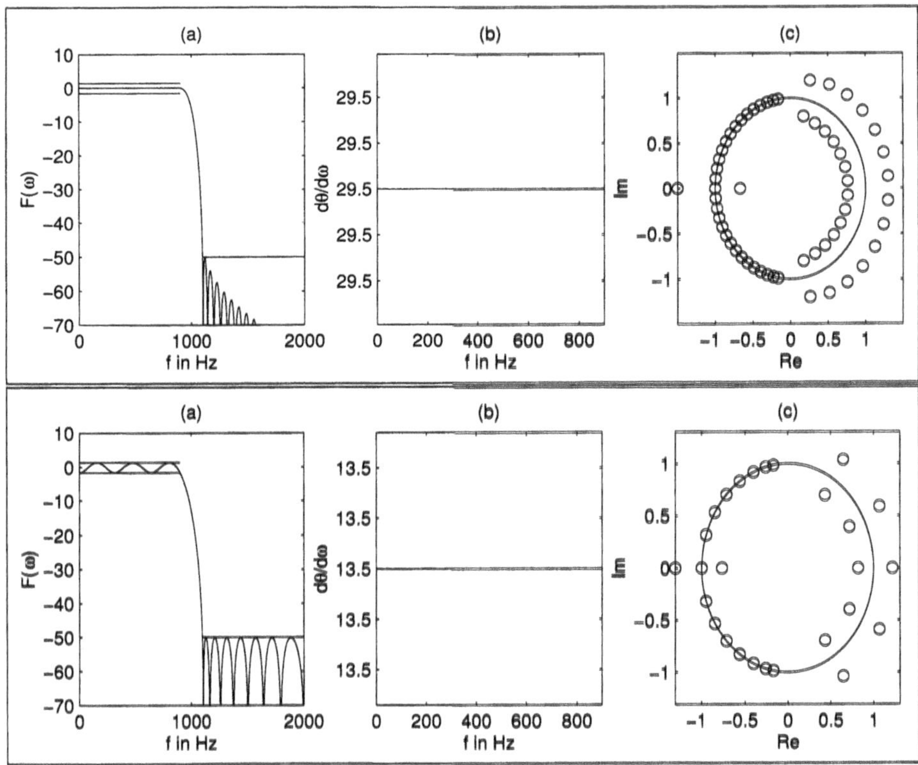

Fig. 3.7. **(upper)** Kaiser window design with $L = 59$. **(lower)** Parks-McClellan design with $L = 27$.
(a) Transfer function. **(b)** Group delay of passband. **(c)** Zero plot.

- Lowpass filter design (in MATLAB use `remez(L,F,A,W)`), with tolerance scheme as shown in Fig. 3.8(a)
- Hilbert filter, i.e., a unit magnitude filter that produces a 90° phase shift for all frequencies in the passband (in MATLAB use `remez(L, F, A, 'Hilbert')`)
- Differentiator filter which has a linear increasing frequency magnitude proportional to ω (in MATLAB use `remez(L,F,A,'differentiator')`)

The equiripple or minimum-maximum algorithm is normally implemented using the *Parks-McClellan iterative method.* The Parks-McClellan method is used to produce a equiripple or minimax data fit in the frequency domain. It's based on the "alternation theorem" which says that there is exactly one polynomial, a Chebyshev polynomial with minimum length, that fits into a given tolerance scheme. Such a tolerance scheme is shown in Fig. 3.8(a), and Fig. 3.8(b) shows a polynomial which fulfills this tolerance scheme. The length of the polynomial, and therefore the filter, can be estimated for a lowpass with

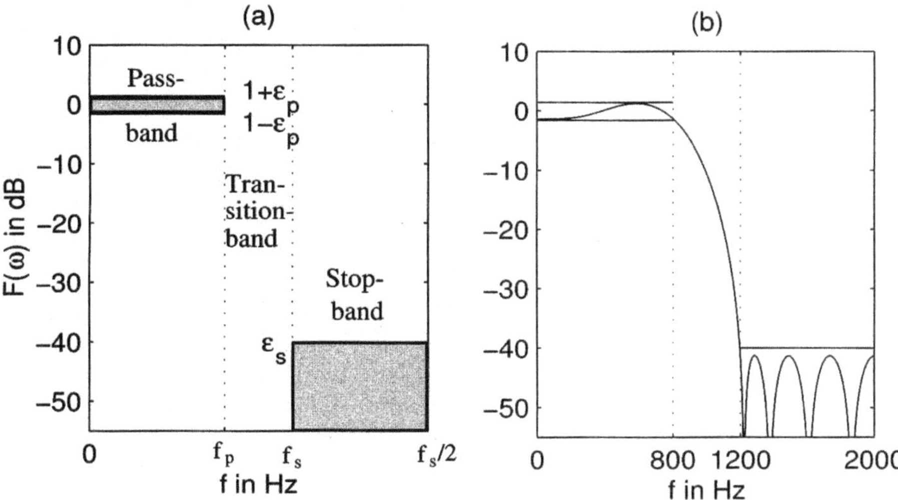

Fig. 3.8. Parameter for the filter design. **(a)** Tolerance scheme **(b)** Example function, which fulfills the scheme.

$$L = \frac{-10\log_{10}(\varepsilon_p \varepsilon_s) - 13}{2.324(\omega_s - \omega_p)} + 1 \tag{3.15}$$

where ε_p is the passband and ε_s the stopband ripple.

The algorithm iteratively finds the location of locally maximum errors that deviate from a nominal value, reducing the size of the maximal error per iteration, until all deviation errors have the same value. Most often, the *Remez method* is used to select the new frequencies by selecting the frequency set with the largest peaks of the error curve between two iterations, see [65, p. 478]. This is the reason why the MATLAB equiripple function is called `remez`.

Compared to the *direct frequency method*, with or without data windows, the advantage of the equiripple design method is that passband and stopband deviations can be specified differently. This may, for instance, be useful in audio applications where the ripple in the passband may be specified to be higher, because the ear only perceives differences larger than 3 dB.

We notice from Fig. 3.7(lower) that the equiripple design having the same tolerance requirements as the Kaiser window design enjoys a considerably reduced filter order, i.e., 27 compared with 59.

3.4 Constant Coefficient FIR Design

There are only a few applications (e.g. adaptive filters) where we need a general programmable filter architecture like the one shown in Example 3.1 (p. 81). In many applications, the filters are LTI (i.e., linear time-invariant) and the coefficients do not change over time. In this case, the hardware effort can essentially be reduced by exploiting the multiplier and adder (trees) needed to implement the FIR filter arithmetic.

With available digital filter design software the production of FIR coefficients is a straightforward process. The challenge remains mapping the FIR design into a suitable architecture. The direct or transposed forms are preferred for maximum speed and lowest resource utilization. Lattice filters are used in adaptive filters because the filter can be enlarged by one section, without the need for recomputation of the previous lattice sections. But this feature only applies to PDSPs and is less applicable to FPGAs. We will therefore focus our attention on the direct and transposed implementations. We will start with possible improvements to the direct form and will then move on to the transposed form. At the end of the section we will discuss an alternative design approach using distributed arithmetic.

3.4.1 Direct FIR Design

The Direct FIR filter shown in Fig. 3.1 (p. 80) can be implemented in VHDL using (sequential) `PROCESS` statements or by "component instantiations" of the adders and multipliers. A `PROCESS` design provides more freedom to the synthesizer, while component instantiation gives full control to the designer. To illustrate this, a length-4 FIR will be presented as a `PROCESS` design. Although a length-4 FIR is far too short for most practical applications, it is easily extended to higher orders and has the advantage of a short compiling time. The linear-phase (therefore symmetric) FIR's impulse response is assumed to be given by

$$f[k] = \{-1.0, 3.75, 3.75, -1.0\}. \tag{3.16}$$

These coefficients can be directly encoded into a 4-bit fraction number. For example, 3.75_{10} would have a 4-bit binary representation 11.11_2 where "." denotes the location of the binary point. Note that it is in general more efficient to implement only *positive* CSD coefficients, because positive CSD coefficients have fewer nonzero terms and we can take the sign of the coefficient into account when the summation of the products is computed. See also the first two steps in the RAG algorithm 3.4 discussed later, p. 97.

In a practical situation, the FIRs are obtained from a computer design tool and presented to the designer as floating-point numbers. The performance of a fixed-point FIR, based on floating-point coefficients, needs to be verified using simulation or algebraic analysis to ensure that design specifications remain satisfied. In the above example, the floating-point numbers are 3.75 and 1.0, which can be represented exactly with fixed-point numbers, and the check can be skipped.

Another issue that must be addressed when working with fixed-point designs is protecting the system from *dynamic range overflow*. Fortunately, the *worst-case* dynamic range growth G of an Lth order FIR is easy to compute and it is:

$$G \leq \log_2 \left(\sum_{k=0}^{L-1} |f[k]| \right) \tag{3.17}$$

The total bit width is then the sum of the input bit width and the bit growth G. For the above filter for (3.16) we have $G = \log_2(9.5) < 4$ which states that the system's internal data registers need to have at least four more integer bits than the input data to insure no overflow. If 8-bit internal arithmetic is used the input data should be bounded by $\pm 128/9.5 = \pm 13$.

Example 3.2: Four-tap Direct FIR Filter

The VHDL design[3] for a filter with coefficients $\{-1, 3.75, 3.75, -1\}$ is shown in the following listing.

```
PACKAGE eight_bit_int IS      -- User defined types
  SUBTYPE BYTE IS INTEGER RANGE -128 TO 127;
  TYPE ARRAY_BYTE IS ARRAY (0 TO 3) OF BYTE;
END eight_bit_int;

LIBRARY work;
USE work.eight_bit_int.ALL;

LIBRARY ieee;
USE ieee.std_logic_1164.ALL;
USE ieee.std_logic_arith.ALL;

ENTITY fir_srg IS                            ------> Interface
  PORT (clk    :   IN  STD_LOGIC;
        x      :   IN  BYTE;
        y      :   OUT BYTE);
END fir_srg;

ARCHITECTURE flex OF fir_srg IS

  SIGNAL tap : ARRAY_BYTE;     -- Tapped delay line of bytes

BEGIN

  p1: PROCESS                ------> Behavioral Style
  BEGIN
    WAIT UNTIL clk = '1';
  -- Compute output y with the filter coefficients weight.
  -- The coefficients are [-1  3.75  3.75  -1].
  -- Multiplication and division for Altera VHDL are only
  -- allowed for powers of two!
    y <= 2 * tap(1) + tap(1) + tap(1) / 2 + tap(1) / 4
         + 2 * tap(2) + tap(2) + tap(2) / 2 + tap(2) / 4
         - tap(3) - tap(0);
```

[3] The equivalent Verilog code `fir_srg.v` for this example can be found in Appendix A on page 358.

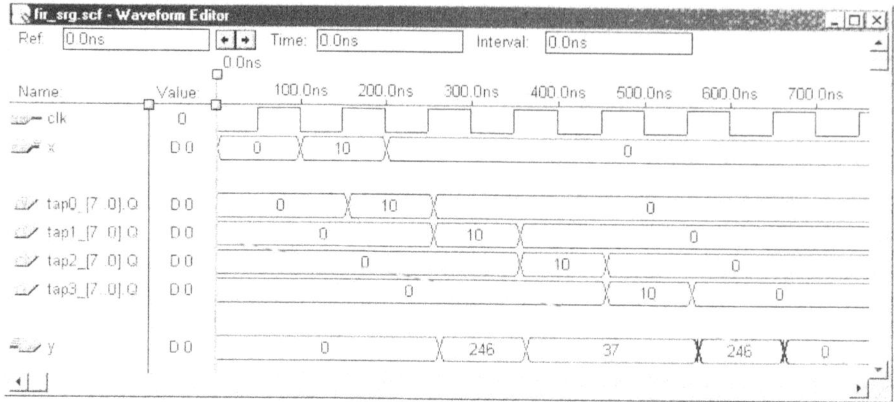

Fig. 3.9. VHDL simulation results of the FIR filter with impulse input 10.

```
    FOR I IN 3 DOWNTO 1 LOOP
      tap(I) <= tap(I-1); -- Tapped delay line: shift one
    END LOOP;
    tap(0) <= x;                -- Input in register 0
  END PROCESS;

END flex;
```

The design is a literal interpretation of the Direct FIR architecture found in Fig. 3.1 (p. 80). The design is applicable to both symmetric and asymmetric filters. The output of each tap of the tapped delay line is multiplied with the appropriately weighted binary value and the results are added. The impulse response y of the filter to an impulse 10 is shown in Fig. 3.9. Note that MaxPlusII displays -10 as unsigned number, i.e., $256 - 10 = 246$. 3.2

There are three obvious actions that can improve this design:

1) Realize each filter coefficient with an optimized CSD code (see Chapter 2, Example 2.1, p. 33).

2) Increase effective multiplier speed by pipelining. The output adder should be arranged in a pipelined balance tree. If the coefficients are coded as "powers-of-two," the pipelined multiplier and the adder tree can be merged. Pipelining has low overhead due to the fact that the LC registers are otherwise often unused. A few additional pipeline registers may be necessary if the number of terms in the tree to be added is not a power of two.

3) For symmetric coefficients, the multiplication complexity can be reduced as shown in Fig. 3.5 (p. 86).

The first two actions are applicable to all FIR filters, while the third applies only to linear-phase (symmetric) filters. These ideas will be illustrated by example designs.

Table 3.3. Improved FIR filter.

Symmetry	no	yes	no	no	yes	yes
CSD	no	no	yes	no	yes	yes
Tree	no	no	no	yes	no	yes
Speed/MHz	18.05	37.17	37.74	100.00	56.81	104.16
Size/LCs	104	80	70	120	64	72

Example 3.3: Improved Four-tap Direct FIR Filter

The design from the previous example can be improved using a CSD code for the coefficients $3.75 = 2^2 - 2^{-2}$. In addition, symmetry and pipelining can also be employed to enhance the filter's performance. Table 3.3 shows the maximum throughput that can be expected for each different design. CSD coding and symmetry result in smaller, more compact designs. Improvements in `Registered Performance` are obtained by pipelining the multiplier and providing an adder tree for the output accumulation. Two additional pipeline registers (i.e., 16 LCs) are necessary, however. The most compact design is archived using symmetry and CSD coding without the use of an adder tree. The partial VHDL code for producing the filter output y is shown below.

```
WAIT UNTIL clk = '1';
t1 <= tap(1) + tap(2); -- Using symmetry
t2 <= tap(0) + tap(3);

y <= 4 * t1 - t1 / 4 - t2; Apply CSD code and add
...
```

The fastest design is obtained when all three enhancements are used. The partial VHDL code, in this case, becomes:

```
WAIT UNTIL clk = '1';
t1 <= tap(1) + tap(2); -- Use symmetry of coefficients
t2 <= tap(0) + tap(3);
t3 <= 4 * t1 - t1 / 4; -- Pipelined multiplier
t4 <= -t2;             -- Build a binary tree and add delay
y <=  t3 + t4;
...
```

3.3

Rephasing Pipelined Multiplier in FIR Filter

Sometimes a single coefficient has more pipeline delay than all the other coefficients. We can model this delay by $f[n]z^{-d}$. If we now add a positive delay with

$$f[n] = z^d f[n] z^{-d} \tag{3.18}$$

the two delays are eliminated. Translating this into a hardware means that for the direct form FIR filter we have to use the ouput of the d position previous register.

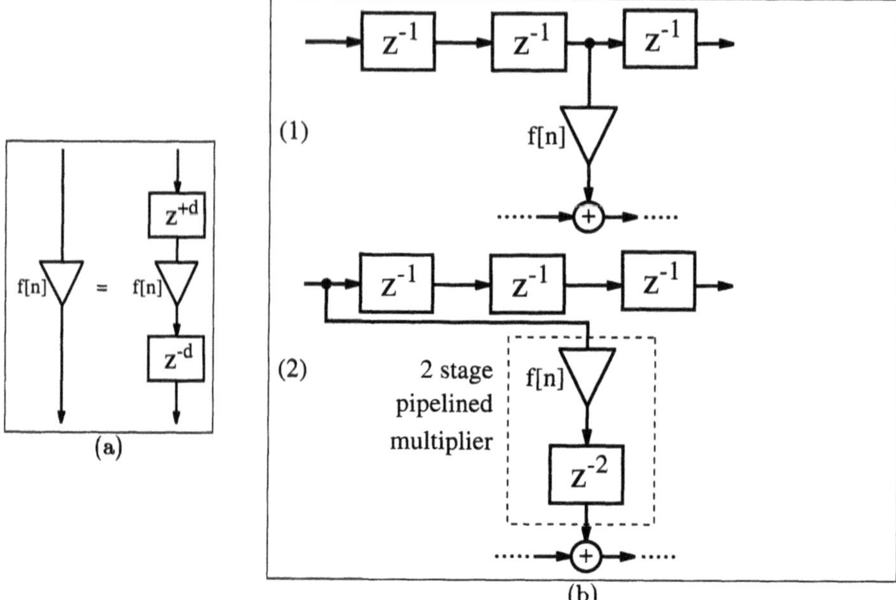

Fig. 3.10. Rephasing FIR filter. **(a)** Principle. **(b)** Rephasing a multiplier. **(1)** Without pipelining. **(2)** With two-stage pipelining.

This principle is shown in Fig. 3.10(a). The Fig. 3.10(b) shows an example for rephasing a pipelined multiplier which has two delays.

3.4.2 FIR Filter with Transposed Structure

A variation of the Direct FIR filter is called the *transposed filter* and has been discussed in Section 3.2.1 (p. 81). The transposed filter enjoys, in the case of a constant coefficient filter, the following two additional improvements, compared with the direct form FIR:

- Multiple use of the repeated coefficients using the Reduced Adder Graph (RAG) algorithm [27, 28, 29, 30]
- Pipeline adders using a carry-save adder

The pipeline adder will increase the speed, at additional adder and register costs, while the RAG principle will reduce the size (i.e., number of LCs) of the filter and sometimes also increase the speed. The pipeline adder principle has been discussed in Chapter 2 and here we will focus on the RAG algorithm.

In Chapter 2 it was noted that it can sometimes be advantageous to implement the factors of a constant coefficient, rather than implement the CSD code directly. For example, the CSD code realization of the constant multiplier coefficient 45 requires 3 adders, while the factors $5 \cdot 9$ only requires two adders. For a transposed FIR filter, the probability is high that all the coefficients will have several factors in common. For instance, the coefficients 9 and 11 can be built using $8 + 1 = 9$ for the first and $11 = 9 + 2$ for the second. This reduces the total effort to one adder. In general, however, finding the optimal reduced adder graph (RAG) is an NP-hard problem. As a result, heuristics must be used. Some obvious actions can, however, reduce the effort. Specifically:

Algorithm 3.4: Reduced Adder Graph

a) Remove the sign of the coefficient because the sign can be realized by a subtraction in the filter's tapped delay line.
b) Remove all coefficients and factors which are a power of two, since they can be implemented by a hard-wired data shift.
c) Realize all cost "1" coefficients.
d) Use cost "1" coefficients in building the multiplier of higher cost.

Step (a)-(c) are straight forward, but step (d) is potentially complex since the number of theoretical graphs increases exponentially. To simplify the process it is helpful to use the CSD coding algorithm shown in Table 2.3 (p. 37). To illustrate the RAG algorithm, consider coding the coefficients defining the F6 halfband FIR filter of Goodman and Carey [66].

Example 3.5: Reduced Adder Graph for F6 Halfband Filter

The halfband filter F6 has four nonzero coefficients, namely $f[0], f[1], f[3]$, and $f[5]$ which are $346, 208, -44$, and 9. For a first cost estimation we convert the decimal values (index 10) into binary representations (index 2) and look-up the cost for the coefficients from Table 2.3 (p. 37). It follows:

		$f[k]$		Cost
$f[0] = 346_{10}$	$=$	$2 \cdot 173$	$= 101011010_2$	4
$f[1] = 208_{10}$	$=$	$2^4 \cdot 13$	$= 11010000_2$	2
$f[3] = -44_{10}$	$=$	$-2^2 \cdot 11$	$= -101100_2$	2
$f[5] = 9_{10}$	$=$	3^2	$= 1001_2$	1
		Total		9

For the direct CSD code realization, 9 adders are required. The RAG algorithms proceeds as follows:

Step	To be realized	Already realized	Action
0)	{346,208,-44,9}	{ – }	Initialization
1)	{346,208,44,9}	{ – }	No negative coefficients
2)	{346,208,44,9}	{ – }	Remove 2^k coefficients
3)	{173,13,11,9}	{ – }	Remove 2^k factors from coefficients
4)	{173,13,11}	{ 9 }	Realize cost 1 coefficients
5)	{173,13,11}	{ 9 }	Other coefficients are primes

Apply the heuristic to the remaining coefficients, starting with the coefficient with the lowest cost and smallest value. It follows that:

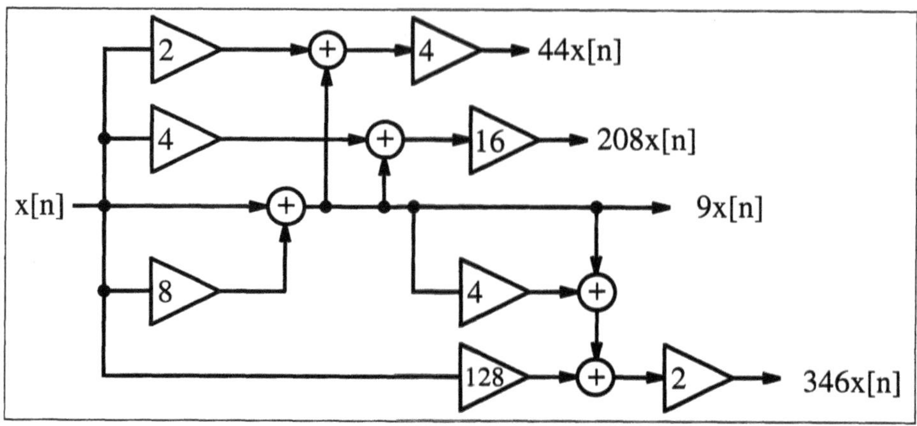

Fig. 3.11. Realization of F6 using RAG algorithm.

Step	Realize	Already realized	Action Find representation
6)	{173,13}	{9,11}	for $11 = 9 + 2$
7)	{173}	{9,11,13}	for $13 = 9 + 4$
8)	{ − }	{9,11,13,173}	for $173 = (4 + 1)9 + 128$

Fig. 3.11 shows the resulting reduced-adder graph. The number of adders is reduced from 9 to 5. The adder path delay is also reduced from 4 to 3. 3.5

3.4.3 FIR Filter Using Distributed Arithmetic

A completely different FIR architecture is based on the distributed arithmetic (DA) concept introduced in Sect. 2.5.1 (p. 60). In contrast to a conventional sum-of-products architecture, in distributed arithmetic we always compute the sum of products of a specific bit b over *all* coefficients in one step. This is computed using a small table and an accumulator with shifter.

To illustrate, consider the three-coefficient FIR found in Example 2.17 (p. 63).

Example 3.6: Distributed Arithmetic Filter as State Machine

A distributed arithmetic filter can be built in VHDL code[4] using the following state machine description:

```
LIBRARY ieee;                     -- Using predefined packages
USE ieee.std_logic_1164.ALL;
USE ieee.std_logic_arith.ALL;
PACKAGE da_package IS       -- User defined component
  COMPONENT case3
```

[4] The equivalent Verilog code **dafsm.v** for this example can be found in Appendix A on page 359.

```
    PORT ( table_in    : IN   STD_LOGIC_VECTOR(2 DOWNTO 0);
           table_out   : OUT  integer RANGE 0 TO 6);
  END COMPONENT;
END da_package;

LIBRARY work;
USE work.da_package.ALL;

LIBRARY ieee;                  -- Using predefined packages
USE ieee.std_logic_1164.ALL;
USE ieee.std_logic_arith.ALL;
ENTITY dafsm IS                          ------> Interface
       PORT (clk  : IN STD_LOGIC;
             x_in0, x_in1, x_in2 :
                          IN  STD_LOGIC_VECTOR(2 DOWNTO 0);
             y    : OUT INTEGER RANGE 0 TO 63);
END dafsm;

ARCHITECTURE flex OF dafsm IS
  TYPE STATE_TYPE IS (s0, s1);
  SIGNAL state    : STATE_TYPE;
  SIGNAL x0, x1, x2, table_in
                           : STD_LOGIC_VECTOR(2 DOWNTO 0);
  SIGNAL table_out : integer RANGE 0 TO 7;
BEGIN

  table_in(0) <= x0(0);
  table_in(1) <= x1(0);
  table_in(2) <= x2(0);

  PROCESS                 ------> DA in behavioral style
    VARIABLE p     : integer RANGE 0 TO 63;-- Temp. register
    VARIABLE count : integer RANGE 0 TO 3; -- Counts shifts
  BEGIN
    WAIT UNTIL clk = '1';
    CASE state IS
      WHEN s0 =>         -- Initialization step
        state <= s1;
        count := 0;
        p := 0;
        x0 <= x_in0;
        x1 <= x_in1;
        x2 <= x_in2;
      WHEN s1 =>             -- Processing step
        IF count = 3 THEN    -- Is sum of product done ?
          y <= p;            -- Output of result to y and
          state <= s0;       -- start next sum of product
        ELSE
          p := p / 2 + table_out * 4;
          x0(0) <= x0(1);
          x0(1) <= x0(2);
          x1(0) <= x1(1);
          x1(1) <= x1(2);
```

```
            x2(0) <= x2(1);
            x2(1) <= x2(2);
            count := count + 1;
            state <= s1;
          END IF;
      END CASE;
    END PROCESS;

    LC_Table0: case3
      PORT MAP(table_in => table_in, table_out => table_out);

END flex;
```

The LC table[5] defined as **CASE** components was generated with the utility program **dagen.exe**. The output is show below.

```
LIBRARY ieee;
USE ieee.std_logic_1164.ALL;
USE ieee.std_logic_arith.ALL;

ENTITY case3 IS
       PORT ( table_in  : IN  STD_LOGIC_VECTOR(2 DOWNTO 0);
              table_out : OUT INTEGER RANGE 0 TO 6);
END case3;

ARCHITECTURE LCs OF case3 IS
BEGIN

-- This is the DA CASE table for
-- the 3 coefficients: 2, 3, 1
-- automatically generated with dagen.exe -- DO NOT EDIT!

  PROCESS (table_in)
  BEGIN
    CASE table_in IS
      WHEN  "000" =>     table_out <=  0;
      WHEN  "001" =>     table_out <=  2;
      WHEN  "010" =>     table_out <=  3;
      WHEN  "011" =>     table_out <=  5;
      WHEN  "100" =>     table_out <=  1;
      WHEN  "101" =>     table_out <=  3;
      WHEN  "110" =>     table_out <=  4;
      WHEN  "111" =>     table_out <=  6;
      WHEN   OTHERS  =>     table_out <=  0;
    END CASE;
  END PROCESS;
END LCs;
```

As suggested in Chapter 2, a shift/accumulator is used, which shifts only one position to the right for each step, instead of shifting k positions to the left. The simulation results, shown in Fig. 3.12, report the correct result

[5] The equivalent Verilog code **case3.v** for this example can be found in Appendix A on page 360.

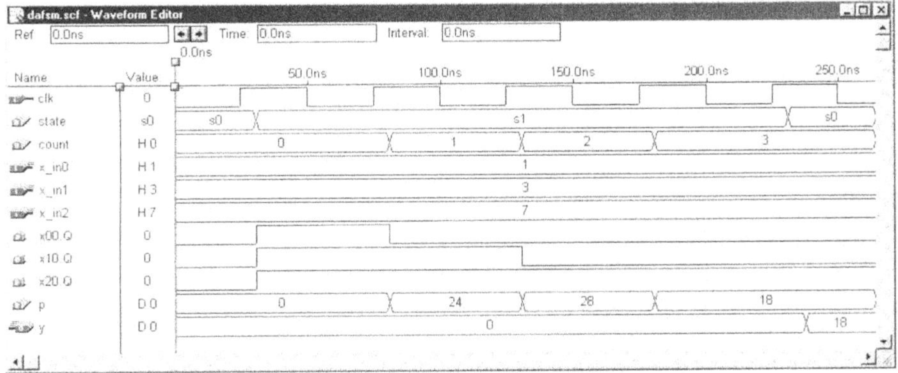

Fig. 3.12. Simulation of the 3-tap FIR filter with input $\{1, 3, 7\}$.

($y = 18$) for an input sequence $\{1, 3, 7\}$. The design runs with `Registered Performance` of 61.72 MHz and uses 37 LCs and no EABs. 3.6

By defining the distributed arithmetic table with a `CASE` statement, the synthesizer will use logic cells to implement the LUT. This will result in a fast and efficient design only if the tables are small. For large tables, alternative means must be found. In this case, we may use the 2Kbit embedded array blocks (EABs), which (as discussed in Chapter 1) can be configured as $2^8 \times 8, 2^9 \times 4, 2^{10} \times 2$ or $2^{11} \times 1$ tables. These two design paths are discussed in more detail in the following.

Distributed Arithmetic Using Logic Cells

The DA implementation of an FIR filter is particularly attractive for low-order cases due to LUT address space limitations (e.g., $L \leq 4$). It should be remembered, however, that FIR filters are linear filters. This implies that the outputs of a collection of low-order filters can be added together to define the output of a high-order FIR, as shown in Fig. 2.19 (p. 67). Based on the LCs found in a FLEX10K device, namely $2^4 \times$ 1-bit tables, a DA table for four coefficients can be implemented. The number of necessary LCs increases exponentially with order. Typically, the number of LCs is much higher than the number of EABs. For example, an EPF10K20RC240-4 contains 1152 LCs but only 6 EABs. Also, EABs can be used to efficiently implement RAMs and FIFOs and other high-valued functions. It is therefore sometimes desirable to use EABs economically. Using EABs in an unregistered mode will decrease the maximal bandwidth of the design. If the design is implemented using larger tables with a $2^b \times b$ `CASE` statement, inefficient designs can result. Even choosing "`Global Project Logic Synthesis`" with the "`Optimize Area`" option, with ''`Reduce Logic`'' and "`Duplicate Logic Extraction`" enabled, which gives optimal area for the `CASE` table, will still result in a

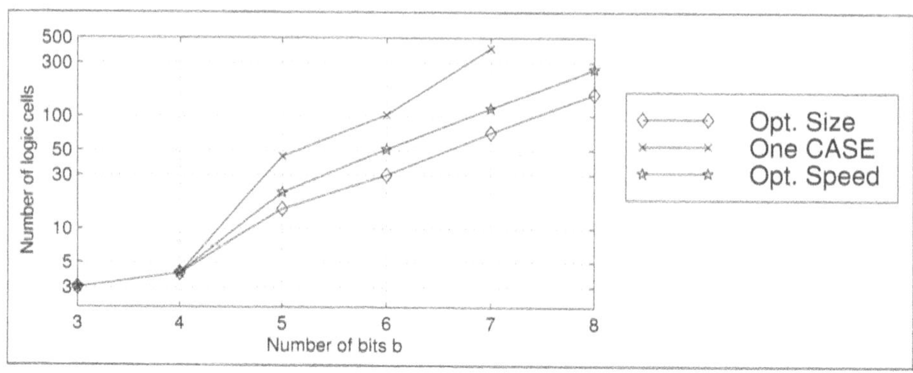

Fig. 3.13. Size comparison of synthesis results for different coding using the **CASE** statement with b input and outputs.

larger-than-expected design. The $2^7 \times 7$ table, for example, required 394 LCs. Fig. 3.13 shows the number of LCs necessary for tables having three to eight inputs using the **CASE** statement. The minimum size curve is obtained from the fact that in the FLEX10K a $4 \rightarrow 1$ multiplexer can (using the **cascade AND** gate) be built with 2 LCs. A $16 \rightarrow 1$ multiplexer can therefore be built with 10 LCs (see help for **busmux** in MaxPlusII).

Synthesizers typically try to optimize the logic equations and are not capable of recognizing structures. It is generally more efficient to realize the 4-input table with **CASE** statements, followed by a (bus) multiplexer. In this model it is also straightforward to add additional pipeline registers to the modular design. For maximum speed, a register must be introduced behind each $2 \rightarrow 1$ multiplexer. This will, however, yield a higher LC count compared to the 2 LC implementation of a $4 \rightarrow 1$ multiplexer. The following example illustrates the structure of a 5-input table.

Example 3.7: Five-input DA Table

The utility program **dagen.exe** accepts filter length and coefficients, and returns the necessary **PROCESS** statements for the 4-input **CASE** table followed by a multiplexer. The VHDL output for an arbitrary set of coefficients, namely $\{1, 3, 5, 7, 9\}$, is given[6] in the following listing:

```
LIBRARY ieee;
USE ieee.std_logic_1164.ALL;
USE ieee.std_logic_arith.ALL;

ENTITY case5p IS
       PORT ( clk        : IN  STD_LOGIC;
              table_in   : IN  STD_LOGIC_VECTOR(4 DOWNTO 0);
              table_out  : OUT INTEGER RANGE 0 TO 25);
END case5p;
```

[6] The equivalent Verilog code **case5p.v** for this example can be found in Appendix A on page 361.

```
ARCHITECTURE LCs OF case5p IS

  SIGNAL lsbs : STD_LOGIC_VECTOR(3 DOWNTO 0);
  SIGNAL msbs0 : STD_LOGIC_VECTOR(1 DOWNTO 0);
  SIGNAL table0out00, table0out01 : INTEGER RANGE 0 TO 25;

BEGIN

-- These are the distributed arithmetic CASE tables for
-- the 5 coefficients: 1, 3, 5, 7, 9
-- automatically generated with dagen.exe -- DO NOT EDIT!

  PROCESS
  BEGIN
    WAIT UNTIL clk = '1';
    lsbs(0) <= table_in(0);
    lsbs(1) <= table_in(1);
    lsbs(2) <= table_in(2);
    lsbs(3) <= table_in(3);
    msbs0(0) <= table_in(4);
    msbs0(1) <= msbs0(0);
  END PROCESS;

  PROCESS      -- This is the final DA MPX stage.
  BEGIN        -- Automatically generated with dagen.exe
    WAIT UNTIL clk = '1';
    CASE msbs0(1) IS
      WHEN  '0' =>    table_out <=  table0out00;
      WHEN  '1' =>    table_out <=  table0out01;
      WHEN  OTHERS  =>    table_out <=  0;
    END CASE;
  END PROCESS;

  PROCESS      -- This is the DA CASE table 00 out of 1.
  BEGIN        -- Automatically generated with dagen.exe
    WAIT UNTIL clk = '1';
    CASE lsbs IS
      WHEN  "0000" =>    table0out00 <=  0;
      WHEN  "0001" =>    table0out00 <=  1;
      WHEN  "0010" =>    table0out00 <=  3;
      WHEN  "0011" =>    table0out00 <=  4;
      WHEN  "0100" =>    table0out00 <=  5;
      WHEN  "0101" =>    table0out00 <=  6;
      WHEN  "0110" =>    table0out00 <=  8;
      WHEN  "0111" =>    table0out00 <=  9;
      WHEN  "1000" =>    table0out00 <=  7;
      WHEN  "1001" =>    table0out00 <=  8;
      WHEN  "1010" =>    table0out00 <=  10;
      WHEN  "1011" =>    table0out00 <=  11;
      WHEN  "1100" =>    table0out00 <=  12;
      WHEN  "1101" =>    table0out00 <=  13;
      WHEN  "1110" =>    table0out00 <=  15;
      WHEN  "1111" =>    table0out00 <=  16;
```

```
        WHEN   OTHERS  =>    table0out00 <=  0;
      END CASE;
    END PROCESS;

    PROCESS     -- This is the DA CASE table 01 out of 1.
    BEGIN       -- Automatically generated with dagen.exe
      WAIT UNTIL clk = '1';
      CASE lsbs IS
        WHEN  "0000" =>    table0out01 <=  9;
        WHEN  "0001" =>    table0out01 <=  10;
        WHEN  "0010" =>    table0out01 <=  12;
        WHEN  "0011" =>    table0out01 <=  13;
        WHEN  "0100" =>    table0out01 <=  14;
        WHEN  "0101" =>    table0out01 <=  15;
        WHEN  "0110" =>    table0out01 <=  17;
        WHEN  "0111" =>    table0out01 <=  18;
        WHEN  "1000" =>    table0out01 <=  16;
        WHEN  "1001" =>    table0out01 <=  17;
        WHEN  "1010" =>    table0out01 <=  19;
        WHEN  "1011" =>    table0out01 <=  20;
        WHEN  "1100" =>    table0out01 <=  21;
        WHEN  "1101" =>    table0out01 <=  22;
        WHEN  "1110" =>    table0out01 <=  24;
        WHEN  "1111" =>    table0out01 <=  25;
        WHEN   OTHERS  =>    table0out01 <=  0;
      END CASE;
    END PROCESS;
  END LCs;
```

The five inputs produce two **CASE** tables and a $2 \rightarrow 1$ bus multiplexer. The multiplexer may also be realized with a component instantiation using the LPM function `busmux`. The program `dagen.exe` writes a VHDL file with the name `caseX.vhd`, where `X` is the filter length that is also the input bit width. The file `caseXp.vhd` is the same table, except with additional pipeline registers. The `component` can be used directly in a state machine design or in an unrolled filter structure. 3.7

Referring to Fig. 3.13, it can be seen that the structured VHDL code improves on the number of required LCs. Fig. 3.14 compares the different design methods in terms of speed. Reducing the LC count also improves the throughput, because the number of LC levels is reduced. Although we get a high `Registered Performance` using seven pipeline stages for a $2^8 \times 8$ table with 88.49 MHz, the design may now be too large for some applications. We may also consider the partitioning technique (Exercise 3.6, p. 114), shown in Fig. 2.18 (p. 66), or implementation with an EAB, discussed next.

DA Using Embedded Array Blocks

As mentioned in the last section, it is not economical to use a 2Kbit EABs for a short FIR filter, mainly because the number of available EABs is lim-

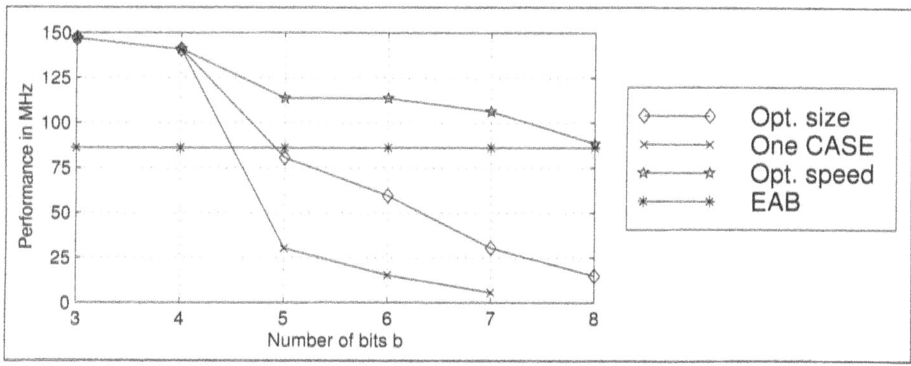

Fig. 3.14. Speed comparison for different coding styles using the **CASE** statement.

ited. Also, the maximum registered speed of a EAB is 76 MHz, and an LC table implementation may be faster. The following example shows the DA implementation using a component instantiation of the EAB.

Example 3.8: Distributed Arithmetic Filter using EABs

The **CASE** table from the last example can be replaced by a EAB ROM. The ROM table is defined by file **da3.mif**. The default input and output configuration of the EAB is given by **"REGISTERED."** If it is not desirable to have a registered configuration, set **LPM_ADDRESS_CONTROL => "UNREGISTERED"** and **LPM_OUTDATA => "UNREGISTERED."** The VHDL code[7] for the state machine design is shown below:

```
LIBRARY lpm;
USE lpm.lpm_components.ALL;

LIBRARY ieee;                    -- Using predefined packages
USE ieee.std_logic_1164.ALL;
USE ieee.std_logic_arith.ALL;
USE ieee.std_logic_unsigned.ALL;  -- Contains conversion
                                  -- VECTOR -> INTEGER
ENTITY darom IS                   ------> Interface
  PORT (clk   : IN STD_LOGIC;
        x_in0, x_in1, x_in2
                          : IN STD_LOGIC_VECTOR(2 DOWNTO 0);
        y     : OUT INTEGER RANGE 0 TO 63);
END darom;

ARCHITECTURE flex OF darom IS
  TYPE STATE_TYPE IS (s0, s1);
  SIGNAL state                     : STATE_TYPE;
  SIGNAL x0, x1, x2, table_in, mem
                           : STD_LOGIC_VECTOR(2 DOWNTO 0);
  SIGNAL table_out                 : integer RANGE 0 TO 7;
BEGIN
```

[7] The equivalent Verilog code **darom.v** for this example can be found in Appendix A on page 363.

```
    table_in(0) <= x0(0);
    table_in(1) <= x1(0);
    table_in(2) <= x2(0);

    PROCESS                          ------> DA in behavioral style
      VARIABLE  p   : integer RANGE 0 TO 63; --Temp. register
      VARIABLE count : integer RANGE 0 TO 3;
    BEGIN                                     -- Counts the shifts
      WAIT UNTIL clk = '1';
      CASE state IS
        WHEN s0 =>              -- Initialization step
          state <= s1;
          count := 0;
          p := 0;
          x0 <= x_in0;
          x1 <= x_in1;
          x2 <= x_in2;
        WHEN s1 =>               -- Processing step
          IF count = 3 THEN      -- Is sum of product done ?
            y <= p;              -- Output of result to y and
            state <= s0;         -- start next sum of product
          ELSE
            p := p / 2 + table_out * 4;
            x0(0) <= x0(1);
            x0(1) <= x0(2);
            x1(0) <= x1(1);
            x1(1) <= x1(2);
            x2(0) <= x2(1);
            x2(1) <= x2(2);
            count := count + 1;
            state <= s1;
          END IF;
      END CASE;
    END PROCESS;

    rom_1: LPM_ROM
      GENERIC MAP ( LPM_WIDTH => 3,
                    LPM_WIDTHAD => 3,
                    LPM_OUTDATA => "UNREGISTERED",
                    LPM_ADDRESS_CONTROL => "UNREGISTERED",
                    LPM_FILE => "darom3.mif")
      PORT MAP ( address => table_in, q => mem);

    table_out <= CONV_INTEGER(mem);

  END flex;
```

Compared with Example 3.6 (p. 98), we have now a component instantiation of the `LPM_ROM`. Because there is a need to convert between the `STD_LOGIC_VECTOR` output of the ROM and the integer, we have used the package `std_logic_unsigned` from the library `ieee`. The latter contains the `CONV_INTEGER` function for unsigned `STD_LOGIC_VECTOR`.

The include file **darom3.mif** was also generated with the program **dagen.exe**. The file has the following contents:

```
-- This is the DA MIF table for the 3 coefficients: 2, 3, 1
-- automatically generated with dagen.exe -- DO NOT EDIT!
WIDTH = 3;
DEPTH = 8;
ADDRESS_RADIX = dec;
DATA_RADIX = dec;
CONTENT BEGIN
      0     :   0;
      1     :   2;
      2     :   3;
      3     :   5;
      4     :   1;
      5     :   3;
      6     :   4;
      7     :   6;
END;
```

The design runs with 32.25 MHz and uses 34 LCs (three less than the **CASE** version) and one EAB (more precisely, 3/8 of a EAB). 3.8

But EABs have only a single address decoder and if we implement a $2^4 \times 8$ table, a complete EAB would be consumed unnecessarily, and it can not be used elsewhere. For longer filters, however, the use of EABs is attractive because:

- EABs have registered throughput at a constant 76 MHz, and
- Routing effort is reduced.

Signed DA FIR Filter

A signed DA filter will require a signed accumulator. The following example shows the VHDL code for the previously studied three-coefficient example, 2.18 from Chapter 2 (p. 64).

Example 3.9: Signed DA FIR Filter

For the signed DA filter, an additional state is required. See the variable **count**[8] to process the sign bit.

```
LIBRARY ieee;                    -- Using predefined packages
USE ieee.std_logic_1164.ALL;
USE ieee.std_logic_arith.ALL;

PACKAGE da_package IS      -- User defined components
COMPONENT case3s
   PORT ( table_in    : IN  STD_LOGIC_VECTOR(2 DOWNTO 0);
          table_out   : OUT INTEGER RANGE -2 TO 4);
END COMPONENT;
```

[8] The equivalent Verilog code **case3s.v** for this example can be found in Appendix A on page 365.

```
END da_package;

LIBRARY work;
USE work.da_package.ALL;

LIBRARY ieee;                  -- Using predefined packages
USE ieee.std_logic_1164.ALL;
USE ieee.std_logic_arith.ALL;

ENTITY dasign IS                           ------> Interface
       PORT (clk  : IN STD_LOGIC;
             x_in0, x_in1, x_in2
                        : IN  STD_LOGIC_VECTOR(3 DOWNTO 0);
             y    : OUT INTEGER RANGE -64 TO 63);
END dasign;

ARCHITECTURE flex OF dasign IS

  TYPE STATE_TYPE IS (s0, s1);
  SIGNAL state      : STATE_TYPE;
  SIGNAL table_in   : STD_LOGIC_VECTOR(2 DOWNTO 0);
  SIGNAL x0, x1, x2 : STD_LOGIC_VECTOR(3 DOWNTO 0);
  SIGNAL table_out  : integer RANGE -2 TO 4;

BEGIN

  table_in(0) <= x0(0);
  table_in(1) <= x1(0);
  table_in(2) <= x2(0);

  PROCESS                  ------> DA in behavioral style
    VARIABLE  p : integer RANGE -64 TO 63;-- Temporary reg.
    VARIABLE count : integer RANGE 0 TO 4; -- Counts the
  BEGIN                                    -- shifts
    WAIT UNTIL clk = '1';
    CASE state IS
      WHEN s0 =>           -- Initialization step
        state <= s1;
        count := 0;
        p := 0;
        x0 <= x_in0;
        x1 <= x_in1;
        x2 <= x_in2;
      WHEN s1 =>             -- Processing step
        IF count = 4 THEN -- Is sum of product done?
          y <= p;       -- Output of result to y and
          state <= s0; -- start next sum of product
        ELSE
          IF count = 3 THEN            -- Subtract for last
          p := p / 2 - table_out * 8; -- accumulator step
          ELSE
          p := p / 2 + table_out * 8;  -- Accumulation for
          END IF;                      -- all other steps
```

```
              FOR k IN 0 TO 2 LOOP     -- Shift bits
                x0(k) <= x0(k+1);
                x1(k) <= x1(k+1);
                x2(k) <= x2(k+1);
              END LOOP;
            count := count + 1;
            state <= s1;
          END IF;
      END CASE;
    END PROCESS;

    LC_Table0: case3s
      PORT MAP(table_in => table_in, table_out => table_out);

  END flex;
```

The LC table (component **case3s.vhd**) was generated using the program **dagen.exe**. The VHDL code[9] is shown below:

```
LIBRARY ieee;
USE ieee.std_logic_1164.ALL;
USE ieee.std_logic_arith.ALL;

ENTITY case3s IS
  PORT ( table_in  : IN   STD_LOGIC_VECTOR(2 DOWNTO 0);
         table_out : OUT  INTEGER RANGE -2 TO 4);
END case3s;

ARCHITECTURE LCs OF case3s IS
BEGIN

-- This is the DA CASE table for
-- the 3 coefficients: -2, 3, 1
-- automatically generated with dagen.exe -- DO NOT EDIT!

  PROCESS (table_in)
  BEGIN
    CASE table_in IS
      WHEN  "000" =>    table_out <=  0;
      WHEN  "001" =>    table_out <=  -2;
      WHEN  "010" =>    table_out <=  3;
      WHEN  "011" =>    table_out <=  1;
      WHEN  "100" =>    table_out <=  1;
      WHEN  "101" =>    table_out <=  -1;
      WHEN  "110" =>    table_out <=  4;
      WHEN  "111" =>    table_out <=  2;
      WHEN   OTHERS  =>    table_out <=  0;
    END CASE;
  END PROCESS;
END LCs;
```

[9] The equivalent Verilog code **case3s.v** for this example can be found in Appendix A on page 366.

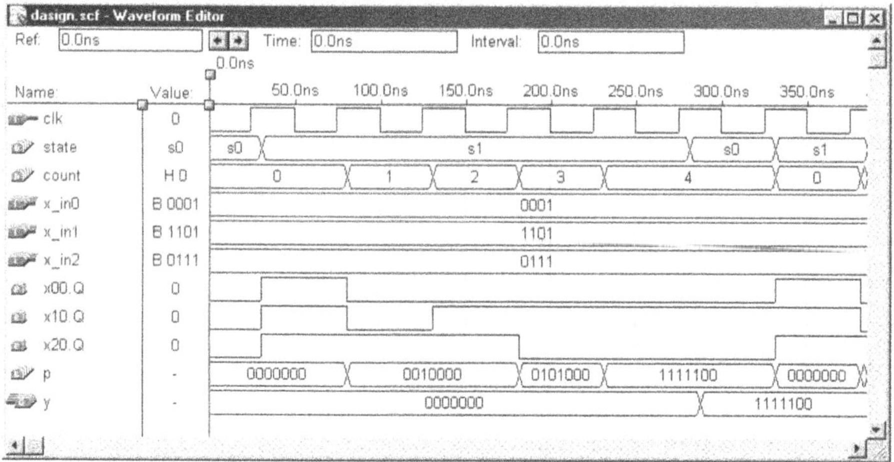

Fig. 3.15. Simulation of the 3-tap signed FIR filter with input $\{1, -3, 7\}$.

Fig. 3.15 shows the simulation for the input sequence $\{1, -3, 7\}$. As expected, the output y is $-4_{10} = 1111100_{2C}$. 3.9

To accelerate a DA filter, unrolled loops can be used. The input is applied sample by sample (one word at a time), in a bit-parallel form. In this case, for each bit of input a separate table is required. While the table size varies (input bit width equals number of filter taps), the contents of the tables are the same. The obvious advantage is a reduction of VHDL code size, if we use a component definition for the LC tables, as previously presented. To demonstrate, the unrolling of the 3-coefficients, 4-bit input example, previously considered, is developed below.

Example 3.10: Loop Unrolling for DA FIR Filter

In a typical FIR application, the input values are processed in word parallel form (i.e., see Fig. 3.16). The following VHDL code[3] illustrates the unrolled DA code, according to Fig. 3.16.

```
LIBRARY ieee;                    -- Using predefined packages
USE ieee.std_logic_1164.ALL;
USE ieee.std_logic_arith.ALL;

PACKAGE da_package IS            -- User defined components
  COMPONENT case3s
    PORT ( table_in    : IN  STD_LOGIC_VECTOR(2 DOWNTO 0);
           table_out   : OUT INTEGER RANGE -2 TO 4);
  END COMPONENT;
END da_package;
```

[3] The equivalent Verilog code **dapara.v** for this example can be found in Appendix A on page 367.

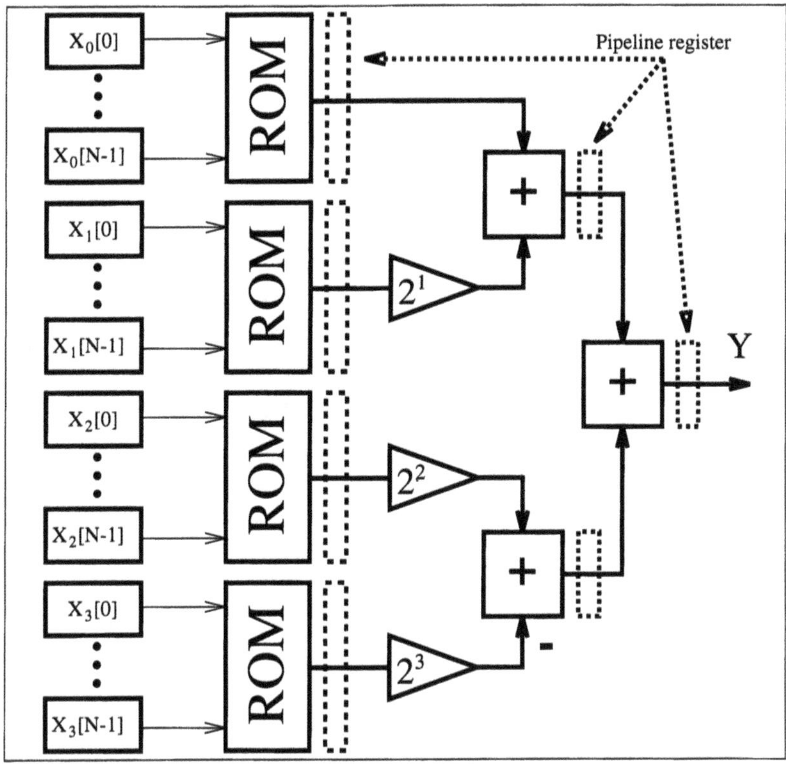

Fig. 3.16. Parallel implementation of a distributed arithmetic FIR filter.

```
LIBRARY work;
USE work.da_package.ALL;

LIBRARY ieee;                    -- Using predefined packages
USE ieee.std_logic_1164.ALL;
USE ieee.std_logic_arith.ALL;

ENTITY dapara IS                           ------> Interface
       PORT (clk  : IN STD_LOGIC;
             x_in : IN  STD_LOGIC_VECTOR(3 DOWNTO 0);
             y    : OUT INTEGER RANGE -46 TO 44);
END dapara;

ARCHITECTURE flex OF dapara IS
  SIGNAL x0, x1, x2, x3 : STD_LOGIC_VECTOR(2 DOWNTO 0);
  SIGNAL y0, y1, y2, y3 : integer RANGE -2 TO 4;
  SIGNAL s0 : integer RANGE -6 TO 12;
  SIGNAL s1 : integer RANGE -10 TO 8;
  SIGNAL t0, t1, t2, t3 : integer RANGE -2 TO 4;

BEGIN
```

```
    PROCESS                              ------> DA in behavioral style
    BEGIN
      WAIT UNTIL clk = '1';
      FOR k IN 0 TO 1 LOOP  -- Shift all four bits
        x0(k) <= x0(k+1);
        x1(k) <= x1(k+1);
        x2(k) <= x2(k+1);
        x3(k) <= x3(k+1);
      END LOOP;
      x0(2) <= x_in(0); -- Load x_in in the
      x1(2) <= x_in(1); -- MSBs of register 2
      x2(2) <= x_in(2);
      x3(2) <= x_in(3);
      y <= y0 + 2 * y1 + 4 * y2 - 8 * y3;
  -- Pipeline register and adder tree
  --  t0 <= y0; t1 <= y1; t2 <= y2; t3 <= y3;
  --  s0 <= t0 + 2 * t1; s1 <= t2 - 2 * t3;
  --  y <= s0 + 4 * s1;
    END PROCESS;

    LC_Table0: case3s
               PORT MAP(table_in => x0, table_out => y0);
    LC_Table1: case3s
               PORT MAP(table_in => x1, table_out => y1);
    LC_Table2: case3s
               PORT MAP(table_in => x2, table_out => y2);
    LC_Table3: case3s
               PORT MAP(table_in => x3, table_out => y3);

  END flex;
```

The design uses 4 tables of size $2^3 \times 4$ and all tables have the *same* content as the table shown in Example 3.9 (p. 107). Fig. 3.17 shows the simulation for the input sequence $\{1, -3, 7\}$. Because the input is applied serial (and bit-parallel) the expected result $-4_{10} = 1111100_{2C}$ is computed at the 400 ns interval. 3.10

The previous design requires 39 LCs and runs at 42.19 MHz. An important advantage of the DA concept, compared with general purpose MAC design, is that pipelining is easy archived. We can add additional pipeline registers at the table output and at the adder-tree output with no costs. To compute y, i.e., instead

```
y <= y0 + 2 * y1 + 4 * y2 - 8 * y3;
```

we use signals `t0` to `t1` for the pipelined version within a `PROCESS` statement

```
t0 <= y0; t1 <= y1; t2 <= y2; t3 <= y3;
s0 <= t0 + 2 * t1; s1 <= t2 - 2 * t3;
y <= s0 + 4 * s1;
```

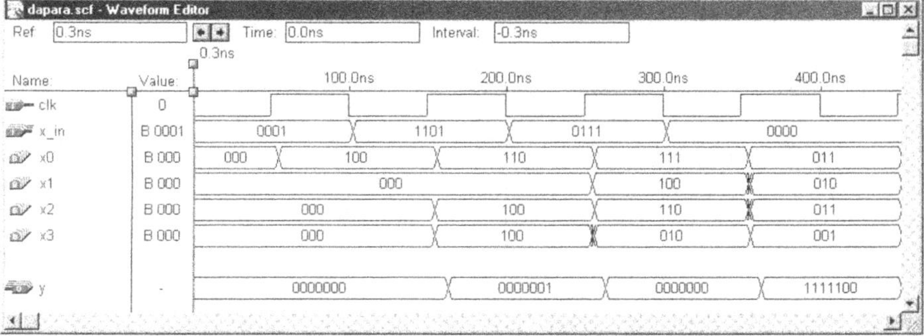

Fig. 3.17. Simulation results for the parallel distributed arithmetic FIR filter.

The size of the design does not increase, because the registers of the LC table and adder are otherwise unused. But the `Registered Performance` increases from 42.19 MHz to 106.38 MHz!

Exercises

3.1: A filter has the following specification: sampling frequency 2 KHz; passband 0-0.4 KHz, stopband 0.5-1 KHz; passband ripple, 3 dB, and stopband ripple, 48 dB. Use the MATLAB software and the "Interactive Lowpass Filter Design" demo for the filter design.
(a1) Design a direct filter with Kaiser window.
(a2) Determine filter length and the absolute ripple in the passband.
(b1) Design an equiripple filter.
(b2) Determine filter length and the absolute ripple in the passband.

Exercises Using MaxPlusII

3.2: **(a)** Compute the RAG for a length-11 halfband filter F5 that has the nonzero coefficients $f[0] = 256, f[\pm 1] = 150, f[\pm 3] = -25, f[\pm 5] = 3$.
(b) What is the minimum output bit width of the filter, if the input bit width is 8 bits?
(c1) Write and compile (with the MaxPlusII compiler) the VHDL code for the filter.
(c2) Simulate the filter with impulse and step response.
(d) Write the VHDL code for the filter in distributed arithmetic, using the state machine approach with the table realized as `LPM_ROM`.

3.3: **(a)** Compute the RAG for length-11 halfband filter F7 that has the nonzero coefficients $f[0] = 521, f[\pm 1] = 302, f[\pm 3] = -53, f[\pm 5] = 7$.
(b) What is the minimum output bit width of the filter, if the input bit width is 8 bits?
(c1) Write and compile (with the MaxPlusII compiler) the VHDL code for the filter.

(c2) Simulate the filter with impulse and step response.

3.4: Hartley [67] has introduced a concept to implement constant coefficient filters, by exploiting common subexpressions across coefficients. For instance, the filter

$$y[n] = \sum_{k=0}^{L-1} a[k]x[n-k] \tag{3.19}$$

with three coefficients $a[k] = \{480, -302, 31\}$. The CSD code of these three coefficients is given by

	512	256	128	64	32	16	8	4	2	1
480 :	1	0	0	0	−1	0	0	0	0	0
−302 :	0	−1	0	−1	0	1	0	0	1	0
31 :	0	0	0	0	1	0	0	0	0	−1

From the table we notice that the pattern $\begin{array}{|cc|}\hline 1 & 0 \\ 0 & -1 \\ \hline\end{array}$ can be found four times. If we therefore build the temporary variable $h[n] = 2x[n] - x[n-1]$, we can compute the filter output with

$$y[n] = 256h[n] - 32h[n] - 16h[n-1] + h[n-1]. \tag{3.20}$$

(a) Verify (3.20) by substituting $h[n] = 2x[n] - x[n-1]$.
(b) How many adders are required to yield the direct CSD implementation of (3.19) and the implementation with subexpression sharing?
(c1) Implement the filter with subexpression sharing with MaxPlusII for 8-bit inputs.
(c2) Simulate the impulse response of the filter.
(c3) Determine LC usage and **Registered Performance**.

3.5: Use the subexpression method from Exercise 3.4 to implement a 4-tap filter with the coefficients $a[k] = \{-1046, -1109, -894, 2072\}$.
(a) Find the CSD code and the subexpression representation for the most frequent pattern.
(b) Substitute for the subexpression a 2 or −2, respectively. Apply the subexpression sharing one more time to the reduced set.
(c) Determine the temporary equations and check by substitution back into (3.19).
(d1) Implement the filter with subexpression sharing with MaxPlusII for 8-bit inputs.
(d2) Simulate the impulse response of the filter.
(d3) Determine LC usage and **Registered Performance**.

3.6: **(a)** Use the program **dagen.exe** to compile a DA table for the coefficients {20, 24, 21, 100, 13, 11, 19, 7} using multiple **CASE** statements.
Synthesize the design for maximum speed and determine the size and **Registered Performance**.
(b) Use the partitioning technique to implement the same table using two sets, namely {20, 24, 21, 100} and {13, 11, 19, 7}, and an additional adder. Synthesize the design for maximum speed and determine the size and **Registered Performance**.
(c) Compare the designs from (a) and (b).

4. Infinite Impulse Response (IIR) Digital Filters

Introduction

In Chapter 3 we introduced the FIR filter. The most important properties that make the FIR attractive (+) or unattractive (−) for selective applications include:

+ FIR linear-phase performance is easily achieved.
+ Multiband filters are possible.
+ The Kaiser window method allows iterative-free design.
+ FIRs have a simple structure for decimators and interpolators (see Chapter 5).
+ Nonrecursive filters are always stable and have no limit cycles.
+ It is easy to get high-speed, pipelined designs.
+ FIRs typically have low coefficient and arithmetic roundoff error budgets, and well-defined quantization noise.
− Recursive FIR filters may be unstable because of imperfect pole/zero annihilation.
− The sophisticated Parks-McClellan algorithms must be available for minimax filter design.
− High filter length requires high implementation effort.

Compared to an FIR filter, an IIR filter can often be much more efficient in terms of attaining certain performance characteristics with a given filter order. This is because the IIR filter incorporates feedback and is capable of realizing both zeros and poles of a system transfer function, whereas the FIR filter is an all-zero filter. In this chapter, the fundamentals of IIR filter design will be developed. The traditional approach to the design of IIR filters involves the transformation of an analog filter, with defined feedback specifications, into the digital domain. This is a reasonable approach, mainly because the art of designing analog filters is highly advanced, and many standard tables are available, i.e., [68]. We will review the four most important classes of these analog prototype filters in this chapter, namely Butterworth, Chebyshev I and II, and elliptic filters.

The IIR will be shown to overcome many of the deficiencies of the FIR, but to have some less desirable properties as well. The general desired (+) and undesired (−) properties of an IIR filter are:

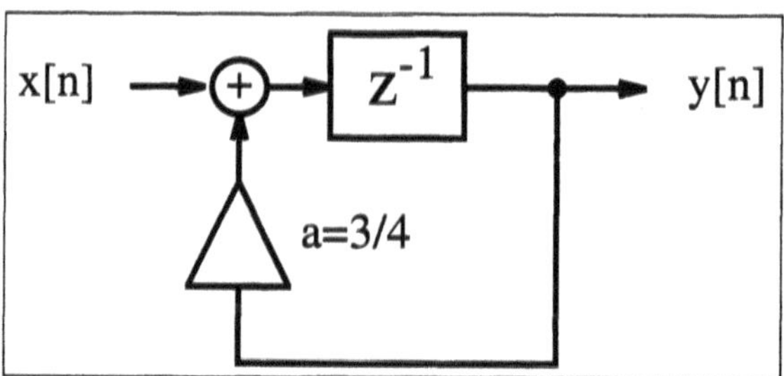

Fig. 4.1. First-order IIR filter used as lossy integrator.

+ Standard design using an analog prototype filter is well understood.
+ Highly selective filters can be realized with low-order designs which can run at high speeds.
+ Design using tables and a pocket calculator is possible.
+ For the same tolerance scheme, filters are short, compared with FIR filters.
+ Closed-loop design algorithms can be used.
– Nonlinear-phase response is typical, i.e., it is difficult to get linear-phase response. (Using an all-pass filter for phase compensation results in twice the complexity.)
– Limit cycles may occur for integer implementation.
– Multiband design is difficult; only low, high, or bandpass filters are designed.
– Feedback can introduce instabilities. (Most often, the mirror pole to the unit circle can be used to produce the same magnitude response, and the filter will be stable.)
– It is more difficult to get high-speed, pipelined designs

To demonstrate the possible benefits of using IIR filters, we will discuss a first-order IIR filter example.

Example 4.1: Lossy Integrator I

One of the basic tasks of a filter may be to smooth a noisy signal. Assume that a signal $x[n]$ is received in the presence of wideband zero-mean random noise. Mathematically, an integrator could be used to suppress the effects of the noise. If the average value of the input signal is to be preserved over a finite time interval, a *lossy* integrator is often used to process the signal with additive noise. Fig. 4.1 displays a simple first-order lossy integrator that satisfies the discrete-time difference equation:

$$y[n+1] = \frac{3}{4}y[n] + x[n]. \tag{4.1}$$

As we can see from the impulse response in Fig. 4.2(a), the same functionality of the first-order lossy integrator can be achieved with a 15-tap FIR filter. The step response to the lossy integrator is shown in Fig. 4.2(b).

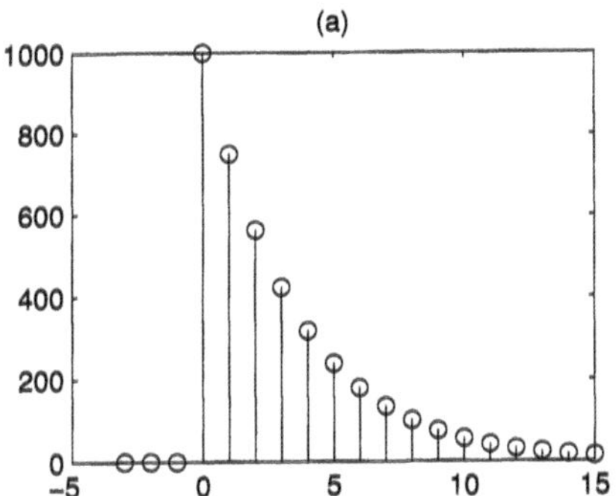

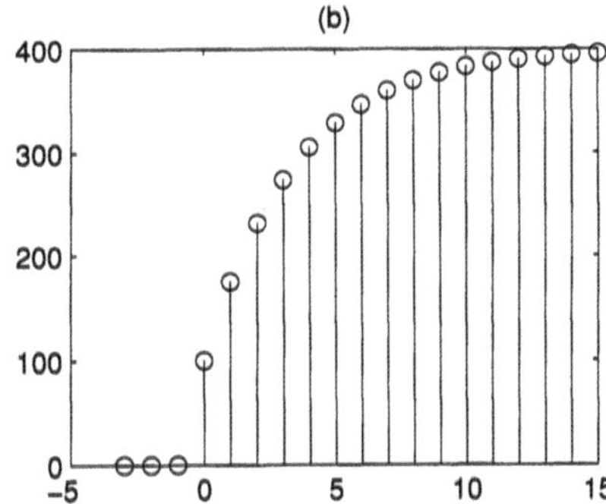

Fig. 4.2. Simulation of lossy integrator with $a = 3/4$. **(a)** Impulse response for $1000\delta[n]$. **(b)** Step response for $100\sigma[n]$.

The following VHDL code[1] shows a possible implementation of this IIR filter.

```
PACKAGE n_bit_int IS                    -- User defined type
  SUBTYPE BITS15 IS INTEGER RANGE -2**14 TO 2**14-1;
END n_bit_int;

LIBRARY work;
USE work.n_bit_int.ALL;

LIBRARY ieee;
USE ieee.std_logic_1164.ALL;
USE ieee.std_logic_arith.ALL;

ENTITY iir IS
  PORT (x_in  : IN  BITS15;        -- Input
        y_out : OUT BITS15;        -- Result
        clk   : IN  STD_LOGIC);
END iir;

ARCHITECTURE flex OF iir IS

  SIGNAL x, y : BITS15;

BEGIN

  PROCESS      -- Use FF for input and recursive part
  BEGIN
  WAIT UNTIL clk = '1';
    x  <= x_in;
    y  <= x + y / 4 + y / 2;
  end process;

  y_out <= y;                -- Connect y to output pins
```

[1] The equivalent Verilog code `iir.v` for this example can be found in Appendix A on page 368.

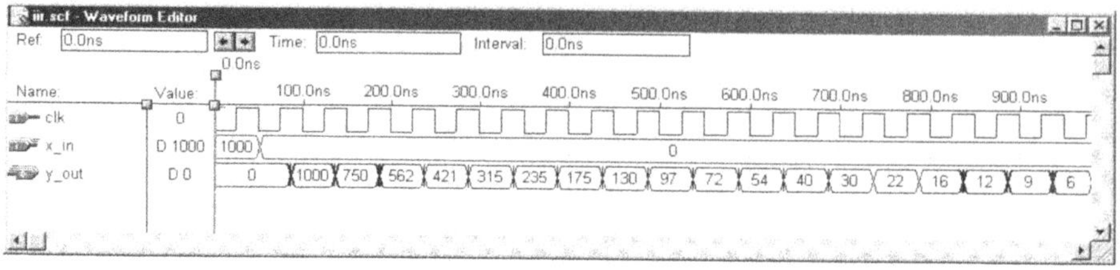

Fig. 4.3. Impulse response for MaxPlusII simulation of the lossy integrator.

```
END flex;
```

Registers have been implemented using a **WAIT** statement inside a **PROCESS** block, while the multiplication and addition is implemented using CSD code. The design uses 31 logic cells and runs at a speed of 55.86 MHz, if synthesized with an `Optimize Speed=10` option. The response of the filter to an impulse, with amplitude 1000 is shown in Fig. 4.3, and agrees with the simulated results presented in Fig. 4.2(a). 4.1

An alternative design approach using a "standard logic vector" data type and `LPM_ADD_SUB` megafunctions is discussed in Exercise 4.4 (p. 140). This second approach will produce longer VHDL code but will have the benefit of direct control, at the bit level, over the sign extension and multiplier.

4.1 IIR Theory

A nonrecursive filter incorporates, as the name implies, no feedback. The impulse response of such a filter is finite, i.e., it is an FIR filter. A recursive filter, on the other hand has feedback, and is expected in general to have an infinite impulse response, i.e., to be an IIR filter. Fig. 4.4(a) shows filters with separate recursive and nonrecursive parts. A *canonical* filter is produced if these recursive and nonrecursive parts are merged together, as shown in Fig. 4.4(b). The transfer function of the filter from Fig. 4.4 can be written as:

$$F(z) = \frac{\sum\limits_{l=0}^{L-1} a[l]z^{-l}}{1 - \sum\limits_{l=1}^{L-1} b[l]z^{-l}} \tag{4.2}$$

The difference equation for such a system yields:

$$y[n] = \sum_{l=0}^{L-1} a[l]x[n-l] + \sum_{l=1}^{L-1} b[l]y[n-k] \tag{4.3}$$

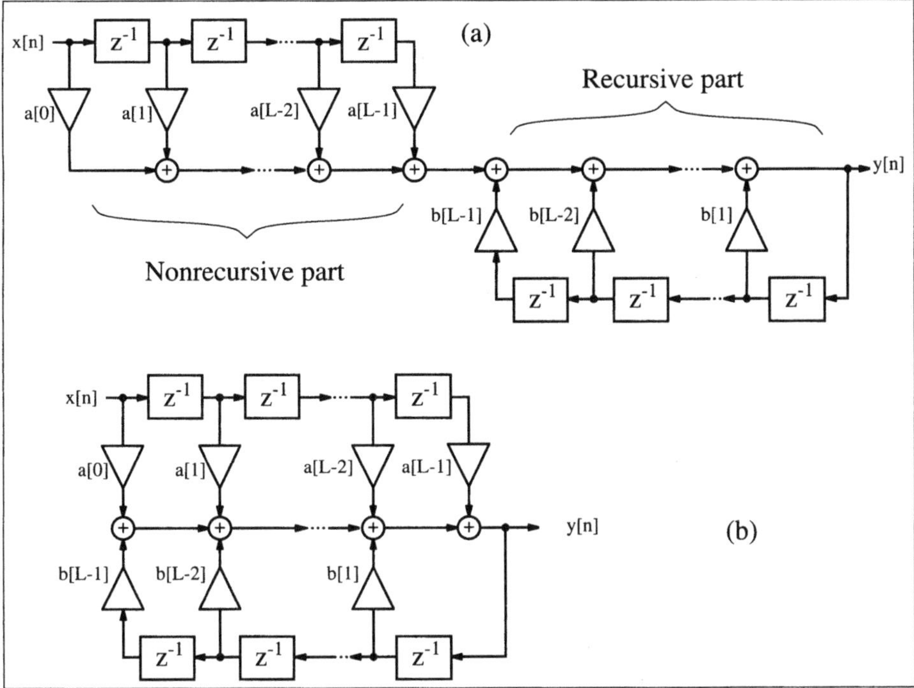

Fig. 4.4. Filter with feedback.

Comparing this with the difference equation for the FIR filter (3.2) on p. 80, we find that the difference equation for recursive systems depends not only on the L previous values of the input sequence $x[n]$, but also on the $L-1$ previous values of $y[n]$.

If we compute poles and zeros of $F(z)$, we see that the nonrecursive part, i.e., the nominator of $F(z)$, produces the *zeros* p_{0l}, while the denominator of $F(z)$ produces the *poles* $p_{\infty l}$.

For the transfer function, the *pole/zero plot* can be used to look up the most important properties of the filter. If we substitute $z = e^{j\omega T}$ in the z-domain transfer function, we can construct the Fourier transfer function

$$F(\omega) = |F(\omega)|e^{\theta(\omega)} = \frac{\prod_{l=0}^{L-2} p_{0l} - e^{j\omega T}}{\prod_{l=0}^{L-2} p_{\infty l} - e^{j\omega T}} = \frac{e^{\sum_l \alpha_l} \prod_{l=0}^{L-2} v_l}{e^{\sum_l \beta_l} \prod_{l=0}^{L-2} u_l} \tag{4.4}$$

by graphical means. This is shown in Fig. 4.5, for a specific amplitude (i.e., gain) and phase value. The gain at a specific frequency ω_0 is the quotient of the zero vectors v_l and the pole vectors u_l. These vectors start at a specific zero or pole, respectively, and end at the frequency point, $e^{j\omega_0 T}$, of interest. The phase gain for the example from Fig. 4.5 becomes $\theta(\omega_0) = \alpha_0 + \alpha_1 - \beta_0$.

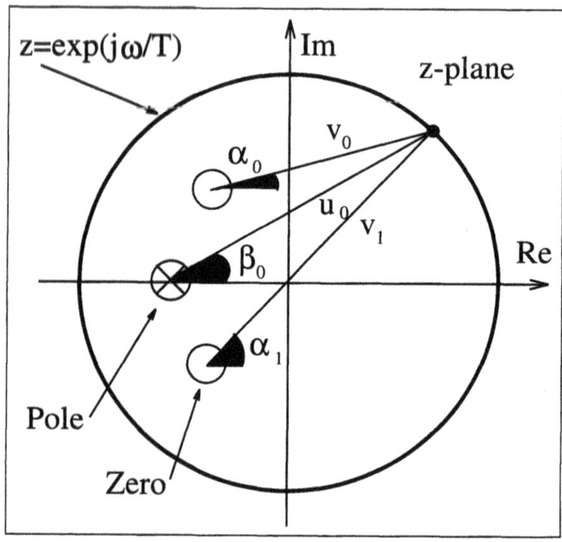

Fig. 4.5. Computation of transfer function using the pole/zero plot. Amplitude gain $= v_0 v_1/u_0$, phase gain $= \alpha_0 + \alpha_1 - \beta_0$.

Using the connection between the transfer function in the Fourier domain and the pole/zero plot, we can already deduce several properties:

1) A *zero on the unit circle* $p_0 = e^{j\omega_0 T}$ (with no annihilating pole) produces a zero in the transfer function in the Fourier domain at the frequency ω_0.
2) A *pole on the unit circle* $p_\infty = e^{j\omega_0 T}$ (and no annihilating zero) produces an infinite gain in the transfer function in the Fourier domain at the frequency ω_0.
3) An independent from the input signal *stable* filter has all poles inside the unit circle.
4) A *real filter* has single poles and zeros on the unit circle, while complex poles and zeros appear always in pairs, i.e., if $a_0 + ja_1$ is a pole or zero, $a_0 - ja_1$ must also be a pole or zero.
5) A *linear-phase* (i.e., constant group delay) filter has all poles and zeros symmetric to the unit circle or at $z = 0$.

If we combine observations 3 and 5, we find that, for a stable linear-phase system, all zeros must be symmetric to the unit circle and only poles at $z = 0$ are permitted.

An IIR filter (with poles $z \neq 0$) can therefore be only approximately linear-phase. To achieve this approximation a well-known principle from analog filter design is used: an allpass has an unit gain, and introduces a nonzero phase gain, which is used to achieve linearization in the frequency range of interest, i.e., the passband.

4.2 IIR Coefficient Computation

In classical IIR design, a digital filter is designed which approximates an ideal filter. The ideal digital filter model specifications are mathematically converted into a set of specifications for an analog filter model using the *bi-linear z-transform* given by:

$$s = \frac{z-1}{z+1}. \tag{4.5}$$

A classic analog Butterworth, Chebyshev, or elliptic model can be synthesized from these specifications, and is then mapped into a digital IIR using this bi-linear z-transform.

An analog *Butterworth* filter has a magnitude-squared frequency response given by:

$$|F(\omega)|^2 = \frac{1}{1+\left(\frac{\omega}{\omega_s}\right)^{2N}}. \tag{4.6}$$

The poles of $|F(\omega)|^2|$ are distributed along a circular arc at locations separated by π/N radians. More specifically, the transfer function is N times differentiable at $\omega = 0$. This results in a locally smooth transfer function around 0 Hz. An example of a Butterworth filter model is shown in Fig. 4.6(upper). Note that the tolerance scheme for this design is the same as for the Kaiser window and equiripple design shown in Fig. 3.7 (p. 90).

An analog *Chebyshev* filter of Type I or II is defined in terms of a Chebyshev polynomial $V_N(\omega) = \cos(N\cos(\omega))$, which forces the filter poles to reside on an ellipse. The magnitude-squared frequency response of a Type I filter is represented by:

$$|F(\omega)|^2 = \frac{1}{1+\varepsilon^2 V_N^2\left(\frac{\omega}{\omega_s}\right)}. \tag{4.7}$$

An example of a typical Type I magnitude frequency and impulse response is shown in Fig. 4.7(upper). Note the ripple in the passband, and smooth stopband behavior.

The Type II magnitude-squared frequency response is modeled as:

$$|F(\omega)|^2 = \frac{1}{1+\left(\varepsilon^2 V_N^2\left(\frac{\omega}{\omega_s}\right)^{-1}\right)}. \tag{4.8}$$

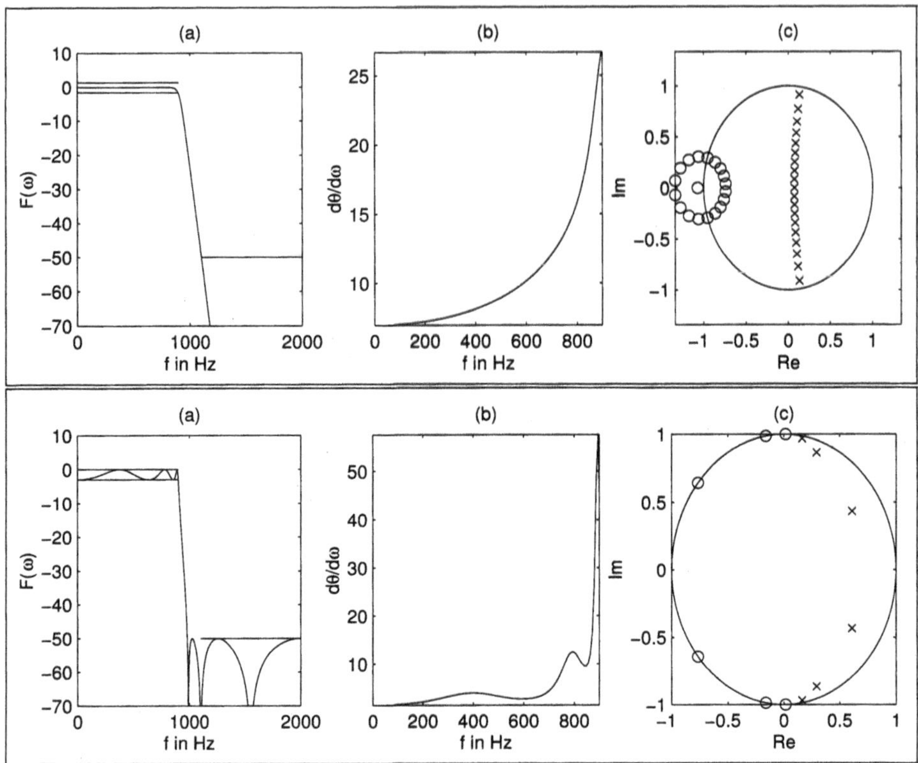

Fig. 4.6. Filter design with MATLAB toolbox. **(upper)** Butterworth filter and **(lower)** elliptic Filter.
(a) Transfer function. **(b)** Group delay of passband. **(c)** Pole/zero plot. ($\times$ = pole; $\circ$ = zero).

An example of a typical Type II magnitude frequency and impulse response is shown in Fig. 4.7(lower). Note that in this case a smooth passband results, and the stopband now exhibits ripple behavior.

An analog *elliptic* prototype filter is defined in terms of the solution to the Jacobian elliptic function, $U_N(\omega)$. The magnitude-squared frequency response is modeled as:

$$|F(\omega)|^2 = \frac{1}{1+\varepsilon^2 U_N^2\left(\frac{\omega}{\omega_s}\right)^{-1}}. \tag{4.9}$$

The magnitude-squared and impulse response of a typical elliptic filter is shown in Fig. 4.6(lower). Observe that the elliptic filter exhibits ripple in both the passband and stopband.

If we compare the four different IIR filter implementations, we find that Butterworth filter has order 19, Chebyshev has order 8, while the elliptic design has order 6, for the same tolerance scheme shown in Fig. 3.8 (p. 91).

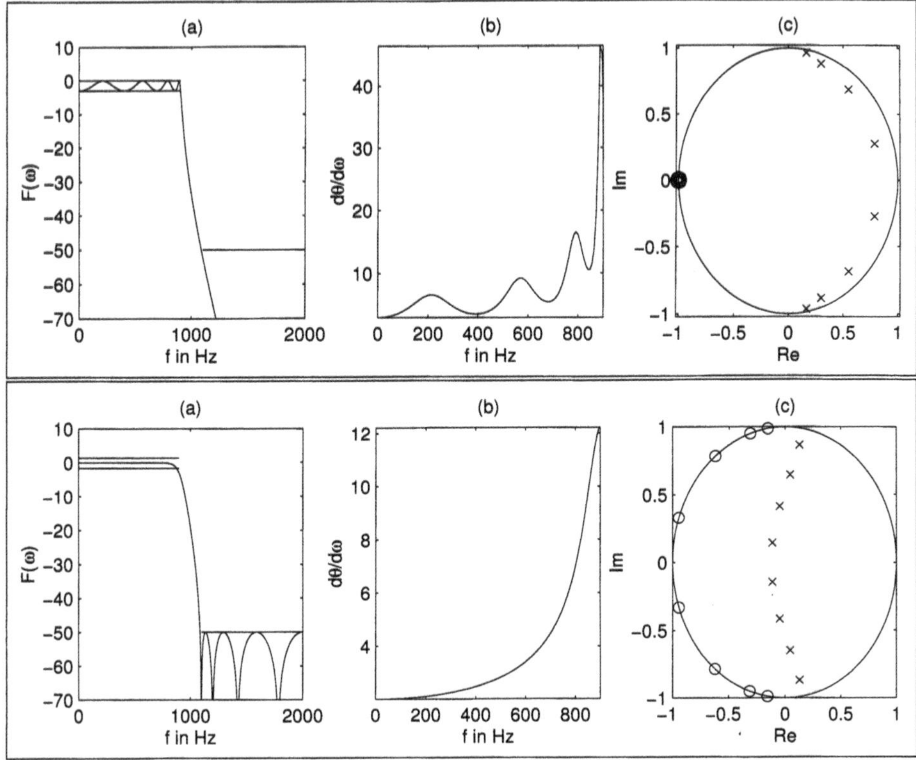

Fig. 4.7. Chebyshev filter design with MATLAB toolbox. Chebyshev I **(upper)** and Chebyshev II **(lower)**.
(a) Transfer function. **(b)** Group delay of passband. **(c)** Pole/zero plot ($\times$ = pole; $\circ$ = zero).

If we compare Fig. 4.6 and 4.7, we find that for the filter with shorter order the ripple increases, and the group delay becomes highly nonlinear. A good compromise is most often the Chebyshev Type II filter with medium order, a flat passband, and tolerable group delay.

4.2.1 Summary of Important IIR Design Attributes

In the previous section, classic IIR types were presented. Each model provides the designer with trade-off choices. The attributes of classic IIR types are summarized as follows:

- **Butterworth:** Maximally-flat passband, flat stopband, wide transition band
- **Chebyshev I:** Equiripple passband, flat stopband, moderate transition band

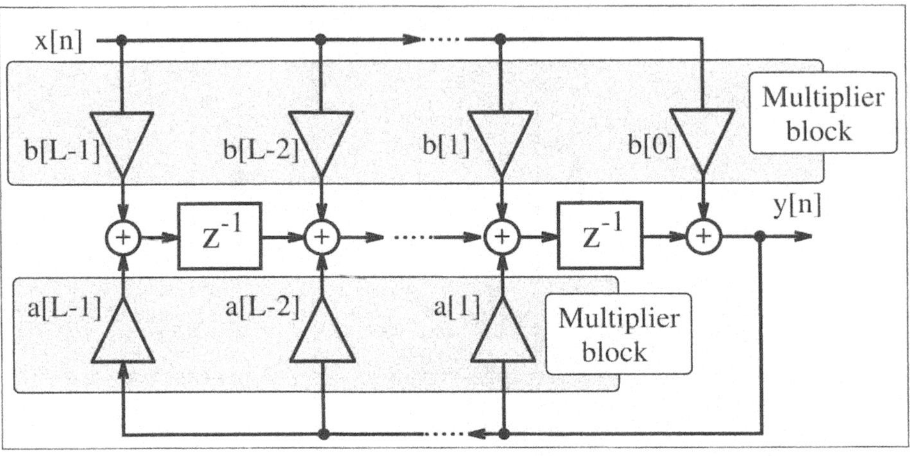

Fig. 4.8. Direct I form IIR filter using multiplier blocks.

- **Chebyshev II:** Flat passband, equiripple stopband, moderate transition band
- **Elliptic:** Equiripple passband, equiripple stopband, narrow transition band

For a given set of filter requirement, the following observations generally hold:

- Filter order
  - Lowest: Elliptic
  - Medium: Chebyshev I or II
  - Highest: Butterworth
- Passband characteristics
  - Equiripple: Elliptic, Chebyshev I
  - Flat: Butterworth, Chebyshev II
- Stopband characteristics
  - Equiripple: Elliptic, Chebyshev II
  - Flat: Butterworth, Chebyshev I
- Transition band characteristics
  - Narrowest: Elliptic
  - Medium: Chebyshev I+II
  - Widest: Butterworth

4.3 IIR Filter Implementation

Obtaining an IIR transfer function is generally considered to be a straightforward exercise, especially if design software like MATLAB is used. IIR filters

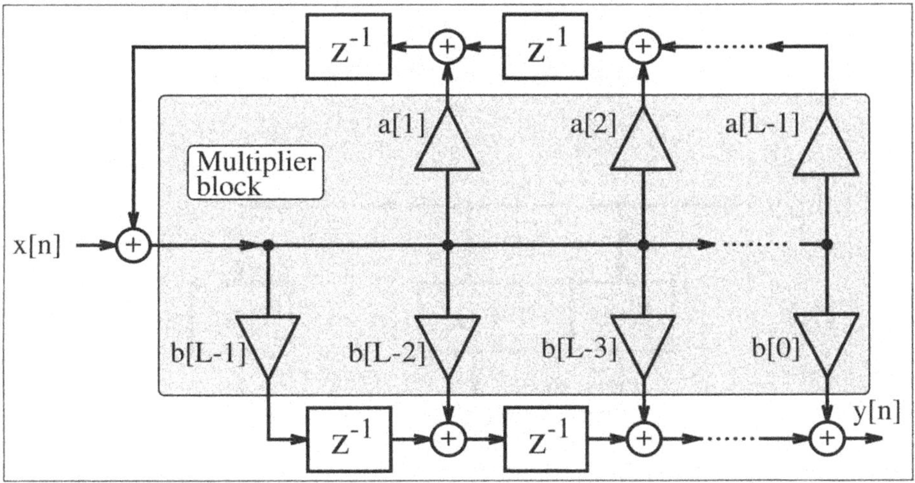

Fig. 4.9. Direct II form IIR filter using multiplier blocks.

can be developed in the context of many architectures. The most important structures are summarized as follows:

- Direct I form (see Fig. 4.8)
- Direct II form (see Fig. 4.9)
- Cascade of first- or second-order systems (see Fig. 4.10(a))
- Parallel implementation of first- or second-order systems (see Fig. 4.10(b)).
- BiQuad implementation of a typical second-order section found in basic cascade or parallel designs (see Fig. 4.11)
- Normal [69], i.e., cascade of first- or second-order state variable systems (see Fig. 4.10(a))
- Parallel normal, i.e., parallel first- or second-order state variable systems (see Fig. 4.10(b))
- Continued fraction structures
- Lattice filter (after Gray-Markel, see Fig. 4.12)
- Wave digital implementation (after Fettweis [70])
- General state space filter

Each architecture serves a unique purpose. Some of the general selection rules are summarized below:

- Speed
  - High: Direct I & II
  - Low: Wave
- Fixed-point arithmetic roundoff error sensitivity
  - High: Direct I & II
  - Low: Normal, Lattice

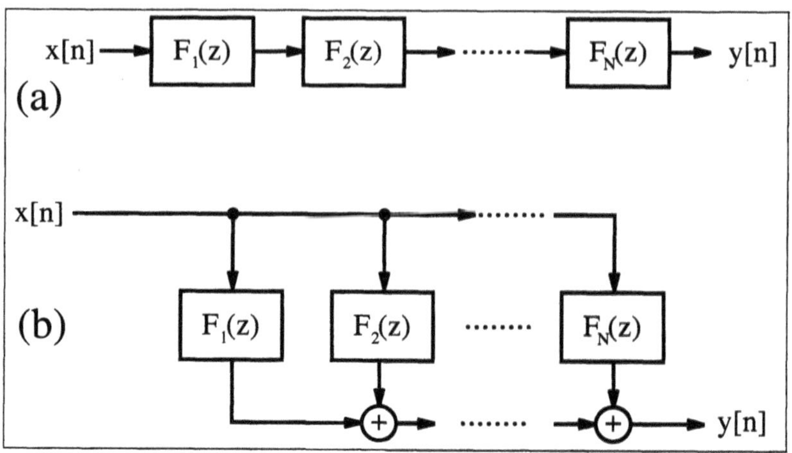

Fig. 4.10. (a) Serial implementation $F(z) = \prod_{k=1}^{N} F_k(z)$. (b) Parallel implementation $F(z) = \sum_{k=1}^{N} F_k(z)$.

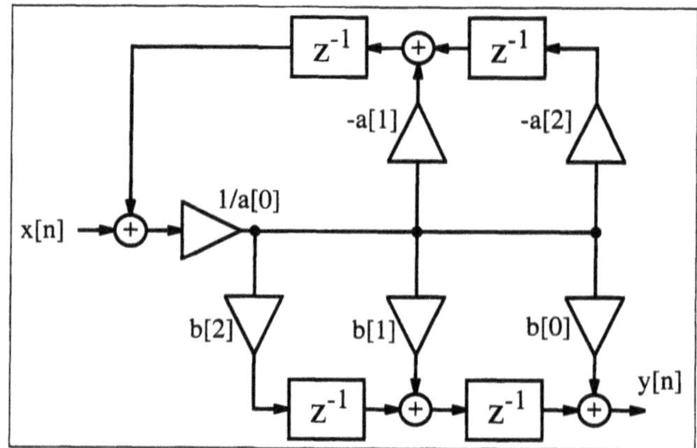

Fig. 4.11. Possible second-order section BiQuad with transfer function $F(z) = (b[0] + b[1]z^{-1} + b[2]z^{-2})/(a[0] + a[1]z^{-1} + a[2]z^{-2})$.

- Fixed-point coefficient roundoff error sensitivity
  - High: Direct I & II
  - Low: Parallel, Wave
- Special properties
  - Orthogonal weight outputs: Lattice
  - Optimized second-order sections: Normal
  - Arbitrary IIR specification: State variable

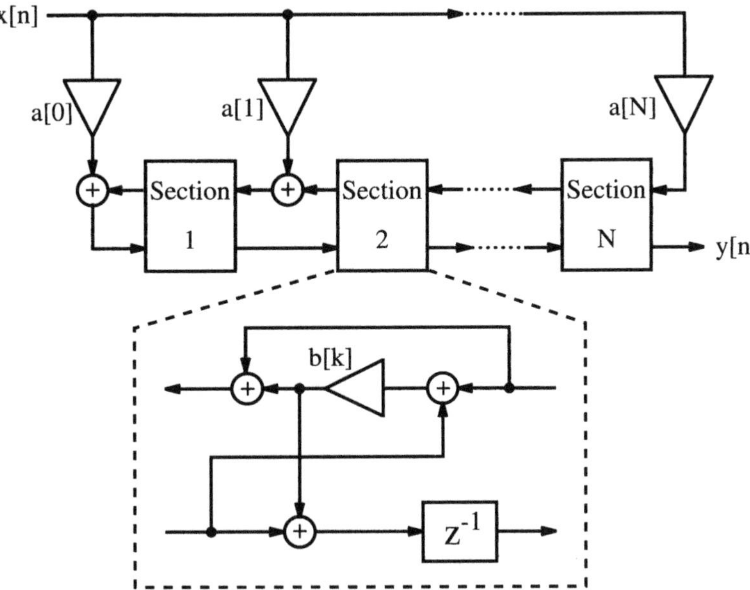

Fig. 4.12. Lattice filter.

With the help of software tools like MATLAB, the coefficients can easily be converted from one architecture to another, as demonstrated by the following example.

Example 4.2: Butterworth Second-Order System

Assume we wish to design a Butterworth filter (order = 10, passband =0.3 Fs) realized by second-order systems. We can use the following MATLAB code to generate the coefficients:

```
N=10;Fp=0.3;
[B,A]=butter(N,Fp)
[sos, gain]=tf2sos(B,A)
```

i.e., we first compute the Butterworth coefficient using the function `butter()`, and then convert this filter coefficient using the "transfer function to second-order section" function **tf2sos** to compute the BiQuad coefficients.
We will get the following results using MATLAB:

$$B = 0.0001, 0.0012, 0.0048, 0.0112, 0.0168,$$
$$0.0168, 0.0112, 0.0048, 0.0012, 0.0001$$
$$A = 1.0000, -3.5863, 6.5587, -7.5520, 5.9363,$$
$$-3.2606, 1.2421, -.03146, 0.0479, -0.0033$$

and for the second-order section we finally get:

$b[0,i]$	$b[1,i]$	$b[2,i]$	$a[0,i]$	$a[1,i]$	$a[2,i]$
1.0000	2.0019	0.9997	1.0000	−0.4709	0.0610
1.0000	2.0762	1.0784	1.0000	−0.4936	0.1121
1.0000	2.0274	1.0296	1.0000	−0.5434	0.2243
1.0000	1.9695	0.9717	1.0000	−0.6310	0.4216
1.0000	1.9250	0.9271	1.0000	−0.7786	0.7541

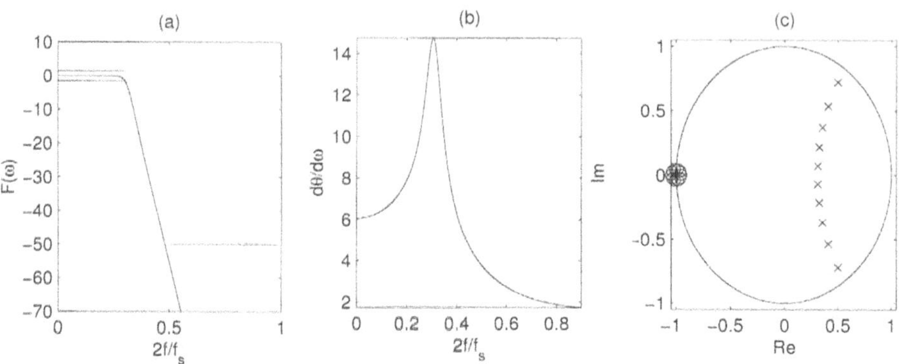

Fig. 4.13. Tenth order Butterworth filter showing **(a)** magnitude, **(b)** phase, and **(c)** group delay response.

Fig. 4.13 shows the transfer function, group delay, and the pole/zero plot of the filter. Note that all zeros are at $z_{0i} = -1$, which can also be seen from the nominator coefficients of the second-order systems. Note also the rounding error in $b[1,i] = 2$ and $b[0,i] = b[2,i] = 1$. 4.2

4.3.1 Finite Wordlength Effects

Crochiere and Oppenheim [71] have shown that the coefficient wordlength required for a digital filter is closely related to the coefficient sensitivities. Implementation of the same IIR filter can therefore lead to a wide range of required wordlengths. To illustrate some of the dynamics of this problem, consider an eighth-order order elliptic filter analyzed by Crochiere and Oppenheim [71]. The resulting eighth-order transfer function was implemented with a Wave, Cascade, Parallel, Lattice, Direct I and II, and Continuous Fraction architecture. The estimated coefficient wordlength to meet a specific maximal passband error criterion was conservatively estimated as shown in the second column of Table 4.1. As a result, it can be seen that the Direct form needs more wordlength than the Wave or Parallel structure. This has led to the conclusion that a Wave structure gives the best complexity (MW) in terms of the bit-width (W) multiplier product (M), as can be seen from column six of Table 4.1.

In the context of FIR filters (see Chapter 3), the reduced adder graph (RAG) technique was introduced in order to simplify the design of a block of several multipliers [72, 73]. Dempster and Macleod have evaluated the eighth-order elliptic filter from above, in the context of RAG multiplier implementation strategies. A comparison is presented in Table 4.2. The second column displays the multiplier block size. For a Direct II architecture, two multiplier blocks, of size 9 and 7, are required. For a Wave architecture, no two

Table 4.1. Data for eighth-order elliptic filter by Crochiere and Oppenheim [71] sorted according the costs $M \cdot W$.

Type	Word-length W	Mults M	Adds	Delays	Cost $M \cdot W$
Wave	11.35	12	31	10	136
Cascade	11.33	13	16	8	147
Parallel	10.12	18	16	8	182
Lattice	13.97	17	32	8	238
Direct I	20.86	16	16	16	334
Direct II	20.86	16	16	8	334
Cont.-frac	22.61	18	16	8	408

coefficients have the same input, and, as a result, no multiplier blocks can be developed. Instead, eleven individual multipliers must be implemented. The third column displays the number of adder/subtractors B for a canonical signed digit (CSD) design required to implement the multiplier blocks. Column four shows the same result for single-optimized multiplier adder graphs (MAG) [74]. Column five shows the result for the reduced adder graph. Column six shows the overall adder/wordwidth product for a RAG design. Table 4.2 shows that Cascade and Parallel forms give comparable or better results, compared with Wave digital filters, because the multiplier block size is an essential criterion when using the RAG algorithms. Delays have not been considered for the FPGA design, because all the logic cells have an associated flip-flop.

4.3.2 Optimization of the Filter Gain Factor

In general, we derive the IIR integer coefficient from floating point filter coefficients by first normalizing to the maximum coefficient, and then multiplying with the desire gain factor, i.e., bit-width $2^{\mathrm{round}(W)}$. However, most often it

Table 4.2. Data for eighth-order elliptic filter implemented using CSD, MAG, and RAG strategies [72].

Type	Block size	CSD B	MAG B	RAG B	RAG $W(B+A)$
Cascade	$4 \times 3, 2 \times 1$	26	26	24	453
Parallel	$11 \times 9, 4 \times 2, 1 \times 1$	31	30	29	455
Wave	11×1	58	63	22	602
Lattice	$1 \times 9, 8 \times 1$	33	31	29	852
Direct I	1×16	103	83	36	1085
Direct II	$1 \times 9, 1 \times 7$	103	83	41	1189
Cont.-frac	18×1	118	117	88	2351

Table 4.3. Variation of the gain factor to minimize filter complexity of the cascade filter.

	CSD	MAG	RAG
Optimal gain	1122	1121	1121
# adders for optimal gain	23	21	18
# adders for gain = 1024	26	26	24
Improvement	12%	19%	25%

is more efficient to select the gain factor within a range, $2^{\lfloor W \rfloor} \ldots 2^{\lceil W \rceil}$. There will be essentially no change in the transfer function, because the coefficients must be rounded anyway, after multiplying with the gain factor. If we apply, for instance, this search in the range $2^{\lfloor W \rfloor} \ldots 2^{\lceil W \rceil}$ for the cascade filter in the Crochiere and Oppenheim design example from above (gain used in Table 4.1 was $2^{\lfloor 11.33 \rfloor - 1} = 1024$), we get the data reported in Table 4.3.

We notice, from the comparison shown in Table 4.3, a substantial improvement in the number of adders required to implement the multiplier. Although the optimal gain factor for MAG and RAG in this case is the same, it can be different.

4.4 Fast IIR Filter

In Chapter 3, FIR filter `Registered Performance` was improved using pipelining (see Fig. 3.6, p. 88). In the case of FIR filters, pipelining can be achieved at essentially no cost. Pipelining IIR filters, however, is more sophisticated and is certainly not free. Strategies reported to improve IIR filter throughput are:

- Look-ahead interleaving in the time domain [75]
- Clustered look-ahead pole/zero assignment [76, 77]
- Scattered look-ahead pole/zero assignment [75, 78]
- IIR decimation filter design [79]
- Parallel processing [80]
- RNS implementation [35, Sect. 4.2][44]

The first five methods are based on filter architecture or signal flow techniques, and the last is based on computer arithmetic (see Chapter 2). These techniques will be demonstrated with examples. To simplify the VHDL representation of each case, only a first-order IIR filter will be considered, but the same ideas can be applied to higher-order IIR filters and can be found in the literature references.

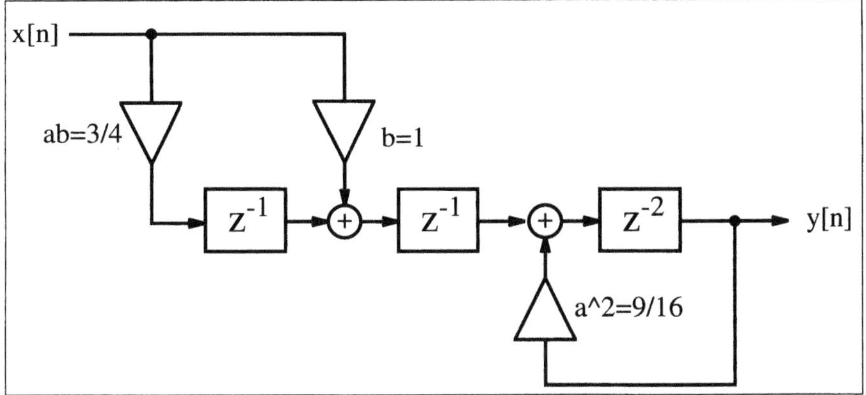

Fig. 4.14. Lossy integrator with look-ahead arithmetic.

4.4.1 Time Domain Interleaving

Consider the differential equation of a first-order IIR system, namely

$$y[n+1] = ay[n] + bx[n]. \tag{4.10}$$

The output of the first-order system, namely $y[n+1]$, can be computed using a look-ahead methodology by substituting $y[n+1]$ into the differential equation for $y[n+2]$. That is

$$y[n+2] = ay[n+1] + bx[n+1] = a^2y[n] + abx[n] + bx[n+1] \tag{4.11}$$

The equivalent system is shown in Fig. 4.14.

This concept can be generalized by applying the look-ahead transform for $(S-1)$ steps, resulting in:

$$y[n+S] = a^S y[n] + \underbrace{\sum_{k=0}^{S-1} a^k bx[n+S-1-k]}_{(\eta)}. \tag{4.12}$$

It can be seen that the term (η) defines an FIR filter having coefficients $\{b, ab, a^2b, \ldots, a^{S-1}b\}$, that can be pipelined using the pipelining techniques presented in Chapter 3 (i.e., pipelined multiplier and pipelined adder trees). The recursive part of (4.12) can now also be implemented with an S-stage pipelined multiplier for the coefficient a^S. We will demonstrate the look-ahead design with the following example.

Example 4.3: Lossy Integrator II

Consider again the lossy integrator from Example 4.1 (p. 116), but now with look-ahead. Fig. 4.14 shows the look-ahead lossy integrator, which is a combination of a nonrecursive part (i.e., FIR filter for x), and a recursive part with delay 2 and coefficient 9/16.

$$y[n+2] = \frac{3}{4}y[n+1] + x[n+1] = \frac{3}{4}(\frac{3}{4}y[n] + x[n]) + x[n+1]$$
$$= \frac{9}{16}y[n] + \frac{3}{4}x[n] + x[n+1] \qquad (4.13)$$

The VHDL code[2] shown below, implements the IIR filter in look-ahead form.

```
PACKAGE n_bit_int IS                    -- User defined type
  SUBTYPE BITS15 IS INTEGER RANGE -2**14 TO 2**14-1;
END n_bit_int;

LIBRARY work;
USE work.n_bit_int.ALL;

LIBRARY ieee;
USE ieee.std_logic_1164.ALL;
USE ieee.std_logic_arith.ALL;

ENTITY iir_pipe IS
  PORT ( x_in  : IN   BITS15;     -- Input
         y_out : OUT  BITS15;     -- Result
         clk   : IN   STD_LOGIC);
END iir_pipe;

ARCHITECTURE flex OF iir_pipe IS

  SIGNAL  x, x3, sx, y, y9 : BITS15;

BEGIN

  PROCESS  -- Use FFs for input, output and pipeline stages
  BEGIN
    WAIT UNTIL clk = '1';
    x   <= x_in;
    x3  <= x / 2 + x / 4;    -- Compute x*3/4
    sx <=  x + x3; -- Sum of x element i.e. output FIR part
    y9  <= y / 2 + y / 16;  -- Compute y*9/16
    y   <= sx + y9;          -- Compute output
  END PROCESS;

  y_out <= y ;     -- Connect register y to output pins

END flex;
```

The pipelined adder and multiplier in this example are implemented in two steps. In the first stage, $\frac{9}{16}y[n]$ is computed. In the second stage, $\frac{3}{4}x[n] + x[n+1]$ and $\frac{9}{16}y[n]$ are added. The design consumes 64 logic cells and runs with a **Registered Performance** of 73.52 MHz. The response of the filter to an impulse with amplitude 1000 is shown in Fig. 4.15.

4.3

[2] The equivalent Verilog code **iir_pipe.v** for this example can be found in Appendix A on page 368.

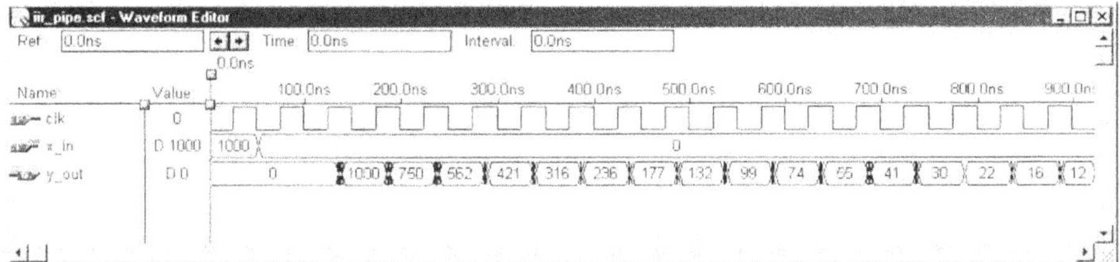

Fig. 4.15. VHDL simulation of impulse response of the look-ahead lossy integrator.

Comparing the look-ahead scheme with the 31 LCs and 55.86 MHz solution reported in Example 4.1 (p. 116), we find that look-ahead pipelining has doubled the complexity of the design, with an attendant speed-up of 31%. The comparison of the two filter's response to the impulse with amplitude 1000 shown in Fig. 4.2 (p. 117) and Fig. 4.15 reveals that the look-ahead scheme has an additional overall delay, and that the quantization effect differs by a ± 2 amount between the two methodologies.

An alternative design approach, using a standard logic vector data type and `LPM_ADD_SUB` megafunctions, is discussed in Exercise 4.5 (p. 140). The second approach will produce longer VHDL code, but will have the benefit of direct control at the bit level of the sign extension and multiplier.

4.4.2 Clustered and Scattered Look-Ahead Pipelining

Clustered and scattered look-ahead pipelining schemes add self-canceling poles and zeros to the design to facilitate pipelining of the recursive portion of the filter. In the *clustered* method, additional pole/zeros are introduced in such a way that in the denominator of the transfer function the coefficients for $z^{-1}, z^{-2}, \ldots, z^{-(S-1)}$ become zero. The following example shows clustering for a second-order filter.

Example 4.4: Clustering Method

A second-order transfer function is assumed to have a pole at 1/2 and 3/4 and a transfer function given by:

$$F(z) = \frac{1}{1 - 1.25z^{-1} + 0.375z^{-2}} = \frac{1}{(1 - 0.5z^{-1})(1 - 0.75z^{-1})} \tag{4.14}$$

Adding a canceling pole/zero at $z = -1.25$ results in a new transfer function

$$F(z) = \frac{1 + 1.25z^{-1}}{1 - 1.1875z^{-2} + 0.4688z^{-3}}. \tag{4.15}$$

The recursive part of the filter can now be implemented with an additional pipeline stage. 4.4

The problem with clustering is that the cancelled pole/zero pair may lie outside the unit circle, as is the case in the previous example (i.e., $z_\infty = -1.25$). This introduces instability into the design if the pole/zero annihilating is not perfect. In general, a second-order system with poles at r_1, r_2

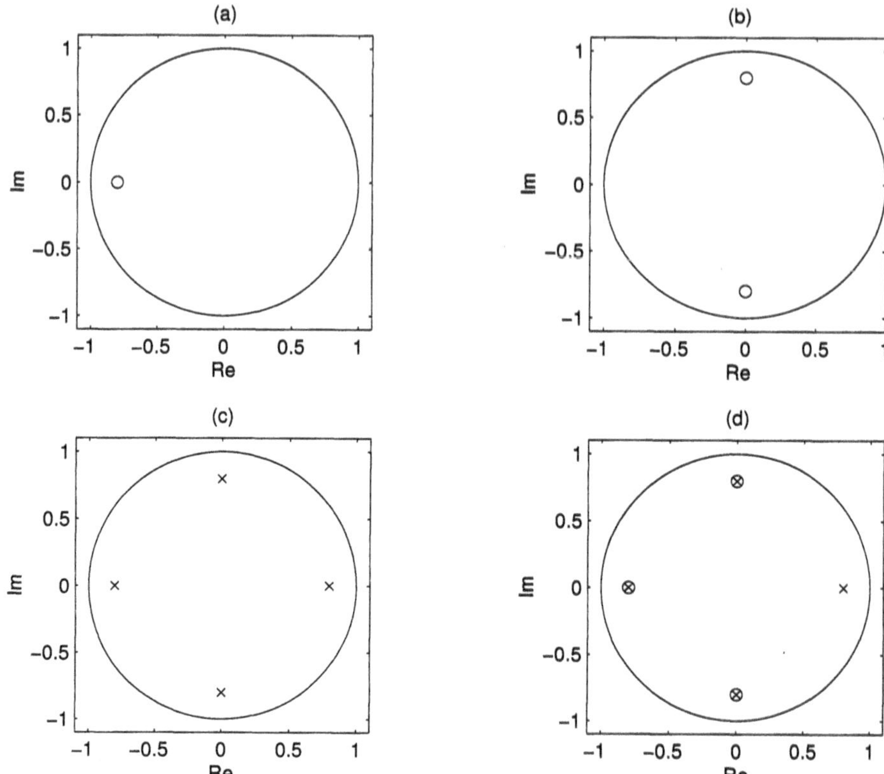

Fig. 4.16. Pole/zero plot for scattered look-ahead first-order IIR filter. **(a)** $F_1(z) = (1 + az^{-1})$. **(b)** $F_2(z) = 1 + a^2 z^{-2}$. **(c)** $F_3(z) = 1/(1 - a^4 z^{-4})$. **(d)** $F(z) = \prod_k F_k(z) = (1 + az^{-1})(1 + a^2 z^{-2})/(1 - a^4 z^{-4}) = 1/(1 - az^{-1})$.

and with one extra canceling pair, has a pole location at $1/(r_1 + r_2)$, which lies outside the unit circle for $(r_1 + r_2) > 1$. Soderstrand et al. [77], have described a stable clustering method, which in general introduces more than one canceling pole/zero pair.

The *scattered look-ahead* approach does not introduce stability problems. It introduces $(S - 1)$ canceling pole/zero pairs located at $z_k = pe^{j\pi k/S}$, for an original filter with a pole located at p. The denominator of the transfer function has, as a result, only zero coefficients associated with the terms z^0, z^S, z^{-2S} etc.

Example 4.5: Scattered Look-Ahead Method

Consider implementing a second-order system having poles located at $z_{\infty 1} = 0.5$ and $z_{\infty 2} = 0.25$ with two additional pipeline stages. A second-order transfer function of a filter with poles at $1/2$ and $1/4$ has the transfer function

$$F(z) = \frac{1}{1 - 3/4z^{-1} + 1/8z^{-2}} \tag{4.16}$$

The scattered look-ahead introduces two additional pipeline stages by adding pole/zero pairs at $0.25e^{\pm j2\pi/3}$ and $0.5e^{\pm j2\pi/3}$. Adding a canceling pole/zero at this location results in

$$
\begin{aligned}
F(z) &= \frac{1}{1+0.5z^{-1}+0.25z^{-2}} \\
&\quad \cdot \frac{(1+0.5z^{-1}+0.25z^{-2})(1+.75z^{-1}+0.5625z^{-2})}{(1+.75z^{-1}+0.5625z^{-2})(1-0.75z^{-1}+0.125z^{-2})} \\
&= \frac{1+1.25z^{-1}+1.1875z^{-2}+0.4687z^{-3}+.1406z^{-4}}{1-0.5469z^{-3}+0.0527z^{-6}} \\
&= \frac{512+640z^{-1}+608z^{-2}+240z^{-3}}{512-280z^{-3}+27z^{-6}}
\end{aligned}
$$

and the recursive part can be implemented with two additional pipeline stages. 4.5

It is interesting to note that for a first-order IIR system, clustered and scattered look-ahead methods result in the same pole/zero canceling pair lying on a circle around the origin with angle differences $2\pi/S$. The nonrecursive part can be realized with a "power-of-two decomposition" according to

$$(1+az^{-1})(1+a^2z^{-2})(1+a^4z^{-4})\cdots \tag{4.17}$$

Fig. 4.16 shows such a pole/zero representation for a first-order section, which enables an implementation with four pipeline stages in the recursive part.

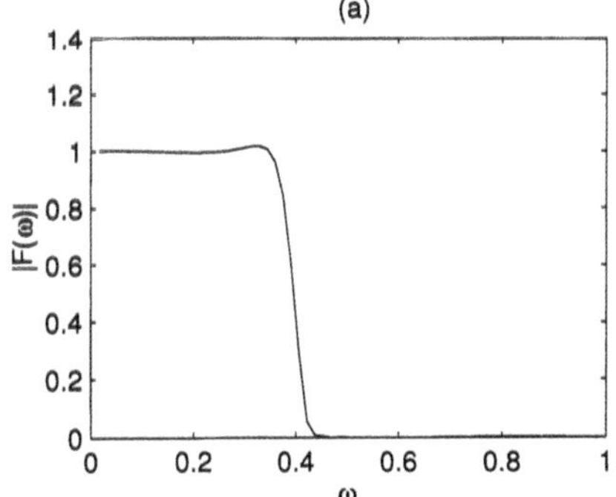

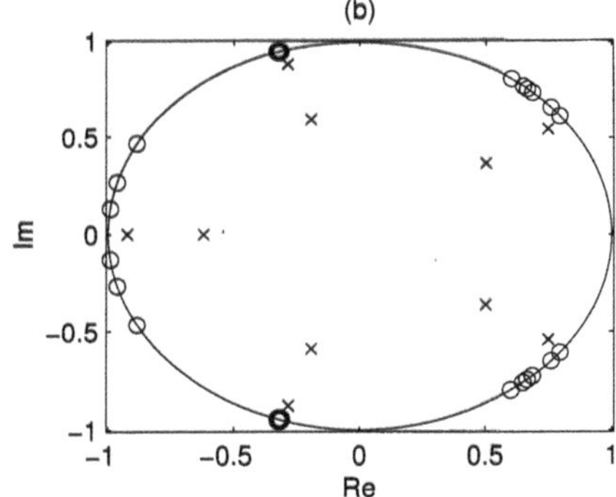

Fig. 4.17. **(a)** Transfer function, and **(b)** pole/zero distribution of a 37 order Martinez-Parks IIR filter with $S = 5$.

4.4.3 IIR Decimator Design

Martinez and Parks [79] have introduced, in the context of decimation filters (see Chapter 5, p. 152), a filter design algorithm based on the minimax method. The resulting transfer function satisfies

$$F(z) = \frac{\sum\limits_{l=0}^{L} b[l]z^{-l}}{1 - \sum\limits_{n=0}^{N/S} a[n]z^{-nS}} \tag{4.18}$$

That is, only every other S coefficient in the denominator is nonzero. In this case, the recursive part (i.e., the denominator) can be pipelined with S stages. It has been found that in the resulting pole/zero distribution, all zeros are on the unit circle, as is usual for an elliptic filter, while the poles lie on circles, whose main axes have a difference in angle of $2\pi/S$, as shown in Fig. 4.17(b).

4.4.4 Parallel Processing

In a parallel-processing filter implementation [80], P parallel IIR paths are formed, each running at a $1/P$ input sampling rate. They are combined at the output using a multiplexer, as shown in Fig. 4.18. Because a multiplexer, in general, will be faster than a multiplier and/or adder, the parallel approach will be faster. Furthermore, each path P has a factor of P more time to compute its assigned output.

To illustrate, consider again a first-order system and $P = 2$. The look-ahead scheme, as in (4.11)

$$y[n+2] = ay[n+1] + x[n+1] = a^2y[n] + ax[n] + x[n+1] \tag{4.19}$$

is now split into even $n = 2k$ and odd $n = 2k-1$ output sequences, obtaining

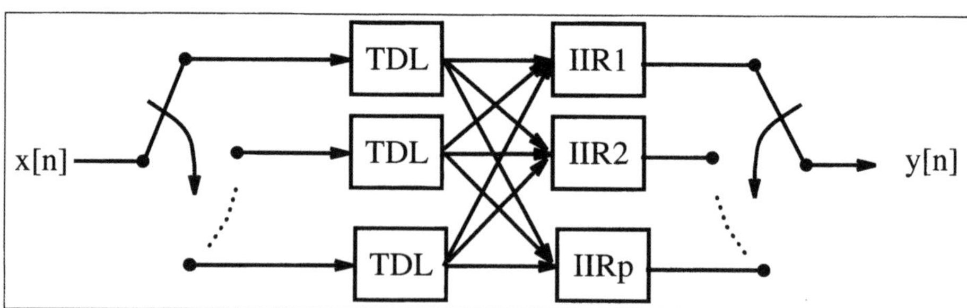

Fig. 4.18. Parallel IIR implementation. The tapped delay lines (TDL) run with a $1/p$ input sampling rate.

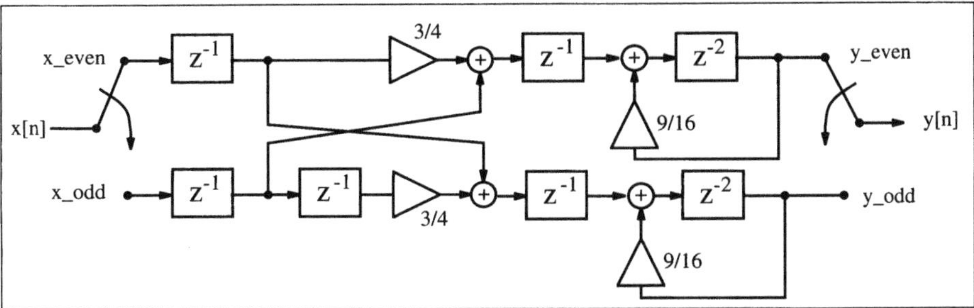

Fig. 4.19. Two-path parallel IIR filter implementation.

$$y[n+2] = \begin{cases} y[2k+2]=a^2y[2k]+ax[2k]+x[2k+1] \\ y[2k+1]=a^2y[2k-1]+ax[2k-1]+x[2k] \end{cases} \quad (4.20)$$

where $n, k \in \mathbb{Z}$. The two equations are the basis for the following parallel IIR filter FPGA implementation.

Example 4.6: Lossy Integrator III

Consider implementing a parallel lossy integrator, with $a = 3/4$, as an extension to the methods presented in Examples 4.1 (p. 116) and 4.3 (p. 131). A two-channel parallel lossy integrator, which is a combination of two non-recursive parts (i.e., an FIR filter for **x**), and two recursive parts with delay 2 and coefficient 9/16, is shown in Fig. 4.19. The VHDL code[3] shown below implements the design.

```
PACKAGE eight_bit_int IS                  -- User defined type
  SUBTYPE BITS15 IS INTEGER RANGE -2**14 TO 2**14-1;
END eight_bit_int;

LIBRARY work;
USE work.eight_bit_int.ALL;

LIBRARY ieee;
USE ieee.std_logic_1164.ALL;
USE ieee.std_logic_arith.ALL;

ENTITY iir_par IS                              ------> Interface
  PORT ( clk       : IN  STD_LOGIC;
         x_in      : IN  BITS15;
         clk2      : OUT STD_LOGIC;
         y_out     : OUT BITS15);
END iir_par;

ARCHITECTURE flex OF iir_par IS

  TYPE STATE_TYPE IS (even, odd);
  SIGNAL  state                            : STATE_TYPE;
  SIGNAL  x_even, x_odd, xd_odd, x_wait : BITS15;
```

[3] The equivalent Verilog code **iir_par.v** for this example can be found in Appendix A on page 369.

```
    SIGNAL  y_even, y_odd, y_wait, y      : BITS15;
    SIGNAL  x_e, x_o, y_e, y_o            : BITS15;
    SIGNAL  sum_x_even, sum_x_odd         : BITS15;
    SIGNAL  clk_div2                      : STD_LOGIC;

  BEGIN

    Multiplex: PROCESS --> Split x into even and odd samples
    BEGIN              --> recombine y at clk rate
      WAIT UNTIL clk = '1';
      CASE state IS
        WHEN even =>
           x_even <= x_in;
           x_odd <= x_wait;
           clk_div2 <= '1';
           y <= y_wait;
           state <= odd;
        WHEN odd =>
           x_wait <= x_in;
           y <= y_odd;
           y_wait <= y_even;
           clk_div2 <= '0';
           state <= even;
        END CASE;
    END PROCESS Multiplex;

    y_out <= y;
    clk2  <= clk_div2;

    Arithmetic: PROCESS
    BEGIN
      WAIT UNTIL clk_div2 = '0';
      sum_x_even <= (x_even * 2 + x_even) /4 + x_odd;
      y_even <= (y_even * 8 + y_even ) /16 + sum_x_even;
      xd_odd <= x_odd;
      sum_x_odd <= (xd_odd * 2 + xd_odd) /4 + x_even;
      y_odd  <= (y_odd * 8 + y_odd) / 16 + sum_x_odd;
    END PROCESS Arithmetic;

  END flex;
```

The design is realized with two **PROCESS** statements. In the first, **PROCESS Multiplex**, **x** is split into even and odd indexed parts, and the output **y** is recombined at the **clk** rate. In addition, the first **PROCESS** statement generates the second clock, running at **clk/2**. The second block implements the filter's arithmetic according to (4.20). Measuring the **Registered Performance** of the design with MaxPlusII software, a problem arises MaxPlusII cannot compute the **Registered Performance** in multiple clock domains. Assuming that the input multiplexer runs at twice the arithmetic speed, an estimated 58.13 MHz yields an input rate of more than 100 MHz. This is the result of the **y_odd** and **y_even** filters running off the **clk_div2** clock. This can be validated by the simulation shown in Fig. 4.20. The **clk** frequency was chosen to be higher than the computed **Registered Performance**. 4.6

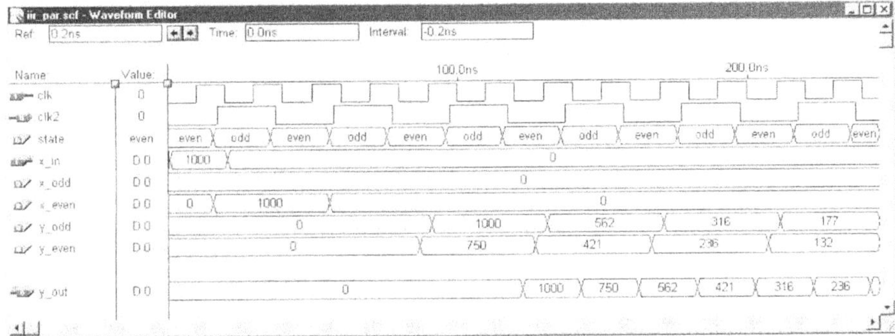

Fig. 4.20. VHDL simulation of the response of the parallel IIR filter to an impulse 1000.

The disadvantage of the parallel implementation, compared with the other methods presented, is the relatively high implementation cost of 213 LCs.

4.4.5 IIR Design Using RNS

Because the Residue Number System (RNS) uses an intrinsically short wordlength, it is an excellent candidate to implement fast (recursive) IIR filters. In a typical IIR-RNS design, a system is implemented as a collection of recursive and nonrecursive systems, each defined in terms of an FIR structure (see Fig. 4.21). Each FIR may be implemented in RNS-DA, using a quarter-square multiplier, or in the index domain, as developed in Chapter 2 (p. 41).

For a stable filter, the recursive part should be scaled to control dynamic range growth. The scaling operation may be implemented with mixed radix conversion, Chinese Remainder Theorem (CRT), or the $\epsilon-$CRT method. For high-speed designs, it is preferable to add an additional pipeline delay based on the clustered or scattered look-ahead pipelining technique [35, Sect. 4-2]. An RNS recursive filter design will be developed in detail in Sect. 5.3. It will be seen that RNS design will improve speed from 50 MHz to more than 70 MHz.

Exercises

Exercises Using MaxPlusII

4.1: **(a)** Implement a first-order IIR filter with a pole at $z_{\infty 0} = 3/8$ and 12-bit input width, using MaxPlusII.
(b) Determine the number of LCs and the `Registered Performance`.
(c) Simulate the design with an input impulse of 100.

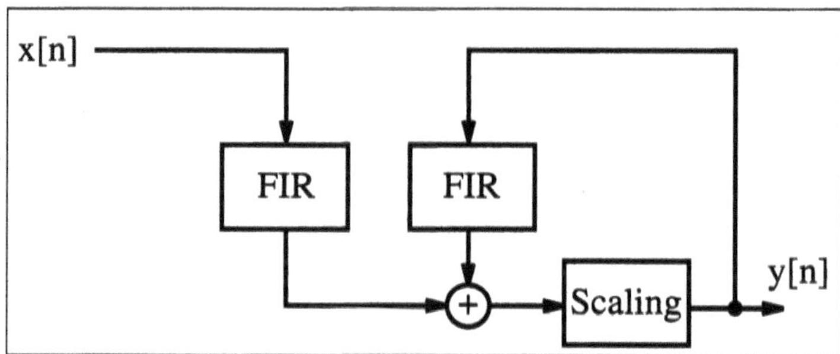

Fig. 4.21. RNS implementation of IIR filters using two FIR sections and scaling.

4.2: (a) Implement a first-order IIR filter with a pole at $z_{\infty 0} = 3/8$, 12-bit input width, and a look-ahead of one step, using MaxPlusII.
(b) Determine the number of LCs and the **Registered Performance**.
(c) Simulate the design with an input impulse of 100.

4.3: (a) Implement a first-order IIR filter with a pole at $z_{\infty 0} = 3/8$, 12-bit input width, and a parallel design with two paths, using MaxPlusII.
(b) Determine the number of LCs and the **Registered Performance**.
(c) Simulate the design with an input impulse of 100.

4.4: (a) Implement a first-order IIR filter as in Example 4.1 (p. 116), using a 15-bit `std_logic_vector`, and implement the adder with two `lpm_add_sub` megafunctions, using MaxPlusII.
(b) Determine the number of LCs and the **Registered Performance**.
(c) Simulate the design with an input impulse of 1000, and compare the results to Fig. 4.3 (p. 118).

4.5: (a) Implement a first-order pipelined IIR filter from Example 4.3 (p. 131) using a 15-bit `std_logic_vector`, and implement the adder with four `lpm_add_sub` megafunctions, using MaxPlusII.
(b) Determine the number of LCs and the **Registered Performance**.
(c) Simulate the design with an input impulse of 1000, and compare the results to Fig. 4.15 (p. 133).

4.6: Shajaan and Sorensen have shown that an IIR Butterworth filter can be efficiently designed by implementing the coefficients as "signed-power-of-two" (SPT) values [81]. The transfer function of a cascade filter with N sections

$$F(z) = \prod_{l=1}^{N} S[l] \frac{b[l,0] + b[l,1]z^{-1} + b[l,2]z^{-2}}{a[l,0] + a[l,1]z^{-1} + a[l,2]z^{-2}} \tag{4.21}$$

should be implemented using the second-order sections shown in Fig. 4.11, p. 126. A tenth-order filter, as discussed in Example 4.2 (p. 127), can be realized with the following SPT filter coefficients [81]:

l	$S[l]$	$1/a[l,0]$	$a[l,1]$	$a[l,2]$
1	2^{-1}	2^{-1}	$-1-2^{-4}$	$1-2^{-2}$
2	2^{-1}	2^{-1}	$-1-2^{-1}$	$1-2^{-5}$
3	2^{-1}	2^{-1}	$-1-2^{-1}$	$2^{-1}+2^{-6}$
4	1	2^{-1}	$-1-2^{-2}$	$2^{-2}+2^{-5}$
5	2^{-1}	2^{-1}	$-1-2^{-1}$	$2^{-2}+2^{-4}$

Because the zeros of the Butterworth are all at $z = -1$, we choose $b[0] = b[2] = 0.5$ and $b[1] = 1$.

(a) Compute and plot the transfer function of the first BiQuad and the complete filter.

(b) Implement and simulate the first BiQuad for 8-bit inputs.

(c) Build and simulate the 5-stage filter with MaxPlusII.

(d) Determine LC usage and `Registered Performance` of the filter.

5. Multirate Signal Processing

Introduction

A frequent task in digital signal processing is to adjust the sampling rate according to the signal of interest. Systems with different sampling rates are referred to as *multirate* systems. In this chapter, two typical examples will illustrate decimation and interpolation in multirate DSP systems. We will then introduce polyphase notation, and will discuss some efficient decimator designs. At the end of the chapter we will discuss filter banks and a quite new, highly celebrated addition to the DSP toolbox: wavelet analysis.

5.1 Decimation and Interpolation

If, after A/D conversion, the signal of interest can be found in a small frequency band (typically, lowpass or bandpass), then it is reasonable to filter with a lowpass or bandpass filter and to reduce the sampling rate. A narrow filter followed by a downsampler is usually referred to as a *decimator* [65]. [1] The filtering, downsampling, and the effect on the spectrum is illustrated in Fig. 5.1.

We can reduce the sampling rate up to the limit called the "Nyquist rate," which says that the sampling rate must be higher than the bandwidth of the signal, in order to avoid aliasing. Aliasing is demonstrated in Fig. 5.2 for a lowpass signal. Aliasing is irreparable, and should avoided at all cost.

For a bandpass signal, the frequency band of interest must fall within an *integer band.* If f_s is the sampling rate, and R is the desired downsampling factor, then the band of interest must fall between

$$k\frac{f_s}{2R} < f < (k+1)\frac{f_s}{2R} \qquad k \in \mathbb{N}. \tag{5.1}$$

If it does not there may be aliasing due to "copies" from the negative frequency bands, although the sampling rate may still be higher than the Nyquist rate, as shown in Fig. 5.3.

Increasing the sampling rate can be useful, in the D/A conversion process, for example. Typically, D/A converters use a sample-and-hold of first-order

[1] Some authors refer to a downsampler as a decimator.

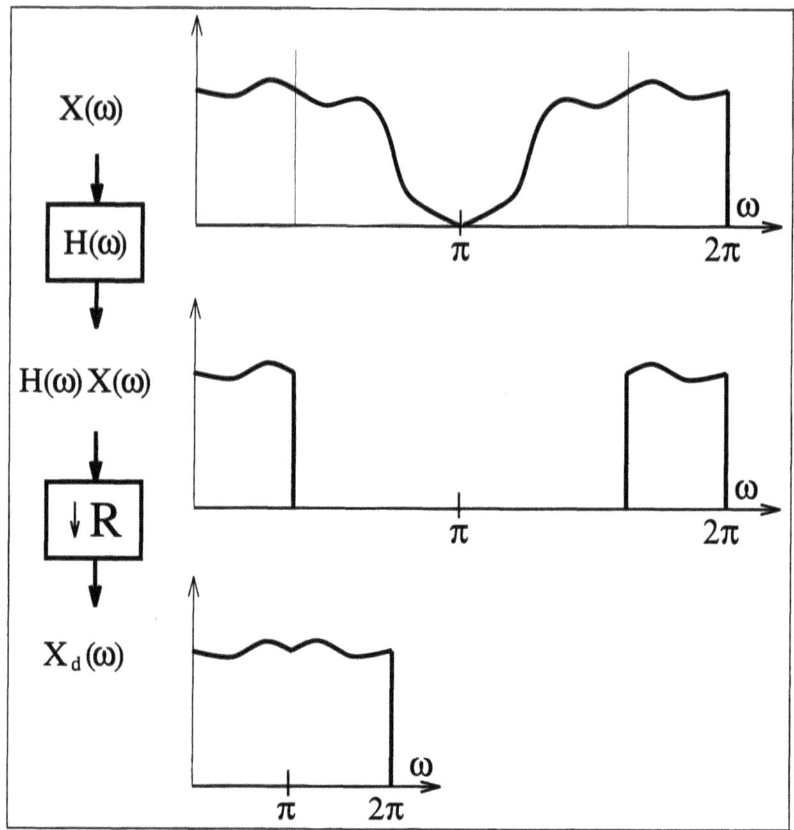

Fig. 5.1. Decimation of signal $x[n]$ ○—● $X(\omega)$.

at the output, which produces a step-like output function. This can be compensated for with an analog $1/\text{sinc}(x)$ compensation filter, but most often a digital solution is more efficient. We can use, in the digital domain, an *expander* and an additional filter to get the desired frequency band. We notice, from Fig. 5.4, that the introduced zeros produce an extra copy of the baseband spectrum which must first be removed before the signal can be processed with the D/A converter. The much smoother output signal of such an *interpolation*[2] can be seen in Fig. 5.5.

5.1.1 Noble Identities

When manipulating signal flow graphs of multirate systems it is sometimes useful to rearrange the filter and downsampler/expander, as shown in Fig. 5.6. These are the so-called "Noble" relations [82]. For the decimator, it follows

[2] Some authors refer to the expander as interpolator.

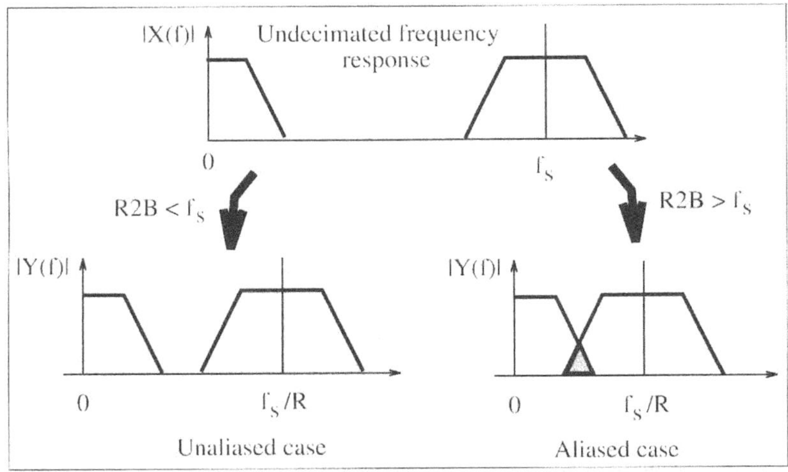

Fig. 5.2. Unaliased and aliased decimation cases..

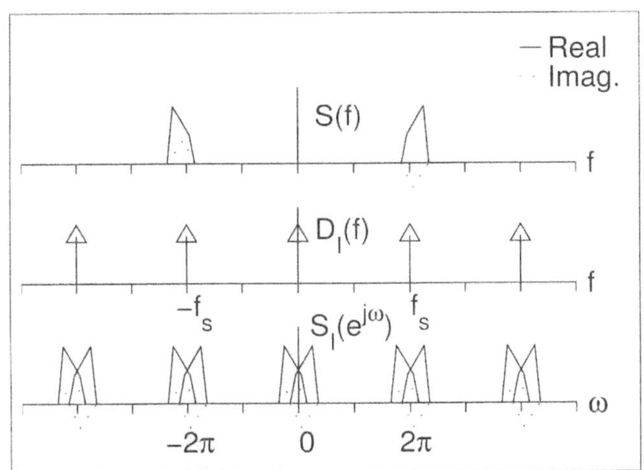

Fig. 5.3. Integer band violation (©1995 VDI Press [4]).

$$(\downarrow R)\ F(z) = F(z^R)\ (\downarrow R). \tag{5.2}$$

i.e., if the downsampling is done first, we can reduce the filter length $F(z^R)$ by a factor of R.

For the interpolator, the Noble relation is defined as

$$F(z)\ (\uparrow R) = (\uparrow R)\ F(z^R) \tag{5.3}$$

i.e., in a interpolation putting the filter before the expander results in an R times shorter filter.

These two identities will become very useful when we discuss polyphase implementation in Sect. 5.2 (p. 147).

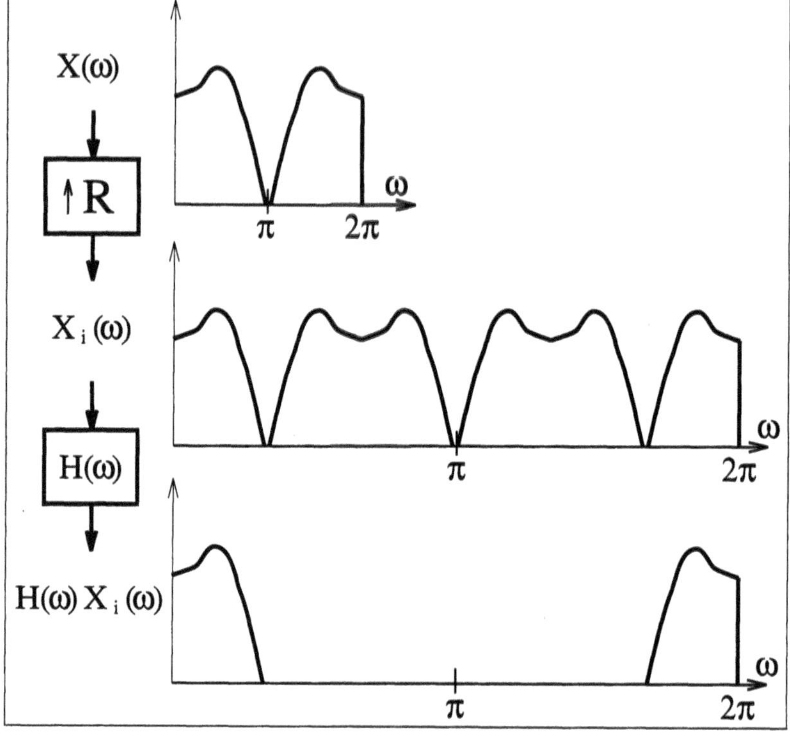

Fig. 5.4. Interpolation example. $R = 3$ for $x[n]$ o—• $X(\omega)$.

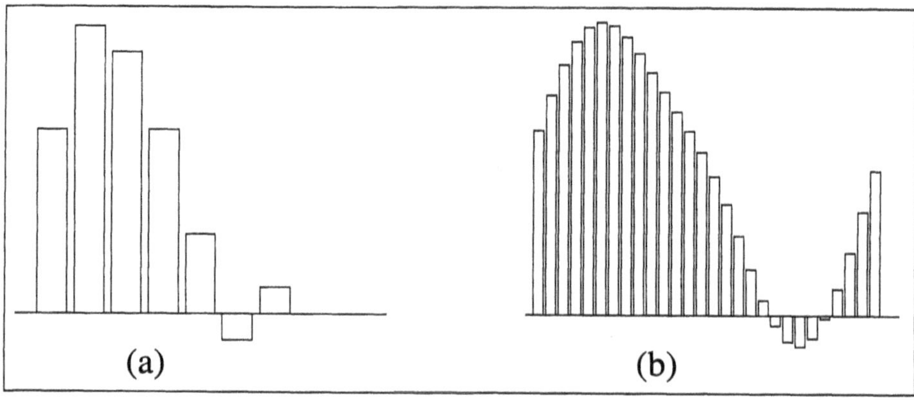

Fig. 5.5. D/A conversion. **(a)** Low oversampling, high degradation. **(b)** High oversampling, low degradation.

5.1.2 Sampling Rate Conversion by Rational Factor

If the input and output rate of a multirate system is *not* an integer factor, then a rational change factor R_1/R_2 in the sampling rate can be used. More

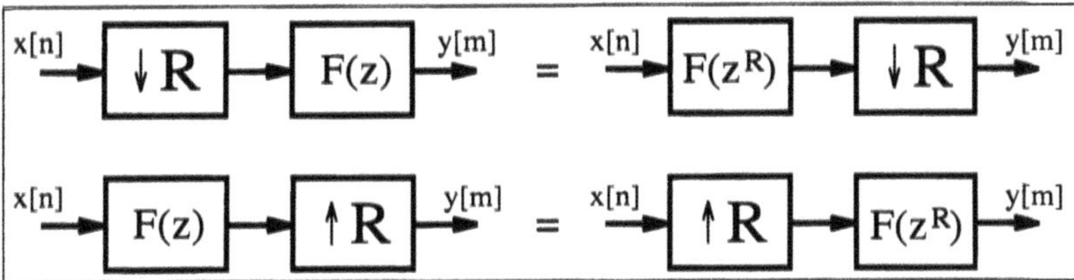

Fig. 5.6. Equivalent multirate systems (Noble relation).

precisely, we first use an interpolator to increase the sampling rate by R_1, and then use a decimator to downsample by R_2. Since the filters used for interpolation and decimation are both lowpass filters, it follows, from the upper configuration in Fig. 5.7, that we only need to implement the lowpass filter with the smaller passband frequency, i.e.,

$$f_p = \min\left(\frac{\pi}{R_1}, \frac{\pi}{R_2}\right). \tag{5.4}$$

This is graphically interpreted in the lower configuration of Fig. 5.7.

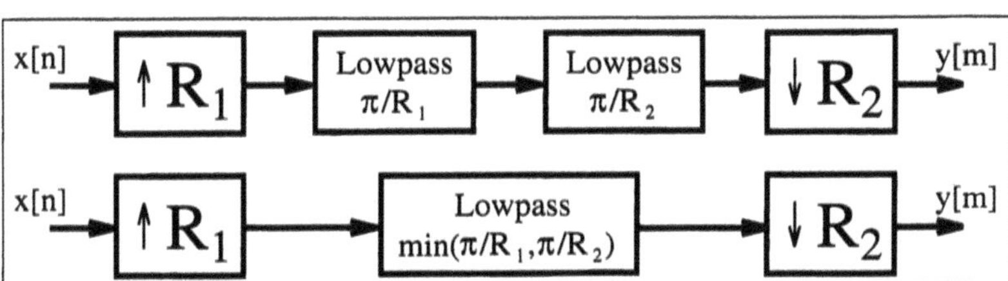

Fig. 5.7. Noninteger decimation system. **(upper)** Cascade of an interpolator and a decimator. **(lower)** Result combining the lowpass filters.

5.2 Polyphase Decomposition

Polyphase decomposition is very useful when implementing decimation or interpolation in IIR or FIR filter and filter banks. To illustrate this, consider the polyphase decomposition of an FIR decimation filter. If we add downsampling by a factor of R to the FIR filter structure shown in Fig. 3.1 (p. 80), we find that we only need to compute the outputs $y[n]$ at time instances

$$y[0], y[R], y[2R], \ldots \tag{5.5}$$

It follows that we do not need to compute not all sums-of-product $x[n]f[n-k]$ of the convolution. For instance, $x[0]$ only needs to be multiplied with

$$f[0], f[R], f[2R], \ldots \tag{5.6}$$

Besides $x[0]$, these coefficients only need to be multiplied with

$$x[R], x[2R], \ldots \tag{5.7}$$

It is therefore reasonable to split the input signal first into R separate sequences according to

$$\begin{aligned}
x[n] &= \sum_{r=0}^{R-1} x_r[n] \\
x_0[n] &= \{x[0], x[R], \ldots\} \\
x_1[n] &= \{x[1], x[R+1], \ldots\} \\
&\vdots \\
x_{R-1}[n] &= \{x[R-1], x[2R-1], \ldots\}
\end{aligned}$$

and also to split the filter $f[n]$ in R sequences

$$\begin{aligned}
f[n] &= \sum_{r=0}^{R-1} f_r[n] \\
f_0[n] &= \{f[0], f[R], \ldots\} \\
f_1[n] &= \{f[1], f[R+1], \ldots\} \\
&\vdots \\
f_{R-1}[n] &= \{f[R-1], f[2R-1], \ldots\}.
\end{aligned}$$

Fig. 5.8 shows a decimator filter implemented using polyphase decomposition. Such a decimator can run R times faster than the usual FIR filter followed by a downsampler. The filters $f_r[n]$ are called polyphase filters, because they all have the same magnitude transfer function, but they are separated by a sample delay, which introduces a phase offset.

A final example illustrates the polyphase decomposition.

Example 5.1: Polyphase Decimator Filter

Consider a Daubechies length-4 filter with $G(z)$ and $R = 2$.

$$G(z) = \left((1+\sqrt{3}) + (3+\sqrt{3})z^{-1} + (3-\sqrt{3})z^{-2} + (1-\sqrt{3})z^{-3}\right)\frac{1}{4\sqrt{2}}$$

$$G(z) = 0.48301 + 0.8365z^{-1} + 0.2241z^{-2} - 0.1294z^{-3}$$

Quantizing the filter to 8 bits of precision results in the following model:

$$G(z) = \left(124 + 214z^{-1} + 57z^{-2} - 33z^{-3}\right)/256$$

$$\begin{aligned}
G(z) &= G_0(z^2) + z^{-1}G_1(z^2) \\
&= \underbrace{\left(\frac{124}{256} + \frac{57}{256}z^{-2}\right)}_{G_0(z^2)} + z^{-1}\underbrace{\left(\frac{214}{256} - \frac{33}{256}z^{-2}\right)}_{G_1(z^2)}
\end{aligned}$$

and it follows

$$G_0(z) = \frac{124}{256} + \frac{57}{256}z^{-1} \qquad G_1(z) = \frac{214}{256} - \frac{33}{256}z^{-1} \tag{5.8}$$

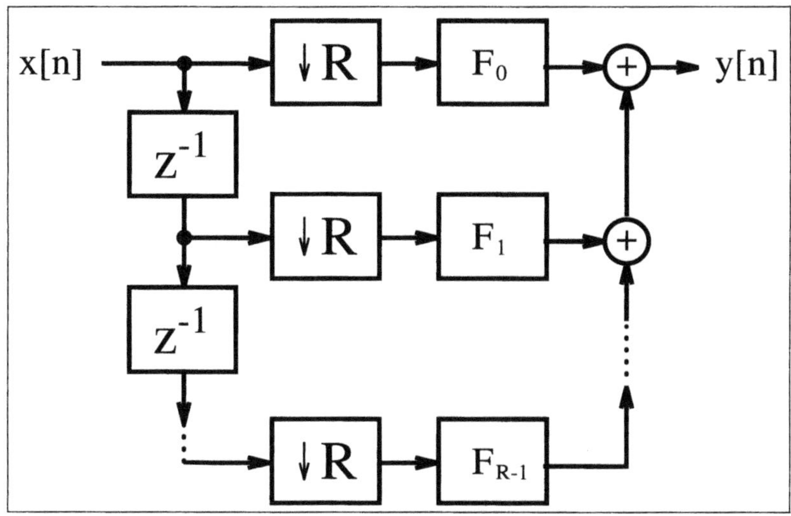

Fig. 5.8. Polyphase realization of a decimation filter.

The following VHDL code[3] shows the polyphase implementation for DB4.

```
PACKAGE n_bits_int IS            -- User defined types
  SUBTYPE BITS8 IS INTEGER RANGE -128 TO 127;
  SUBTYPE BITS9 IS INTEGER RANGE -2**8 TO 2**8-1;
  SUBTYPE BITS17 IS INTEGER RANGE -2**16 TO 2**16-1;
  TYPE ARRAY_BITS17_4 IS ARRAY (0 TO 3) of BITS17;
END n_bits_int;

LIBRARY work;
USE work.n_bits_int.ALL;

LIBRARY ieee;
USE ieee.std_logic_1164.ALL;
USE ieee.std_logic_arith.ALL;
USE ieee.std_logic_signed.ALL;

ENTITY db4poly IS                          ------> Interface
  PORT (clk              : IN  STD_LOGIC;
        x_in             : IN  BITS8;
        clk2             : OUT STD_LOGIC;
        x_e, x_o, g0, g1 : OUT BITS17;
        y_out            : OUT BITS9);
END db4poly;

ARCHITECTURE flex OF db4poly IS

  TYPE STATE_TYPE IS (even, odd);
  SIGNAL state                   : STATE_TYPE;
```

[3] The equivalent Verilog code db4poly.v for this example can be found in Appendix A on page 374.

```
    SIGNAL x_odd, x_even, x_wait : BITS8;
    SIGNAL clk_div2              : STD_LOGIC;
    -- Arrays for multiplier and taps:
    SIGNAL r  : ARRAY_BITS17_4 ;
    SIGNAL x33, x99, x107, y     : BITS17;

  BEGIN

    Multiplex: PROCESS    ------> Split into even and odd
    BEGIN                          -- samples at clk rate
      WAIT UNTIL clk = '1';
      CASE state IS
        WHEN even =>
          x_even <= x_in;
          x_odd  <= x_wait;
          clk_div2 <= '1';
          state <= odd;
        WHEN odd =>
          x_wait <= x_in;
          clk_div2 <= '0';
          state <= even;
      END CASE;
    END PROCESS Multiplex;

    AddPolyphase: PROCESS (clk_div2,x_odd,x_even)
    VARIABLE m  : ARRAY_BITS17_4 ;
    BEGIN
  -- Compute auxiliary multiplications of the filter
      x33  <= x_odd * 32 + x_odd;
      x99  <= x33 * 2 + x33;
      x107 <= x99 + 8 * x_odd;
  -- Compute all coefficients for the transposed filter
      m(0) := 4 * (32 * x_even - x_even);         -- m[0] = 127
      m(1) := 2 * x107;                           -- m[1] = 214
      m(2) := 8 * (8 * x_even - x_even) + x_even;-- m[2] = 57
      m(3) := x33;                                -- m[3] = -33
  ------> Compute the filters and infer registers
      IF clk_div2'event and (clk_div2 = '0') THEN
  ------------ Compute filter G0
        r(0) <=  r(2) + m(0);     -- g[0] = 127
        r(2) <=  m(2);            -- g[2] = 57
  ------------ Compute filter G1
        r(1) <=  -r(3) + m(1);    -- g[1] = 214
        r(3) <=  m(3);            -- g[3] = -33
  ------------ Add the polyphase components
        y <= r(0) + r(1);
      END IF;
    END PROCESS AddPolyphase;

    x_e <= x_even; -- Provide some test signal as outputs
    x_o <= x_odd;
    clk2 <= clk_div2;
    g0 <= r(0);
```

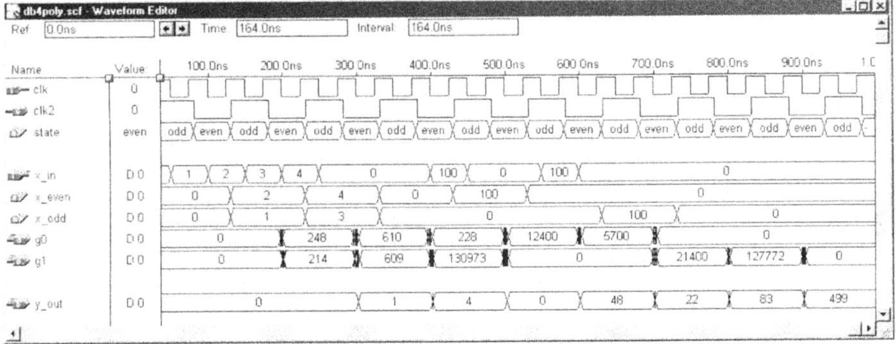

Fig. 5.9. VHDL simulation of the polyphase implementation of the length-4 Daubechies filter.

```
    g1 <= r(1);

    y_out <= y / 256; -- Connect to output

  END flex;
```

The first **PROCESS** is the FSM, which includes the control flow and the splitting of the input stream at the sampling rate into even and odd samples. The second **PROCESS** includes the reduced adder graph (RAG) multiplier, and the last **PROCESS** hosts the two filters in a transposed structure. Although the output is scaled, there is potential growth by the amount $\sum |g_k| = 1.673 < 2^1$. Therefore the output **y_out** was chosen to have an additional guard bit. The design uses 208 LCs and runs with 90.90 MHz **Registered Performance**. A simulation of the filter is shown in Fig. 5.9. The first four input samples are a triangle function to demonstrate the splitting into even and odd samples. Impulses with an amplitude of 100 are used to verify the coefficients of the two polyphase filters. Note that the filter is *not* shift invariant! 5.1

From the VHDL simulation shown in Fig. 5.9, it can be seen that such a decimator is no longer shift invariant, resulting in a technically nonlinear system. This can be validated by applying a single impulse. Initializing at an even indexed sample, the response is $G_0(z)$, while for an odd indexed sample, the response is $G_1(z)$.

5.2.1 Recursive IIR Decimator

It is also possible to apply polyphase decomposition to recursive filters and to get the speed benefit, if we follow the idea from Martinez and Parks [79], in the transfer function

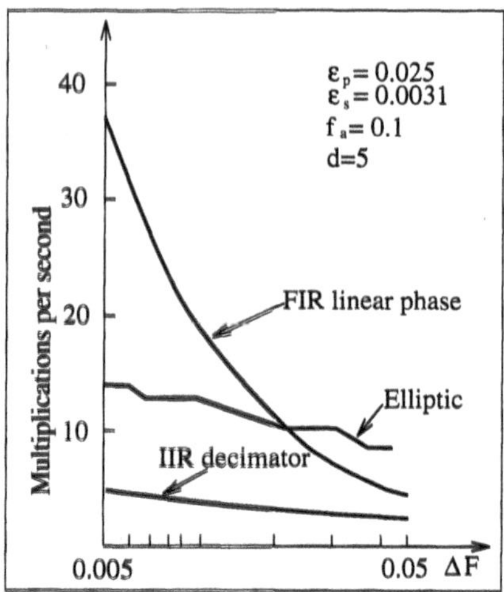

Fig. 5.10. Comparison of computational effort for decimators $\Delta F = f_p - f_s$.

$$F(z) = \frac{\sum_{l=0}^{L-1} a[l]z^{-l}}{1 - \sum_{l=1}^{K-1} b[l]z^{-lR}}. \tag{5.9}$$

i.e., the recursive part has only each Rth coefficient. We have already discussed such a design in the context of IIR filters (Fig. 4.17, p. 135). Fig. 5.10 shows that, depending on the transition width ΔF of the filter, an IIR decimator offers substantial savings compared with an FIR decimator.

5.2.2 Fast-Running FIR Filter

An interesting application of polyphase decomposition is the so-called *fast-running* FIR filter. The basic idea of this filter is the following: If we decompose the input signal $x[n]$ into R polyphase components, we can use Winograd's short convolution algorithms to implement a fast filter. Let us demonstrate this with an example for $R = 2$.

Example 5.2: Fast-Running FIR filter

We decompose the input signal $X(z)$ and filter $F(z)$ into even and odd polyphase components, i.e.,

$$X(z) = \sum_n x[n]z^{-n} = X_0(z^2) + z^{-1}X_1(z^2) \tag{5.10}$$

$$F(z) = \sum_n f[n]z^{-n} = F_0(z^2) + z^{-1}F_1(z^2). \tag{5.11}$$

The convolution in the time domain of $x[n]$ and $f[n]$ yields a polynomial multiply in the z-domain. It follows for the output signal $Y(z)$ that

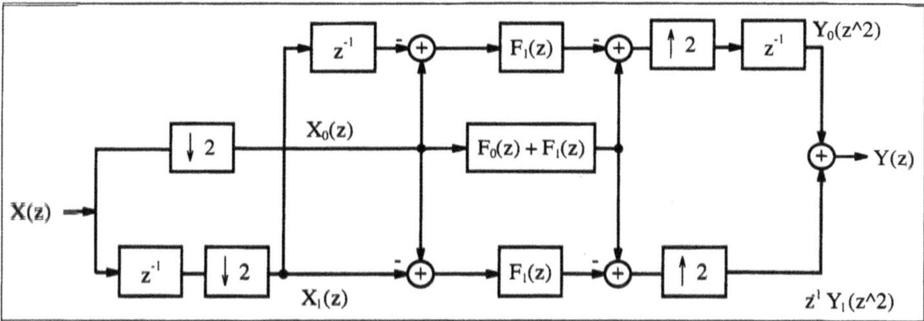

Fig. 5.11. Fast-running FIR filter with $R = 2$.

$$Y(z) = Y_0(z^2) + z^{-1}Y_1(z^2) \tag{5.12}$$
$$= (X_0(z^2) + z^{-1}X_1(z^2))(F_0(z^2) + z^{-1}F_1(z^2)). \tag{5.13}$$

If we split (5.13) into the polyphase components $Y_0(z)$ and $Y_1(z)$ we get

$$Y_0(z) = X_0(z)F_0(z) + z^{-1}X_1(z)F_1(z) \tag{5.14}$$
$$Y_1(z) = X_1(z)F_0(z) + X_0(z)F_1(z). \tag{5.15}$$

If we now compare (5.13) with a 2×2 linear convolution

$$A(z) \cdot B(z) = (a[0] + z^{-1}a[1])(b[0] + z^{-1}b[1]) \tag{5.16}$$
$$= a[0]b[0] + z^{-1}(a[0]b[1] + a[1]b[0]) + a[1]b[1]z^{-2}, \tag{5.17}$$

we notice that the factors for z^{-1} are the same, but for $Y_0(z)$ we must compute an extra addition to get the right phase relation. Winograd [83] has compiled a list of short convolution algorithms, and a linear 2×2 convolution can be computed using three multiplications and six adds with

$$\begin{array}{lll} a[0] = x[0] - x[1] & a[1] = x[0] & a[2] = x[1] - x[0] \\ b[0] = f[0] - f[1] & b[1] = f[0] & b[2] = f[1] - f[0] \\ c[k] = a[k]b[k] & k = 0,1,2 & \\ y[0] = c[1] + c[2] & y[1] = c[1] - c[0]. & \end{array} \tag{5.18}$$

With the help of this short convolution algorithm, we can now define the fast-running filter as follows:

$$\begin{bmatrix} Y_0 \\ Y_1 \end{bmatrix} = \begin{bmatrix} 0 & 1 & -1 \\ -1 & 1 & 0 \end{bmatrix} \begin{bmatrix} F_0 & 0 & 0 \\ 0 & F_0 + F_1 & 0 \\ 0 & 0 & F_1 \end{bmatrix} \begin{bmatrix} 1 & -1 \\ 1 & 0 \\ 1 & z^{-1} \end{bmatrix} \begin{bmatrix} X_0 \\ X_1 \end{bmatrix}. \tag{5.19}$$

Fig. 5.11 shows the graphical interpretation. 5.2

If we compare the direct filter implementation with the fast-running FIR filter we must distinguish between hardware effort and average number of adder and multiplier operations. A direct implementation would have L multipliers and $L-1$ adders running at full speed. For the fast-running filter we have three filters of length $L/2$ running at half speed. This results in $3L/4$ multiplications per output sample and $(2+2)/2+3/2(L/2-1) = 3L/4+1/2$ additions for the whole filter, i.e., the arithmetic count is about 25% better than in the direct implementation. From an implementation standpoint, we need $3L/2$ multipliers and $4+3(L/2-1) = 3L/2+1$ adders, i.e., the effort is about 50% higher than in the direct implementation. The important feature in Fig. 5.11 is that the fast-running filter basically runs at twice the speed of the direct implementation. Using a higher number R of decomposition may further increase the maximum throughput. The general methology for R polyphase signals with f_a as input rate is now as follows:

Algorithm 5.3: Fast-Running FIR Filter

1) Decompose the input signal into R polyphase signals, using A_e adders to form R sequences at a rate of f_a/R.
2) Filter the R sequences with R filters of length L/R.
3) Use A_a additions to compute the polyphase representation of the output $Y_k(z)$. Use a final output multiplexer to generate the output signal $Y(z)$.

Notice that the computed partial filter of length L/R may again be decomposed, using Algorithm 5.3. Then the question arises: When should we stop the iterative decomposition? Mou and Duhamel [84] have compiled a table with the goal of minimizing the average arithmetic count. Table 5.1 shows the optimal decomposition. The criterion used was a minimum total number of multiplications and additions, which is typical for a MAC-based design. In Table 5.1, all partial filters which should be implemented based on Algorithm 5.3 are underlined.

For a larger length than 60, a fast convolution using the FFT is more efficient, and will be discussed in Chapter 6.

Table 5.1. Computational effort for the recursive FIR decomposition [84, 85].

L	Factors	$M+A$	$\frac{M+A}{L}$
2	direct	6	3
3	direct	15	5
4	$\underline{2} \times 2$	26	6.5
5	direct	45	9
6	$\underline{3} \times 2$	56	9.33
8	$\underline{2^2} \times 2$	94	11.75
9	$\underline{3} \times 3$	120	13.33
10	$\underline{5} \times 2$	152	15.2
12	$\underline{2} \times \underline{3} \times 2$	192	16
14	$\underline{7} \times 2$	310	22.1
15	$\underline{5} \times 3$	300	20
16	$\underline{2^3} \times 2$	314	19.63
18	$\underline{2} \times \underline{3} \times 3$	396	22
20	$\underline{5} \times \underline{2} \times 2$	472	23.6
21	$\underline{7} \times 3$	591	28.1
22	$\underline{11} \times 2$	668	30.4
24	$\underline{2^2} \times \underline{3} \times 2$	624	26
25	$\underline{5} \times 5$	740	29.6
26	$\underline{13} \times 2$	750	28.9
27	$\underline{3^2} \times 3$	810	30
30	$\underline{5} \times \underline{3} \times 2$	912	30.4
32	$\underline{2^4} \times 2$	1006	31.44
33	$\underline{11} \times 3$	1248	37.8
35	$\underline{7} \times 5$	1405	40.1
36	$\underline{2^2} \times \underline{3} \times 3$	1260	35
39	$\underline{13} \times 3$	1419	36.4
55	$\underline{11} \times 5$	2900	52.7
60	$\underline{5} \times \underline{2} \times \underline{3} \times 2$	2784	46.4
65	$\underline{13} \times 5$	3345	51.46

5.3 Hogenauer CIC Filters

A very efficient architecture for a high decimation-rate filter is the "cascade integrator comb" (CIC) filter introduced by Hogenauer [86]. The CIC (also known as the Hogenauer filter), has proven to be an effective element in high-decimation or interpolation systems. One application is in wireless communications, where signals, sampled at RF or IF rates, need to be reduced to baseband. For narrowband applications (e.g., cellular radio), decimation rates in excess of 1,000 are routinely required. Such systems are sometimes referred to as channelizers [87]. Another application area is in $\Sigma\Delta$ data converters [88].

CIC filters are based on the fact that perfect pole/zero canceling can be achieved. This is only possible with exact integer arithmetic. Both two's complement and the residue number system have the ability to support error-free arithmetic. In the case of two's complement, arithmetic is performed modulo 2^b, and, in the case of the RNS, modulo M.

An introductory case study will be used to demonstrate.

5.3.1 Single-Stage CIC Case Study

Fig. 5.12 shows a first-order CIC filter without decimation in 4-bit arithmetic. The filter consists of a (recursive) integrator (I-section), followed by a 4-bit differentiator or comb (C-section). The filter is realized with 4-bit values, which are implemented in two's complement arithmetic, and the values are bounded by $-8_{10} = 1000_{2C}$ and $7_{10} = 0111_{2C}$.

Fig. 5.13 shows the impulse response of the filter. Although the filter is recursive, the impulse response is finite, i.e., it is a *recursive* FIR filter. This is unusual because we generally expect recursive filter to be an IIR filter. The impulse response shows that the filter computes the sum

$$y[n] = \sum_{k=0}^{D-1} x[n-k], \tag{5.20}$$

where D is the delay found in the comb section. The filter's response is a *moving average* defined over D contiguous sample values. Such a moving average is a very simple form of a lowpass filter. The same moving-average

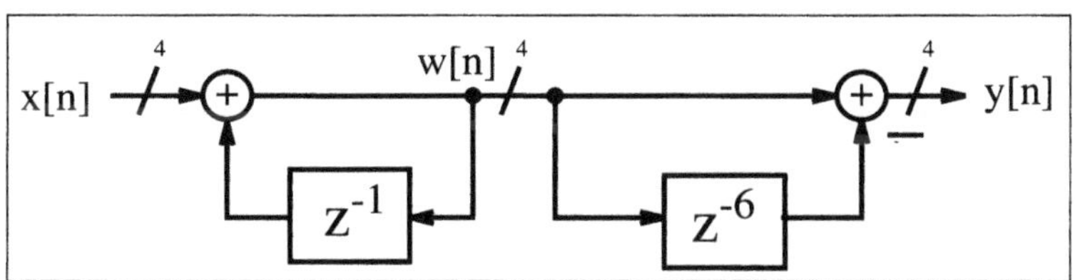

Fig. 5.12. Moving average in 4-bit arithmetic.

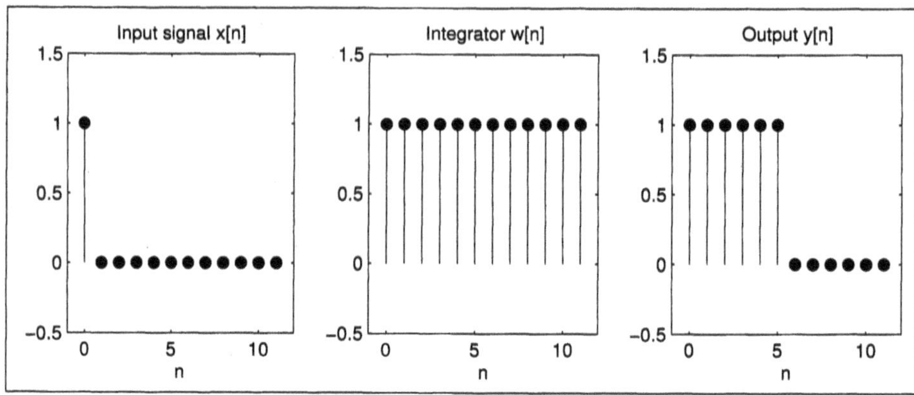

Fig. 5.13. Impulse response of the filter from Fig. 5.12.

filter implemented as a nonrecursive FIR filter, would require five adders, compared with one adder and one subtractor for the CIC design.

A recursive filter having a known pole location has its largest steady-state sinusoidal output when the input is an "eigenfrequency" signal, one whose pole directly coincides with a pole of the recursive filter. For the CIC section, the eigenfrequency corresponds to the frequency $\omega = 0$, i.e., a step input. The step response of the first-order moving average given by (5.20) is a ramp for the first D samples, and a constant $y[n] = D = 6$ thereafter, as shown in Fig. 5.14. Notice that although the integrator $w[n]$ shows frequent overflows, the output is still correct. This is because the comb subtraction also uses two's complement arithmetic, e.g., at the time of the first wrap-around, the actual integrator signal is $w[n] = -8_{10} = 1000_{2C}$, and the delay signal is $w[n-6] = 2_{10} = 0010_{2C}$. This results in $y[n] = -8_{10} - 2_{10} = 1000_{2C} - 0010_{2C} = 0110_{2C} = 6_{10}$, as expected. The accumulator would continue to count upward until $w[n] = -8_{10} = 1000_{2C}$ is again reached. This pattern would continue as long as the step input is present. In fact, as long as the output $y[n]$ is a valid 4-bit two's complement number in the range $[-8, 7]$, the exact arithmetic of the two's complement system will automatically compensate for the integrator overflows.

In general, 4-bit filter width is usually much too small for a typical application. The Harris IC HSP43220, for instance, has five stages and uses 66-bit integrator width. To reduce the adder latency, it is therefore reasonable to use a multibase RNS system. If we use, for instance, the set $\mathbb{Z}_{30} = \{2, 3, 5\}$, it can be seen from Table 5.2 that a total of $2 \cdot 3 \cdot 5 = 30$ unique values can be represented. The mapping is unique (bijective) and is proven by the Chinese Remainder Theorem.

Fig. 5.15 displays the step response of the illustrated RNS implementation. The filter's output, $y[n]$, has been reconstructed using data from Table 5.2. The output response is identical with the sample value obtained in the

Table 5.2. RNS mapping for the set $(2, 3, 5)$.

$a =$	0	1	2	3	4	5	6	7	8	9	10	11	12	13	14	15
$a \bmod 2$	0	1	0	1	0	1	0	1	0	1	0	1	0	1	0	1
$a \bmod 3$	0	1	2	0	1	2	0	1	2	0	1	2	0	1	2	0
$a \bmod 5$	0	1	2	3	4	0	1	2	3	4	0	1	2	3	4	0
$a =$	16	17	18	19	20	21	22	23	24	25	26	27	28	29	30	
$a \bmod 2$	0	1	0	1	0	1	0	1	0	1	0	1	0	1	0	
$a \bmod 3$	1	2	0	1	2	0	1	2	0	1	2	0	1	2	0	
$a \bmod 5$	1	2	3	4	0	1	2	3	4	0	1	2	3	4	0	

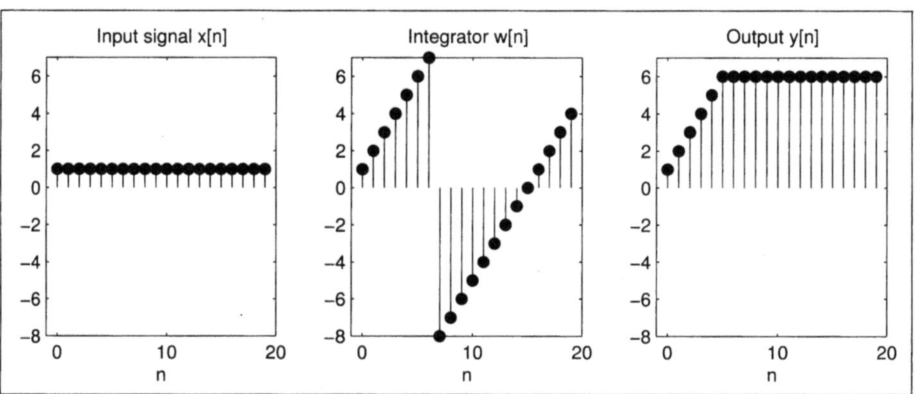

Fig. 5.14. Step response (eigenfrequency test) of the filter from Fig. 5.12.

two's complement case (see Fig. 5.14). A mapping that preserves the structure is called *homomorph*. A bijective homomorphism is called an *isomorphism* (notation $\cong$), which can be expressed as:

$$\mathbb{Z}_{30} \cong \mathbb{Z}_2 \times \mathbb{Z}_3 \times \mathbb{Z}_5. \tag{5.21}$$

5.3.2 Multistage CIC Filter Theory

The transfer function of a general CIC system consisting of S stages is given by:

$$F(z) = \left(\frac{1 - z^{-RD}}{1 - z^{-1}} \right)^S, \tag{5.22}$$

where D is the number of delays in the comb section, and R the down-sampling (decimation) factor.

It can be seen from (5.22) that $F(z)$ is defined with respect to RDS zeros and S poles. The RD zeros generated by the numerator term $(1 - z^{-RD})$ are located on $2\pi/(RD)$-radian centers beginning at $z = 1$. Each distinct

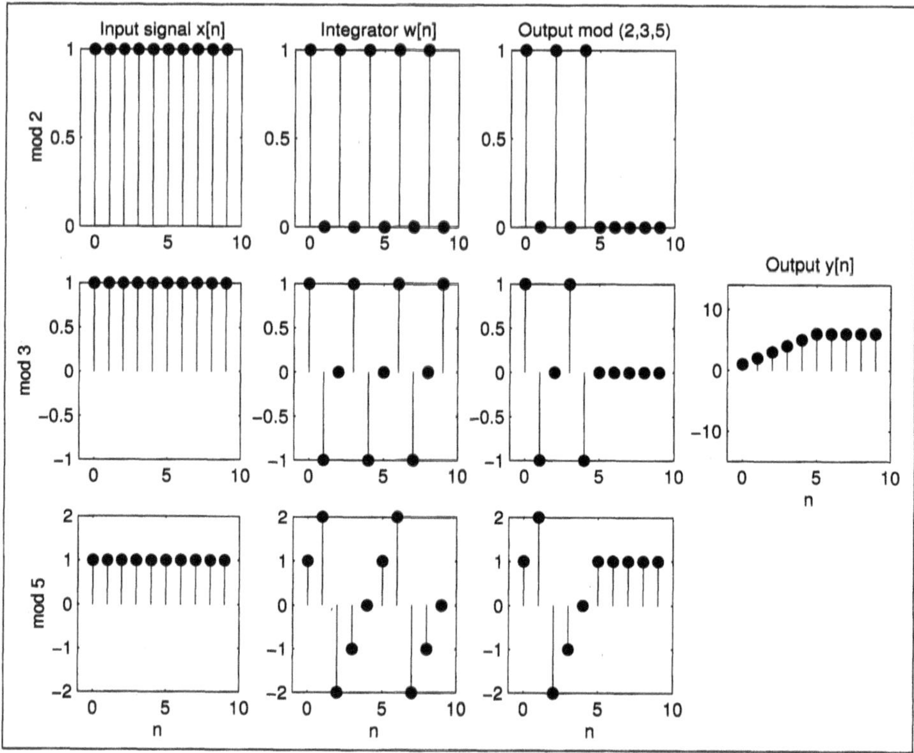

Fig. 5.15. Step response of the first-order CIC in RNS arithmetic.

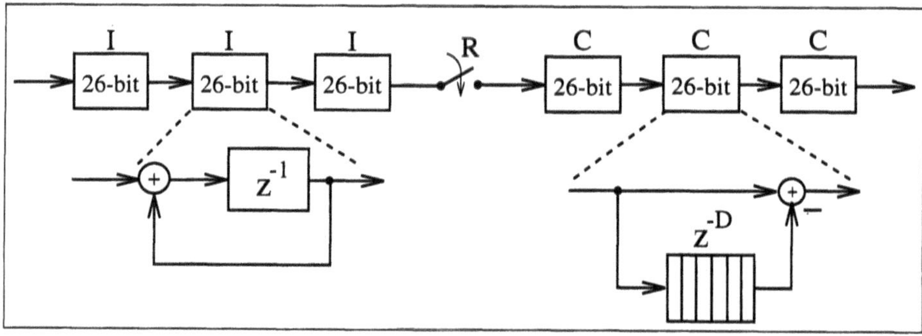

Fig. 5.16. CIC filter. Each stage 26-bit.

zero appears with multiplicity S. The S poles of $F(z)$ are located at $z = 1$, i.e., at the zero frequency (DC) location. It can immediately be seen that they are annihilated by S zeros of the CIC filter. The result is an S-stage moving average filter. The maximum dynamic range growth occurs at the DC frequency (i.e., $z = 1$). The maximum dynamic range growth is

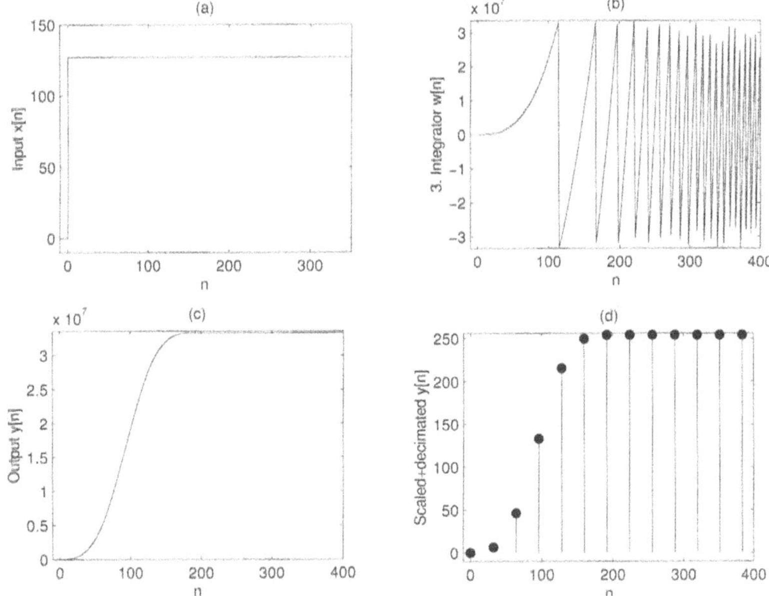

Fig. 5.17. Simulation of the three-stage CIC filter shown in Fig. 5.16.

$$B_{\mathbf{grow}} = (RD)^S \quad \text{or} \quad b_{\mathbf{grow}} = \log_2 (B_{\mathbf{grow}}) \text{ bits.} \tag{5.23}$$

Knowledge of this value is important when designing a CIC filter, since the need for exact arithmetic was in the single-state CIC example. In practice, the worst-case gain can be substantial, as evidenced by a 66-bit dynamic range built into commercial CIC filters (e.g., the Harris HSP43220 [87] channelizer), typically designed using two's complement arithmetic.

Fig. 5.16 shows a three-stage CIC filter that consists of a three-stage integrator, a sampling rate reduction by R, and a three-stage comb. Note that all integrators are implemented first, then the decimator, and finally the comb sections. The rearrangement saves a factor R of delay elements in the comb sections. The number of delays D for a high-decimation rate filter is typically one or two.

A three-stage RNS CIC filter with an input wordwidth of eight bits, along with $D = 2$, $R = 32$, or $DR = 2 \cdot 32 = 64$, would require an internal wordwidth of $W = 8 + 3 \log_2(64) = 26$ bits to ensure that run-time overflow would not occur. The output wordwidth would normally be a value significantly less than W, such as nine bits. If the ratio of signal bandwidth to sampling frequency is, for instance, 1/32, then the aliasing suppression is 89.6 dB, and the passband attenuation is 0.17 [86, Table I+II].

Example 5.4: Three-Stage CIC Decimator I

The worst-case gain condition can be forced by supplying a step (DC) signal to the CIC filter. Fig. 5.17(a) shows a step input signal with amplitude 127.

Fig. 5.17(b) displays the output found at the third integrator section. Observe that run-time overflows occur at a regular rate. The CIC output shown in Fig. 5.17(c) is interpolated (smoothed) for display at the input sampling rate. The output shown in Fig. 5.17(d) is scaled to 9-bit precision and displayed at the decimated sample rate.
The following VHDL code[5] shows the CIC example design.

```
PACKAGE n_bit_int IS    -- User defined types
  SUBTYPE word26 IS INTEGER RANGE 0 TO 2**26-1;
END n_bit_int;

LIBRARY work;
USE work.n_bit_int.ALL;

LIBRARY ieee;
USE ieee.std_logic_1164.ALL;
USE ieee.std_logic_arith.ALL;
USE ieee.std_logic_unsigned.ALL;

ENTITY cic3r32 IS
       PORT ( clk   :   IN  STD_LOGIC;
              x_in  :   IN  STD_LOGIC_VECTOR(7 DOWNTO 0);
              y_out :   OUT STD_LOGIC_VECTOR(8 DOWNTO 0));
END cic3r32;

ARCHITECTURE flex OF cic3r32 IS
  TYPE    STATE_TYPE IS (hold, sample);
  SIGNAL  state     : STATE_TYPE ;
  SIGNAL  count      : integer RANGE 0 TO 31;
  SIGNAL  clk2       : STD_LOGIC;
  SIGNAL  x : STD_LOGIC_VECTOR(7 DOWNTO 0);
                                          -- Registered input
  SIGNAL  sxtx : STD_LOGIC_VECTOR(25 DOWNTO 0);
                                          -- Sign extended input
  SIGNAL  i0, i1 , i2 : word26;   -- I section  0, 1, and 2
  SIGNAL  i2d1, i2d2, i2d3, i2d4, c1, c0 : word26;
                                       -- I and COMB section 0
  SIGNAL  c1d1, c1d2, c1d3, c1d4, c2 : word26;      -- COMB 1
  SIGNAL  c2d1, c2d2, c2d3, c2d4, c3 : word26;      -- COMB 2

BEGIN

  FSM: PROCESS
  BEGIN
    WAIT UNTIL clk = '0';
    CASE state IS
      WHEN hold =>
        IF count < 31 THEN
           state <= hold;
        ELSE
           state <= sample;
        END IF;
```

[5] The equivalent Verilog code cic3r32.v for this example can be found in Appendix A on page 370.

```
      WHEN OTHERS =>
        state <= hold;
    END CASE;
  END PROCESS FSM;

  sxt: PROCESS (x)
  BEGIN
    sxtx(7 DOWNTO 0) <= x;
    FOR k IN 25 DOWNTO 8 LOOP
      sxtx(k) <= x(x'high);
    END LOOP;
  END PROCESS sxt;

  Int: PROCESS
  BEGIN
    WAIT UNTIL clk = '1';
      x    <= x_in;
      i0   <= i0 + CONV_INTEGER(sxtx);
      i1   <= i1 + i0 ;
      i2   <= i2 + i1 ;
    CASE state IS
      WHEN sample =>
        c0    <= i2;
        count <= 0;
      WHEN OTHERS =>
        count <= count + 1;
    END CASE;
    IF (count > 8) and (count <16) THEN
      clk2  <= '1';
    ELSE
      clk2  <= '0';
    END IF;
  END PROCESS Int;

  Comb: PROCESS
  BEGIN
    WAIT UNTIL clk2 = '1';
      i2d1 <= c0;
      i2d2 <= i2d1;
      c1   <= c0 - i2d2;
      c1d1 <= c1;
      c1d2 <= c1d1;
      c2   <= c1  - c1d2;
      c2d1 <= c2;
      c2d2 <= c2d1;
      c3   <= c2  - c2d2;
  END PROCESS Comb;

  y_out <= CONV_STD_LOGIC_VECTOR(c3 / 2**17 , 9);

END flex;
```

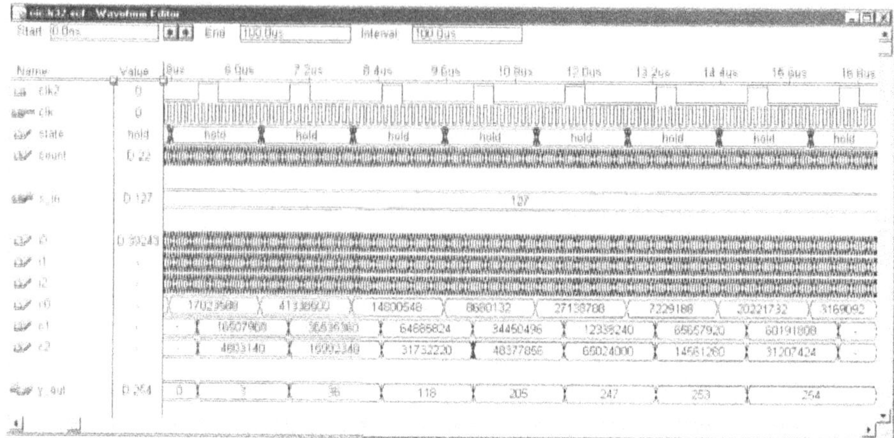

Fig. 5.18. VHDL simulation of the three-stage CIC filter shown in Fig. 5.16.

The designed filter includes a finite state machine (FSM); a sign extension, `sxt: PROCESS`, and two "arithmetic" `PROCESS` blocks. The `Int: PROCESS` realizes the three integrators and the clock divider for the comb section. The `Comb: PROCESS` includes the three comb filters, each having a delay of two samples. The filter runs at 38.46 MHz and uses 332 LCs. Note that the filter would not fit in the target device without the early downsampling. The early downsampling saves $3 \cdot 32 \cdot 27 = 2592$ registers or LCs!
If we compare the filter outputs (Fig. 5.18 shows the VHDL output `y_out` and the response $y[n]$ from the MATLAB simulation shows in Fig. 5.17(d) we see that the filter behaves as expected. 5.4

Hogenauer [86] noted, based on a careful analysis, that some of the lower significant bits from early stages can be eliminated without sacrificing system integrity. Fig. 5.19 displays the system's magnitude frequency response for a design using full (worst-case) wordwidth in all stages, and using the wordlength "pruning" policy suggested by Hogenauer.

5.3.3 Amplitude and Aliasing Distortion

The transfer function of an S-stage CIC filter was reported to be

$$F(z) = \left(\frac{1 - z^{RD}}{1 - z^{-1}}\right)^{S}. \tag{5.24}$$

The amplitude distortion and the maximum aliasing component can be computed in the frequency domain by evaluating $F(z)$ along the arc $z = e^{j2\pi fT}$. The magnitude response becomes

$$|F(f)| = \left(\frac{\sin(2\pi fTRD/2)}{\sin(2\pi fT/2)}\right)^{S}. \tag{5.25}$$

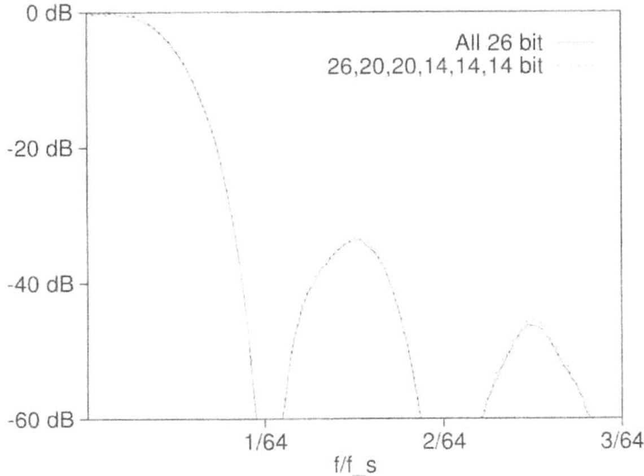

Fig. 5.19. CIC transfer function (f_s is sampling frequency at the input).

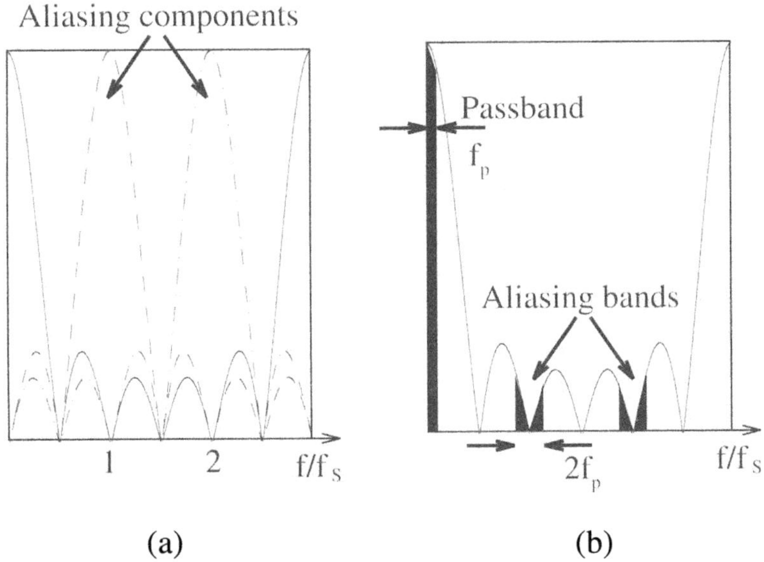

Fig. 5.20. Transfer function of a three-stage CIC decimator. Note that f_s is the sampling frequency at the lower rate.

which can be used to directly compute the *amplitude distortion* at the pass-band edge ω_p. Fig. 5.20 shows $|F(f - k\frac{1}{2R})|$ for a three-stage CIC filter with $R = 3$, $D = 2$, and $RD = 6$. Observe that several copies of the CIC filter's low-frequency response are aliased in the baseband.

It can be seen that the maximum aliasing component can be computed from $|F(f)|$ at the frequency

$$f|_{\text{Aliasing has maximum}} = 1/(2R) - f_p. \tag{5.26}$$

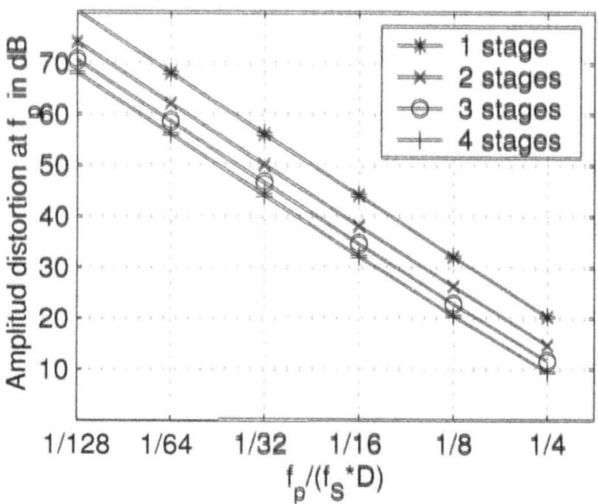

Fig. 5.21. Amplitude distortion for the CIC decimator.

Most often, only the first aliasing component is taken into consideration, because the second component is smaller. Fig. 5.21 shows the amplitude distortion at f_p for different ratios of $f_p/(Df_s)$.

Fig. 5.22 shows, for different values of S, R, and D, the maximum aliasing component for a special ratio of passband frequency and sampling frequency, f_p/f_s.

It may be argued that the amplitude distortion can be corrected with a cascaded FIR compensation filter, which has a transfer function $1/|F(z)|$ in the passband, but the aliasing distortion can *not* be repaired. Therefore, the acceptable aliasing distortion is most often the dominant design parameter.

5.3.4 Hogenauer Pruning Theory

The total internal wordwidth is defined as the sum of the input wordwidth and the maximum dynamic growth requirement (5.23), or algebraically:

$$B_{\text{intern}} = B_{\text{input}} + B_{\text{growth}} \tag{5.27}$$

If the CIC filter is designed to perform exact arithmetic with this wordwidth at all levels, no run-time overflow will occur at the output. In general, input and output bit width of a CIC filter are in the same range. We find then that quantization introduced through pruning in the output is in general larger than quantization introduced by also pruning some LSBs at previous stages. If $\sigma^2_{T,2S+1}$ is the quantization noise introduced through pruning in the output, Hogenauer suggested to set it equal to the sum of the noise σ_k^2 introduced by all previous sections. For a CIC filter with S integrator and S comb sections, it follows that:

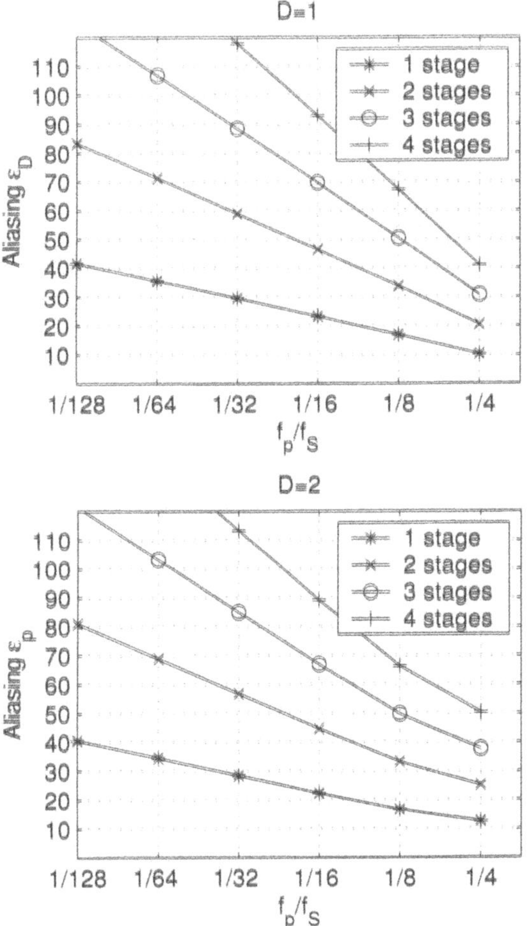

Fig. 5.22. Maximum aliasing for one- to four stage CIC decimator.

$$\sum_{k=1}^{2S} \sigma_{T,k}^2 = \sum_{k=1}^{2N} \sigma_k^2 P_k^2 \leq \sigma_{T,2S+1}^2 \tag{5.28}$$

$$\sigma_{T,k}^2 = \frac{1}{2S} \sigma_{T,2S+1}^2 \tag{5.29}$$

$$P_k^2 = \sum_n (h_k[n])^2 \quad k = 1, 2, \ldots, 2S, \tag{5.30}$$

where P_k^2 is the power gain from stage k to the output. Compute next the number of bits B_k, which should be pruned by

$$B_k = \left\lfloor 0,5 \log_2 \left(P_k^{-2} \cdot \frac{6}{N} \cdot \sigma_{T,2S+1}^2 \right) \right\rfloor \tag{5.31}$$

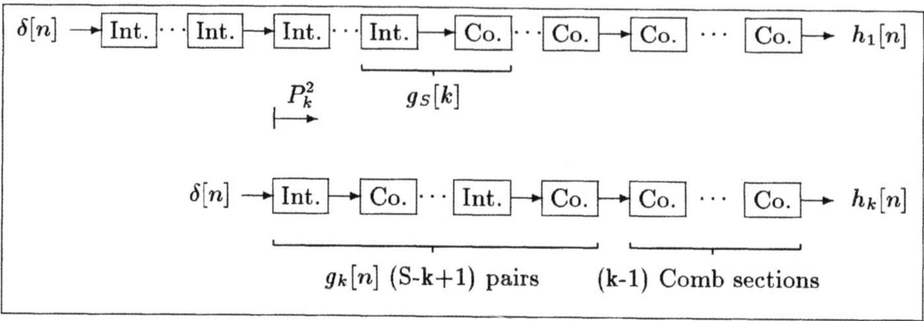

Fig. 5.23. Rearrangement to simplify the computation of P_k^2 (©1995 VDI Press [4]).

$$\sigma_{T,k}^2\Big|_{j=2N+1} = \frac{1}{12}2^{2B_k} = \frac{1}{12}2^{2(B_{\text{in}}-B_{\text{out}}+B_{\text{growth}})} \tag{5.32}$$

The power gain $P_k^2, k = S+1, \ldots, 2S$ for the comb sections can be computed using the binomial coefficient

$$H_k(z) = \sum_{m=0}^{2S+1-k} (-1)^n \binom{2S+1-k}{n} z^{-kRD}$$
$$k = S, S+1, \ldots, 2S. \tag{5.33}$$

For computation of the first factor P_k^2 for $k = 1, 2, \ldots, S$, it is useful to keep in mind that each integrator/comb pair produces a finite (moving average) impulse response. The resulting system for stage k is therefore a series of $S-k+1$ integrator/comb pairs followed by $k-1$ comb sections. Fig. 5.23 shows this rearrangement for a simplified computation of P_k^2.

The program `cic.exe` (included on the CD-ROM under `book/util`) computes this CIC pruning. The program produces a file `cicXX.dat`, where `XX` must be specified. The following design example explains the results.

Example 5.5: Three-Stages CIC Decimator II

Garcia et al. [44], have designed an I/Q demodulator which uses a three-stage CIC filter. The row data of the decimator were: $B_{\text{input}} = 8, B_{\text{output}} = 9$, Bit $R = 32$, and $D = 2$. Obviously, the bit growth is

$$B_{\text{growth}} = \lceil \log_2(RD^S) \rceil = \log_2(64^3) \lceil 3 \cdot 6 \rceil = 18 \tag{5.34}$$

and the total internal bit width becomes

$$B_{\text{intern}} = B_{\text{input}} + B_{\text{growth}} = 8 + 18 = 26. \tag{5.35}$$

The program `cic.exe` shows the following results:

```
-- ---------------------------------------------------------
-- Program for the design of a CIC decimator.
-- ---------------------------------------------------------
--          Input bit width        Bin  =    8
--          Output bit width       Bout =    9
--          Number of stages       S    =    3
--          Decimation factor      R    =   32
--          COMB delay             D    =    2
```

```
--       Frequency resolution DR    =     64
--       Passband freq. ratio P     =      4
-- ---------------------------------------------------------
-- ----------------- Results of the Design ----------------
-- ---------------------------------------------------------
-- -------- Computed bit width:
-- -------- Maximum bit growth over all stages      =     18
-- -------- Maximum bit width including sign Bmax+1 =     26
-- Stage   1 INTEGRATOR. Bit width :  26
-- Stage   2 INTEGRATOR. Bit width :  21
-- Stage   3 INTEGRATOR. Bit width :  16
-- Stage   1 COMB.       Bit width :  14
-- Stage   2 COMB.       Bit width :  13
-- Stage   3 COMB.       Bit width :  12
-- ------- Maximum aliasing component : 0.009582 = 40.37 dB
-- ------- Amplitude distortion       : 0.258012 = 11.77 dB
```

5.5

The design charts shown in Figs. 5.21 and 5.22 may also be used to compute the maximum aliasing component and the amplitude distortion.

The following design example demonstrates the detailed bit-width design, using MaxPlusII.

Example 5.6: Three-Stage CIC Decimator III

The data for the design should be the same as for Example 5.4 (p. 159), but we now consider the pruning as computed in Example 5.5 (p. 166).
The following VHDL code[6] shows the CIC example design with pruning.

```
PACKAGE n_bit_int IS               -- User defined types
  SUBTYPE word26 IS INTEGER RANGE -2**25 TO 2**25-1;
  SUBTYPE word21 IS INTEGER RANGE -2**20 TO 2**20-1;
  SUBTYPE word16 IS INTEGER RANGE -2**15 TO 2**15-1;
  SUBTYPE word14 IS INTEGER RANGE -2**14 TO 2**14-1;
  SUBTYPE word13 IS INTEGER RANGE -2**13 TO 2**13-1;
  SUBTYPE word12 IS INTEGER RANGE -2**12 TO 2**12-1;
END n_bit_int;

LIBRARY work;
USE work.n_bit_int.ALL;

LIBRARY ieee;
USE ieee.std_logic_1164.ALL;
USE ieee.std_logic_arith.ALL;
USE ieee.std_logic_unsigned.ALL;


ENTITY cic3s32 IS
  PORT ( clk   :   IN  STD_LOGIC;
         x_in  :   IN  STD_LOGIC_VECTOR(7 DOWNTO 0);
```

[6] The equivalent Verilog code cic3s32.v for this example can be found in Appendix A on page 372.

```
        y_out   :   OUT STD_LOGIC_VECTOR(8 DOWNTO 0));
END cic3s32;

ARCHITECTURE flex OF cic3s32 IS
  TYPE    STATE_TYPE IS (hold, sample);
  SIGNAL  state      : STATE_TYPE ;
  SIGNAL  count      : integer RANGE 0 TO 31;
  SIGNAL  clk2       : STD_LOGIC;
  SIGNAL  x          : STD_LOGIC_VECTOR(7 DOWNTO 0);
                                              -- Registered input
  SIGNAL  sxtx : STD_LOGIC_VECTOR(25 DOWNTO 0);
                                           -- Sign extended input
  SIGNAL  i0 :  word26;                            -- I section 0
  SIGNAL  i1 :  word21;                            -- I section 1
  SIGNAL  i2 :  word16;                            -- I section 2
  SIGNAL  i2d1, i2d2, i2d3, i2d4, c1, c0 : word14;
                                         -- I and COMB section 0
  SIGNAL  c1d1, c1d2, c1d3, c1d4, c2 : word13;   -- COMB  1
  SIGNAL  c2d1, c2d2, c2d3, c2d4, c3 : word12;   -- COMB  2

BEGIN

  FSM: PROCESS
  BEGIN
    WAIT UNTIL clk = '0';
    CASE state IS
      WHEN hold =>
        IF count < 31 THEN
           state <= hold;
        ELSE
           state <= sample;
        END IF;
      WHEN OTHERS =>
        state <= hold;
    END CASE;
  END PROCESS FSM;

  Sxt: PROCESS (x)
  BEGIN
    sxtx(7 DOWNTO 0) <= x;
    FOR k IN 25 DOWNTO 8 LOOP
      sxtx(k) <= x(x'high);
    END LOOP;
  END PROCESS Sxt;

  Int: PROCESS
  BEGIN
  WAIT
    UNTIL clk = '1';
      x   <= x_in;
      i0  <= i0 + CONV_INTEGER(sxtx);
      i1  <= i1 + i0 / 32;
      i2  <= i2 + i1 / 32;
```

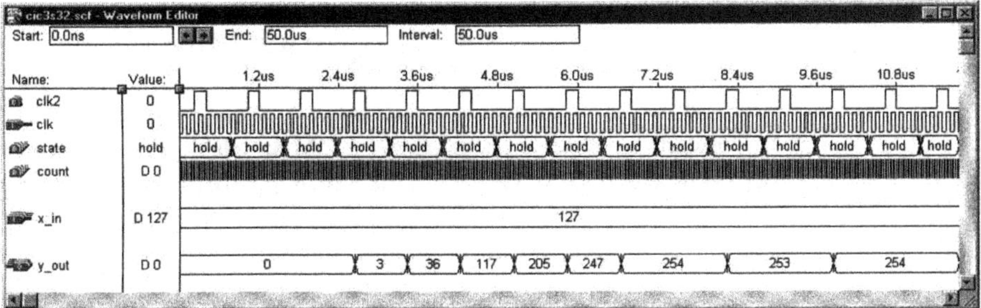

Fig. 5.24. VHDL simulation of the three-stage CIC filter, implemented with bit pruning.

```
      CASE state IS
        WHEN sample =>
          c0    <= i2 /4;
          count <= 0;
        WHEN OTHERS =>
          count <= count + 1;
      END CASE;
      IF (count > 8) and (count <16) THEN
        clk2  <= '1';
      ELSE
        clk2  <= '0';
      END IF;
    END PROCESS Int;

    Comb: PROCESS
    BEGIN
      WAIT UNTIL clk2 = '1';
        i2d1 <= c0;
        i2d2 <= i2d1;
        c1   <= c0 - i2d2;
        c1d1 <= c1 / 2;
        c1d2 <= c1d1;
        c2   <= c1 / 2 - c1d2;
        c2d1 <= c2 / 2;
        c2d2 <= c2d1;
        c3   <= c2 / 2 - c2d2;
    END PROCESS Comb;

    y_out <= CONV_STD_LOGIC_VECTOR(c3 / 8, 9);

  END flex;
```

The design has the same architecture as the unscaled CIC shown in Example 5.4 (p. 159). The design consists of a finite state machine (FSM), a sign extension `sxt: PROCESS`, and two "arithmetic" `PROCESS` blocks. The `Int: PROCESS` realizes the three integrators and the clock divider for the comb sections. The `Comb: PROCESS` includes the three comb sections, each having a delay of two. But now, all integrator and comb sections are designed with the bit width suggested by Hogenauer's pruning technique. This reduces the size of the design to 199 LCs and the design now runs now at 41.66 MHz. 5.6

This design improved the speed from 38 to 41 MHz and also saved 133 LCs, or 40%, compared with the design considered in Example 5.4 (p. 159). Comparing the filter output of the VHDL simulations, shown in Fig. 5.24 and Fig. 5.18 (p. 162), different LSB quantization behavior can be noted (see Exercise 5.11, p. 207). In the pruned design, "noise" possesses the asymptotic behavior of the LSB ($253 \leftrightarrow 254$).

5.3.5 CIC RNS Design

The design of a CIC filter using the RNS was proposed by Garcia, Meyer-Bäse, and Taylor [44]. A three-stage CIC filter, with 8-bit input, 10-bit output, $D = 2$, and $R = 32$ was implemented. The maximum wordwidth was 26 bits. For the RNS implementation, the 4-moduli set $(256, 63, 61, 59)$, i.e., one 8-bit two's complement and three 6-bit moduli, covers this range (see Fig. 5.25). The output was scaled using an ε-CRT requiring eight tables and three two's complement adders [39, Fig. 1], or (as shown in Fig. 5.26) using a base removal scaling (BRS) algorithm based on two 6-bit moduli (after [38]), and an ε-CRT for the remaining two moduli, for a total of five modulo adders and nine ROM tables, or seven tables (if multiplicative inverse ROM and the ε-CRT are combined). The following table shows the speed in MSPS and the number of LEs and EABs used for the three scaling schemes.

Type	ε-CRT	BRS ε-CRT (Speed data for BRS m_4 only)	BRS ε-CRT combined ROM
MSPS	58.8	70.4	58.8
#LE	34	87	87
#Table (EAB)	8	9	7

The decrease in speed to 58.8 MSPS, for the scaling schemes 1 and 3, is the result of the need for a 10-bit ε-CRT. It should be noted that this does not reduce the *system* speed, since scaling is applied at the lower (output) sampling rate. For the BRS ε-CRT, it is assumed that only the BRS m_4 part (see Fig. 5.13) must run at the input sampling rate, while BRS m_3 and ε-CRT run at the output sampling rate.

Some resources can be saved if a scaling scheme, similar to Example 5.5 (p. 166), and illustrated in Fig. 5.25, is used. With this scheme, the BRS ε-CRT scheme must be applied to reduce the bit width in the earlier sections of the filter. The early use of ROMs decreases the possible throughput from 76.3 to 70.4 MSPS, which is the maximum speed of the BRS with m_4. At the output, the efficient ε-CRT scheme was applied.

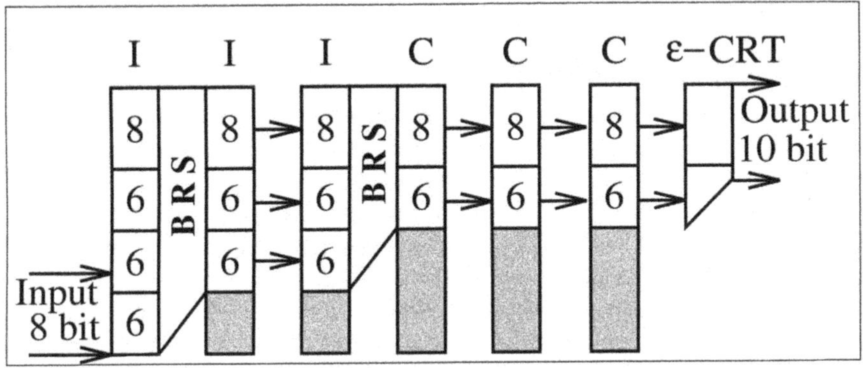

Fig. 5.25. CIC filter. Detail of design with base removal scaling (BRS).

The following table summarizes the three implemented filter designs, without including the scaling data.

Type	2C 26-bit	RNS 8, 6, 6, 6-bit	Detailed bit width RNS design
MSPS	49.3	76.3	70.4
#LEs	343	559	355

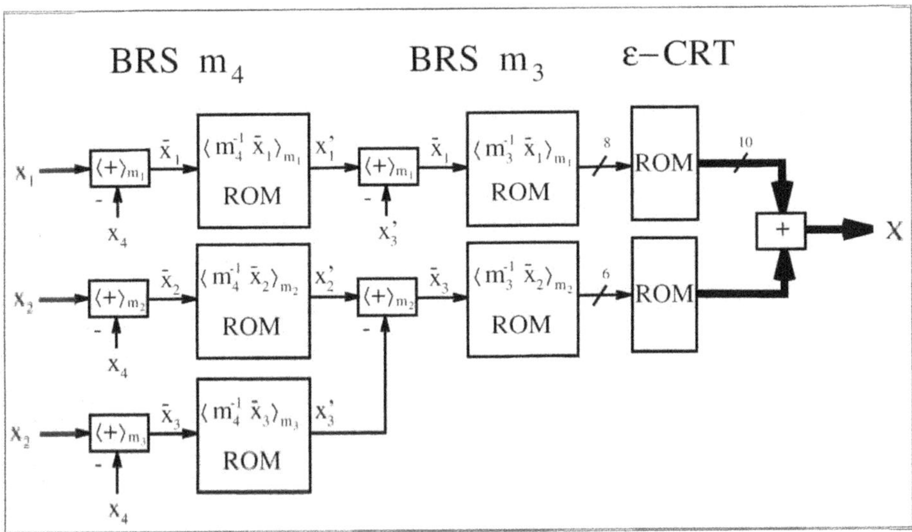

Fig. 5.26. BRS and ε-CRT conversion steps.

5.4 Multistage Decimator

If the decimation rate R is large it can be shown that a multistage design can be realized with less effort than a single-stage converter. In particular, S stages, each having a decimation capability each of R_k, are designed to have an overall downsampling rate of $R = R_1 R_2 \cdots R_S$. Unfortunately, passband imperfections, such as ripple deviation, accumulate from stage to stage. As a result, a passband deviation target of ε_p must normally be tightened on the order of $\varepsilon_p' = \varepsilon_p/S$ to meet overall system specifications. This is obviously a worst-case assumption, in which all short filters have the maximum ripple at the same frequencies, which is in general too pessimistic. It is often more reasonable to try an initial value near the given passband specification ε_p, and then selectively reduce it if necessary.

5.4.1 Multistage Decimator Design Using Goodman-Carey Halfband Filters

Goodman and Carey [66] proposed to develop multi-stage systems based on the use of CIC and halfband filters. As the name implies, a *halfband filter* has a passband and stopband located at $\omega_s = \omega_p = \pi/2$, or midway in the baseband. A halfband filter can therefore be used to change the sampling rate by a factor of two. If the halfband filter has *point symmetry* relative to $\omega = \pi/2$, then all even coefficients (except the center tap) become zero.

Definition 5.7: **Halfband Filter**

The impulse response of a halfband filter is symmetric to $k = d$ and obeys the following rule

$$f[k] = 0 \qquad k = \text{even without } k = d. \tag{5.36}$$

The same condition transformed in the z-domain reads

$$F(z) + F(-z) = c \cdot z^{-d}, \tag{5.37}$$

where $c \in \mathbb{C}$ is a constant and $d \in \mathbb{N}_0$.

Table 5.3. Coefficients of the halfband filter F1 to F9 from Goodman and Carey [66].

Name	L	Ripple	$f[0]$	$f[1]$	$f[3]$	$f[5]$	$f[7]$	$f[9]$
F1	3	–	1	1				
F2	3	–	2	1				
F3	7	–	16	9	−1			
F4	7	36 dB	32	19	−3			
F5	11	–	256	150	−25	3		
F6	11	49 dB	346	208	−44	9		
F7	11	77 dB	521	302	−53	7		
F8	15	65 dB	802	490	−116	33	−6	
F9	19	78 dB	8192	5042	−1277	429	−116	18

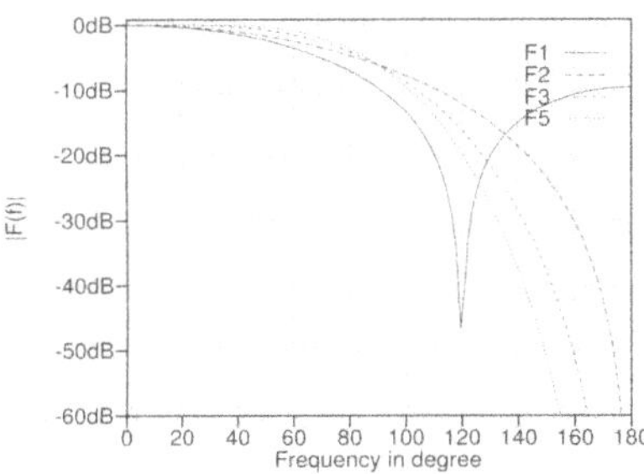

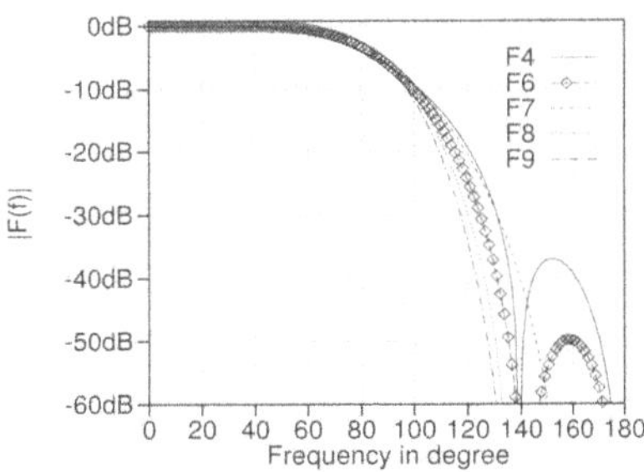

Fig. 5.27. Transfer function of the halfband filter F1 to F9.

Goodman and Carey [66] have compiled a list of integer halfband filters which, with increased length, have smaller amplitude distortions. Table 5.3 shows the coefficients of these halfband filters. To simplify the representation, the coefficients were noted with a center tap located at $d = 0$. F1 is the moving-average filter of length L, i.e., it is Hogenauer's CIC filter, and may therefore be used in the first stage also, to change the rate with a factor other than two. Fig. 5.27 shows the transfer function of the nine different filters. Note that in the logarithmic plot of Fig. 5.27, the point symmetry (as is usual for halfband filters) cannot be observed.

The basic idea of the Goodman and Carey multistage decimator design is that, in the first stages, filters with larger ripple and less complexity can be applied, because the passband-to-sampling frequency ratio is relatively small. As the passband-to-sampling frequency ratio increases, we must use filters with less distortion. The algorithm stops at $R = 2$. For the final decimation ($R = 2$ to $R = 1$), a longer halfband filter must be designed.

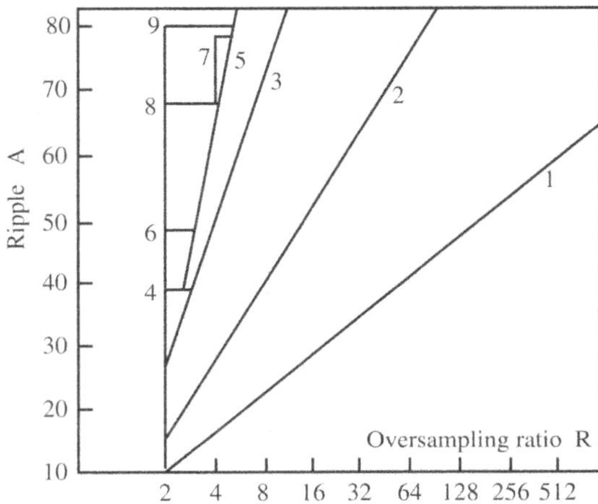

Fig. 5.28. Goodman and Carey design chart [66].

Goodman and Carey have provided the design chart shown in Fig. 5.28. Initially, the input oversampling factor R and the necessary attenuation in the passband and stopband $A = A_p = A_s$ must be computed. From this starting point, the necessary filters for $R, R/2, R/4, \ldots$ can be drawn as a horizontal line (at the same stopband attenuation). The filters F4 and F6-F9 have ripple in the passband (see Exercise 5.8, p. 207), and if several such filters are used it may be necessary to adjust ε_p. We may, therefore, consider the following adjustment

$$A = -20\log_{10}\varepsilon_p \quad \text{for} \quad \text{F1-F3, F5} \tag{5.38}$$

$$A = -20\log_{10}\min\left(\frac{\varepsilon_p}{S'}, \varepsilon_s\right) \quad \text{for} \quad \text{F4, F6-F9.} \tag{5.39}$$

where S' is the number of stages with ripple.

We will demonstrate the multistage design with the following example.

Example 5.8: Multistage Halfband Filter Decimator

We wish to develop a decimator with $R = 160, \varepsilon_p = 0.015$, and $\varepsilon_s = 0.031 =$ 30 dB, using the Goodman and Carey design approach.

At a first glance, we can conclude that we need a total of five filters and mark the starting point at $R = 160$ and 30 dB in Fig. 5.29(a). From 160 to 32, we use a CIC filter of length $L = 5$. This CIC filter is followed by two F2 filter and one F3 filter to reach $R = 8$. Now we need a filter with ripple. It follows that

$$A = -20\log_{10}\min\left(\frac{0.015}{1}, 0.031\right) = 36.48 \text{ dB}. \tag{5.40}$$

From Fig. 5.28, we conclude that for 36 dB the filter F4 is appropriate. We may now compute the whole filter transfer function $F(\omega)|$ by using the Noble relation (see Fig. 5.6, p. 147) $F(z) = \mathrm{F1}(z)\mathrm{F2}(z^5)\mathrm{F2}(z^{10})\mathrm{F3}(z^{20})\mathrm{F4}(z^{40})$, which is shown in Fig. 5.29(b). Fig. 5.29(a) shows the design algorithm, using the design chart from Fig. 5.28. 5.8

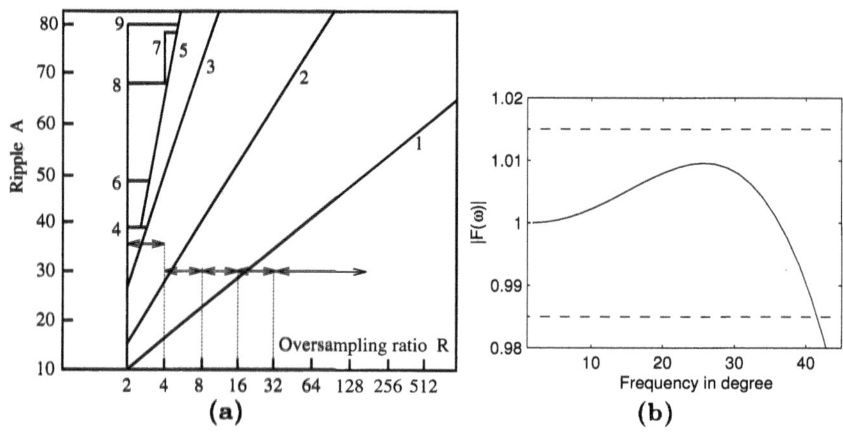

Fig. 5.29. Design example for Goodman and Carey halfband filter. **(a)** Design chart. **(b)** Transfer function $|F(\omega)|$.

Example 5.8 shows that considering only the filter with ripple in (5.39) was sufficient. Using a more pessimistic approach, with $S = 6$, we would have gotten $A = -20\log(0.015/6) = 52$ dB, and we would have needed filter F8, with essentially higher effort. It's therefore better to start with an optimistic assumption and possibly correct this later.

5.5 Frequency Sampling Filters as Bandpass Decimators

The CIC filters discussed in Sect. 5.3 (p. 155) belong to a larger class of systems called *frequency sampling filters* (FSFs). Frequency sampling filters can be used, as channelizer or decimating filter, to decompose the information spectrum into a set of discrete subbands, such as those found in multi-user communication systems. A classic FSF consists of a comb filter cascaded with a bank of frequency selective resonators [4, 53]. The resonators independently produce a collection of poles that selectively annihilate the zeros produced by the comb prefilter. Gain adjustments are applied to the output of the resonators to shape the resulting magnitude frequency response of the overall filter. An FSF can also be created by cascading all-pole filter sections with all-zero filter (comb) sections, as suggested in Fig. 5.30. The delay of the comb-section, $1 \pm z^{-D}$, is chosen so that its zeros cancel the poles of the all-pole prefilter as shown in Fig. 5.31. Wherever there is a complex pole, there is also an annihilating complex zero that results in an all-zero FIR, with the usual linear-phase and constant group-delay properties.

Frequency sampling filters are of interest to designers of multirate filter banks due, in part, to their intrinsic low complexity and linear-phase behavior. FSF designs rely on exact pole-zero annihilation and are often found in embedded applications. Exact FSF pole-zero annihilation, can be guaranteed by using polynomial filters defined over an *integer ring* using the two's complement or the residue number system (RNS). The poles of an FSF filter developed in this manner can reside on the periphery of the unit circle. This conditionally unstable location is acceptable, due to the guarantee of exact pole-zero cancellation. Without this guarantee, the designer would have to locate the poles of the resonators within the unit circle, with a loss in performance. In addition, by allowing the FSF poles and zeros to reside on the unit circle, a multiplier-less FSF can be created, with an attendant reduction in complexity and an increase in data bandwidth.

Consider the filter shown in Fig. 5.30. It can be shown that first-order filter sections (with integer coefficients) produce poles at angles of 0° and 180°. Second-order sections, with integer coefficients, can produce poles at angles of 60°, 90°, and 120°, according to the relationship $2\cos(2\pi K/D)$=1, 0, and −1. The frequency selectivity of higher-order sections is shown in Table 5.4. The angular frequencies for all polynomials having integer coefficients with roots on the unit circle, up to order six, are reported. The building blocks

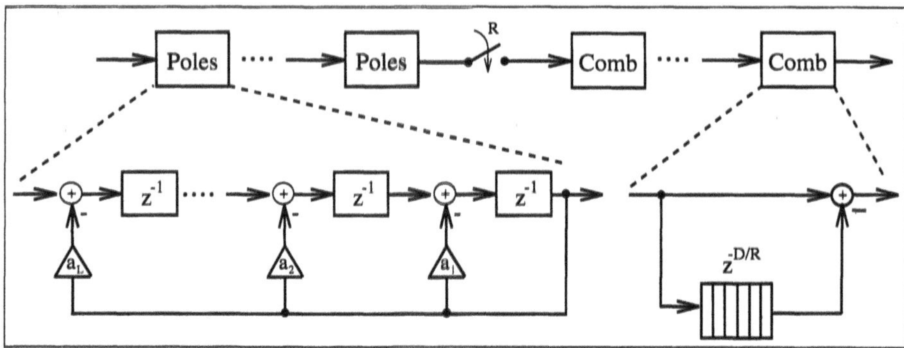

Fig. 5.30. Cascading of frequency sampling filters to save a factor of R delays for multirate signal processing [4, Sect. 3.4].

Table 5.4. Filters with integer coefficients producing unique angular pole locations up to order six. Shown are the filter coefficients and nonredundant angular locations of the roots on the unit circle.

$C_k(z)$	Order	a_0	a_1	a_2	a_3	a_4	a_5	a_6	θ_1	θ_2	θ_3
$-C_1(z)$	1	1	-1						0°		
$C_2(z)$	1	1	1						180°		
$C_6(z)$	2	1	-1	1					60°		
$C_4(z)$	2	1	0	1					90°		
$C_3(z)$	2	1	1	1					120°		
$C_{12}(z)$	4	1	0	-1	0	1			30°	150°	
$C_{10}(z)$	4	1	-1	1	-1	1			36°	108°	
$C_8(z)$	4	1	0	0	0	1			45°	135°	
$C_5(z)$	4	1	1	1	1	1			72°	144°	
$C_{16}(z)$	6	1	0	0	-1	0	0	1	20.00°	100.00°	140.00°
$C_{14}(z)$	6	1	-1	1	-1	1	-1	1	25.71°	77.14°	128.57°
$C_7(z)$	6	1	1	1	1	1	1	1	51.42°	102.86°	154.29°
$C_9(z)$	6	1	0	0	1	0	0	1	40.00°	80.00°	160.00°

listed in Table 5.4 can be used to efficiently design and implement such FSF filters. For example, a two's complement (i.e., RNS single modulus) filter bank was developed for use as a constant-Q speech processing filter bank. It covers a frequency range from 900 to 8,000 Hz [89, 90], using 16 KHz sampling frequency. An integer coefficient halfband filter HB6 [66] anti-aliasing filter and a third-order multiplier-free CIC filter (also known as Hogenauer filter [86] see Sect. 5.3, p. 155), was then added to the design to suppress unwanted frequency components, as shown in Fig. 5.32. The bandwidth of each resonator can be independently tuned by the number of stages and delays in the comb-section. The number of stages and delays is optimized to meet

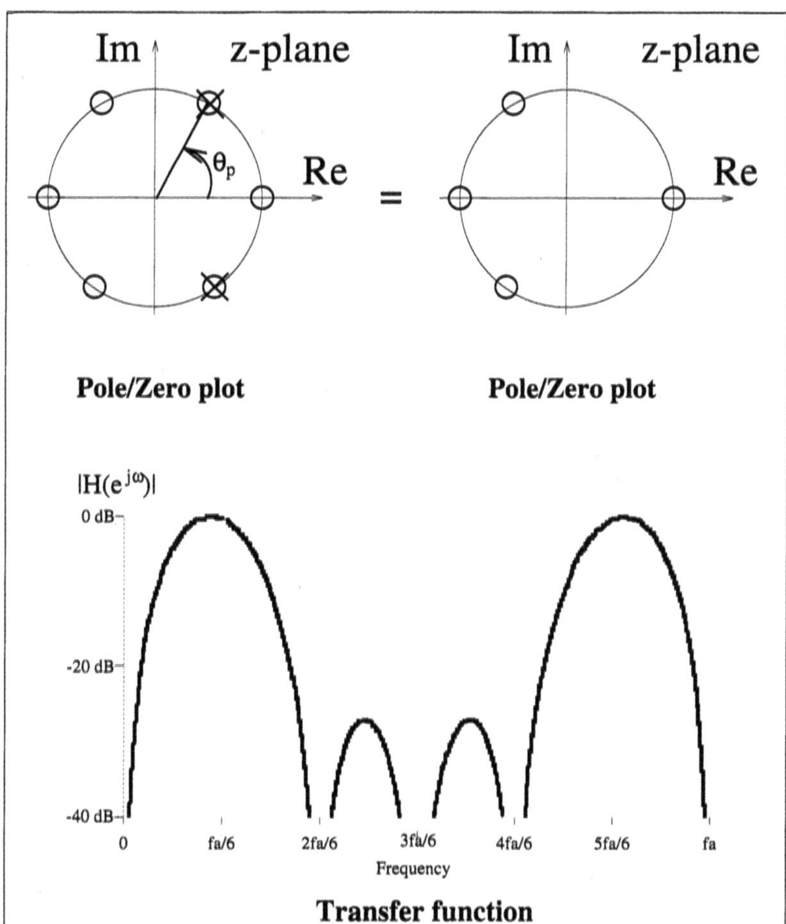

Fig. 5.31. Example of pole/zero-compensation for a pole-angle of 60° and Comb-delay $D = 6$.

the desired bandwidth requirements. All frequency-selective filters have two stages and delays.

The filter bank was prototyped using a Xilinx XC4000 FPGA with the complexity reported in Table 5.5. Using high-level design tools (XBLOCKS from Xilinx), the number of used CLBs was typically 20% higher than the theoretical prediction obtained by counting adders, flip-flops, ROMs and RAMs.

The design of an FSF can be manipulated by changing the comb delay, channel amplitude, or the number of sections. For example, adaptation of the comb delay can easily be achieved because the CLBs are used as 32×1 memory cells, and a counter realizes specific comb delays with the CLB used as a memory cell.

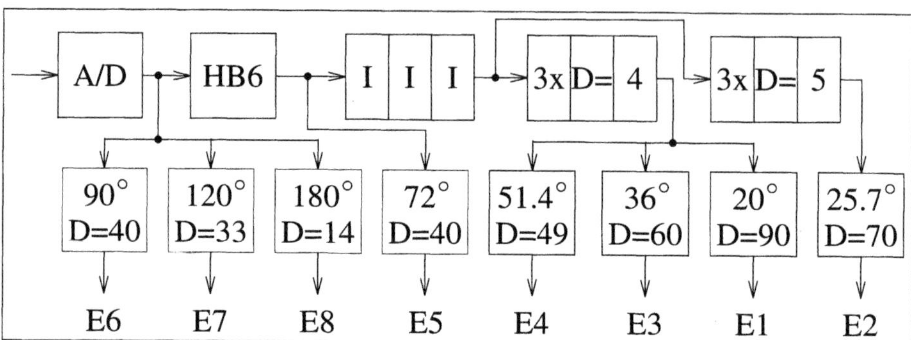

Fig. 5.32. Design of an filter bank consisting of a half-band and CIC prefilter and FSF comb-resonator sections.

Table 5.5. Number of used CLBs of Xilinx XC4000 FPGAs (Notation: F20D90 means filter pole-angle 20.00° delay Comb $D = 90$). Total: Actual 1572 CLBs, nonrecursive FIR: 11292 CLBs

	F20D90	F25D70	F36D60	F51D49	F72D40	F90D40
Theory	122	184	128	164	124	65
Practice	160	271	190	240	190	93
Nonre. FIR	2256	1836	1924	1140	1039	1287

	F120D33	F180D14	HB6	III	D4	D5
Theory	86	35	122	31	24	24
Practice	120	53	153	36	33	33
Nonre. FIR	1260	550				

5.6 Filter Banks

A *digital filter bank* is a collection of filters having a common input or output, as shown in Fig. 5.33. One common application of the *analysis filter bank* shown in 5.33(a) is spectrum analysis, i.e., to split the input signal into R different so-called subband signals. The combination of several signals into a common output signal, shown in Fig. 5.33(b), is called a *synthesis filter bank.* The analysis filter may be nonoverlapping, slightly overlapping, or substantially overlapping. Fig. 5.34 shows an example of a slightly overlapping filter bank, which is the most common case.

Another important characteristic that distinguishes different classes of filter banks is the bandwidth and spacing of the center frequencies of the filters. A popular example of a *nonuniform filter bank* is the octave-spaced or *wavelet filter bank*, which will be discussed in Sect. 5.7 (p. 197). In *uniform filter banks*, all filters have the same bandwidth and sampling rates. From the implementation standpoint, uniform, maximal decimating filter banks

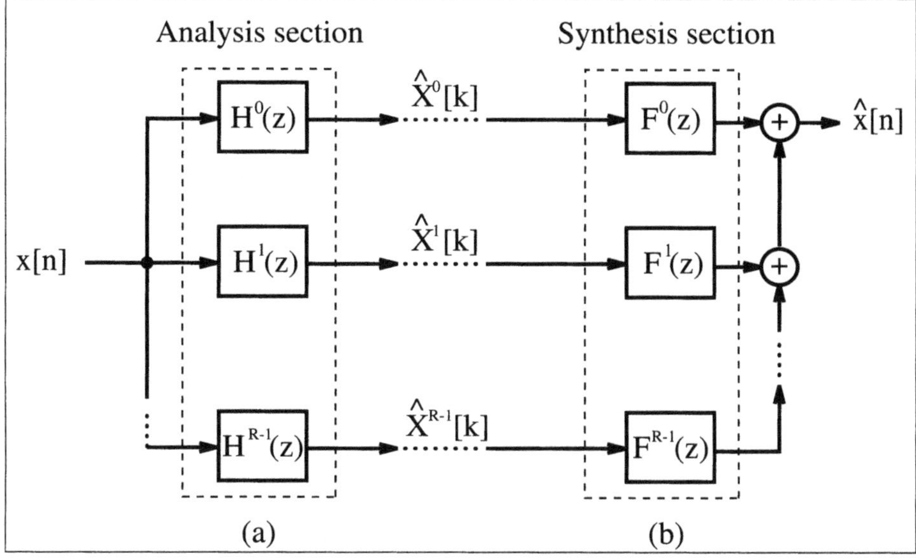

Fig. 5.33. Typical filter bank decomposition system showing **(a)** analysis, and **(b)** synthesis filters.

are often preferred, because they can be realized with the help of an FFT algorithm, as shown in the next section.

5.6.1 Uniform DFT Filter Bank

In a maximal decimating, or critically sampled filter bank, the decimation or interpolation R is equal to the number of bands K. We call it a *DFT filter bank* if the rth band filter $h^r[n]$ is computed from the "modulation" of a single prototype filter $h[n]$, according to

$$h^r[n] = h[n]W_R^{rn} = h[n]e^{-j2\pi rn/R} \tag{5.41}$$

An efficient implementation of the R channel filter bank can be generated if we use polyphase decomposition (see Sect. 5.2, p. 147) of the filter $h^r[n]$

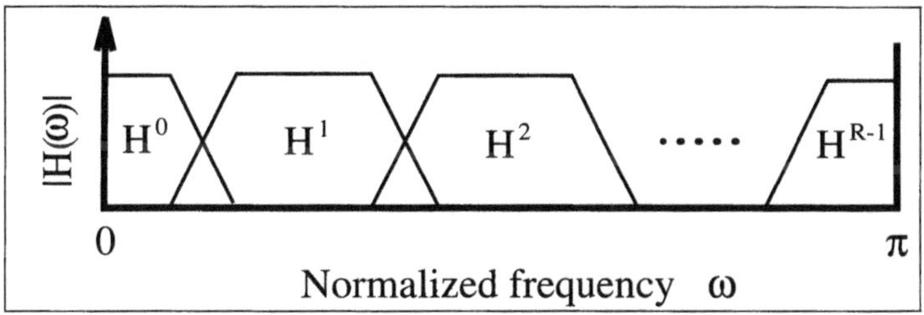

Fig. 5.34. R channel filter bank, with slight amount of overlapping.

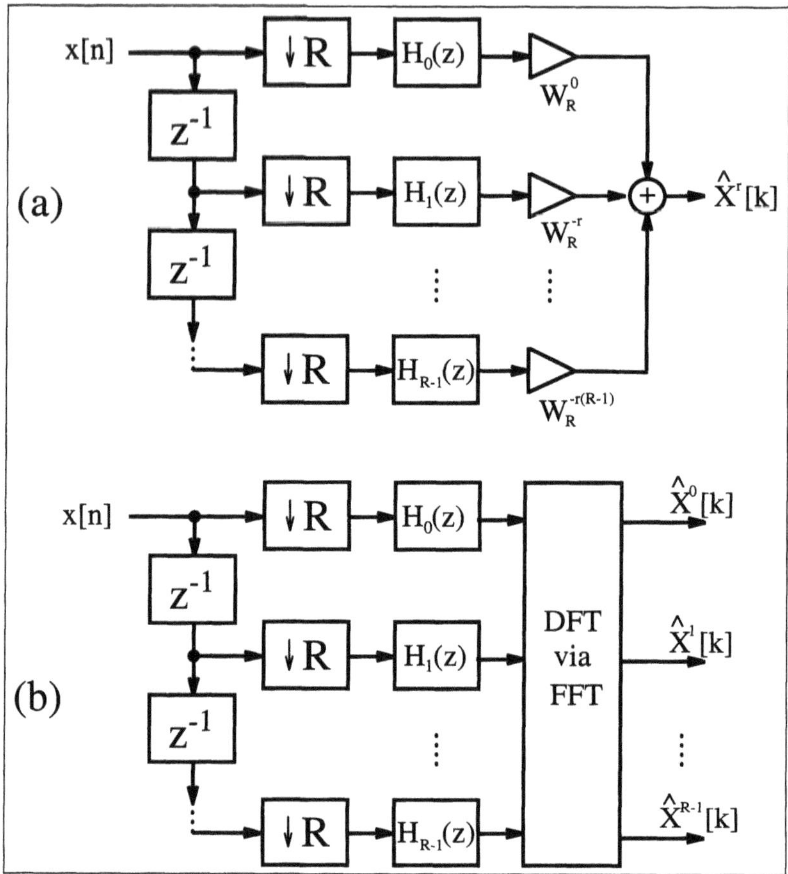

Fig. 5.35. **(a)** Analysis DFT filter bank for channel k. **(b)** Complete analysis DFT filter bank.

and the input signal $x[n]$. Because each of these bandpass filters is critically sampled, we use a decomposition with R polyphase signals according to

$$h[n] = \sum_{k=0}^{R-1} h_k[n] \leftrightarrow h_k[m] = h[mR-k] \tag{5.42}$$

$$x[n] = \sum_{k=0}^{R-1} x_k[n] \leftrightarrow x_k[m] = x[mR-k] \tag{5.43}$$

If we now substitute (5.42) in (5.41), we find that all bandpass filters $h^r[n]$ share the same polyphase filter $h_k[n]$, while the "twiddle factors" for each filter are different. This structure is shown in Fig. 5.35(a) for the rth filter $h^r[n]$. It is now obvious that this "twiddle multiplication" for $h^r[n]$ corresponds to the rth DFT component, with an input vector of $\hat{x}_0[n], \hat{x}_1[n], \ldots, \hat{x}_{R-1}[n]$. The computation for the whole analysis band can be reduced to filtering with R polyphase filters, followed by a DFT (or FFT) of these R filtered components, as shown in Fig. 5.35(b). This is obviously much more efficient than direct computation using the filter defined in (5.41) (see Exercise 5.6, p. 206).

The polyphase filter bank for the uniform DFT *synthesis bank* can be developed as an inverse operation to the analysis bank, i.e., we can use the R spectral components $\hat{X}^r[k]$ as input for the inverse DFT (or FFT), and reconstruct the output signal using a polyphase interpolator structure, shown in Fig. 5.36. The reconstruction bandpass filter becomes

$$f^r[n] = \frac{1}{R} f[n] W_R^{-rn} = f[n] e^{j2\pi rn/R}. \tag{5.44}$$

If we now combine the analysis and synthesis filter banks, we can see that the DFT and IDFT annihilate each other, and *perfect reconstruction* occurs if the convolution of the included polyphase filter gives an unit sample function, i.e.,

$$h_r[n] * f_r[n] = \begin{cases} 1 & n = d \\ 0 & \text{else.} \end{cases} \tag{5.45}$$

In other words, the two polyphase functions must be inverse filters of each other, i.e.,

$$H_r(z) \cdot F_r(z) = z^{-d}$$

$$F_r(z) = \frac{z^{-d}}{H_r(z)}.$$

where we allow a delay d in order to have causal (realizable) filters. In a practical design, these ideal conditions cannot be met exactly by two FIR filters. We can use approximation for the two FIR filters, or we can combine an FIR and IIR, as shown in the following example.

Example 5.9: DFT Filter Bank

The *lossy integrator* studied in Example 4.3 (p. 131) should be interpreted in the context of a DFT filter bank with $R = 2$. The difference equation was

$$y[n+1] = \frac{3}{4} y[n] + x[n]. \tag{5.46}$$

The impulse response of this filter in the z-domain is

$$F(z) = \frac{z^{-1}}{1 - 0.75 z^{-1}} \tag{5.47}$$

In order to get two polyphase filters, we use a similar scheme as for the "scattered look-ahead" modification (see Example 4.5, p. 134), i.e., we introduce an additional pole/zero pair at the mirror position. Multiplying nominator and denominator with $(1 + 0.75z^{-1})$ yields

$$F(z) = \underbrace{\frac{0.75 z^{-2}}{1 - 0.75^2 z^{-2}}}_{H_0(z^2)} + z^{-1} \underbrace{\frac{1}{1 - 0.75^2 z^{-2}}}_{H_1(z^2)} \tag{5.48}$$

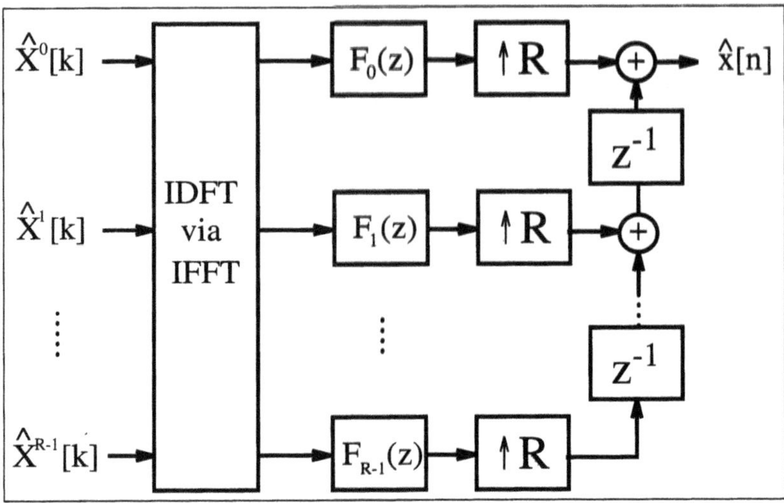

Fig. 5.36. DFT synthesis filter bank.

$$= H_0\left(z^2\right) + z^{-1}H_1\left(z^2\right) \tag{5.49}$$

which gives the two polyphase filters:

$$H_0(z) = \frac{0.75z^{-1}}{1-0.75^2z^{-1}} = 0.75z^{-1} + 0.4219z^{-2} + 0.2373z^{-3} + \ldots \tag{5.50}$$

$$H_1(z) = \frac{1}{1-0.75^2z^{-1}} = 1 + 0.5625z^{-1} + 0.3164z^{-2} + \ldots \tag{5.51}$$

We can approximate these impulse responses with a nonrecursive FIR, but to get less than 1% error we must use about 16 coefficients. It is therefore much more efficient if we use the two recursive polyphase IIR filters defined by (5.50) and (5.51). After decomposition with the polyphase filters, we then apply a 2-point DFT, which is given by

$$\boldsymbol{W} = \begin{bmatrix} 1 & 1 \\ 1 & -1 \end{bmatrix}$$

The whole analysis filter bank can now be constructed as shown in 5.37(a). For the synthesis bank, we first compute the inverse DFT using

$$\boldsymbol{W}^{-1} = \frac{1}{2}\begin{bmatrix} 1 & 1 \\ 1 & -1 \end{bmatrix}.$$

In order to get a perfect reconstruction we must find the inverse polyphase filter to $h_0[n]$ and $h_1[n]$. This is not difficult, because the $H_r(z)'s$ are single-pole IIR filters, and $F_r(z) = z^{-d}/H_r(z)$ must therefore be two-tap FIR filters. Using (5.50) and (5.51), we find that $d = 1$ is already sufficient to get causal filters, and it is

$$F_0[n] = \frac{4}{3}\left(1 - 0.75^2z^{-1}\right) \tag{5.52}$$

$$F_1[n] = z^{-1} - 0.75^2z^{-2}. \tag{5.53}$$

The synthesis bank is graphically interpreted in Fig. 5.37(b).

5.9

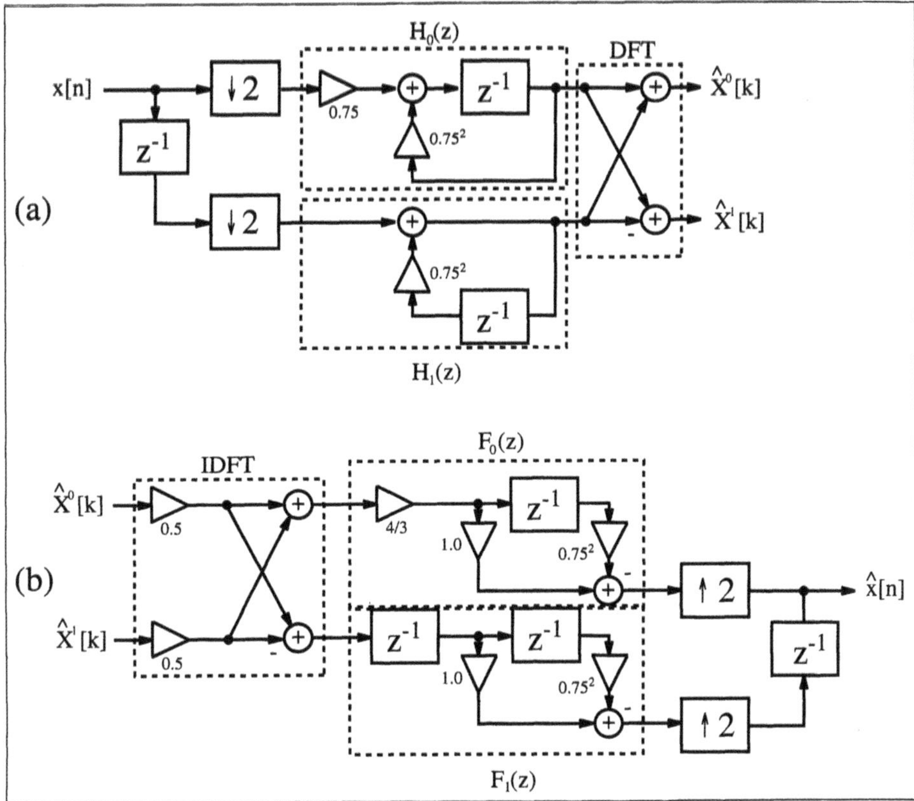

Fig. 5.37. Critically sampled uniform DFT filter bank for $R = 2$. **(a)** Analysis filter bank. **(b)** Synthesis filter bank.

5.6.2 Two-Channel Filter Banks

Two-channel filter banks are an important tool for the design of general filter banks and wavelets. Fig. 5.38 shows an example of a two-channel filter bank that splits the input $x[n]$ using lowpass ($G(z)$) and highpass ($H(z)$) "analysis" filters. The resulting signal $\hat{x}[n]$ is reconstructed using low and highpass "synthesis" filters. Between the analysis and synthesis sections are decimation and interpolation by 2 units. The signal between the decimators and interpolators is often quantized, and nonlinearly processed for enhancement, or compressed.

It is common practice to define only the lowpass filter $G(z)$, and to use its definition to specify the highpass filter $H(z)$. The construction rule is normally given by

$$h[n] = (-1)^n g[n] \quad \circ\!\!-\!\!\bullet \quad H(z) = G(-z). \tag{5.54}$$

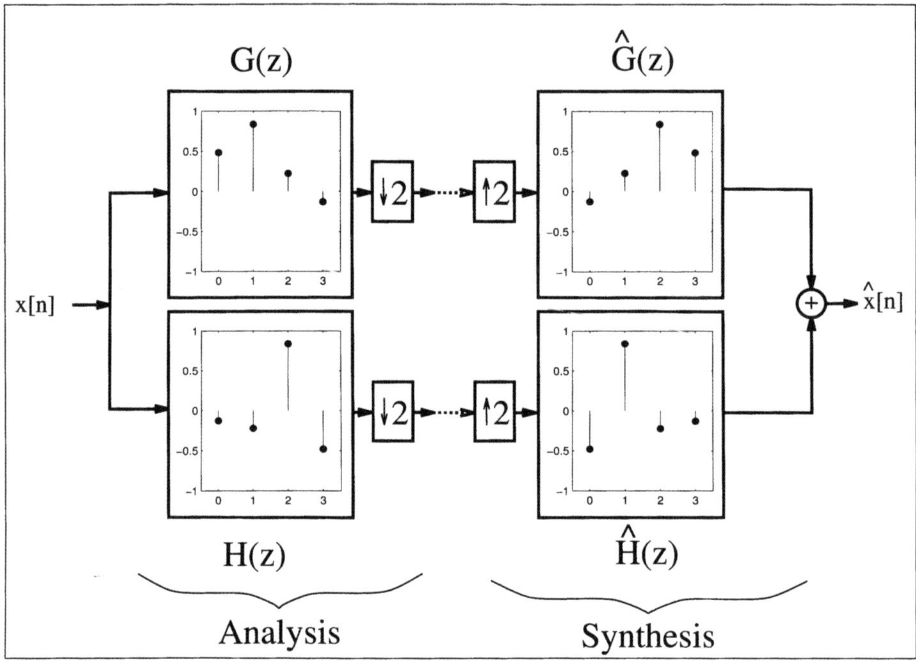

Fig. 5.38. Two-channel filter bank using Daubechies filter of length-4.

which defines the filters to be mirrored pairs. Specifically, in the frequency domain, $|H(e^{j\omega})| = |G(e^{j(\omega-\pi)})|$. This is a *Quadrature Mirror Filter (QMF)* bank, because the two filters have mirror symmetry to $\pi/2$.

For the synthesis shown in Fig. 5.38, we first use an expander (a sampling rate increase of 2), and then two separate reconstruction filters, $\hat{G}(z)$ and $\hat{H}(z)$, to reconstruct $\hat{x}[n]$. A challenging question now is, can the input signal be perfectly reconstructed, i.e., can we satisfy

$$\hat{x}[n] = x[n-d]? \tag{5.55}$$

That is, a perfectly reconstructed signal has the same shape as the original, up to a phase (time) shift. Because $G(z)$ and $H(z)$ are not ideal rectangular filters, achieving perfect reconstruction is not a trivial problem. Both filters produce essential aliasing components after the downsampling by 2, as shown in Fig. 5.38. The simple orthogonal filter bank which satisfies (5.55) is attributed to Alfred Haar (circa 1910) [91].

Example 5.10: Two-Channel Haar Filter Bank I

The filter transfer functions of the two-channel QMF filter bank from Fig. 5.39 are[8]

[8] Sometimes the amplitude factors are chosen in such a way that *orthonormal* filters are obtained, i.e., $\sum_n |h[n]|^2 = 1$. In this case, the filters have an amplitude factor of $1/\sqrt{2}$. This will complicate a hardware design significantly.

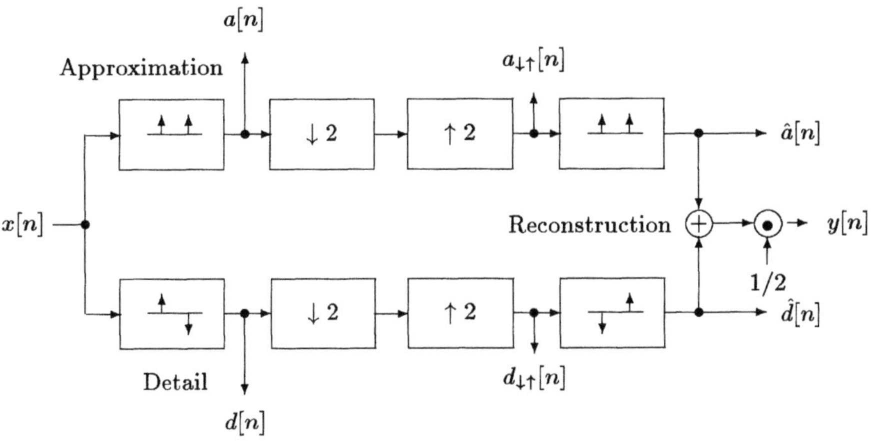

	Time step n				
	0	1	2	3	4
$x[n]$	$x[0]$	$x[1]$	$x[2]$	$x[3]$	$x[4]$
$a[n]$	$x[0]$	$x[0]+x[1]$	$x[1]+x[2]$	$x[2]+x[3]$	$x[3]+x[4]$
$d[n]$	$x[0]$	$x[1]-x[0]$	$x[2]-x[1]$	$x[3]-x[2]$	$x[4]-x[3]$
$a_{\downarrow\uparrow}[n]$	0	$x[0]+x[1]$	0	$x[2]+x[3]$	0
$d_{\downarrow\uparrow}[n]$	0	$x[1]-x[0]$	0	$x[3]-x[2]$	0
$\hat{a}[n]$	0	$x[0]+x[1]$	$x[0]+x[1]$	$x[2]+x[3]$	$x[2]+x[3]$
$\hat{d}[n]$	0	$x[0]-x[1]$	$x[1]-x[0]$	$x[2]-x[3]$	$x[3]-x[2]$
$\hat{x}[n]$	0	$x[0]$	$x[1]$	$x[2]$	$x[3]$

Fig. 5.39. Two-channel Haar-QMF bank (©1999 Springer Press [5]).

$$G(z) = 1 + z^{-1} \qquad H(z) = 1 - z^{-1}$$
$$\hat{G}(z) = \frac{1}{2}(1 + z^{-1}) \qquad \hat{H}(z) = \frac{1}{2}(-1 + z^{-1}).$$

Using data found in the table in Fig. 5.39, it can be verified that the filter produces a perfect reconstruction of the input. The input sequence $x[0], x[1], x[2], \ldots$, processed by $G(z)$ and $H(z)$, yields the sum $x[n]+x[n-1]$ and difference $x[n]-x[n-1]$, respectively. The downsampling followed by upsampling forces every second value to zero. After applying the synthesis filter and combining the output we again get the input sequence delayed by one, i.e., $\hat{x}[n] = x[n-1]$, a *perfect* reconstruction with $d = 1$. 5.10

In the following we will discuss the general relationships the four filters must obey to get a perfect reconstruction. It is useful to remember that decimation and interpolation by 2 of a signal $s[k]$ is equivalent to multiplying $S(z)$ by the sequence $\{1, 0, 1, 0, \ldots,\}$. This translates, in the z-domain, to

$$S_{\downarrow\uparrow}(z) = \frac{1}{2}\left(S(z) + S(-z)\right). \tag{5.56}$$

If this signal is applied to the two-channel filter bank, the lowpass path $X_{\downarrow\uparrow\mathrm{G}}(z)$ and highpass path $X_{\downarrow\uparrow\mathrm{H}}(z)$ become

$$X_{\downarrow\uparrow\mathrm{G}}(z) = \frac{1}{2}\left(X(z)G(z) + X(-z)G(-z)\right), \tag{5.57}$$

$$X_{\downarrow\uparrow\mathrm{H}}(z) = \frac{1}{2}\left(X(z)H(z) + X(-z)H(-z)\right). \tag{5.58}$$

After multiplication with the synthesis filter $\hat{G}(z)$ and $\hat{H}(z)$, and summation of the results, we get $\hat{X}(z)$ as

$$\begin{aligned} \hat{X}(z) &= X_{\downarrow\uparrow\mathrm{G}}(z)\hat{G}(z) + X_{\downarrow\uparrow\mathrm{H}}(z)\hat{H}(z) \\ &= \frac{1}{2}\left(G(z)\hat{G}(z) + H(z)\hat{H}(z)\right) X(z) \\ &\quad + \frac{1}{2}\left(G(-z)\hat{G}(z) + H(-z)\hat{H}(z)\right) X(-z). \end{aligned} \tag{5.59}$$

The factor of $X(-z)$ shows the aliasing component, while the term at $X(z)$ shows the amplitude distortion. For a *perfect* reconstruction this translates into the following:

> **Theorem 5.11: Perfect Reconstruction**
>
> A perfect reconstruction for a two-channel filter bank, as shown in Fig. 5.38, is achieved if
>
> **1)** $G(-z)\hat{G}(z) + H(-z)\hat{H}(z) = 0$, i.e., the reconstruction is free of aliasing.
>
> **2)** $G(z)\hat{G}(z) + H(z)\hat{H}(z) = 2z^{-d}$, i.e., the amplitude distortion has amplitude one.

Let us check this condition for the Haar filter bank.

Example 5.12: Two-Channel Haar Filter bank II

The filters of the two-channel Haar QMF bank were defined by

$$G(z) = 1 + z^{-1} \quad H(z) = 1 - z^{-1}$$

$$\hat{G}(z) = \frac{1}{2}(1 + z^{-1}) \quad \hat{H}(z) = \frac{1}{2}(-1 + z^{-1}).$$

The two conditions from Theorem 5.11 can be proved with:

1) $G(-z)\hat{G}(z) + H(-z)\hat{H}(z)$

$$= \frac{1}{2}((1 - z^{-1})(1 + z^{-1}) + \frac{1}{2}((1 + z^{-1})(-1 + z^{-1})$$

$$= \frac{1}{2}((1 - z^{-2}) + \frac{1}{2}((-1 + z^{-2}) = 0 \quad \checkmark$$

2) $G(z)\hat{G}(z) + H(z)\hat{H}(z)$

$$= \frac{1}{2}(1 + z^{-1})^2 + \frac{1}{2}(1 - z^{-1})(-1 + z^{-1})$$

$$= \frac{1}{2}\left((1 + 2z^{-1} + z^{-2}) + (-1 + 2z^{-1} - z^{-2})\right) = 2z^{-1} \quad \checkmark$$

5.12

For the proof using Theorem 5.11, it can be noted that the perfect reconstruction condition does not change if we switch the analysis and synthesis filters.

In the following we will discuss some restrictions that can be made in the filter design to fulfill the condition from 5.11 more easily.

First, we limit the filter choice by using the following:

Theorem 5.13: Aliasing-Free Two-Channel Filter Bank

A two-channel filter bank is *aliasing-free* if

$$G(-z) = -\hat{H}(z) \quad \text{and} \quad H(-z) = \hat{G}(z). \tag{5.60}$$

This can be checked if we use (5.60) for the first condition of Theorem 5.11. Using a length-4 filter, these two conditions can be interpreted as follows:

$$\begin{aligned} g[n] &= \{g[0], g[1], g[2], g[3]\} \to \hat{h}[n] = \{-g[0], g[1], -g[2], g[3]\} \\ h[n] &= \{h[0], h[1], h[2], h[3]\} \to \hat{g}[n] = \{h[0], -h[1], h[2], -h[3]\}. \end{aligned}$$

With the restriction of the filters as in Theorem 5.13, we can now simplify the second condition in Theorem 5.11. It is useful to define first an auxiliary *product filter* $F(z) = G(z)\hat{G}(z)$. The second condition from Theorem 5.11 becomes

$$G(z)\hat{G}(z) + H(-z)\hat{H}(-z) = F(z) + \hat{G}(-z)G(-z) = F(z) - F(-z) \tag{5.61}$$

and we finally get

$$\boxed{F(z) - F(-z) = 2z^{-d}.} \tag{5.62}$$

i.e., the product filter must be a *halfband filter*.[9] The construction of a perfect reconstruction filter bank uses the following three simple steps:

Algorithm 5.14: Perfect-Reconstruction Two-Channel Filter Bank

1) Define a normalized halfband filter according to Equation (5.62).
2) Factor the filter $F(z)$ in $F(z) = G(z)\hat{G}(z)$.
3) Compute $H(z)$ and $\hat{H}(z)$ using Equations (5.60), i.e., $\hat{H}(z) = -G(-z)$ and $H(z) = \hat{G}(-z)$.

We wish to demonstrate Algorithm 5.14 with the following example. To simplify the notation we will, in the following example, write a combination of a length L filter for $G(z)$, and length N for $\hat{G}(z)$, as an L/N filter.

Example 5.15: Perfect-Reconstructing Filter Bank Using F3

The (normalized) halfband filter F3 (Table 5.3, p. 172) of length 7 has the following z-domain transfer function

$$\mathrm{F3}(z) = \frac{1}{16}\left(-1 + 9z^{-2} + 16z^{-3} + 9z^{-4} - z^{-6}\right) \tag{5.63}$$

[9] For the definition of a halfband filter, see p. 172.

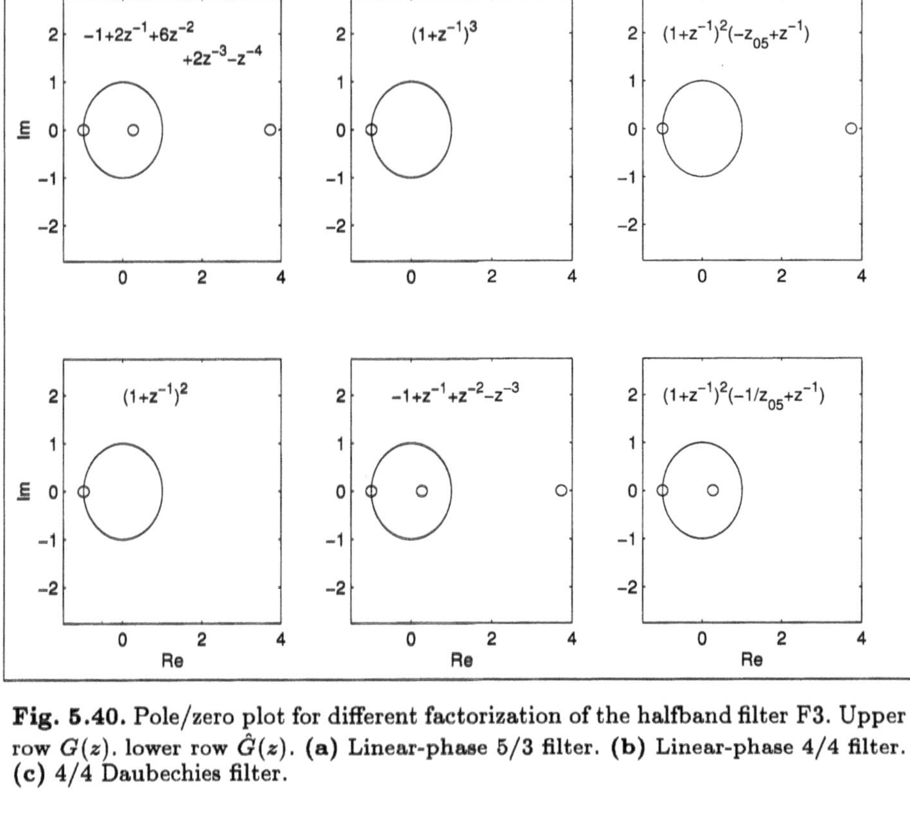

Fig. 5.40. Pole/zero plot for different factorization of the halfband filter F3. Upper row $G(z)$. lower row $\hat{G}(z)$. **(a)** Linear-phase 5/3 filter. **(b)** Linear-phase 4/4 filter. **(c)** 4/4 Daubechies filter.

The zeros of the transfer function are at $z_{01-4} = -1$, $z_{05} = 2+\sqrt{3} = 3.7321$, and $z_{06} = 2-\sqrt{3} = 0.2679 = 1/z_{05}$. There are different choices for factoring $F(z) = G(z)\hat{G}(z)$. A 5/3 filter is, for instance,
a) $G(z) = (-1+2z^{-1}+6z^{-2}+2z^{-3}-z^{-4})/8$ and $\hat{G}(z) = (1+2z^{-1}+z^{-2})/2$.
We may design a 4/4 filter as:
b) $G(z) = \frac{1}{4}(1+z^{-1})^3$ and $\hat{G}(z) = \frac{1}{4}(-1+3z^{-1}+3z^{-2}-z^{-3})$.
Another configuration of the 4/4 configuration uses the Daubechies filter configuration, which is often found in wavelet applications and has the form:
c) $G(z) = \frac{1-\sqrt{3}}{4\sqrt{2}}(1+z^{-1})^2(-z_{05}+z^{-1})$ and $\hat{G}(z) = -\frac{1+\sqrt{3}}{4\sqrt{2}}(1+z^{-1})^2(-z_{06}+z^{-1})$.
Fig. 5.40 shows these three combinations, along with their pole/zero plots.

5.15

For the Daubechies filter, the condition $H(z) = -z^{-N}G(-z^{-1})$ holds in addition, i.e., high and lowpass polynoms are mirror versions of each other. This is a typical behavior in *orthogonal* filter banks.

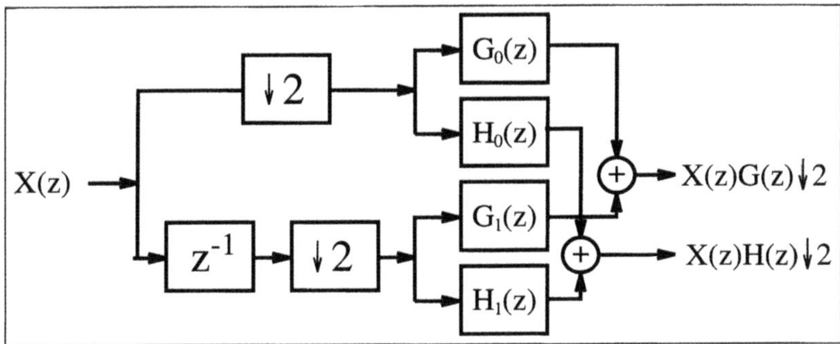

Fig. 5.41. Polyphase implementation of the two-channel filter bank.

From the pole/zero plots shown in Fig. 5.40, for $F(z) = G(z)\hat{G}(z)$ the following conclusions can be made:

Corollary 5.16: Factorization of a Halfband Filter

1) To construct a *real* filter, we must always group the conjugate symmetric zeros at (z_0 and z_0^*) in the same filter.

2) For *linear-phase* filters, the pole/zero plot must be symmetrical to the unit circle ($z = 1$). Zero pairs at (z_0 and $1/z_0$) must be assigned to the *same* filter.

3) To have *orthogonal* filters which are mirror polynoms of each other, ($F(z) = U(z)U(z^{-1})$), all pairs z_0 and $1/z_0$ must be assigned to *different* filters.

We notice that some of the above conditions can *not* be fulfilled at the same time. In particular, rules 2 and 3 represent a contradiction. Orthogonal, linear-phase filters are in general not possible, except when all zeros are on the unit circle, as in the case of the Haar filter bank.

If we classify the filter banks from Example 5.15, we find that configurations (a) and (b) are real linear-phase filters, while (c) is an real orthogonal filter.

Implementing Two-Channel Filter Banks

We will now discuss different options for implementing two-channel filter banks. We will first discuss the general case, and then special simplifications which are possible if the filters are QMF, linear-phase, or orthogonal. We will only discuss the analysis filter bank, as synthesis may be achieved with graph transposition.

Polyphase two-channel filter banks. In the general case, with two filters $G(z)$ and $H(z)$, we can realize each filter as a polyphase filter

$$H(z) = H_0(z^2) + z^{-1}H_1(z^2) \qquad G(z) = G_0(z^2) + z^{-1}G_1(z^2) \tag{5.64}$$

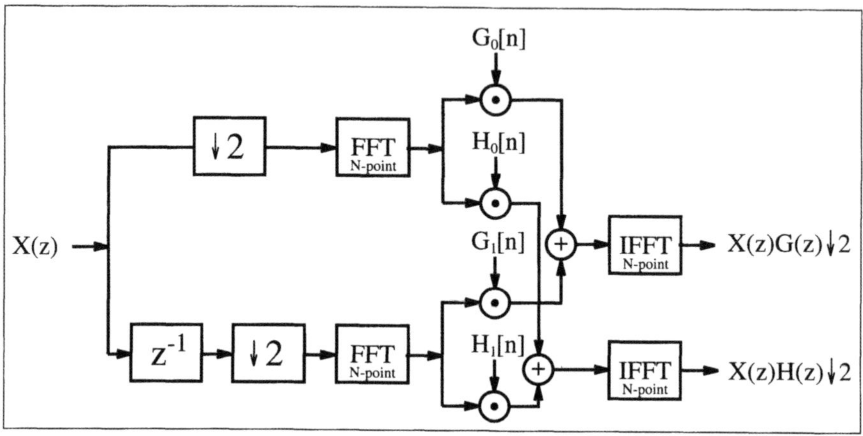

Fig. 5.42. Two-channel filter bank with polyphase decomposition and fast convolution using the FFT (©1999 Springer Press [5]).

which is shown in Fig. 5.41. This does not reduce the hardware effort ($2L$ multipliers and $2(L-1)$ adders are still used), but the design can be run with twice the usual sampling frequency, $2f_s$.

These four polyphase filters have only half the length of the original filters. We may implement these length $L/2$ filters directly or with one of the following methods:

1) Run-length filter using short Winograd convolution algorithms [84], discussed in Sect. 5.2.2, p. 152.
2) Fast convolution using FFT (discussed in Chapter 6) or NTTs (discussed in Chapter 7).
3) Using advanced arithmetic concepts discussed in Chapter 3, such as distribute arithmetic, reduced adder graph, or residue number system.

Using the fast convolution FFT/NTT techniques has the additional benefit that the forward transform for each polyphase filter need only be done once, and also, the inverse transform can be applied to the spectral sum of the two components, as shown in Fig. 5.42. But in general FFT methods only give improvements for longer filters, typically, larger than 32; however, the typical two-channel filter length is less than 32.

Lifting. Another general approach to constructing fast and efficient two-channel filter banks is the *lifting* scheme introduced recently by Swelden and Herley [92, 93]. The basic idea is the use of cross terms (called lifting and dual lifting), as in a lattice filter, to construct a longer filter from a short filter, while preserving the perfect reconstruction conditions. The basic structure is shown in Fig. 5.43.

Designing a lifting scheme typically starts with the "lazy filter bank," with $G(z) = \hat{H}(z) = 1$ and $H(z) = \hat{G}(z) = z^{-1}$. This channel bank fulfills

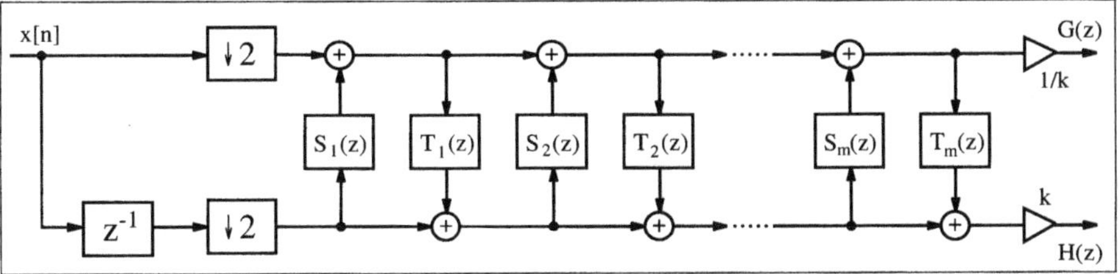

Fig. 5.43. Two-channel filter implementation using lifting and dual lifting steps.

both conditions from Theorem 5.11 (p. 186), i.e., it is a perfect reconstruction filter bank. The following question arises: if we keep one filter fixed, what are filters $S(z)$ and $T(z)$ such that the filter bank is still a perfect reconstruction? The answer is important, and not trivial:

$$\text{Lifting:} \quad H'(z) = H(z) + G(-z)S(z^2) \quad \text{for any } S(z^2). \tag{5.65}$$

$$\text{Dual Lifting:} \quad G'(z) = G(z) + H(-z)T(z^2) \quad \text{for any } T(z^2). \tag{5.66}$$

To check, if we substitute the lifting equation into the perfect reconstruction condition from Theorem 5.11 (p. 186), and we see that both conditions are fulfilled if $\hat{G}(z)$ and $\hat{H}(z)$ still meet the conditions of Theorem 5.13 (p. 187) for the aliasing free filter bank (Exercise 5.9, p. 207).

The conversion of the Daubechies length-4 filter bank into lifting steps demonstrates the design.

Example 5.17: Lifting Implementation of the DB4 Filter

One filter configuration in Example 5.15 (p. 187) was the Daubechies length-4 filter [94, p. 195]. The filter coefficients were

$$G(z) =$$
$$\left((1+\sqrt{3}) + (3+\sqrt{3})z^{-1} + (3-\sqrt{3})z^{-2} + (1-\sqrt{3})z^{-3}\right)\frac{1}{4\sqrt{2}}$$
$$H(z) =$$
$$\left(-(1-\sqrt{3}) + (3-\sqrt{3})z^{-1} - (3+\sqrt{3})z^{-2} + (1+\sqrt{3})z^{-3}\right)\frac{1}{4\sqrt{2}}$$

A possible implementation uses two lifting steps and one dual-lifting step. The differential equations that produce a two-channel filter bank based on to the above equation are

$$h_1[n] = x[2n+1] - \sqrt{3}x[2n]$$
$$g_1[n] = x[2n] + \frac{\sqrt{3}}{4}h_1[n] + \frac{\sqrt{3}-2}{4}h_1[n-1]$$
$$h_2[n] = h_1[n] + g_1[n+1]$$
$$g[n] = \frac{\sqrt{3}+1}{\sqrt{2}}g_1[n]$$
$$h[n] = \frac{\sqrt{3}-1}{\sqrt{2}}h_1[n]$$

Notice that the early decimation and splitting of the input into even $x[2n]$ and odd $x[2n-1]$ sequences allows the filter to run with $2f_s$. This structure can be

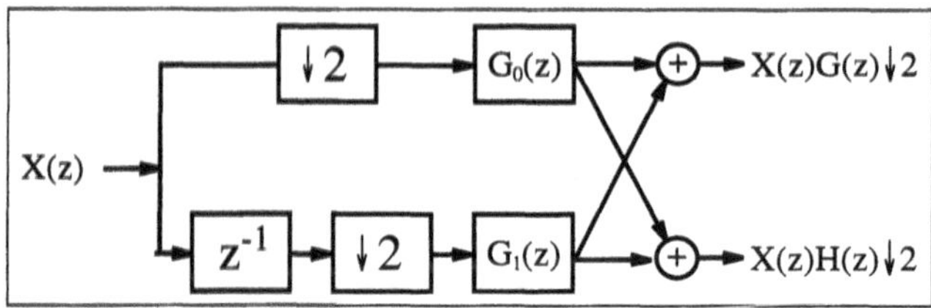

Fig. 5.44. Polyphase realization of the two-channel QMF bank (©1999 Springer Press [5]).

directly translated into hardware and can be implemented using MaxPlusII (Exercise 5.10, p. 207). The implementation will use five multiplications and four adders. The reconstruction filter bank can be constructed based on graph transposition, which is, in the case of the differential equations, a reversing of the operations and flipping of the signs. 5.17

Daubechies et al. [95], have shown that *any* (bi)orthogonal wavelet filter bank can be converted into a sequence of lifting and dual lifting steps. The number of multipliers and adders required then depends on the number of lifting steps (more steps gives less complexity) and can reach up to 50% compared with the direct polyphase implementation. This approach seems especially promising if the bit width of the multiplier is small [96]. On the other hand, the lattice-like structure does not allow use of reduced adder graph (RAG) techniques, and for longer filters the direct polyphase approach will often be more efficient.

Although the techniques (polyphase decomposition and lifting) discussed so far improve speed or size and cover all types of two-channel filters, additional savings can be achieved if the filters are QMF, linear-phase, or orthogonal. This will be discussed in the following.

QMF implementation. For QMF [97] we have found that according to (5.54),

$$h[n] = (-1)^n g[n] \quad \circ\!\!-\!\!\bullet \quad H(z) = G(-z). \tag{5.67}$$

But this implies that the polyphase filters are the same (except the sign), i.e.,

$$G_0(z) = H_0(z) \quad G_1(z) = -H_1(z). \tag{5.68}$$

Instead of the four filters from Fig. 5.41, for QMF we only need two filters and an additional "Butterfly," as shown in Fig. 5.44. This saves about 50%. For the QMF filter we need:

$$L \text{ real adders} \qquad L \text{ real multipliers,} \tag{5.69}$$

and the filter can run with twice the usual input-sampling rate.

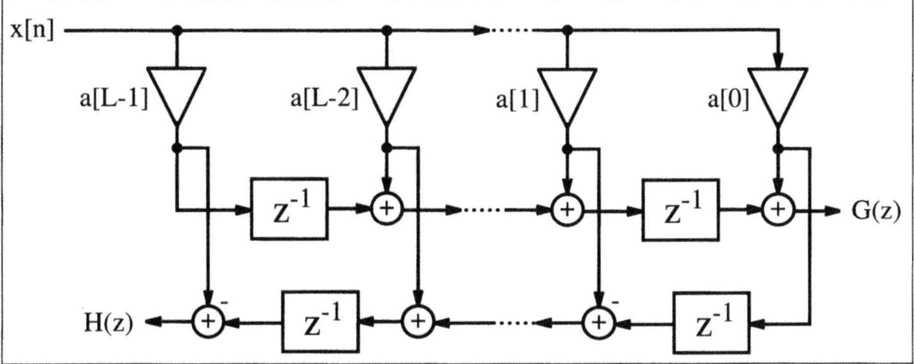

Fig. 5.45. Orthogonal two-channel filter bank using the transposed FIR structure.

Orthogonal filter banks. An orthogonal filter pair[10] obeys the conjugate mirror filter (CQF) [98] condition, defined by

$$H(z) = z^{-N}\, G(-z^{-1}). \tag{5.70}$$

If we use the transposed FIR filter shown in Fig. 5.45, we need only half the number of multipliers. The disadvantage is that we can *not* benefit from polyphase decomposition to double the speed.

Another alternative is realization of the CQF bank using the lattice filter shown in Fig. 5.46. The following example demonstrates the conversion of the direct FIR filter into a lattice filter.

Example 5.18: Lattice Daubechies $L = 4$ Filter Implementation

One filter configuration in Example 5.15 (p. 187) was the Daubechies length-4 filter [94, p. 195]. The filter coefficients were

$$\begin{aligned} G(z) &= \frac{(1+\sqrt{3}) + (3+\sqrt{3})z^{-1} + (3-\sqrt{3})z^{-2} + (1-\sqrt{3})z^{-3}}{4\sqrt{2}} \\ &= 0.48301 + 0.8365z^{-1} + 0.2241z^{-2} - 0.1294z^{-3} \end{aligned} \tag{5.71}$$

$$\begin{aligned} H(z) &= \frac{-(1-\sqrt{3}) + (3-\sqrt{3})z^{-1} - (3+\sqrt{3})z^{-2} + (1+\sqrt{3})z^{-3}}{4\sqrt{2}} \\ &= 0.1294 + 0.2241z^{-1} - 0.8365z^{-2} + 0.48301z^{-3}. \end{aligned} \tag{5.72}$$

The transfer function for a two-channel lattice with two stages is

$$G(z) = \left(1 + a[0]z^{-1} - a[0]a[1]z^{-2} + a[1]z^{-3}\right) s \tag{5.73}$$

$$H(z) = \left(-a[1] - a[0]a[1]z^{-1} - a[0]z^{-2} + z^{-3}\right) s. \tag{5.74}$$

If we now compare (5.71) with (5.73) we find

$$s = \frac{1+\sqrt{3}}{4\sqrt{2}} \qquad a[0] = \frac{3+\sqrt{3}}{4\sqrt{2}s} \qquad a[1] = \frac{1-\sqrt{3}}{4\sqrt{2}s}. \tag{5.75}$$

[10] The orthogonal filter name comes from the fact that the scalar product of the filters, for a shift by two (i.e., $\sum g[k]h[k-2l] = 0, k, l \in \mathbb{Z}$), is zero.

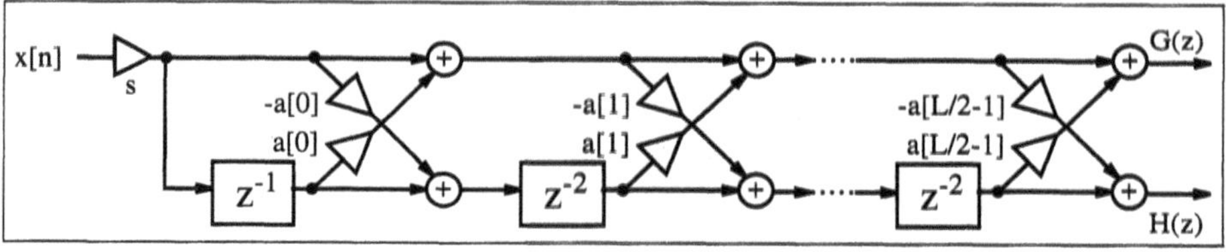

Fig. 5.46. Lattice realization for the orthogonal two-channel filter bank (©1999 Springer Press [5]).

We can now translate this structure direct into hardware and implement the filter bank with MaxPlusII as shown in the following VHDL[11] code.

```
PACKAGE n_bits_int IS               -- User defined types
  SUBTYPE BITS8 IS INTEGER RANGE -128 TO 127;
  SUBTYPE BITS9 IS INTEGER RANGE -2**8 TO 2**8-1;
  SUBTYPE BITS17 IS INTEGER RANGE -2**16 TO 2**16-1;
  TYPE ARRAY_BITS17_4 IS ARRAY (0 TO 3) OF BITS17;
END n_bits_int;

LIBRARY work;
USE work.n_bits_int.ALL;

LIBRARY ieee;
USE ieee.std_logic_1164.ALL;
USE ieee.std_logic_arith.ALL;
USE ieee.std_logic_unsigned.ALL;

ENTITY db4latti IS                             ------> Interface
  PORT (clk       : IN  STD_LOGIC;
        clk2      : OUT STD_LOGIC;
        x_in      : IN  BITS8;
        x_e, x_o  : OUT BITS17;
        g, h      : OUT BITS9);
END db4latti;

ARCHITECTURE flex OF db4latti IS

  TYPE STATE_TYPE IS (even, odd);
  SIGNAL state                    : STATE_TYPE;
  SIGNAL sx_up, sx_low, x_wait    : BITS17;
  SIGNAL clk_div2                 : STD_LOGIC;
  SIGNAL sxa0_up, sxa0_low        : BITS17;
  SIGNAL up0, up1, low0, low1     : BITS17;

BEGIN

  Multiplex: PROCESS ------> Split into even and odd
  BEGIN                    --    samples at clk rate
    WAIT UNTIL clk = '1';
```

[11] The equivalent Verilog code **db4latti.v** for this example can be found in Appendix A on page 376.

```
    CASE state IS
      WHEN even =>
        -- Multiply with 256*s=124
        sx_up   <= 4 * (32 *   x_in - x_in);
        sx_low  <= 4 * (32 * x_wait - x_wait);
        clk_div2 <= '1';
        state <= odd;
      WHEN odd =>
        x_wait <= x_in;
        clk_div2 <= '0';
        state <= even;
    END CASE;
  END PROCESS;

---------- Multipy a[0] = 1.7321
  sxa0_up  <= (2*sx_up  - sx_up /4)
                                   - (sx_up /64 + sx_up/256);
  sxa0_low <= (2*sx_low - sx_low/4)
                                  - (sx_low/64 + sx_low/256);
---------- First stage -- FF in lower tree
  up0  <= sxa0_low + sx_up;
  LowerTreeFF: PROCESS
  BEGIN
    WAIT UNTIL clk_div2 = '0';
    low0 <= sx_low - sxa0_up;
  END PROCESS;

---------- Second stage  a[1]=0.2679
  up1  <= (up0 - low0/4) - (low0/64 + low0/256);
  low1 <= (low0 + up0/4) + (up0/64  +  up0/256);

  x_e  <= sx_up;    -- Provide some extra test signals
  x_o  <= sx_low;
  clk2 <= clk_div2;

  OutputScale: PROCESS
  BEGIN
    WAIT UNTIL clk_div2 = '0';
    g <=  up1 / 256;
    h <= low1 / 256;
  END PROCESS;

END flex;
```

This VHDL code is a direct translation of the lattice shown in Fig. 5.46. The incoming stream is multiplied by $s = 0.48 \approx 124/256$. Next, the cross term product multiplications with $a[0] = 1.73 \approx (2 - 2^{-2} - 2^{-6} - 2^{-8})$ of the first stage are computed. It follows that the stage one additions and the lower tree signal must be delayed by one sample. In the second stage, the cross multiplication with $a[1] = 0.27 \approx (2^{-2} + 2^{-6} + 2^{-8})$ and the final output addition are implemented. The design uses 334 LCs and runs at 60.60 MHz.

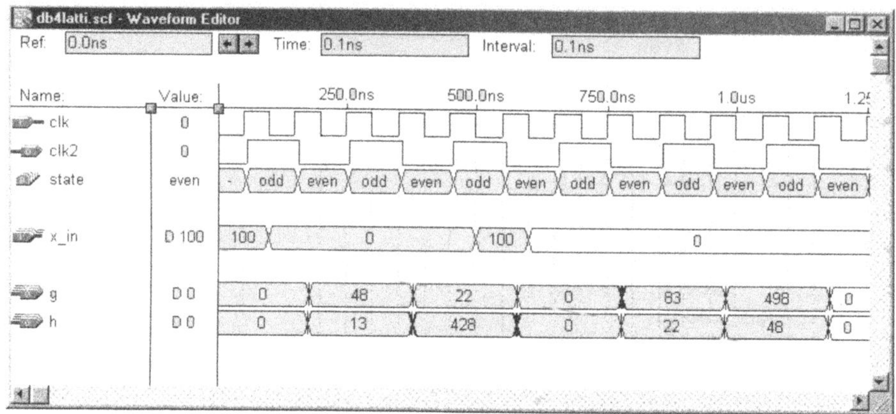

Fig. 5.47. VHDL simulation of the Daubechies length-4 lattice filter bank.

The VHDL simulation is shown in Fig. 5.47. The simulation shows the response to an impulse with amplitude 100 at even and odd positions for the filters $G(z)$ and $H(z)$, respectively. 5.18

If we compare the size of the lattice with the direct polyphase implementation of $G(z)$ shown in Example 5.1 on p. 148 (LCs multiplied by two), we notice that both designs have about the same size ($208 \cdot 2 = 416$ LCs, versus 334 LCs). Although the lattice implementation needs only five multipliers, compared with eight multipliers for the polyphase implementation, we notice that in the polyphase implementation we can use RAG to implement the coefficients of the transposed filter, while in the lattice we must implement single multipliers, which, in general, are less efficient.

Linear-phase two-channel filter bank. We have already seen in Chapter 3 that if a linear filter has even or odd symmetry, 50% of multiplier resources can be saved. The same symmetry also applies for polyphase decomposition of the filters if the filters, have even length. In addition, these filters may run at twice the speed.

If $G(z)$ and $H(z)$ have the same length, another implementation using lattice filters can further decrease the implementation effort, as shown in Fig. 5.48. Notice that the lattice is different from the lattice used for the orthogonal filter bank shown in Fig. 5.46.

The following example demonstrates how to convert a direct architecture into a lattice filter.

Example 5.19: Lattice for $L = 4$ Linear-Phase Filter

One filter configuration in Example 5.15 (p. 187) was a linear-phase filter pair, with both filters of length 4. The filters are

$$G(z) = \frac{1}{4}\left(1 + 3z^{-1} + 3z^{-2} + 1z^{-3}\right) \tag{5.76}$$

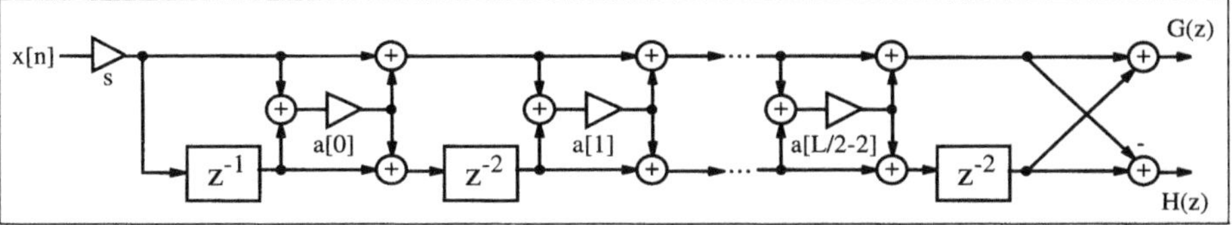

Fig. 5.48. Lattice filter to realize linear-phase two-channel filter bank (©1999 Springer Press [5]).

$$H(z) = \frac{1}{4}\left(-1 + -3z^{-1} + 3z^{-2} + 1z^{-3}\right). \tag{5.77}$$

The transfer functions for the two-channel length-4 linear-phase lattice filters are:

$$G(z) = \left((1 + a[0]) + a[0]z^{-1} + a[0]z^{-2} + (1 + a[0])z^{-3}\right)s \tag{5.78}$$

$$H(z) = \left(-(1 + a[0]) - a[0]z^{-1} + a[0]z^{-2} + (1 + a[0])z^{-3}\right)s. \tag{5.79}$$

Comparing (5.76) with (5.78), we find

$$s = -2 \qquad a[0] = -1,5. \tag{5.80}$$

5.19

Notice that, compared with the direct implementation, only about one quarter of the multipliers are required.

The disadvantage of the linear-phase lattice is that not all linear-phase filters can be implemented. Specifically, $G(z)$ must be even symmetric, $H(z)$ must be odd symmetric, and both filters must be of the same length, with an even number of samples.

Comparison of implementation options. Finally, Table 5.6 compares the different implementation options, which include the general case and special types like QMF, linear-phase and orthogonal.

The Table 5.6 shows the required number of multipliers and adders, the reference figure, the maximum input rate, and the structurally important question of whether the coefficients can be implemented using reduced adder graph technique, or occur as single-multiplier coefficients. For shorter filters, the lattice structure seems to be attractive, while for longer filters, RAG will most often produce smaller and faster designs. Note that the number of multipliers and adders in Table 5.6 are an estimate of the hardware effort required for the filter, and *not* the typical number found in the literature for the computational effort per input sample in a PDSP/μP solution [84, 99].

Excellent additional literature about two-channel filter banks is available (see [82, 99, 96, 100]).

5.7 Wavelets

A time-frequency representation of signals processed through transform methods has proven beneficial for audio and image processing [101, 96, 102].

Table 5.6. Effort to compute two-channel filter banks if both filter are of length L.

Type	Number of real multipliers	Number of real adders	see Fig.	Speed	Can use RAG ?
Polyphase with any coefficients					
Direct FIR filtering	$2L$	$2L-2$	5.41	$2f_s$	✓
Lifting	$\approx L$	$\approx L$	5.43	$2f_s$	–
Quadrature mirror filter (QMF)					
Identical polyphase filter	L	L	5.44	$2f_s$	✓
Orthogonal filter					
Transposed FIR filter	L	$2L-2$	5.45	f_s	✓
Lattice	$L+1$	$3L/4$	5.46	$2f_s$	–
Linear-phase filter					
Symmetric filter	L	$2L-2$	3.5	$2f_s$	✓
Lattice	$L/2$	$3L/2-1$	5.48	$2f_s$	–

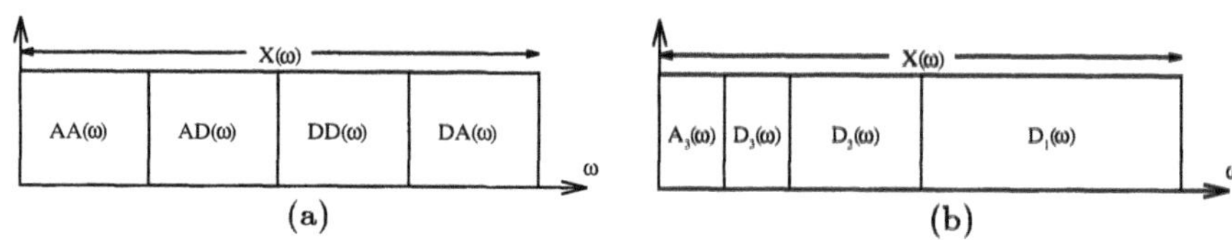

Fig. 5.49. Frequency distribution for **(a)** Fourier (constant bandwidth) and **(b)** constant Q.

Many signals subject to analysis are known to have statistically constant properties for only short time frames (e.g., speech or audio signals). It's therefore reasonable to analyze such signals in a short window, compute the signal parameter, and slide the window forward to analyze the next frame. If this analysis is based on Fourier transforms, it is called a *short term Fourier transform* (STFT).

A short-time Fourier transform (STFT) is formally defined by

$$X(\tau, f) = \int_{-\infty}^{\infty} x(t)\, w(t-\tau)\, e^{-j2\pi f t}\, dt. \tag{5.81}$$

i.e., it slides a window function $w(t-\tau)$ over the signal $x(t)$, and produces a continuous time-frequency map. The window should taper smoothly to zero, both in frequency and time, to ensure localization in frequency Δ_f and time Δ_t of the mapping. One weight function, the Gaussian function ($g(t) = e^{-t^2}$), is optimal in this sense, and provides the minimum (Heisenberg principle) product $\Delta_f \Delta_t$ (i.e., best localization), as proposed by Gabor

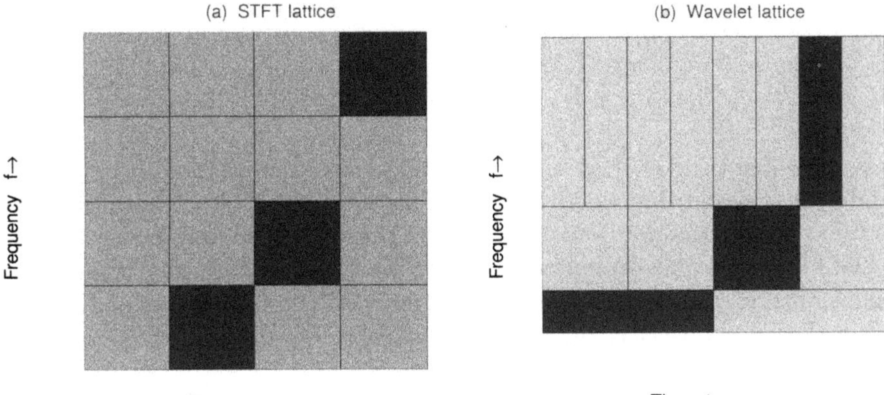

Fig. 5.50. Time frequency grids for a chirp signal. **(a)** Short term Fourier transform. **(b)** Wavelet transform.

in 1949 [103]. The discretization of the Gabor transform leads to the discrete Gabor Transform (DGT). The Gabor transform uses identical resolution windows throughout the time and frequency plane (see Fig. 5.50(a)). Every rectangle in Fig. 5.50(a) has exactly the same shape, but often a constant Q (i.e., the quotient of bandwidth to center frequency) is desirable, especially in audio and image processing. That is for high frequencies we wish to have broadband filters and short sampling intervals, while for low frequencies, the bandwidth should be small and the intervals larger. This can be accomplished with the continuous wavelet transform (CWT), introduced by Morlet [104],

$$\mathrm{CWT}(\tau, f) = \int_{-\infty}^{\infty} x(t)\, h\left(\frac{t-\tau}{s}\right)\, \mathrm{d}t \tag{5.82}$$

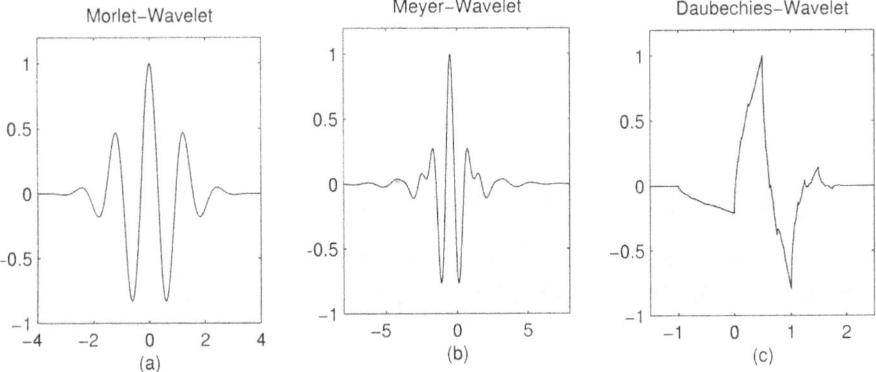

Fig. 5.51. Some typical wavelets from Morlet, Meyer, and Daubechies.

where $h(t)$, known from the Heugens principle in physics, is called a small wave or *wavelet*. Some typical wavelets are displayed in Fig. 5.51.

If we use now as a wavelet

$$h(t) = \left(e^{j2\pi kt} - e^{k^2/2}\right) e^{-t^2/2} \tag{5.83}$$

we still enjoy the "optimal" properties of the Gaussian window, but now with different scales in time and frequency. This so-called Morlet transform is also subject to quantization, and is then called the discrete Morlet Transformation (DMT) [105]. In the discrete case the lattice points in time and frequency as shown in Fig. 5.50(b). The exponential term $e^{-k^2/2}$ in (5.83) was introduced such that the wavelet is DC free. The following examples show the excellent performance of the Gaussian window.

Example 5.20: Analysis of a Chirp Signal

Fig. 5.52 shows the analysis of a constant amplitude signal with increasing frequency. Such signals are called chirp signals. If we applied the Fourier transform we would get a uniform spectrum, because all frequencies are present. The Fourier spectrum does not preserve time-related information. If we use instead an STFT with a Gaussian window, i.e., the Morlet transform, as shown in Fig. 5.52(a), we can clearly see the increasing frequency. But the Gaussian window shows the best localization of all windows. On the other hand, with a Haar window we would have less computational effort, but, as can be seen from Fig. 5.52(b), the Haar window will achieve less precise time-frequency localization of the signal. 5.20

Both DGT and DMT provide good localization by using a Gaussian window, but both are computationally intensive. An efficient multiplier-free implementation is based on two ideas. First, the Gaussian window can be sufficiently approximated by a convolution of (≥ 3) rectangular functions, and second, single-passband frequency-sampling filters (FSF) can be efficiently implemented by defining algebraic integers over polynomial rings, as introduced in [105].

In the following, we wish to focus our attention on a newly popular analysis method called the *discrete wavelet transform*, which better exploits the auditory and visual human perception mode (i.e., constant Q), and also can often be more efficiently computed, using $\mathcal{O}(n)$ complexity algorithms.

5.7.1 The Discrete Wavelet Transformation

A discrete-time version of the analog model leads to the *discrete wavelet transform* (DWT). In practical applications, the DWT is restricted to the discrete time *dyadic DWT* with $a = 2$, and will be considered in the following. The DWT achieves the constant Q bandwidth distribution shown in Fig. 5.49(b) and Fig. 5.50(b) by always applying the two-channel filter bank in a filter tree to the lowpass signal, as shown in Fig. 5.53.

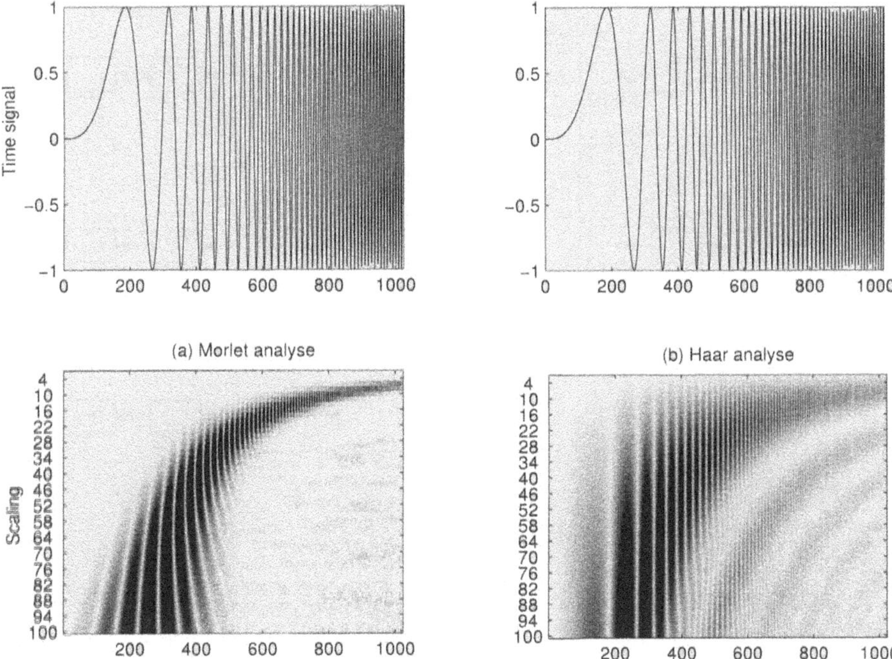

Fig. 5.52. Analysis of a chirp signal with **(a)** Discrete Morlet transform. **(b)** Haar transform.

We now wish to focus on what conditions for the CWT wavelet allow it to be realized with a two-channel DWT filter bank. We may argue that if we sample a continuous wavelet at an appropriate rate (above the Nyquist rate), we may call the sampled version a DWT. But in general, only those continuous wavelet transforms which can be realized with a two-channel filter bank are called DWT.

Closely related to whether a continuous wavelet $\psi(t)$ can be realized with a two-channel DWT, is the question of whether the *scaling equation*

$$\phi(t) = \sum_n g[n]\, \phi(2t - n) \tag{5.84}$$

exists, where the actual wavelet is computed with

$$\psi(t) = \sum_n h[n] \phi(2t - n) \tag{5.85}$$

where $g[n]$ is a low-pass, and $h[n]$ a high-pass filter. Note that $\phi(t)$ and $\psi(t)$ are continuous functions, while $g[n]$ and $h[n]$ are sample sequences (but still may also be IIR filters). Note that (5.84) is similar to the *self similarity*

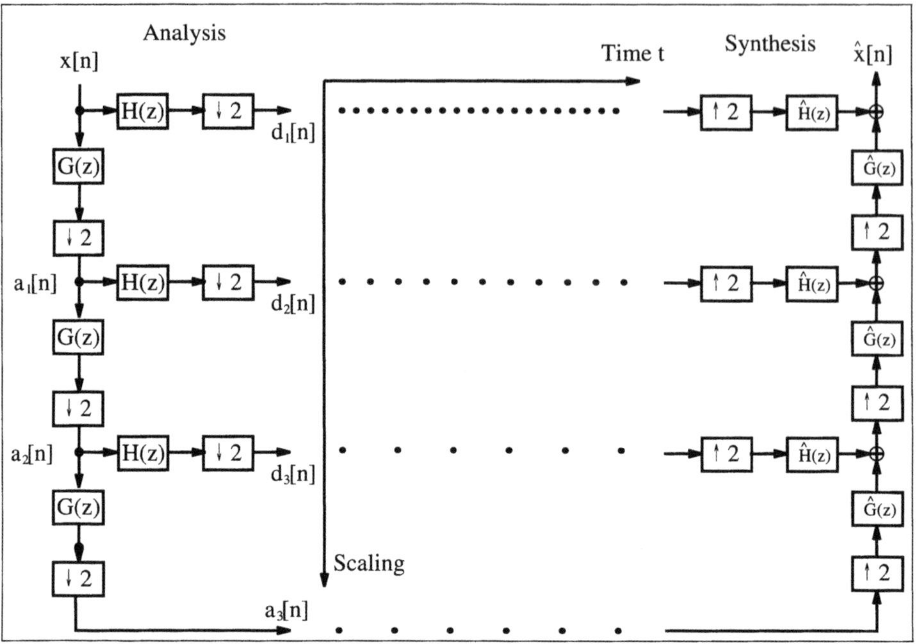

Fig. 5.53. Wavelets tree decomposition in three octaves (©1999 Springer Press [5]).

($\phi(t) = \phi(at)$) exhibited by fractals. In fact, the scaling equation may iterate to a fractal, but that is, in general, not the desired case, because most often a smooth wavelet is desired. The smoothness can be improved if we use a filter with maximal numbers of zeros at π.

We consider now backwards reconstruction: we start with the filter $g[n]$, and construct the corresponding wavelet. That is the most common case, especially if we use the halfband design from Algorithm 5.14 (p. 187) to generate perfect reconstruction filter pairs of the desired length and property.

To get a graphical interpretation of the wavelet, we start with a rectangular function (box function) and build, according to (5.84), the following graphical iteration:

$$\phi^{(k+1)}(t) = \sum_n g[n]\, \phi^{(k)}(2t - n). \tag{5.86}$$

If this coverges to a stable $\phi(t)$, the (new) wavelet is found. This iteration obviously converges for the Haar filter $\{1, 1\}$ immediately after the first iteration, because the sum of two box function scaled and added is again a box function, i.e.,

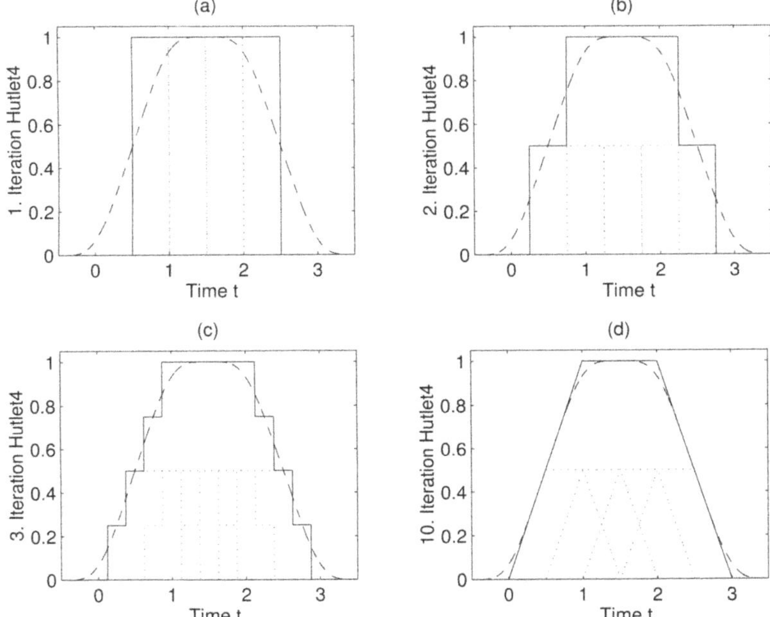

Fig. 5.54. Iteration steps 1, 2, 3, and 10 for Hutlet4. (solid line: $\phi^{(k+1)}(t)$; dotted: $\phi^{(k)}(2t-n)$; and ideal Hut-function: dashed)

r(t) = r(2t)+r(2t-1)

Let us now graphically construct the wavelet that belongs to the filter $g[n] = \{1, 1, 1, 1\}$, which we will call Hutlet4 [106].

Example 5.21: Hutlet of Length-4

We start with four box functions weighted by $g[n] = \{1, 1, 1, 1\}$. The sum shown in Fig. 5.54(a) is the starting $\phi^{(1)}(t)$. This function is scaled by two, and the sum gives a two-step function. After 10 iterations we already get a very smooth trapezoid function. If we now use the QMF relation, from (5.54) (p. 183), to construct the actual wavelet, we get the Hutlet4, which has two triangles as shown in Fig. 5.55. 5.21

We note that $g[n]$ is the impulse response of the moving-average filter, and can be implemented as an one-stage CIC filter [107]. Fig. 5.55 shows all scaling functions and wavelets for this type of wavelet with even length coefficients.

As noted before, the iteration defined by (5.86) may also converge to a fractal. Such an example is shown in Fig. 5.56, which is the wavelet for the

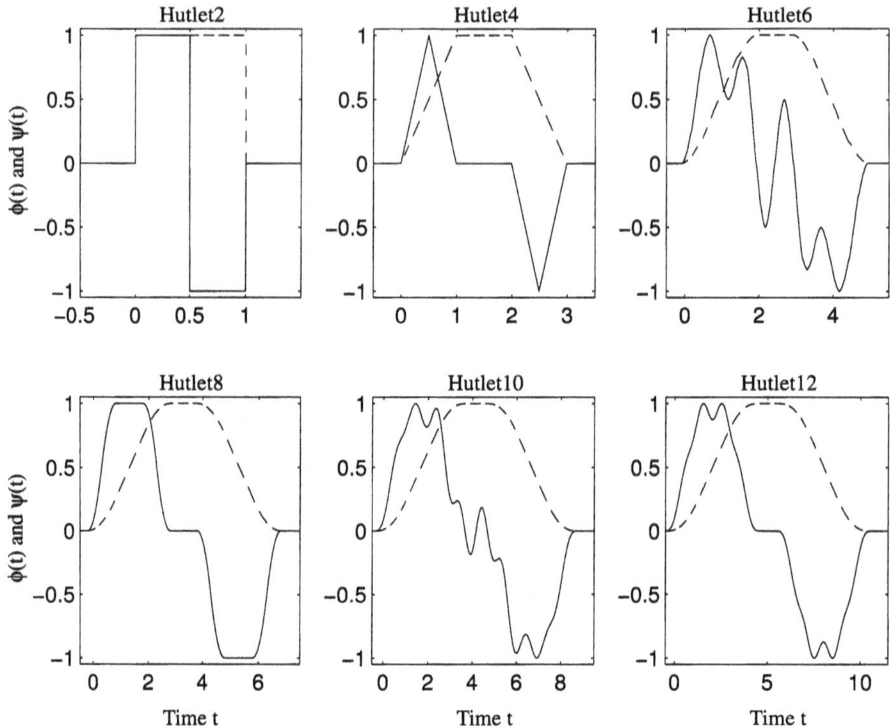

Fig. 5.55. The Hutlet wavelet family (solid line) and scaling function (dashed line) after 10 iterations (©1999 Springer Press [5]).

length-5 "moving average filter." This indicates the challenge of the filter selection $g[n]$: it may converge to a smooth or, totally chaotic function, depending only on an apparently insignificant property like the length of the filter!

We still have not explained why the two-scale equation (5.84) is so important for the DWT. This can be better understood if we rearrange the downsampler (compressor) and filter in the analysis part of the DWT, using the "Noble" relation

$$(\downarrow M)\ H(z) = H(z^M)\ (\downarrow M). \tag{5.87}$$

which was introduced in Sect. 5.1.1, p. 144. The results for a three-level filter bank are shown in Fig. 5.57. If we compute the impulse response of the cascade sequences, i.e.,

$$\begin{aligned} H(z) &\leftrightarrow d_1[k/2] \\ G(z)H(z^2) &\leftrightarrow d_2[k/4] \\ G(z)G(z^2)H(z^4) &\leftrightarrow d_3[k/8] \\ G(z)G(z^2)G(z^4) &\leftrightarrow a_3[k/8] \end{aligned}$$

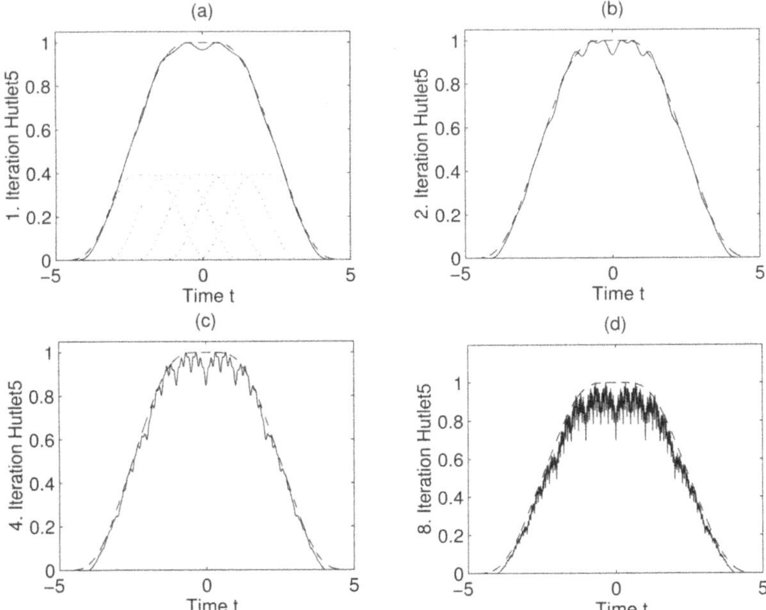

Fig. 5.56. Iteration step 1,2,4, and 8 for Hutlet5. The sequence converges to a fractal!

we find that a_3 is an approximation to the scaling function, while d_3 gives an approximation to the mother wavelet, if we compare the graphs with the continuous wavelet shown in Fig. 5.51 (p. 199).

This is not always possible. For instance, for the Morlet Wavelet shown in Fig. 5.51 (p. 199), no scaling function can be found, and a realization using the DWT is *not* possible.

Two-channel DWT design examples for the Daubechies length-4 filter have already been discussed, in combination with polyphase representation (Example 5.1, p. 148), and the lattice implementation of orthogonal filters in Example 5.18 (p. 193).

Exercises

5.1: Let $F(z) = 1 + z^{-d}$. For which d do we have a halfband filter according to Definition 5.7 (p. 172)?

5.2: Let $F(z) = 1 + z^{-5}$ be a halfband filter.
(a) Draw $|F(\omega)|$. What kind of symmetry does this filter have?
(c) Use Algorithm 5.14 (p. 187) to compute a perfectly reconstructing real filter bank. What is the total delay of the filter bank?

5.3: Use the halfband filter F3 from Example 5.15 (p. 187) to build a perfect-reconstruction filter bank, using Algorithm 5.14 (p. 187), of length

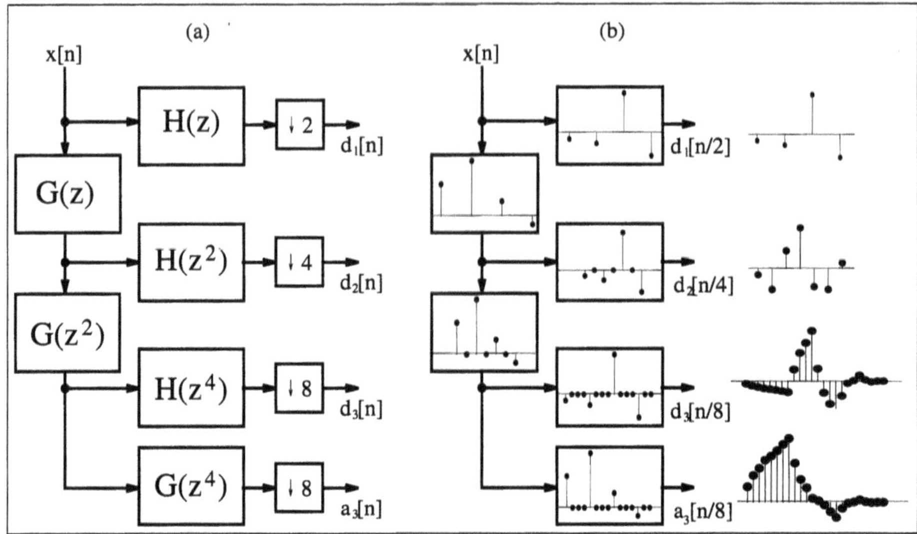

Fig. 5.57. DWT filter bank rearrange using Noble relations. **(a)** Transfer function in the z-domain. **(b)** Impulse response for the length-4 Daubechies filters.

(a) 1/7.
(b) 2/6.

5.4: How many different filter pairs can be built, using F3 from Example 5.15 (p. 187), if both filters are
(a) Complex.
(b) Real.
(c) Linear-phase.
(d) Orthogonal filter bank.

5.5: Use the halfband filter $F2(z) = 1 + z^{-1} + z^{-2}$ to compute, based on Algorithm 5.14 (p. 187), all possible perfect-reconstructing filter banks.

5.6: **(a)** Compute the number of real additions and multiplications for a direct implementation of the critically sampled uniform DFT filter bank shown in Fig. 5.33 (p. 179). Assume the length L analysis and synthesis filters have complex coefficients, and the inputs are real valued.
(b) Assume an FFT algorithm is used which needs $(15N\log_2(N))$ real additions and multiplications. Compute the total effort for a uniform DFT filter bank, using the polyphase representation from Figs. 5.35 (p. 180) and 5.36 (p. 182), for R length L complex filters.
(c) Using the results from (a) and (b) compute the effort for a critically sampled DFT filter bank with $L = 64$ and $R = 16$.

5.7: Use the lossy integrator from Example 5.9 (p. 181) to implement an $R = 4$ uniform DFT filter bank.
(a) Compute the analysis polyphase filter $H_k(z)$.
(b) Determine the synthesis filter $F_k(z)$ for perfect reconstruction.
(c) Determine the 4×4 DFT matrix. How many real additions and multiplications are used to compute the DFT?

(d) Compute the total computational effort of the whole filter bank, in terms of real additions and multiplications per input sample.

5.8: Analyze the frequency response of each Goodman and Carey halfband filter from Table 5.3 (p. 172). Zoom in on the passband to estimate the ripple of the filter.

5.9: Prove the perfect reconstruction for the *lifting* and *dual lifting* scheme from (5.65) and (5.66) on p. 191.

Exercises Using MaxPlusII

5.10: **(a)** Implement the Daubechies length-4 filter using the lifting scheme from Example 5.17 (p. 191), with 8-bit input and coefficient, and 9-bit output quantization.
(b) Simulate the design with two impulses of amplitude 100, similar to Fig. 5.47 (p. 196).
(c) Determine LC utilization and the `Registered Performance`.
(d) Compare the lifting design with the direct polyphase implementation (Example 5.1, p. 148) and with the lattice implementation (Example 5.18, p. 193), in terms of size and speed.

5.11: Use component instantiation of the two designs from Example 5.4 (p. 159) and Example 5.6 (p. 167) to compute the difference of the two filter outputs. Determine the maximum positive and negative deviation.

5.12: **(a)** Use the reduced adder graph design from Fig. 3.11 (p. 98) to build a halfband filter F6 (see Table 5.3, p. 172) for 8-bit inputs using MaxPlusII. Use the transposed FIR structure (Fig. 3.3, p. 81) as the filter architecture.
(b) Determine size in LCs, and `Registered Performance`, of the F6 design.

5.13: **(a)** Compute the polyphase representation for F6 from Table 5.3, p. 172.
(b) Implement the polyphase filter F6 with decimation $R = 2$ for 8-bit inputs with MaxPlusII.
(c) Determine size in LCs, and `Registered Performance`, of the polyphase design.
(d) What are the advantages and disadvantages of the polyphase design, when compared with the direct implementation from Exercise 5.12 (p. 207), in terms of size and speed.

6. Fourier Transforms

The Discrete Fourier Transform (DFT) and its fast implementation, the Fast Fourier Transform (FFT), have played a central role in digital signal processing.

DFT and FFT algorithms have been invented (and reinvented) in many variations. As Heideman et al. [108] pointed out, we know that Gauss used an FFT-type algorithm that today we call the Cooley-Tukey FFT. In this chapter we will discuss the most important algorithms summarized in Fig. 6.1.

We will follow the terminology introduced by Burrus [109], who classified FFT algorithms simply by the (multidimensional) index maps of their input and output sequences. We will therefore call all algorithms which do *not* use an multidimensional index map, DFT algorithms, although some of them, such as the Winograd DFT algorithms, enjoy an essential reduced computational effort. DFT and FFT algorithms do not "stand alone": the most efficient implementations often result in a combination of DFT and FFT al-

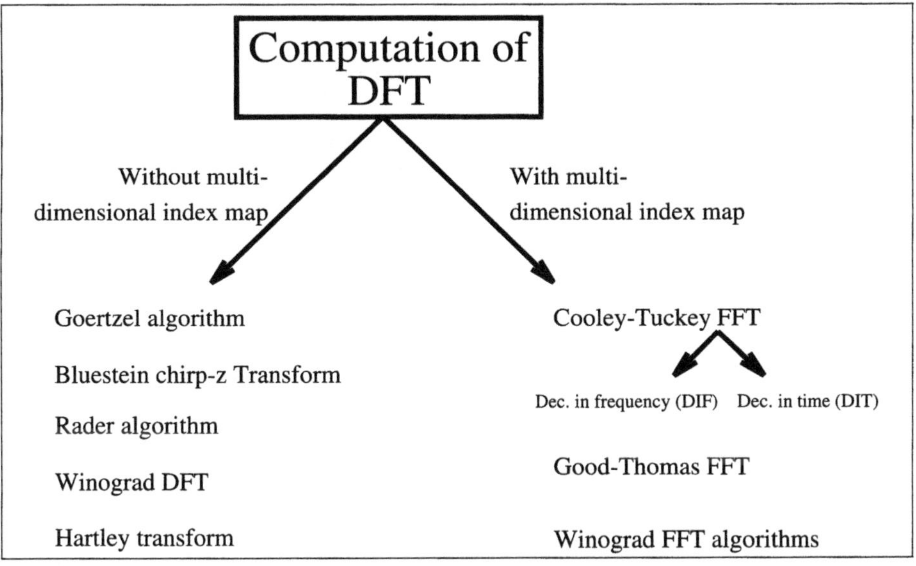

Fig. 6.1. Classifications of DFT and FFT algorithms.

gorithms. For instance, the combination of the Rader prime algorithm and the Good-Thomas FFT results in excellent VLSI implementations. The literature provides many FFT design examples. We find implementations with PDSPs and ASICs [110, 111, 112, 113, 114, 115]. FFTs have also been developed using FPGAs for 1-D [116, 117, 118] and 2-D transforms [42, 119].

We will discuss in this chapter the four most important DFT algorithms and the three most often used FFT algorithms, in terms of computational effort, and will compare the different implementation issues. At the end of the chapter, we will discuss Fourier-related transforms, such as the DCT, which is an important tool in image compression (e.g., JPEG, MPEG). We start with a short review of definitions and the most important properties of the DFT.

For more detailed study, students should be aware that DFT algorithms are covered in basic DSP books [120, 65, 121, 5], and a wide variety of FFT books are also available [122, 123, 124, 125, 126, 54].

6.1 The Discrete Fourier Transform Algorithms

We will start with a review of the most important DFT properties and will then review basic DFT algorithms introduced by Bluestein, Goertzel, Rader, and Winograd.

6.1.1 Fourier Transform Approximations Using the DFT

The Fourier transform pair is defined by

$$X(f) = \int_{-\infty}^{\infty} x(t) e^{-j2\pi ft}\, \mathrm{d}t \longleftrightarrow x(t) = \int_{-\infty}^{\infty} X(f) e^{j2\pi ft}\, \mathrm{d}f. \quad (6.1)$$

The formulation assumes a continuous signal of infinite duration and bandwidth. For practical representation, we must sample in time and frequency, and amplitudes must be quantized. From an implementation standpoint, we prefer to use a finite number of samples in time and frequency. This leads to the *discrete Fourier Transform* (DFT) where N samples are used in time and frequency, according to

$$X[k] = \sum_{n=0}^{N-1} x[n] e^{-j2\pi kn/N} = \sum_{n=0}^{N-1} x[n] W_N^{kn} \quad (6.2)$$

and the inverse DFT (IDFT) is defined as

$$x[n] = \frac{1}{N} \sum_{k=0}^{N-1} X[k] e^{j2\pi kn/N} = \frac{1}{N} \sum_{k=0}^{N-1} X[k] W_N^{-kn} \quad (6.3)$$

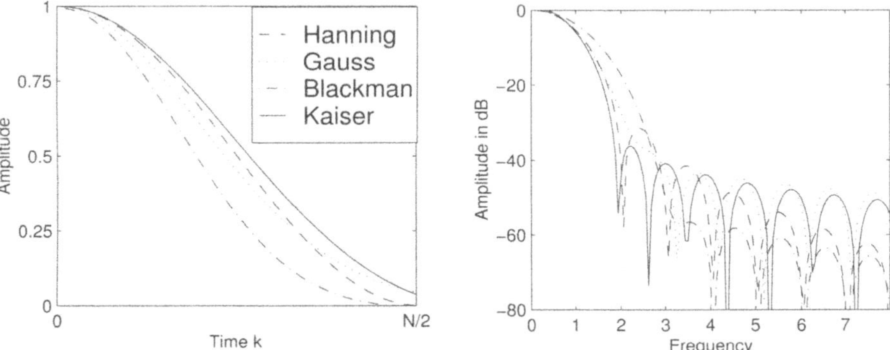

Fig. 6.2. Window functions in time and frequency.

or, in vector/matrix notation

$$\boldsymbol{X} = \boldsymbol{W}\boldsymbol{x} \leftrightarrow \boldsymbol{x} = \frac{1}{N}\boldsymbol{W}^*\boldsymbol{X}. \tag{6.4}$$

If we use the DFT to approximate the Fourier spectrum, we must remember the effect of sampling in time and frequency, namely:

- By sampling in *time,* we get a periodic spectrum with the sampling frequency f_S. The approximation of a Fourier transform by a DFT is reasonable only if the frequency components of $x(t)$ are concentrated on a smaller range than the Nyquist frequency $f_S/2$, as stated in the "Shannon sampling theorem."
- By sampling in the *frequency* domain, the time function becomes periodic, i.e., the DFT assumes the time series to be periodic. If an N-sample DFT is applied to a signal that does not complete an integer number of cycles within an N-sample window, a phenomenon called *leakage* occurs. Therefore if possible, we should choose the sampling frequency and the analysis window in such a way that it covers an integer number of periods of $x(t)$, if $x(t)$ is periodic.

A more practical alternative for decreasing leakage is the use of a window function which tapers smoothly to zero on both sides. Such window functions were already discussed in the context of FIR filter design in Chapter 3 (see Table 3.2, p. 89). Fig. 6.2 shows the time and frequency behavior of some typical windows [87], [127].

An example illustrates the use of a window function.

Example 6.1: Windowing

Fig. 6.3(a) shows a sinusoidal signal that does not complete an integer number of periods in its sample window. The Fourier transform of the signal should ideally include only the two Dirac functions at $\pm\omega_0$, as diplayed in Fig. 6.3(b). Figs. 6.3(c) and (d) show the DFT analysis with different windows. We notice that the analysis with the box function has somewhat more ripple than the

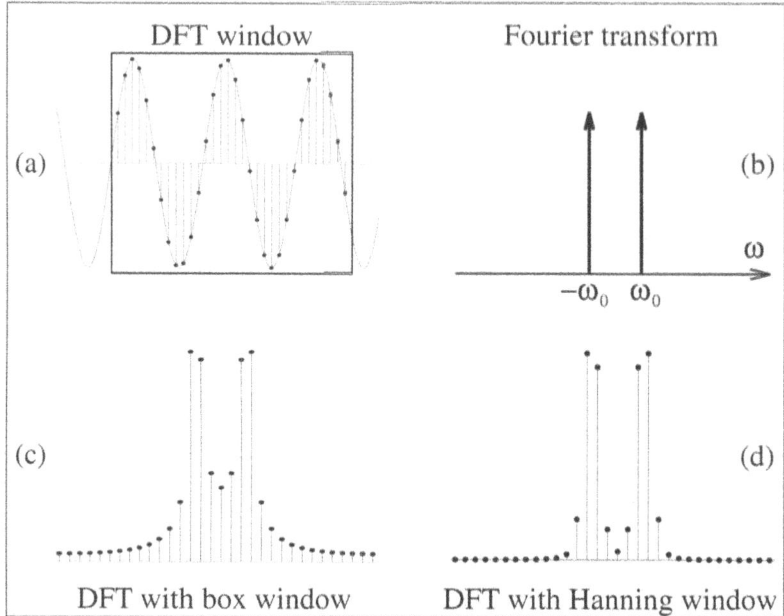

Fig. 6.3. Analysis of periodic function through the DFT, using window functions.

analysis with the Hanning window. An exact analysis would also show that the main lope width with Hanning analysis is larger than the width achieved with the box function, i.e., no window. 6.1

6.1.2 Properties of the DFT

The most important properties of the DFT are summarized in Table 6.1. Many properties are identical with the Fourier transform, e.g., the transform is unique (bijective), the superposition applies, and real and imaginary parts are related through the Hilbert transform.

The similarity of the forward and inverse transform leads to an alternative inversion algorithm. Using the vector/matrix notation (6.4) of the DFT

$$\boldsymbol{X} = \boldsymbol{W}\boldsymbol{x} \leftrightarrow \boldsymbol{x} = \frac{1}{N}\boldsymbol{W}^*\boldsymbol{X} \tag{6.5}$$

we can conclude

$$\boldsymbol{x} = \frac{1}{N}\left(\boldsymbol{W}^*\boldsymbol{X}\right)^* = \boldsymbol{W}\boldsymbol{X}^* \tag{6.6}$$

i.e., we can use the DFT of $\boldsymbol{X}^*$ scaled by $1/N$ to compute the inverse DFT.

Table 6.1. Theorems of the DFT

Theorem	$x[n]$	$X[k]$
Transform	$x[n]$	$\sum_{n=0}^{N-1} x[n]e^{-j2\pi nk/N}$
Inverse Transform	$\frac{1}{N}\sum_{n=0}^{N-1} X[k]e^{j2\pi nk/N}$	$X[k]$
Superposition	$s_1x_1[n]+s_2x_2[n]$	$s_1X_1[k]+s_2X_2[k]$
Time reversal	$x[-n]$	$X[-k]$
Conjugate complex	$x^*[n]$	$X^*[-k]$
Split		
Real part	$\Re(x[n])$	$(X[k]+X^*[-k])/2$
Imaginary part	$\Im(x([n])$	$(X[k]+X^*[-k])/(2j)$
Real even part	$x_e[n]=(x[n]+x[-n])/2$	$\Re(X[k])$
Real odd part	$x_o[n]=(x[n]-x[-n])/2$	$j\Im(X[k])$
Symmetry	$X[n]$	$Nx[-k]$
Cyclic Convolution	$x[n] \circledast f[n]$	$X[k]F[k]$
Multiplication	$x[n]\cdot f[n]$	$\frac{1}{N}X[k] \circledast F[k]$
Periodic shift	$x[n-d \bmod N]$	$X[k]e^{-j2\pi dk/N}$
Parseval theorem	$\sum_{n=0}^{N-1} \lvert x[n]\rvert^2$	$\frac{1}{N}\sum_{k=0}^{N-1} \lvert X[k]\rvert^2$

DFT of a Real Sequence

We now turn to some additional computational savings for DFT (and FFT) computations, when the input sequence is real. In this case, we have two options: we can compute with one N-point DFT the DFT of two N-point sequences, or we can compute with an N-point DFT a length $2N$ DFT of a real sequence.

If we use the Hilbert property from Table 6.1, i.e., a real sequence has an even-symmetric real spectrum and an odd imaginary spectrum, the following algorithms can be synthesized [122].

Algorithm 6.2: Length $2N$ Transform with N-point DFT

The algorithm to compute the $2N$-point DFT $X[k] = X_r[k] + jX_i[k]$ from the time sequence $x[n]$ is as follows:

1) Build an N-point sequence $y[n] = x[2n] + jx[2n+1]$ with $n = 0, 1, \ldots N-1$.
2) Compute $y[n] \circ\!\!-\!\!\bullet\, Y[k] = Y_r[k] + jY_i[k]$. where $\Re(Y[k]) = Y_r[k]$ is the real and $\Im(Y[k]) = Y_i[k]$ is the imaginary part of $Y[k]$, respectively.
3) Compute

$$X_r[k] = \frac{Y_r[k] + Y_r[-k]}{2} + \cos(\pi k/N)\frac{Y_i[k] + Y_i[-k]}{2} - \sin(\pi k/N)\frac{Y_r[k] - Y_r[-k]}{2}$$

$$X_i[k] = \frac{Y_i[k] - Y_i[-k]}{2} - \sin(\pi k/N)\frac{Y_i[k] + Y_i[-k]}{2} - \cos(\pi k/N)\frac{Y_r[k] - Y_r[-k]}{2}$$

with $k = 0, 1, \ldots N-1$.

The computational effort therefore, besides an N-point DFT (or FFT), is 4 N real additions and multiplications, from the twiddle-factors $\pm \exp(j\pi k/N)$.

To transform two length N sequences with a length N DFT, we use the fact (see Table 6.1) that a real sequence has an even spectrum, while the spectrum of a purely imaginary sequence is odd. This is the basis for the following algorithm.

Algorithm 6.3: Two Length N Transforms with one N-point DFT

The algorithm to compute the N-point DFT $g[n] \circ\!\!-\!\!\bullet\, G[k]$ and $h[n] \circ\!\!-\!\!\bullet\, H[k]$ is as follows:

1) Build an N-point sequence $y[n] = h[n] + jg[n]$ with $n = 0, 1, \ldots N-1$.
2) Compute $y[n] \circ\!\!-\!\!\bullet\, Y[k] = Y_r[k] + jY_i[k]$, where $\Re(Y[k]) = Y_r[k]$ is the real and $\Im(Y[k]) = Y_i[k]$ is the imaginary part of $Y[k]$, respectively.
3) Compute finally

$$H[k] = \frac{Y_r[k] + Y_r[-k]}{2} + j\frac{Y_i[k] - Y_i[-k]}{2}$$

$$G[k] = \frac{Y_i[k] + Y_i[-k]}{2} - j\frac{Y_r[k] - Y_r[-k]}{2}$$

with $k = 0, 1, \ldots N-1$.

The computational effort therefore, besides an N-point DFT (or FFT), is 2 N real additions, to form the correct two N-point DFTs.

Fast Convolution Using DFT

One of the most frequent applications of the DFT (or FFT) is the computation of convolutions. As with the Fourier transform, the convolution in time is

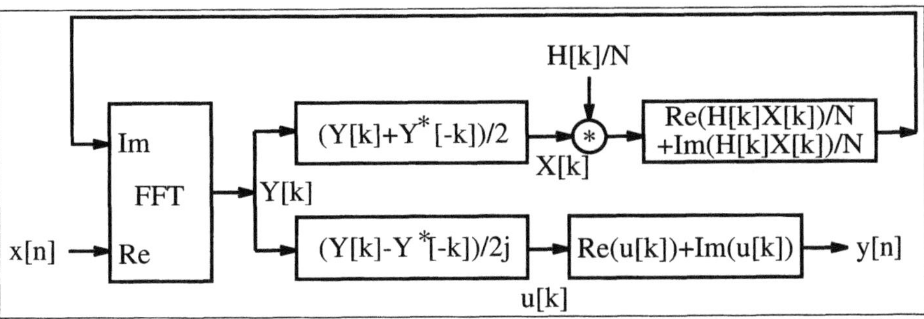

Fig. 6.4. Real convolution using a complex FFT [54].

done by multiplying the two transformed sequences: the two time sequences are transformed in the frequency domain, we compute a (scalar) pointwise product, and we transform the product back into the time domain. The main difference, compared with the Fourier transform, is that now the DFT computes a *cyclic*, and not a linear, convolution. This must be considered when implementing fast convolution with the FFT. This leads to two methods called "overlap save" and "overlap add." In the overlap save method, we basically discharge the samples at the border which are corrupted by the cyclic convolution. In the overlap add method, we zero-pad the filter and signal in such a way that we can directly add the partial sequences to a common product stream.

Most often the input sequences for the fast convolution are real. An efficient convolution may therefore be accomplished with a real transform, such as the Hartley transform discussed in Exercise 6.16, p. 253. We may also construct an FFT-like algorithm for the Hartley transform, and can get about twice the performance compared with a complex transform [128].

If we wish to utilize an available FFT program, we may use one of the previously discussed Algorithms, 6.2 or 6.3, for real sequences. An alternative approach is shown in Fig. 6.4. It shows a similar approach to Algorithm 6.2, where we implemented two N-point transforms with one N-point DFT, but in this case we use the "real" part for a DFT, and the imaginary part for the IDFT, which is needed for the back transformation, according to the convolution theorem.

It is assumed that the DFT of the real-valued filter (i.e., $F[k] = F[-k]^*$) has been computed offline and, in addition, in the frequency domain we need only $N/2$ multiplications to compute $X[k]F[k]$.

6.1.3 The Goertzel Algorithm

A single spectral component $X[k]$ in the DFT computation is given by

$$X[k] = x[0] + x[1]W_N^k + x[2]W_N^{2k} + \ldots + x[N-1]W_N^{(N-1)k}$$

We can combine all x[n] with the same common factor W_N^k, and get

$$X[k] = x[0] + W_N^k \left(x[1] + W_N^k \left(x[2] + \ldots + W_N^k x[N-1]\right) \ldots\right)\right).$$

It can be noted that this results in a possible recursive computation of $X[k]$. This is called the Goertzel algorithm, and is graphically interpreted by Fig. 6.5. The computation of $y[n]$ starts with the last value of the input sequence $x[N-1]$. After step three, a spectrum value of $X[k]$ is available at the output.

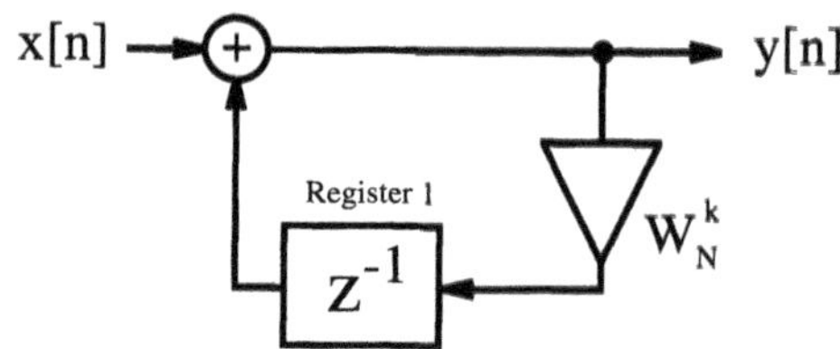

Step	$x[n]$	Register 1	$y[n]$
0	$x[3]$	0	$x[3]$
1	$x[2]$	$W_4^k x[3]$	$x[2] + W_4^k x[3]$
2	$x[1]$	$W_4^k x[2] + W_4^{2k} x[3]$	$x[1] + W_4^k x[2] + W_4^{2k} x[3]$
3	$x[0]$	$W_4^k x[1]$ $+W_4^{2k} x[2] + W_4^{3k} x[3]$	$x[0] + W_4^k x[1]$ $+W_4^{2k} x[2] + W_4^{3k} x[3]$

Fig. 6.5. The length-4 Goertzel algorithm.

If we have to compute several spectral components, we can reduce the complexity if we combine factors of the type $e^{\pm j2\pi n/N}$. This will result in second-order systems having a denominator according to

$$z^2 - 2z\cos\left(\frac{2\pi n}{N}\right) + 1$$

which is discussed in Exercise 6.4 (p. 252). All complex multiplications are then reduced to real multiplications.

In general, the Goertzel algorithm can be attractive if only a few spectral components have to be computed. For the whole DFT, the effort is of order N^2, and therefore yields no advantage compared with the direct DFT computation.

6.1.4 The Bluestein Chirp-z Transform

In the Bluestein chirp-z transform (CZT) algorithms, the DFT exponent nk is quadratic expanded to

$$nk = -(k-n)^2/2 + n^2/2 + k^2/2. \tag{6.7}$$

The DFT therefore becomes

$$X[k] = W^{k^2/2} \sum_{n=0}^{N-1} \left(x[n] W^{n^2/2} \right) W^{-(k-n)^2/2}. \tag{6.8}$$

This algorithm is graphically interpreted in Fig. 6.6. This results in the following

Algorithm 6.4:	**Bluestein Chirp-z Algorithm**
The computation of the DFT is done in three steps, namely	
1) N multiplication of $x[n]$ with $W_N^{n^2/2}$.	
2) Linear convolution of $x[n]W_N^{n^2/2} * W_N^{n^2/2}$.	
3) N multiplications with $W_N^{k^2/2}$.	

For a complete transform, we therefore need a length N convolution and $2N$ complex multiplications. The advantage, compared with the Rader algorithms, is that there is no restriction to primes in the transform length N. CZT can be defined for every length.

Narasimha [129] and others have noticed that in the CZT algorithm many coefficients of the FIR filter part are trivial or identical. For instance, the length-8 CZT has an FIR filter of length 16, but there are only four different complex coefficients as graphically interpreted in Fig. 6.7. These four coefficients are $1, j$, and $\pm e^{22.5^\circ}$, i.e., we have only two nontrivial real coefficients to implement.

It may be of general interest what the *maximum* DFT length for a fixed number C_N of (complex) coefficients is. This is shown in the following table.

DFT length	8	12	16	24	40	48	72	80	120	144	168	180	240	360	504
C_N	4	6	7	8	12	14	16	21	24	28	32	36	42	48	64

As mentioned before, the number of different complex coefficients does not directly correspond to the implementation effort, because some coefficients

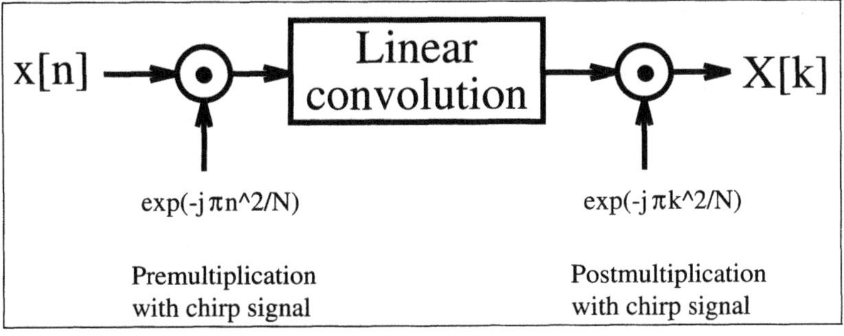

Fig. 6.6. The Bluestein chirp-z algorithm.

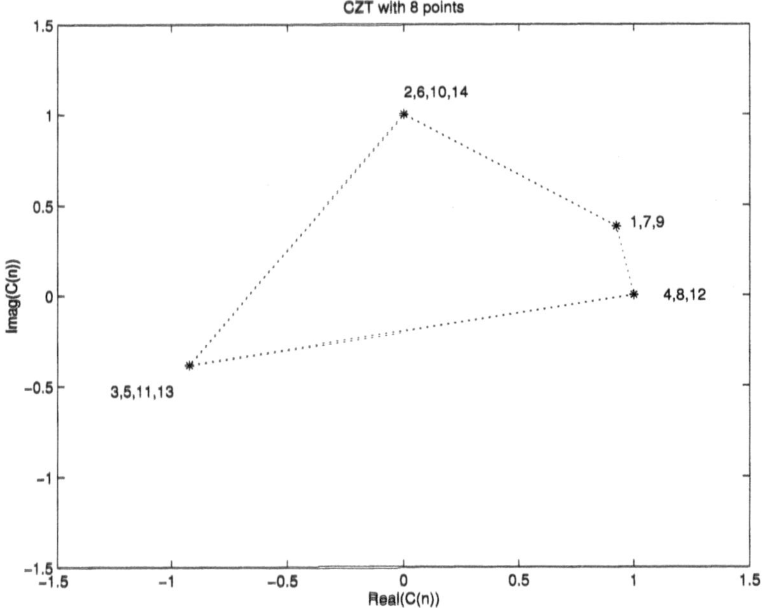

Fig. 6.7. CZT coefficients $C(n) = e^{j2\pi \frac{n^2/2 \bmod 8}{8}}; n = 1, 2, \ldots, 16.$

may be trivial (i.e., ± 1 or $\pm j$) or may show symmetry. In particular, the power-of-two length transform enjoys many symmetries, as can be seen from Fig. 6.8. If we compute the maximum DFT length for a specific number of nontrivial real coefficients, we find as maximum length transforms:

DFT length	10	16	20	32	40	48	50	80	96	160	192
sin/cos	2	3	5	6	8	9	10	11	14	20	25

Length 16 and 32 are therefore the maximum length DFTs with only 3 and 6 real multipliers, respectively!

In general, power-of-two lengths are popular FFT building blocks, and the following table therefore shows, for length $N = 2^n$, the effort when implementing the CZT filter in transposed form.

DFT length	8	16	32	64	128	256	512	1024
C_N	4	7	12	23	44	87	172	343
sin/cos	2	3	6	11	22	43	86	171
CSD	7	13	24	48	90	188	355	741
MAG	7	10	21	40	77	149	295	586
RAG	7	11	13	23	41	63	95	169

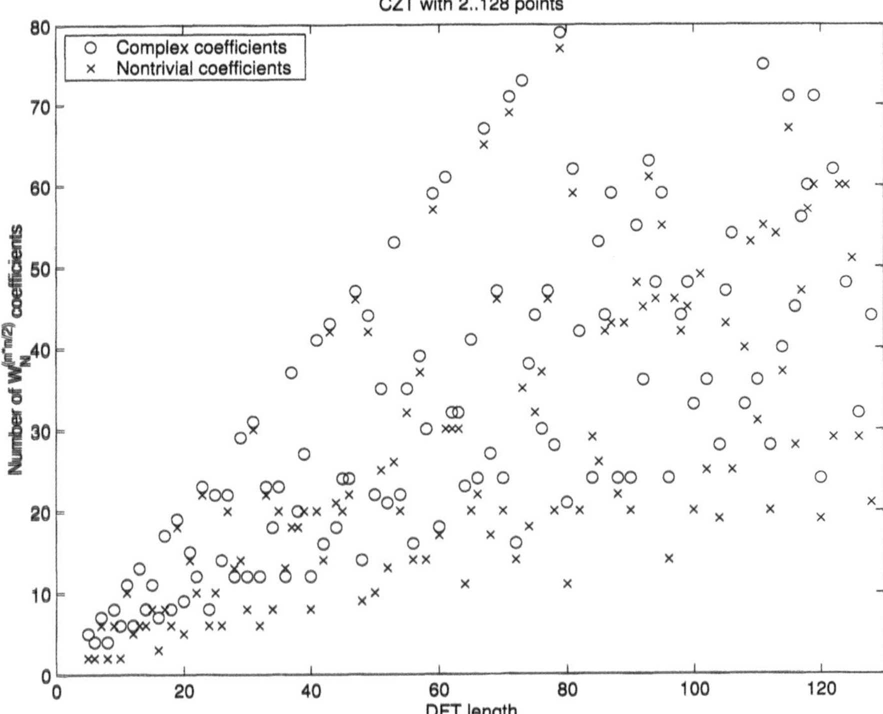

Fig. 6.8. Number of complex coefficient and nontrivial real multiplications for the CZT.

The first row shows the DFT length N. The second row shows the total number of complex exponential C_N. The worst-case effort for C_N complex coefficients is that $2C_N$ real, nontrivial coefficients must be implemented. The actual number of different nontrivial real coefficients is shown in row three. We notice when comparing of rows two and three, that for power-of-two lengths the symmetry and trivial coefficients reduce the number of nontrivial coefficients. The last three rows show, for CZT DFTs up to length 256, the effort (i.e., number of adders) for an 16-bit (15-bit unsigned plus sign bit) coefficient precision implementation, using CSD, MAG, or RAG algorithms (discussed in Chapter 2), respectively. We notice that the RAG algorithm when compared with CSD can essentially reduce the effort for DFT length larger than 32.

6.1.5 The Rader Algorithm

The Rader algorithm [130, 131] to compute a DFT,

$$X[k] = \sum_{n=0}^{N-1} x[n] W_N^{nk} \qquad k, n \in \mathbb{Z}_N; \quad \mathrm{ord}(W_N) = N \tag{6.9}$$

is defined only for prime length N. We first compute the DC component with

$$X[0] = \sum_{n=0}^{N-1} x[n]. \tag{6.10}$$

Because $N = p$ is a prime, we know from the discussion in Chapter 2 (p. 41) that there is a primitive element, a *generator* g, that generates all elements of n and k in the field $\mathbb{Z}_p$, excluding zero, i.e., $g^k \in \mathbb{Z}_p/\{0\}$. We substitute n with $g^n \bmod N$ and k through $g^k \bmod N$, and get the following index transform:

$$X[g^k \bmod N] - x[0] = \sum_{n=0}^{N-2} x[g^n \bmod N] W_N^{g^{n+k \bmod (N-1)}} \tag{6.11}$$

for $k \in \{1, 2, 3, \ldots, N-1\}$. We notice that the right side of (6.11) is a cyclic convolution, i.e.,

$$\begin{aligned} &\left[x[g^0 \bmod N], x[g^1 \bmod N], \ldots, x[g^{N-2} \bmod N]\right] \\ &\quad \circledast \left[W_N, W_N^g, \ldots, W_N^{g^{N-2 \bmod (N-1)}}\right]. \end{aligned} \tag{6.12}$$

An example with $N = 7$ demonstrates the Rader algorithms.

Example 6.5: Rader Algorithms for $N = 7$

For $N = 7$, we know that $g = 3$ is a primitive element (see, for instance, [5], Table B.7), and the index transform is

$$[3^0, 3^1, 3^2, 3^3, 3^4, 3^5] \bmod 7 \equiv [1, 3, 2, 6, 4, 5]. \tag{6.13}$$

We first compute the DC component

$$X[0] = \sum_{n=0}^{6} x[n] = x[0] + x[1] + x[2] + x[3] + x[4] + x[5] + x[6]$$

and in the second step, the cyclic convolution of $X[k] - x[0]$

$$[x[1], x[3], x[2], x[6], x[4], x[5]] \circledast [W_7, W_7^3, W_7^2, W_7^6, W_7^4, W_7^5].$$

or in matrix notation

$$\begin{bmatrix} X[1] \\ X[3] \\ X[2] \\ X[6] \\ X[4] \\ X[5] \end{bmatrix} = \begin{bmatrix} W_7^1 & W_7^3 & W_7^2 & W_7^6 & W_7^4 & W_7^5 \\ W_7^3 & W_7^2 & W_7^6 & W_7^4 & W_7^5 & W_7^1 \\ W_7^2 & W_7^6 & W_7^4 & W_7^5 & W_7^1 & W_7^3 \\ W_7^6 & W_7^4 & W_7^5 & W_7^1 & W_7^3 & W_7^2 \\ W_7^4 & W_7^5 & W_7^1 & W_7^3 & W_7^2 & W_7^6 \\ W_7^5 & W_7^1 & W_7^3 & W_7^2 & W_7^6 & W_7^4 \end{bmatrix} \begin{bmatrix} x[1] \\ x[3] \\ x[2] \\ x[6] \\ x[4] \\ x[5] \end{bmatrix} + \begin{bmatrix} x[0] \\ x[0] \\ x[0] \\ x[0] \\ x[0] \\ x[0] \end{bmatrix} \tag{6.14}$$

This is graphically interpreted using an FIR filter in Fig. 6.9.

We now verify the $p = 7$ Rader DFT formula, using a test triangular signal $x[n] = 10\lambda[n]$ (i.e., a triangle with step size 10). Directly interpreting (6.14), one obtains

$$\begin{bmatrix} X[1] \\ X[3] \\ X[2] \\ X[6] \\ X[4] \\ X[5] \end{bmatrix} = \begin{bmatrix} W_7^1 & W_7^3 & W_7^2 & W_7^6 & W_7^4 & W_7^5 \\ W_7^3 & W_7^2 & W_7^6 & W_7^4 & W_7^5 & W_7^1 \\ W_7^2 & W_7^6 & W_7^4 & W_7^5 & W_7^1 & W_7^3 \\ W_7^6 & W_7^4 & W_7^5 & W_7^1 & W_7^3 & W_7^2 \\ W_7^4 & W_7^5 & W_7^1 & W_7^3 & W_7^2 & W_7^6 \\ W_7^5 & W_7^1 & W_7^3 & W_7^2 & W_7^6 & W_7^4 \end{bmatrix} \begin{bmatrix} 20 \\ 30 \\ 40 \\ 50 \\ 60 \\ 70 \end{bmatrix} + \begin{bmatrix} 10 \\ 10 \\ 10 \\ 10 \\ 10 \\ 10 \end{bmatrix}$$

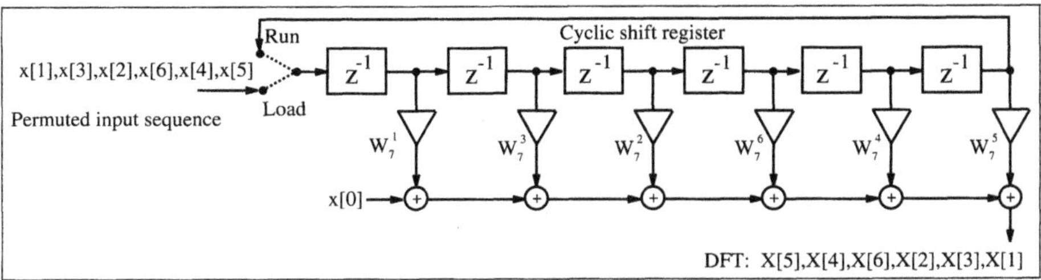

Fig. 6.9. Length $p = 7$ Rader Prime Factor DFT implementation.

$$= \begin{bmatrix} -35 + j72 \\ -35 + j8 \\ -35 + j28 \\ -35 - j72 \\ -35 - j8 \\ -35 - 28 \end{bmatrix}.$$

The value of $X[0]$ is the sum of the time series, which is $10+20+\cdots+70 = 280$.

6.5

In addition, in the Rader algorithms we may use the symmetries of the complex pairs $e^{\pm j2k\pi/N}, k \in [0, N/2]$, to build more efficient FIR realizations (Exercise 6.6, p. 252). Implementing a Rader prime factor DFT is equivalent to implementing an FIR filter, which we discussed in Chapter 3. In order to implement a fast FIR filter, a fully pipelined DA or the transposed filter structure, using the RAG algorithms, is attractive. The RAG FPGA implementation is illustrated in the following example.

Example 6.6: Rader FPGA Implementation

A RAG implementation of the length-7 Rader algorithms is accomplished as follows. The first step is quantizing the coefficients. Assuming that the input values and coefficients are to be represented as a signed 8-bit word, the quantized coefficients are:

$k =$	0	1	2	3	4	5	6
$\mathrm{Re}\{256 \cdot W_7^k\}$	256	160	−57	−231	−231	−57	160
$\mathrm{Im}\{256 \cdot W_7^k\}$	0	−200	−250	−111	111	250	200

A direct form implementation of all the individual coefficients would (consulting Table 2.3, p. 37) consume 24 adders for the constant coefficient multipliers. Using the transposed structure, the individual coefficient implementation effort is reduced to 11 adders, by exploiting the fact that several coefficients differ only in sign. Optimizing further (reduced adder graph, see Fig. 2.3, p. 36), the number of adders reaches a minimum value of 7 (see `Factor: PROCESS` and `Coeffs: PROCESS` below). This is more than a three times improvement over the direct FIR architecture. The following VHDL code[1] illustrates a possible implementation of the length-7 Rader DFT, using transposed FIR filters.

---

[1] The equivalent Verilog code **rader7.v** for this example can be found in Appendix A on page 378.

```
PACKAGE B_bit_int IS    ------> User defined types
  SUBTYPE WORD8 IS INTEGER RANGE -2**7 TO 2**7-1;
  SUBTYPE WORD11 IS INTEGER RANGE -2**10 TO 2**10-1;
  SUBTYPE WORD19 IS INTEGER RANGE -2**18 TO 2**18-1;
  TYPE ARRAY_WORD IS ARRAY (0 to 5) OF WORD19;
END B_bit_int;

LIBRARY work;
USE work.B_bit_int.ALL;

LIBRARY ieee;
USE ieee.std_logic_1164.ALL;
USE ieee.std_logic_arith.ALL;
USE ieee.std_logic_unsigned.ALL;

ENTITY rader7 IS                            ------> Interface
  PORT ( clk            : IN  STD_LOGIC;
         x_in           : IN  WORD8;
         y_real, y_imag : OUT WORD11);
END rader7;

ARCHITECTURE flex OF rader7 IS

  SIGNAL  count     : integer RANGE 0 TO 15;
  TYPE    STATE_TYPE IS (Start, Load, Run);
  SIGNAL  state     : STATE_TYPE ;
  SIGNAL  accu      : WORD11;              -- Signal for X[0]
  SIGNAL  real, imag : ARRAY_WORD;
                                        -- Tapped delay line array
  SIGNAL  x57, x111, x160, x200, x231, x250 : WORD19 ;
                         -- The (unsigned) filter coefficients
  SIGNAL  x5, x25, x110, x125, x256  : WORD19 ;
                                 -- Auxiliary filter coefficients
  SIGNAL  x, x_0 : WORD8;  -- Signals for x[0]

BEGIN

  States: PROCESS    ------> State machine for RADER filter
  BEGIN
    WAIT UNTIL clk = '1';
    CASE state IS
      WHEN Start =>            -- Initialization step
        state <= Load;
        count <= 1;
        x_0 <= x_in;           -- Save x[0]
        accu <= 0 ;            -- Reset accumulator for X[0]
         y_real  <= 0;
         y_imag  <= 0;
      WHEN Load =>    -- Apply x[5],x[4],x[6],x[2],x[3],x[1]
        IF count = 8 THEN      -- Load phase done ?
          state <= Run;
        ELSE
          state <= Load;
```

```
        accu <= accu + x ;
      END IF;
      count <= count + 1;
    WHEN Run  => -- Apply again x[5],x[4],x[6],x[2],x[3]
      IF count = 15 THEN     -- Run phase done ?
        y_real  <= accu;     -- X[0]
        y_imag  <= 0;  -- Only re inputs i.e. Im(X[0])=0
        state <= Start;      -- Output of result
      ELSE                   -- and start again
        y_real  <= real(0) / 256 + x_0;
        y_imag  <= imag(0) / 256;
        state <= Run;
      END IF;
      count <= count + 1;
  END CASE;
END PROCESS States;

Structure: PROCESS      -- Structure of the two FIR
BEGIN                   -- filters in transposed form
  WAIT UNTIL clk = '1';
  x <= x_in;
  -- Real part of FIR filter in transposed form
  real(0) <= real(1) + x160  ;    -- W^1
  real(1) <= real(2) - x231  ;    -- W^3
  real(2) <= real(3) - x57   ;    -- W^2
  real(3) <= real(4) + x160  ;    -- W^6
  real(4) <= real(5) - x231  ;    -- W^4
  real(5) <= -x57  ;              -- W^5

  -- Imaginary part of FIR filter in transposed form
  imag(0) <= imag(1) - x200  ;    -- W^1
  imag(1) <= imag(2) - x111  ;    -- W^3
  imag(2) <= imag(3) - x250  ;    -- W^2
  imag(3) <= imag(4) + x200  ;    -- W^6
  imag(4) <= imag(5) + x111  ;    -- W^4
  imag(5) <= x250;                -- W^5
END PROCESS Structure;

Coeffs: PROCESS       -- Note that all signals
BEGIN                 -- are globally defined
  WAIT UNTIL clk = '1';
-- Compute the filter coefficients and use FFs
  x160    <= x5 * 32;
  x200    <= x25 * 8;
  x250    <= x125 * 2;
  x57     <= x25 + x * 32;
  x111    <= x110 + x;
  x231    <= x256 - x25;
END PROCESS Coeffs;

Factors: PROCESS (x, x5, x25)     -- Note that all signals
BEGIN                             -- are globally defined
-- Compute the auxiliary factor for RAG without an FF
```

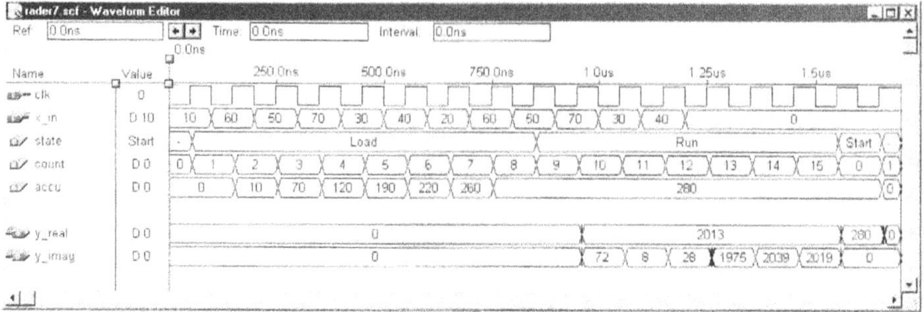

Fig. 6.10. VHDL simulation of a 7-point Rader algorithm.

```
      x5     <= x * 4 + x;
      x25    <= x5 * 4 + x5;
      x110   <= x25 * 4 + x5 * 2;
      x125   <= x25 * 4 + x25;
      x256   <= x * 256;
    END PROCESS Factors;

  END flex;
```

The design consists of four blocks of statements within the four `PROCESS` statements. The first – "`Stages: PROCESS`" – is the state machine, which distinguishes the three processing phases, `Start`, `Load`, and `Run`. The second – "`Structure: PROCESS`" – defines the two FIR filter paths, real and imaginary, respectively. The third item implements the multiplier block using the reduced adder graph. The forth block – "`Factor: PROCESS`" – implements the unregistered factors of the RAG algorithm. It can be seen that all coefficients are realized by using six adders and one subtractors. The design consumes 494 LCs, and runs at 29.94 MHz `Registered Performance`. Fig. 6.10 displays simulation results using MaxPlusII for a triangle input sequence $x[n] = \{10, 20, 30, 40, 50, 60, 70\}$. Notice that the input and output sequences, starting at 950 ns, occur in the permuted order, and negative results appear as unsigned positive numbers. Finally, at 1.55 μs, $X[0]$ is forwarded to the output and `rader7` is ready to process the next input frame. 6.6

Because the Rader algorithm is restricted to prime lengths there is less symmetry in the coefficients, compared with the CZT. The following table shows, for primes length $2^n \pm 1$, the implementation effort of the circular filter in transposed form.

DFT length #Coeffs	7	17	31	61	127	257
sin/cos	6	16	30	60	124	253
CSD	28	76	126	265	553	1075
MAG	24	57	98	219	443	873
RAG	14	36	45	73	131	253

The first row shows, the cyclic convolution length N, which is also the number of complex coefficients. Comparing row two and the worst case with $2N$ real sin/cos coefficients, we see that symmetry and trivial coefficients reduce the number of nontrivial coefficients by a factor of 2. The last three rows show the effort for an 16-bit coefficient precision implementation using CSD, MAG, or RAG algorithms, respectively. Note the advantage of RAG for longer filters. It can be seen from the above table that the effort for CSD-type filters can be estimated with $BN/4$, where B is the coefficient bit width (16 in this table) and N is the filter length. For RAG, the effort (i.e., number of adders) is only N, i.e., a factor $B/4$ improvement over CSD for longer filters (for $B = 16$, a factor 16/4=4 of improvement). For longer filters, RAG needs only one additional adder for each additional coefficient, because the already-synthesized coefficient produces a "dense" grid of small coefficients.

6.1.6 The Winograd DFT Algorithm

The first algorithm with a reduced number of necessary multiplications we wish to discuss is the *Winograd DFT* algorithm. The Winograd algorithm is a combination of the Rader algorithm (which translates the DFT into a cyclic convolution), and Winograd's [83] short convolution algorithms, which we have already used to implement fast running FIR filters (see Sect. 5.2.2, p. 152).

The length is therefore restricted to primes or powers of primes. Table 6.2 gives an overview of the necessary number of arithmetic operations.

The following example for $N = 5$ demonstrates the steps to build a Winograd DFT algorithm.

Example 6.7: $N = 5$ Winograd DFT Algorithm

An alternative representation of the Rader algorithm, using $X[0]$ instead of $x[0]$, is given by [5]

$$X[0] = \sum_{n=0}^{4} x[n] = x[0] + x[1] + x[2] + x[3] + x[4]$$

$$X[k] - X[0] = [x[1], x[2], x[4], x[3]] \circledast [W_5 - 1, W_5^2 - 1, W_5^4 - 1, W_5^3 - 1] \quad k = 1, 2, 3, 4$$

If we implement the cyclic convolution of length 4 with a Winograd algorithm which costs only five nontrivial multiplications, we get the following algorithm:

Table 6.2. Effort for the Winograd DFT with real inputs. Trivial multiplications are those by ± 1 or $\pm j$. For complex inputs, the number of operation is twice as large.

Block length	Total number of real multiplications	Total number nontrivial multiplications	Total number of real additions
2	2	0	2
3	3	2	6
4	4	0	8
5	6	5	17
7	9	8	36
8	8	2	26
9	11	10	44
11	21	20	84
13	21	20	94
16	18	10	74
17	36	35	157
19	39	38	186

$$X[k] = \sum_{n=0}^{4} x[n] e^{-j2\pi kn/5} \qquad k = 0, 1, \ldots, 4$$

$$\begin{bmatrix} X[0] \\ X[1] \\ X[2] \\ X[3] \\ X[4] \end{bmatrix} = \begin{bmatrix} 1 & 0 & 0 & 0 & 0 & 0 \\ 1 & 1 & 1 & 1 & 0 & -1 \\ 1 & 1 & -1 & 1 & 1 & 0 \\ 1 & 1 & -1 & -1 & -1 & 0 \\ 1 & 1 & 1 & -1 & 0 & 1 \end{bmatrix}$$

$$\cdot \mathrm{diag}(1, \frac{1}{2}(\cos(2\pi/5) + \cos(4\pi/5)) - 1,$$

$$\frac{1}{2}(\cos(2\pi/5) - \cos(4\pi/5)), j\sin(2\pi/5),$$

$$j(-\sin(2\pi/5) + \sin(4\pi/5)), j(\sin(2\pi/5) + \sin(4\pi/5)))$$

$$\cdot \begin{bmatrix} 1 & 1 & 1 & 1 & 1 \\ 0 & 1 & 1 & 1 & 1 \\ 0 & 1 & -1 & -1 & 1 \\ 0 & 1 & -1 & 1 & -1 \\ 0 & 1 & 0 & 0 & -1 \\ 0 & 0 & -1 & 1 & 0 \end{bmatrix} \begin{bmatrix} x[0] \\ x[1] \\ x[2] \\ x[3] \\ x[4] \end{bmatrix}.$$

The total computational effort is therefore only 5 or 10 real multiplications for real or imaginary input sequences $x[n]$, respectively. 6.7

It is quite convenient to use a matrix notation for the Winograd DFT algorithm, and so we get

$$\boldsymbol{W}_{N_l} = \boldsymbol{C}_l \cdot \boldsymbol{B}_l \cdot \boldsymbol{A}_l. \tag{6.15}$$

where $\boldsymbol{A}_l$ incorporates the input addition, $\boldsymbol{B}_l$ is the diagonal matrix with the Fourier coefficients, and $\boldsymbol{C}_l$ includes the output additions. The only disadvantage is that now it is not as easy to define the exact steps of the short convolution algorithms, because the sequence, in which input and output additions are computed is lost with this matrix representation.

This combination of Rader algorithms and a short Winograd convolution, known as the Winograd DFT algorithm, will be used later, together with index mapping to introduce the Winograd FFT algorithm. This is the FFT algorithm with the least number of real multiplications among all known FFT algorithms.

6.2 The Fast Fourier Transform (FFT) Algorithms

As mentioned in the introduction of this chapter, we use the terminology introduced by Burrus [109], who classified all FFT algorithms simply by different (multidimensional) index maps of the input and output sequences. These are based on a transform of the length N DFT (6.2)

$$X[k] = \sum_{n=0}^{N-1} x[n]\, W_N^{nk} \tag{6.16}$$

into a multidimensional $N = \prod_l N_l$ representation. It is, in general, sufficient to discuss only the two-factor case, because higher dimensions can be built simply by iteratively replacing again one of these factors. To simplify our representation we will therefore discuss the three FFT algorithms presented only in terms of a two-dimensional index transform.

We transform the (time) index n with

$$n = An_1 + Bn_2 \bmod N \qquad \begin{cases} 0 \leq n_1 \leq N_1 - 1 \\ 0 \leq n_2 \leq N_2 - 1 \end{cases} \tag{6.17}$$

where $N = N_1 N_2$, and $A, B \in \mathbb{Z}$ are constants which must be defined later. Using this index transform, a two dimensional mapping $f : \mathbb{C}^N \to \mathbb{C}^{N_1 \times N_2}$ of the data is built, according to

$$\begin{aligned} &[x[0]\ x[1]\ x[2] \cdots x[N-1]] \\ &= \begin{bmatrix} x[0,0] & x[0,1] & \cdots & x[0,N_2-1] \\ x[1,0] & x[1,1] & \cdots & x[1,N_2-1] \\ \vdots & \vdots & \ddots & \vdots \\ x[N_1-1,0] & x[N_1-1,1] & \cdots & x[N_1-1,N_2-1] \end{bmatrix} \end{aligned} \tag{6.18}$$

Applying another index mapping k to the output (frequency) domain yields

$$k = Ck_1 + Dk_2 \bmod N \qquad \begin{cases} 0 \leq k_1 \leq N_1 - 1 \\ 0 \leq k_2 \leq N_2 - 1 \end{cases} \tag{6.19}$$

where $C, D \in \mathbb{Z}$ are constants which must be defined later. Because the DFT is bijective, we must choose A, B, C, and D in such a way that the transform representation is still unique, i.e., a bijective projection. Burrus [109] has determined the general conditions for how to choose A, B, C, and D for specific N_1 and N_2 such that the mapping is bijective (see Exercises 6.8 and 6.9, p. 252). The transforms given in this chapter are all unique.

An important point in distinguishing different FFT algorithms is the question of whether N_1 and N_2 are allowed to have a common factor, i.e., $\gcd(N_1, N_2) > 1$, or whether the factors must be co-prime. Sometimes algorithms with $\gcd(N_1, N_2) > 1$ are referred to as *common factor algorithms* (CFAs), and algorithms with $\gcd(N_1, N_2) = 1$ are called *prime factor algorithms* (PFAs). A CFA algorithm discussed in the following is the Cooley-Tukey FFT, while the Good-Thomas and Winograd FFTs are of the PFA type. It should be emphasized that the Cooley-Tukey algorithm may indeed realize FFTs with two factors, $N = N_1 N_2$, which are co-prime, and that for a PFA the factors N_1 and N_2 must only be co-prime, i.e., they must *not* be primes themselves. A transform of length $N = 12$ factored with $N_1 = 4$ and $N_2 = 3$, for instance, can therefore be used for both CFA FFTs and PFA FFTs!

6.2.1 The Cooley-Tukey FFT Algorithm

The Cooley-Tukey FFT is the most universal of all FFT algorithms, because any factorization of N is possible. The most popular Cooley-Tukey FFTs are those where the transform length N is a power of a basis r, i.e., $N = r^\nu$. These algorithms are often referred to as radix-r algorithms.

The index transform suggested by Cooley and Tukey (and earlier by Gauss) is also the most simple index mapping. Using (6.17) we have $A = N_2$ and $B = 1$, and the following mapping results

$$n = N_2 n_1 + n_2 \qquad \begin{cases} 0 \leq n_1 \leq N_1 - 1 \\ 0 \leq n_2 \leq N_2 - 1. \end{cases} \tag{6.20}$$

From the valid range of n_1 and n_2, we conclude that the modulo reduction given by (6.17) need not be explicitly computed.

For the inverse mapping from (6.19) Cooley and Tukey, choose $C = 1$ and $D = N_1$, and the following mapping results

$$k = k_1 + N_1 k_2 \qquad \begin{cases} 0 \leq k_1 \leq N_1 - 1 \\ 0 \leq k_2 \leq N_2 - 1. \end{cases} \tag{6.21}$$

The modulo computation can also be skipped in this case. If we now substitute n and k in W_N^{nk} according to (6.20) and (6.21), respectively, we find

$$W_N^{nk} = W_N^{N_2 n_1 k_1 + N_1 N_2 n_1 k_2 + n_2 k_1 + N_1 n_2 k_2}. \tag{6.22}$$

Because W is of order $N = N_1 N_2$, it follows that $W_N^{N_1} = W_{N_2}$ and $W_N^{N_2} = W_{N_1}$. This simplifies (6.22) to

$$W_N^{nk} = W_{N_1}^{n_1 k_1} W_N^{n_2 k_1} W_{N_2}^{n_2 k_2}. \tag{6.23}$$

If we now substitute (6.23) in the DFT from (6.16) it follows that

$$X[k_1,k_2] = \sum_{n_2=0}^{N_2-1} W_{N_2}^{n_2 k_2} \underbrace{\left(W_N^{n_2 k_1} \underbrace{\sum_{n_1=0}^{N_1-1} x[n_1,n_2] W_{N_1}^{n_1 k_1}}_{N_1\text{-point transform}} \right)}_{\bar{x}[n_2,k_1]} \tag{6.24}$$

$$= \underbrace{\sum_{n_2=0}^{N_2-1} W_{N_2}^{n_2 k_2}\, \bar{x}[n_2,k_1]}_{N_2\text{-point transform}} \tag{6.25}$$

We now can define the complete Cooley-Tukey algorithm

Algorithm 6.8: Cooley-Tukey Algorithm

An $N = N_1 N_2$-point DFT can be done using the following steps:

1) Compute an index transform of the input sequence according to (6.20).
2) Compute the N_2 DFTs of length N_1.
3) Apply the twiddle-factors $W_N^{n_2 k_1}$ to the output of the first transform stage.
4) Compute N_1 DFTs of length N_2.
5) Compute an index transform of the output sequence according to (6.21).

The following length-12 transform demonstrates these steps.

Example 6.9: Cooley-Tukey FFT for $N = 12$

Assume $N_1 = 4$ and $N_2 = 3$. It follows then that $n = 3n_1 + n_2$ and $k = k_1 + 4k_2$, and we can compute the following tables for the index mappings:

n_2	n_1 = 0	1	2	3
0	$x[0]$	$x[3]$	$x[6]$	$x[9]$
1	$x[1]$	$x[4]$	$x[7]$	$x[10]$
2	$x[2]$	$x[5]$	$x[8]$	$x[11]$

k_2	k_1 = 0	1	2	3
0	$X[0]$	$X[1]$	$X[2]$	$X[3]$
1	$X[4]$	$X[5]$	$X[6]$	$X[7]$
2	$X[8]$	$X[9]$	$X[10]$	$X[11]$

With the help of this transform we can construct the signal flow graph shown in Fig. 6.11. It can be seen that first we must compute three DFTs with four points each, followed by the multiplication with the twiddle-factors, and finally we compute four DFTs each having length 3. 6.9

For direct computation of the 12-point DFT, a total of $12^2 = 144$ complex multiplications and $11^2 = 121$ complex additions are needed. To compute the Cooley-Tukey FFT with the same length we need a total of 12 complex multiplication for the twiddle-factors, of which 8 are trivial (i.e., ± 1 or $\pm j$)

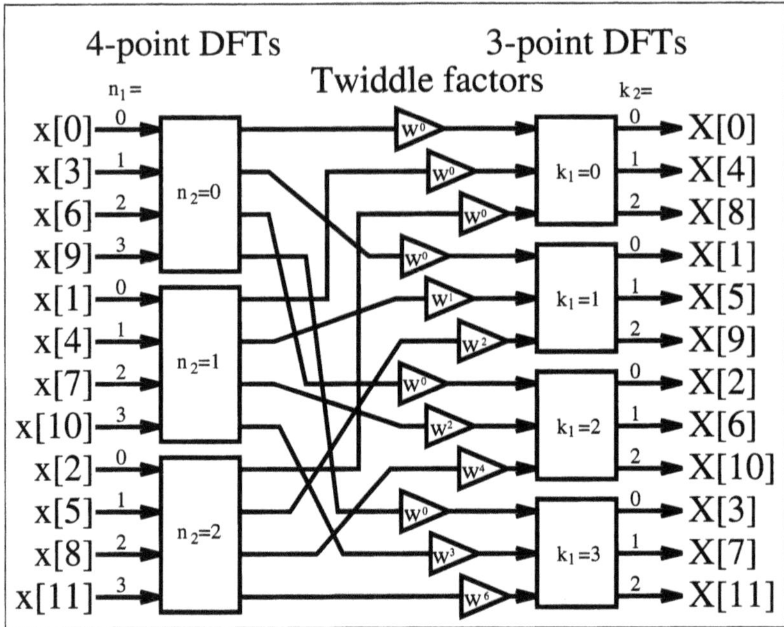

Fig. 6.11. Cooley-Tukey FFT for $N = 12$.

multiplications. According to Table 6.2 (p. 226), the length-4 DFTs can be computed using 8 real additions and no multiplications. For the length-3 DFTs, we need 4 multiplications and 3 additions. If we implement the (fixed coefficient) complex multiplications using 3 additions and 3 multiplications (see Algorithm 6.10, p. 233), the total effort for the 12-point Cooley-Tukey FFT is given by

$$3 \cdot 16 + 4 \cdot 3 + 4 \cdot 12 = 108 \quad \text{real additions and}$$
$$4 \cdot 3 + 4 \cdot 4 = 28 \quad \text{real multiplications.}$$

For the direct implementation we would need $2 \cdot 11^2 + 12^2 \cdot 3 = 674$ real additions and $12^2 \cdot 3 = 432$ real multiplications. It is now obvious why the Cooley-Tukey algorithm is called the "Fast Fourier Transform" (FFT).

Radix-r Cooley-Tukey Algorithm

One important fact that distinguishes the Cooley-Tukey algorithm from other FFT algorithms is that the factors for N can be chosen arbitrarily. It is therefore possible to use a radix-r algorithm in which $N = r^S$. The most popular algorithms are those of basis $r = 2$ or $r = 4$, because the necessary basic DFTs can, according to Table 6.2 (p. 226), be implemented without any multiplications. For $r = 2$ and S stages, for instance, the following index mapping results

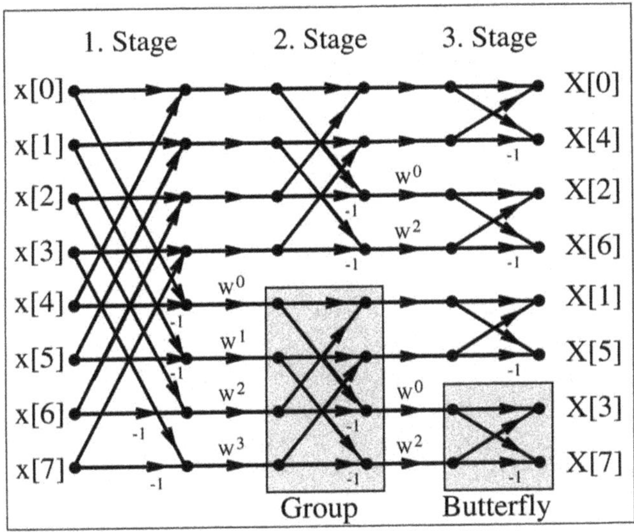

Fig. 6.12. Decimation-in-frequency algorithm of length 8 for radix-2.

$$n = 2^{S-1}n_1 + \cdots + 2n_{S-1} + n_S \tag{6.26}$$
$$k = k_1 + 2k_2 + \cdots + 2^{S-1}k_S. \tag{6.27}$$

For $S > 2$ a common practice is that in the signal flow graph a 2-point DFT is represented with a *Butterfly*, as shown in Fig. 6.12 for an 8-point transform. The signal flow graph representation has been simplified by using the fact that all arriving arrows at a node are added, while the constant coefficient multiplications are symbolized through a factor at an arrow. A radix-r algorithm has $\log_r(N)$ *stages*, and for each *group* the same type of twiddle-factor occurs.

It can be seen from the signal flow graph in Fig. 6.12 that the computation can be done *"in-place,"* i.e., the memory location used by a butterfly can be overwritten, because the data are no longer needed in the next computational steps. The total number of twiddle-factor multiplications for the radix-2 transform is given by

$$\log_2(N)N/2 \tag{6.28}$$

because only every second arrow has a twiddle-factor.

Because the algorithm shown in Fig. 6.12 starts in the frequency domain to split the original DFT into shorter DFTs, this algorithm is called a decimation-in-frequency (DIF) algorithm. The input values typically occur in natural order, while the index of the frequency values is in bit-reversed order. Table 6.3 shows the characteristic values of a DIF radix-2 algorithm.

We may also construct an algorithm with decimation in time (DIT). In this case, we start by splitting the input (time) sequence, and we find that all frequency values will appear in natural order (Exercise 6.11, p. 253).

Table 6.3. Radix-2 FFT with frequency decimation.

	Stage 1	Stage 2	Stage 3	⋯	Stage $\log_2(N)$
Number of groups	1	2	4	⋯	$N/2$
Butterflies per group	$N/2$	$N/4$	$N/8$	⋯	1
Increment exponent twiddle-factors	1	2	4	⋯	$N/2$

The necessary index transform for index 41, for an radix-2 and radix-4 algorithm, is shown in Fig. 6.13. For a radix-2 algorithm, a reversing of the bit sequence, a *bitreverse*, is necessary. For a radix-4 algorithm we must first build "digits" of two bits, and then reverse the order of these digits. This operation is called *digitreverse*.

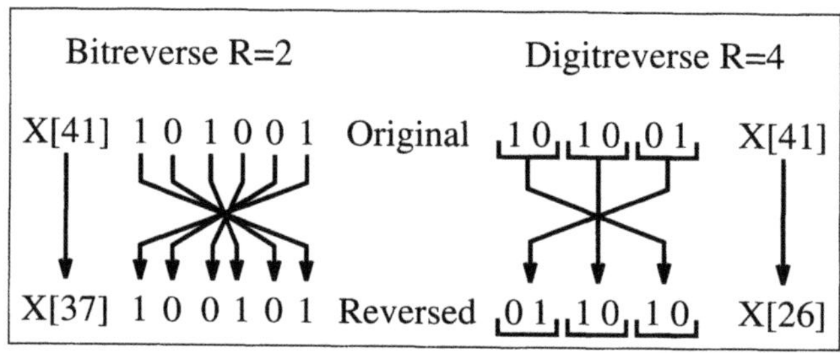

Fig. 6.13. Bitreverse and digitreverse.

Radix-2 Cooley-Tukey Algorithm Implementation

A radix-2 FFT can be efficiently implemented using a butterfly processor which includes, besides the butterfly itself, an additional complex multiplier for the twiddle-factors.

A radix-2 butterfly processor consists of a complex adder, a complex subtraction, and a complex multiplier for the twiddle factors. The complex multiplication with the twiddle-factor is often implemented with four real multiplications and two add/subtract operations. However, it is also possible to build the complex multiplier with only three real multiplications and three add/subtract operations, because one operand is precomputed. The algorithm works as follows:

Algorithm 6.10: Efficient Complex Multiplier

The complex twiddle-factor multiplication $R+j\cdot I = (X+j\cdot Y)(C+j\cdot S)$ can be simplified, because C and S are precomputed and stored in a table. It is therefore also possible to store the following three coefficients

$$C, \quad C+S, \quad \text{and} \quad C-S. \tag{6.29}$$

With these three precomputed factors we first compute

$$E = X - Y, \quad \text{and then} \quad Z = C \cdot E = C \cdot (X - Y). \tag{6.30}$$

We can then compute the final product using

$$R = (C - S) \cdot Y + Z \tag{6.31}$$

$$I = (C + S) \cdot X - Z. \tag{6.32}$$

To check:

$$\begin{aligned} R &= (C-S)Y + C(X-Y) \\ &= CY - SY + CX - CY = CX - SY \checkmark \\ I &= (C+S)X - C(X-Y) \\ &= CX + SX - CX + CY = CY + SX. \checkmark \end{aligned}$$

The algorithm uses three multiplications, one addition, and two subtractions, at the cost of an additional, third table.

The following example demonstrates the implementation of this twiddle-factor complex multiplier.

Example 6.11: Twiddle-Factor Multiplier

Let us first choose some concrete design parameters for the twiddle factor multiplier. Let's assume we have 8-bit input data, the coefficients should have 8 bits plus sign, and we wish to multiply by $e^{j\pi/9} = e^{j20^\circ}$. Quantized to 8 bits, the twiddle factor becomes $C + jS = 256 \cdot e^{j\pi/9} = 121 + j39$. If we use an input value of $70 + j50$, then the expected result is

$$\begin{aligned} (70 + j50)e^{j\pi/9} &= (70 + j50)(121 + j39)/256 \\ &= (6520 + j8780)/256 = 25 + j34. \end{aligned}$$

If we use Algorithm 6.10 to compute the complex multiplication, the three factors become:

$$C = 121, \quad C + S = 160, \quad \text{and} \quad C - S = 82.$$

We note from the above that, in general, the tables $C + S$ and $C - S$ must have one more bit of precision than the C and S tables.

The following VHDL code[2] implements the twiddle factor multiplier.

```
LIBRARY lpm;
USE lpm.lpm_components.ALL;

LIBRARY ieee;
USE ieee.std_logic_1164.ALL;
USE ieee.std_logic_arith.ALL;

ENTITY ccmul IS
  GENERIC (W2   : INTEGER := 17;     -- Multiplier bit width
```

[2] The equivalent Verilog code ccmul.v for this example can be found in Appendix A on page 380.

```
            W1   : INTEGER := 9;      -- Bit width c+s sum
            W    : INTEGER := 8);     -- Input bit width
    PORT (clk   : STD_LOGIC;  -- Clock for the output register
          x_in, y_in, c_in                              -- Inputs
                      : IN  STD_LOGIC_VECTOR(W-1 DOWNTO 0);
          cps_in, cms_in                                -- Inputs
                     : IN  STD_LOGIC_VECTOR(W1-1 DOWNTO 0);
          r_out, i_out                                 -- Results
                     : OUT STD_LOGIC_VECTOR(W-1 DOWNTO 0));
  END ccmul;

  ARCHITECTURE flex OF ccmul IS

    SIGNAL x, y, c : STD_LOGIC_VECTOR(W-1 DOWNTO 0);
                                           -- Inputs and outputs
    SIGNAL r, i, cmsy, cpsx, xmyc                     -- Products
                         : STD_LOGIC_VECTOR(W2-1 DOWNTO 0);
    SIGNAL xmy, cps, cms, sxtx, sxty                   -- x-y etc.
                         : STD_LOGIC_VECTOR(W1-1 DOWNTO 0);

  BEGIN
      x   <= x_in;    -- x
      y   <= y_in;    -- j * y
      c   <= c_in;    -- cos
      cps <= cps_in; -- cos + sin
      cms <= cms_in; -- cos - sin

    PROCESS
    BEGIN
      WAIT UNTIL clk='1';
      r_out <= r(W2-3 DOWNTO W-1);     -- Scaling and FF
      i_out <= i(W2-3 DOWNTO W-1);     -- for output
    END PROCESS;
  ---------- ccmul with 3 mul. and 3 add/sub ---------------
    sxtx  <= x(x'high) & x;         -- Possible growth for
    sxty  <= y(y'high) & y;         -- sub_1 -> sign extension

    sub_1: lpm_add_sub                          -- Sub:  x - y;
      GENERIC MAP (  LPM_WIDTH => W1, LPM_DIRECTION => "SUB",
                     LPM_REPRESENTATION => "SIGNED")
      PORT MAP (dataa => sxtx, datab => sxty, result => xmy);

    mul_1: lpm_mult                   -- Multiply  (x-y)*c = xmyc
      GENERIC MAP ( LPM_WIDTHA => W1, LPM_WIDTHB => W,
                    LPM_WIDTHP => W2, LPM_WIDTHS => W2,
                    LPM_REPRESENTATION => "SIGNED")
      PORT MAP ( dataa => xmy, datab => c, result => xmyc);

    mul_2: lpm_mult                    -- Multiply (c-s)*y = cmsy
      GENERIC MAP ( LPM_WIDTHA => W1, LPM_WIDTHB => W,
                    LPM_WIDTHP => W2, LPM_WIDTHS => W2,
                    LPM_REPRESENTATION => "SIGNED")
      PORT MAP ( dataa => cms, datab => y, result => cmsy);
```

```
  mul_3: lpm_mult                      -- Multiply (c+s)*x = cpsx
    GENERIC MAP ( LPM_WIDTHA => W1, LPM_WIDTHB => W,
                  LPM_WIDTHP => W2, LPM_WIDTHS => W2,
                  LPM_REPRESENTATION => "SIGNED")
    PORT MAP ( dataa => cps, datab => x, result => cpsx);

  sub_2: lpm_add_sub          -- Sub: i <= (c-s)*x - (x-y)*c;
    GENERIC MAP ( LPM_WIDTH => W2, LPM_DIRECTION => "SUB",
                  LPM_REPRESENTATION => "SIGNED")
    PORT MAP ( dataa => cpsx, datab => xmyc, result => i);

  add_1: lpm_add_sub          -- Add: r <= (x-y)*c + (c+s)*y;
    GENERIC MAP ( LPM_WIDTH => W2, LPM_DIRECTION => "ADD",
                  LPM_REPRESENTATION => "SIGNED")
    PORT MAP ( dataa => cmsy, datab => xmyc, result => r);

END flex;
```

The twiddle-factor multiplier is implemented using component instantiations of three `lpm_mult` and three `lpm_add_sub` modules. The output is scaled such that it has the same data format as the input. This is reasonable, because multiplication with a complex exponential $e^{j\phi}$ should not change the magnitude of the complex input. To ensure short latency (for an in-place FFT), the complex multiplier only has output registers, with no internal pipeline registers. With only one output register, it's impossible to determine the **Registered Performance** of the design, but from the simulation results in Fig. 6.14, it can be estimated. The design uses 493 LCs and may run faster, if the `lpm_mult` components can be pipelined (see Fig. 2.15, p. 58). 6.11

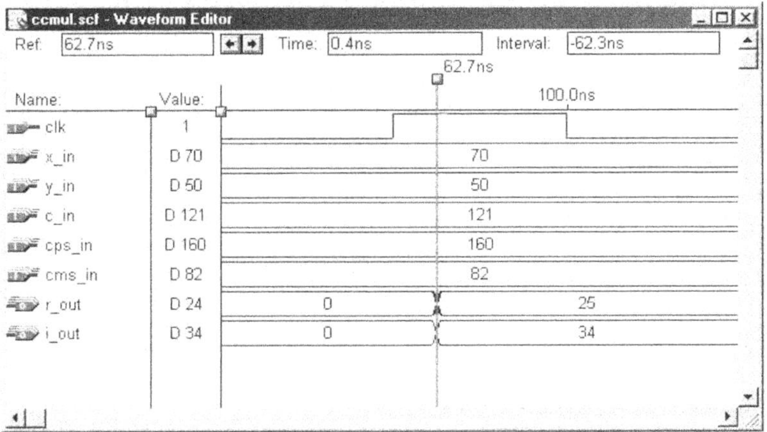

Fig. 6.14. VHDL simulation of a twiddle-factor multiplier.

An in-place implementation, i.e., with only *one* data memory, is now possible, because the butterfly processor is designed without pipeline stages. If we introduce additional pipeline-stages (one for the butterfly and three for the multiplier) the size of the design will increases little (see Exercise 6.24, p. 255) but speed increases. The price for this pipeline design is the cost for extra data memory for the whole FFT, because data read and write memories must now be separated, i.e., no in-place computation can be done.

Using the twiddle-factor multiplier introduced above, it is now possible to design a butterfly processor for a radix-2 Cooley-Tukey FFT.

Example 6.12: Butterfly Processor

To prevent overflow in the arithmetic, the butterfly processor computes the two (scaled) butterfly equations

$$D_{\mathrm{re}} + j \cdot D_{\mathrm{im}} = ((A_{\mathrm{re}} + j \cdot A_{\mathrm{im}}) + (B_{\mathrm{re}} + j \cdot B_{\mathrm{im}}))/2$$

$$E_{\mathrm{re}} + j \cdot E_{\mathrm{im}} = ((A_{\mathrm{re}} + j \cdot A_{\mathrm{im}}) - (B_{\mathrm{re}} + j \cdot B_{\mathrm{im}}))/2$$

Then the temporary result $E_{\mathrm{re}} + j \cdot E_{\mathrm{im}}$ must be multiplied by the twiddle-factor.

The VHDL code[3] of the whole butterfly processor is shown in the following.

```
LIBRARY lpm;
USE lpm.lpm_components.ALL;

LIBRARY ieee;
USE ieee.std_logic_1164.ALL;
USE ieee.std_logic_arith.ALL;

PACKAGE mul_package IS     -- User defined components
  COMPONENT ccmul
    GENERIC (W2   : INTEGER := 17;    -- Multiplier bit width
             W1   : INTEGER := 9;     -- Bit width c+s sum
             W    : INTEGER := 8);    -- Input bit width
    PORT
    (clk    : IN STD_LOGIC; -- Clock for the output register
     x_in, y_in, c_in: IN  STD_LOGIC_VECTOR(W-1 DOWNTO 0);
                                                    -- Inputs
     cps_in, cms_in   : IN  STD_LOGIC_VECTOR(W1-1 DOWNTO 0);
                                                    -- Inputs
     r_out, i_out     : OUT STD_LOGIC_VECTOR(W-1 DOWNTO 0));
                                                   -- Results
  END COMPONENT;
END mul_package;

LIBRARY work;
USE work.mul_package.ALL;


LIBRARY ieee;
USE ieee.std_logic_1164.ALL;
USE ieee.std_logic_arith.ALL;
```

[3] The equivalent Verilog code `bfproc.v` for this example can be found in Appendix A on page 382.

```
LIBRARY lpm;
USE lpm.lpm_components.ALL;

LIBRARY ieee;
USE ieee.std_logic_1164.ALL;
USE ieee.std_logic_arith.ALL;
USE ieee.std_logic_unsigned.ALL;

ENTITY bfproc IS
  GENERIC (W2  : INTEGER := 17;    -- Multiplier bit width
           W1  : INTEGER := 9;     -- Bit width c+s sum
           W   : INTEGER := 8);    -- Input bit width
  PORT
  (clk                 : STD_LOGIC;
   Are_in, Aim_in, c_in,                     -- 8 bit inputs
   Bre_in, Bim_in      : IN  STD_LOGIC_VECTOR(W-1 DOWNTO 0);
   cps_in, cms_in      : IN  STD_LOGIC_VECTOR(W1-1 DOWNTO 0);
                                       -- 9 bit coefficients
   Dre_out, Dim_out,                         -- 8 bit results
   Ere_out, Eim_out    : OUT STD_LOGIC_VECTOR(W-1 DOWNTO 0));
END bfproc;

ARCHITECTURE flex OF bfproc IS

  SIGNAL  dif_re, dif_im                          -- Bf out
                         : STD_LOGIC_VECTOR(W-1 DOWNTO 0);
  SIGNAL  Are, Aim, Bre, Bim : integer RANGE -128 TO 127;
                                       -- Inputs as integers
  SIGNAL  c             : STD_LOGIC_VECTOR(W-1 DOWNTO 0);
                                                  -- Input
  SIGNAL  cps, cms      : STD_LOGIC_VECTOR(W1-1 DOWNTO 0);
                                                -- Coeff in
  SIGNAL  Cre, Cim      : STD_LOGIC_VECTOR(W-1 DOWNTO 0);
                                                -- Results

BEGIN

  PROCESS   -- Compute the additions of the butterfly using
  BEGIN      -- integers and store inputs in flip-flops
    WAIT UNTIL clk = '1';
    Are     <= CONV_INTEGER(Are_in);
    Aim     <= CONV_INTEGER(Aim_in);
    Bre     <= CONV_INTEGER(Bre_in);
    Bim     <= CONV_INTEGER(Bim_in);
    c       <= c_in;                  -- Load from memory cos
    cps     <= cps_in;            -- Load from memory cos+sin
    cms     <= cms_in;            -- Load from memory cos-sin
    Dre_out <= CONV_STD_LOGIC_VECTOR( (Are + Bre )/2, W);
    Dim_out <= CONV_STD_LOGIC_VECTOR( (Aim + Bim )/2, W);
  END PROCESS;

  -- No FF because butterfly difference "diff" is not an
  PROCESS (Are, Bre, Aim, Bim)              -- output port
```

```
  BEGIN
    dif_re <= CONV_STD_LOGIC_VECTOR(Are/2 - Bre/2, 8);
    dif_im <= CONV_STD_LOGIC_VECTOR(Aim/2 - Bim/2, 8);
  END PROCESS;

---- Instantiate the complex twiddle factor multiplier ----
  ccmul_1: ccmul                        -- Multiply (x+jy)(c+js)
    GENERIC MAP ( W2 => W2, W1 => W1, W => W)
    PORT MAP  ( clk => clk, x_in => dif_re, y_in => dif_im,
                c_in => c, cps_in => cps, cms_in => cms,
                r_out => Ere_out, i_out => Eim_out);

END flex;
```

The butterfly processor is implemented using one adder, one subtraction, and the twiddle-factor multiplier instantiated as a component. Flip-flops have been implemented for input A, B, the three table values, and the output port D, in order to have single input/output registered design. The design uses 533 LCs and runs at 18.38 MHz `Registered Performance`. Fig. 6.15 shows the simulation for the zero-pipeline design, for the inputs $A = 100 + j110$, $B = -40 + j10$, and $W = e^{j\pi/9}$. 6.12

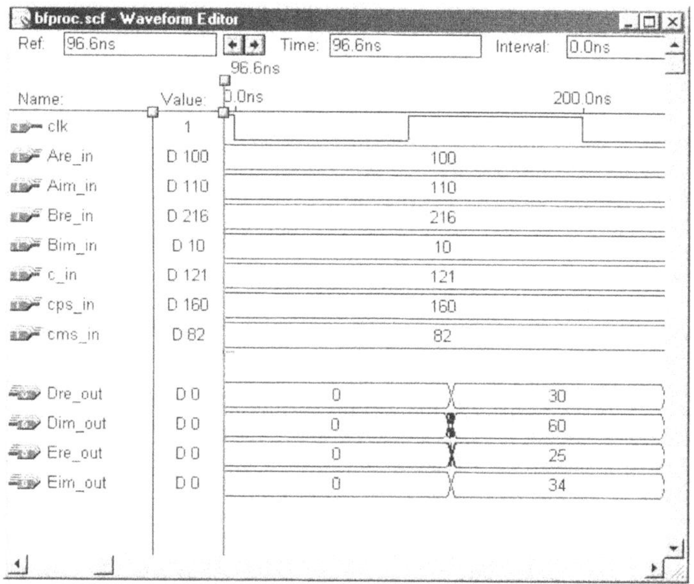

Fig. 6.15. VHDL simulation of a radix-2 butterfly processor.

6.2.2 The Good-Thomas FFT Algorithm

The index transform suggested by Good [132] and Thomas [133] transforms a DFT of length $N = N_1 N_2$ in an "actual" two-dimensional DFT, i.e., there are *no* twiddle-factors as in the Cooley-Tukey FFT. The price we pay for the twiddle-factor free flow is that the factors must be co-prime (i.e., $\gcd(N_k, N_l) = 1$ for $k \neq l$), and we have a somewhat more complicated index mapping, as long as the index computation is done "on-line" and no precomputed tables are used for the index mapping.

If we try to eliminate the twiddle-factors introduced through the index mapping of n and k according to (6.17) and (6.19), respectively, it follows with

$$\begin{aligned} W_N^{nk} &= W_N^{(An_1+Bn_2)(Ck_1+Dk_2)} \\ &= W_N^{ACn_1k_1+ADn_1k_2+BCk_1n_2+BDn_2k_2} \\ &= W_N^{N_2n_1k_1} W_N^{N_1k_2n_2} = W_{N_1}^{n_1k_1} W_{N_2}^{k_2n_2}, \end{aligned} \tag{6.33}$$

that we must fulfill all the following necessary condition at the same time:

$$\langle AD \rangle_N = \langle BC \rangle_N = 0 \tag{6.34}$$

$$\langle AC \rangle_N = N_2 \tag{6.35}$$

$$\langle BD \rangle_N = N_1. \tag{6.36}$$

The mapping suggested by Good [132] and Thomas [133] fulfills this condition and is given by

$$A = N_2 \quad B = N_1 \quad C = N_2 \langle N_2^{-1} \rangle_{N_1} \quad D = N_1 \langle N_1^{-1} \rangle_{N_2}. \tag{6.37}$$

To check: Because the factors AD and BC both include the factor $N_1 N_2 = N$, it follows that (6.34) is checked. With $\gcd(N_1, N_2) = 1$ and a theorem due to Euler, we can write the inverse as $N_2^{-1} \bmod N_1 = N_2^{\phi(N_1)-1} \bmod N_1$ where ϕ is the Euler totient function. The condition (6.35) can now be rewritten as

$$\langle AC \rangle_N = \langle N_2 N_2 \langle N_2^{\phi(N_1)-1} \rangle_{N_1} \rangle_N. \tag{6.38}$$

We can now solve the inner modulo reduction, and it follows with $\nu \in \mathbb{Z}$ and $\nu N_1 N_2 \bmod N = 0$ finally

$$\langle AC \rangle_N = \langle N_2 N_2 (N_2^{\phi(N_1)-1} + \nu N_1) \rangle_N = N_2. \tag{6.39}$$

The same argument can be applied for the condition (6.36), and we have shown that all three conditions from (6.34)-(6.36) are fulfilled if the Good-Thomas mapping (6.37) is used. □

In conclusion, we can now define the following theorem

Theorem 6.13: Good-Thomas Index Mapping

The index mapping suggest by Good and Thomas for n is

$$n = N_2 n_1 + N_1 n_2 \bmod N \qquad \begin{cases} 0 \leq n_1 \leq N_1 - 1 \\ 0 \leq n_2 \leq N_2 - 1 \end{cases} \tag{6.40}$$

and as index mapping for k results

$$k = N_2 \langle N_2^{-1} \rangle_{N_1} k_1 + N_1 \langle N_1^{-1} \rangle_{N_2} k_2 \bmod N \begin{cases} 0 \leq k_1 \leq N_1 - 1 \\ 0 \leq k_2 \leq N_2 - 1 \end{cases} \tag{6.41}$$

The transform from (6.41) is identical to the Chinese Remainder Theorem 2.2.12 (p. 40). It follows, therefore, that k_1 and k_2 can simply be computed via a modulo reduction, i.e., $k_l = k \bmod N_l$.

If we now substitute the Good-Thomas index map in the equation for the DFT matrix (6.16), it follows

$$X[k_1, k_2] = \sum_{n_2=0}^{N_2-1} W_{N_2}^{n_2 k_2} \underbrace{\left(\underbrace{\sum_{n_1=0}^{N_1-1} x[n_1, n_2] W_{N_1}^{n_1 k_1}}_{N_1\text{-point transform}} \right)}_{\bar{x}[n_2, k_1]} \tag{6.42}$$

$$= \underbrace{\sum_{n_2=0}^{N_2-1} W_{N_2}^{n_2 k_2} \; \bar{x}[n_2, k_1]}_{N_2\text{-point transform}}, \tag{6.43}$$

i.e., as claimed at the beginning, it's an "actual" two-dimensional DFT transform without the twiddle-factor introduced by the mapping suggested by Cooley and Tukey. It follows that the Good-Thomas algorithm, although similar to the Cooley-Tukey Algorithm 6.8, has a different index mapping and no twiddle-factors.

Algorithm 6.14: Good-Thomas FFT Algorithm

An $N = N_1 N_2$-point DFT can be computed according to the following steps:

1) Index transform of the input sequence, according to (6.40).
2) Computation of N_2 DFTs of length N_1.
3) Computation of N_1 DFTs of length N_2.
4) Index transform of the output sequence, according to (6.41).

An $N = 12$ transform shown in the following example demonstrates the steps.

Example 6.15: Good-Thomas FFT Algorithm for $N = 12$

Suppose we have $N_1 = 4$ and $N_2 = 3$. Then a mapping for the input index according to $n = 3n_1 + 4n_2 \bmod 12$, and $k = 9k_1 + 4k_2 \bmod 12$ for the output index results, and we can compute the following index mapping tables

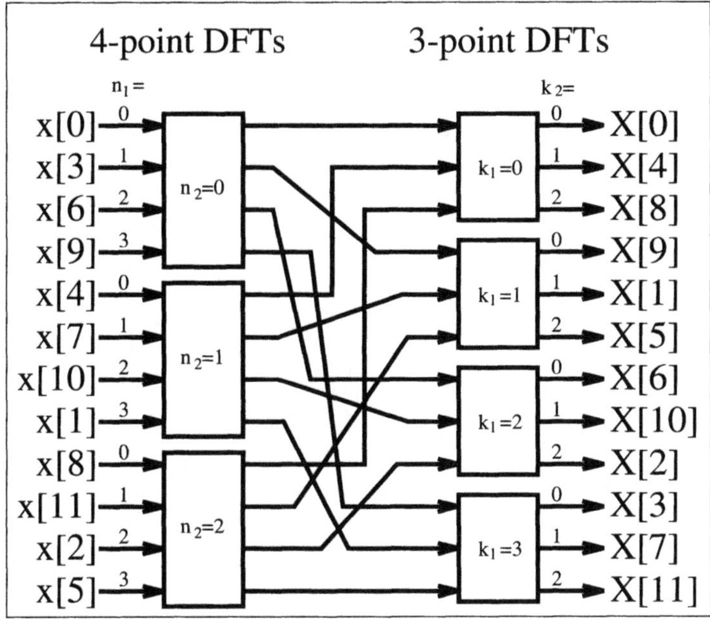

Fig. 6.16. PFA FFT for $N = 12$.

n_2 \ n_1	0	1	2	3
0	$x[0]$	$x[3]$	$x[6]$	$x[9]$
1	$x[4]$	$x[7]$	$x[10]$	$x[1]$
2	$x[8]$	$x[11]$	$x[2]$	$x[5]$

k_2 \ k_1	0	1	2	3
0	$X[0]$	$X[9]$	$X[6]$	$X[3]$
1	$X[4]$	$X[1]$	$X[10]$	$X[7]$
2	$X[8]$	$X[5]$	$X[2]$	$X[11]$

Using these index transforms we can construct the signal flow graph shown in Fig. 6.16. We realize that the first stage has three DFTs each having four points and the second stage four DFTs each of length 3. Multiplication with twiddle-factors between the stages is not necessary. 6.15

6.2.3 The Winograd FFT Algorithm

The Winograd FFT algorithm [83] is based on the observation that the inverse DFT matrix (6.4) (without pre-factor N^{-1}) of dimension $N_1 \times N_2$, with $\gcd(N_1, N_2) = 1$, i.e.,

$$x[n] = \sum_{k=0}^{N-1} X[k] W^{-nk} \qquad (6.44)$$

$$\boldsymbol{x} = \boldsymbol{W}_N^* \boldsymbol{X} \qquad (6.45)$$

can be rewritten using the *Kronecker product*[4] with two quadratic IDFT matrices each, with dimension N_1 and N_2, respectively. As with the index

[4] A Kronecker product is defined by

mapping for the Good-Thomas algorithm, we must write the indices of $X[k]$ and $x[n]$ in a two-dimensional scheme and then read out the indices row by row. The following example for $N = 12$ demonstrates the steps.

Example 6.16: IDFT using Kronecker Product for $N = 12$

Let $N_1 = 4$ and $N_2 = 3$. Then we have the output index transform $k = 9k_1 + 4k_2 \bmod 12$ according to the Good-Thomas index mapping:

$$\begin{bmatrix} X[0] \\ X[1] \\ X[2] \\ X[3] \\ X[4] \\ X[5] \\ X[6] \\ X[7] \\ X[8] \\ X[9] \\ X[10] \\ X[11] \end{bmatrix} \rightarrow \begin{array}{c|cccc} k_2 & & k_1 & & \\ & 0 & 1 & 2 & 3 \\ \hline 0 & X[0] & X[9] & X[6] & X[3] \\ 1 & X[4] & X[1] & X[10] & X[7] \\ 2 & X[8] & X[5] & X[2] & X[11] \end{array} \rightarrow \begin{bmatrix} X[0] \\ X[9] \\ X[6] \\ X[3] \\ X[4] \\ X[1] \\ X[10] \\ X[7] \\ X[8] \\ X[5] \\ X[2] \\ X[11] \end{bmatrix}$$

We can now construct a length-12 IDFT with

$$\begin{bmatrix} x[0] \\ x[9] \\ x[6] \\ x[3] \\ x[4] \\ x[1] \\ x[10] \\ x[7] \\ x[8] \\ x[5] \\ x[2] \\ x[11] \end{bmatrix} = \begin{bmatrix} W_{12}^{0} & W_{12}^{0} & W_{12}^{0} \\ W_{12}^{0} & W_{12}^{-4} & W_{12}^{-8} \\ W_{12}^{0} & W_{12}^{-8} & W_{12}^{-4} \end{bmatrix} \otimes \begin{bmatrix} W_{12}^{0} & W_{12}^{0} & W_{12}^{0} & W_{12}^{0} \\ W_{12}^{0} & W_{12}^{-3} & W_{12}^{-6} & W_{12}^{-9} \\ W_{12}^{0} & W_{12}^{-6} & W_{12}^{0} & W_{12}^{-6} \\ W_{12}^{0} & W_{12}^{-9} & W_{12}^{-6} & W_{12}^{-3} \end{bmatrix} \begin{bmatrix} X[0] \\ X[9] \\ X[6] \\ X[3] \\ X[4] \\ X[1] \\ X[10] \\ X[7] \\ X[8] \\ X[5] \\ X[2] \\ X[11] \end{bmatrix}$$

6.16

So far we have used the Kronecker product to (re)define the IDFT. Using the short hand notation $\tilde{\boldsymbol{x}}$ for the permuted sequence $\boldsymbol{x}$, we may use the following matrix/vector notation:

$$\tilde{\boldsymbol{x}} = \boldsymbol{W}_{N_1} \otimes \boldsymbol{W}_{N_2} \tilde{\boldsymbol{X}}. \tag{6.46}$$

For these short DFTs we now use the Winograd DFT Algorithm 6.7 (p. 225), i.e.,

$$\boldsymbol{W}_{N_l} = \boldsymbol{C}_l \cdot \boldsymbol{B}_l \cdot \boldsymbol{A}_l. \tag{6.47}$$

$$\boldsymbol{A} \otimes \boldsymbol{B} = [a[i,j]]\boldsymbol{B} = \begin{bmatrix} a[0,0]\boldsymbol{B} & \cdots & a[0,L-1]\boldsymbol{B} \\ \vdots & & \vdots \\ a[K-1,0]\boldsymbol{B} & \cdots & a[K-1,L-1]\boldsymbol{B} \end{bmatrix}$$

where $\boldsymbol{A}$ is a $K \times L$ matrix.

where $\boldsymbol{A}_l$ incorporate the input additions, $\boldsymbol{B}_l$ is a diagonal matrix with the Fourier coefficients, and $\boldsymbol{C}_l$ includes the output additions. If we now substitute (6.47) into (6.46), and use the fact that we can change the sequence of matrix multiplications and Kronecker product computation (see for instance [5, App. D], we get

$$\begin{aligned} \boldsymbol{W}_{N_1} \otimes \boldsymbol{W}_{N_2} &= (\boldsymbol{C}_1 \cdot \boldsymbol{B}_1 \cdot \boldsymbol{A}_1) \otimes (\boldsymbol{C}_2 \cdot \boldsymbol{B}_2 \cdot \boldsymbol{A}_2) \\ &= (\boldsymbol{C}_1 \otimes \boldsymbol{C}_2)(\boldsymbol{B}_1 \otimes \boldsymbol{B}_2)(\boldsymbol{A}_1 \otimes \boldsymbol{A}_2). \end{aligned} \qquad (6.48)$$

Because the matrices $\boldsymbol{A}_l$ and $\boldsymbol{C}_l$ are simple addition matrices, the same applies for its Kronecker products, $\boldsymbol{A}_1 \otimes \boldsymbol{A}_2$ and $\boldsymbol{C}_1 \otimes \boldsymbol{C}_2$. The Kronecker product of two quadratic diagonal matrices of dimension N_1 and N_2, respectively, obviously also gives a diagonal matrix of dimension N_1N_2. The total number of necessary multiplications is therefore identical to the number of diagonal elements of $\boldsymbol{B} = \boldsymbol{B}_1 \otimes \boldsymbol{B}_2$, i.e., M_1M_2, if M_1 and M_2, respectively, are the number of multiplications used to compute the smaller Winograd DFTs according to Table 6.2 (p. 226).

We can now combine the different steps to construct a Winograd FFT.

Theorem 6.17: Winograd FFT Design

A $N = N_1N_2$-point transform with co-primes N_1 and N_2 can be constructed as follows:

1) Index transform of the input sequence according to the Good-Thomas mapping (6.40), followed by a row read of the indices.
2) Factorization of the DFT matrix using the Kronecker product.
3) Substitute the length N_1 and N_2 DFT matrices through the Winograd DFT algorithm.
4) Centralize the multiplications.

After successful construction of the Winograd FFT algorithm, we can compute the Winograd FFT using the following three steps:

Theorem 6.18: Winograd FFT Algorithm

1) Compute the pre-additions $\boldsymbol{A}_1$ and $\boldsymbol{A}_2$.
2) Compute M_1M_2 multiplications according to the matrix $\boldsymbol{B}_1 \otimes \boldsymbol{B}_2$.
3) Compute post-additions according to $\boldsymbol{C}_1$ and $\boldsymbol{C}_2$.

Let us now look at a construction of a Winograd FFT of length-12, in detail in the following example.

Example 6.19: Winograd FFT of Length 12

To build a Winograd FFT, we have, according to Theorem 6.17, to compute the necessary matrices used in the transform. For $N_1 = 3$ and $N_2 = 4$ we have the following matrices:

$$\boldsymbol{A}_1 \otimes \boldsymbol{A}_2 = \begin{bmatrix} 1 & 1 & 1 \\ 0 & 1 & 1 \\ 0 & 1 & -1 \end{bmatrix} \otimes \begin{bmatrix} 1 & 1 & 1 & 1 \\ 1 & -1 & 1 & -1 \\ 1 & 0 & -1 & 0 \\ 0 & 1 & 0 & -1 \end{bmatrix} \qquad (6.49)$$

$$\boldsymbol{B}_1 \otimes \boldsymbol{B}_2 = \mathrm{diag}(1, -3/2, \sqrt{3}/2) \otimes \mathrm{diag}(1, 1, 1, -i) \qquad (6.50)$$

$$C_1 \otimes C_2 = \begin{bmatrix} 1 & 0 & 0 \\ 1 & 1 & i \\ 1 & 1 & -i \end{bmatrix} \otimes \begin{bmatrix} 1 & 0 & 0 & 0 \\ 0 & 1 & 0 & 0 \\ 1 & 0 & -1 & 0 \\ 0 & 0 & 1 & -1 \end{bmatrix} \tag{6.51}$$

Combining these matrices according to (6.48) results in the Winograd FFT algorithm. Input and output additions can be realized multiplier free, and the total number of real multiplication becomes $2 \cdot 3 \cdot 4 = 24$. 6.19

So far we have used the Winograd FFT to compute the IDFT. If we now want to compute the DFT with the help of an IDFT, we can use a technique we used in (6.6) on p. 212 to compute the IDFT with help of the DFT. Using matrix/vector notation we find

$$\boldsymbol{x}^* = (\boldsymbol{W}_N^* \boldsymbol{X})^* \tag{6.52}$$

$$\boldsymbol{x}^* = \boldsymbol{W}_N \boldsymbol{X}^*, \tag{6.53}$$

if $\boldsymbol{W}_N = [e^{2\pi jnk/N}]$ with $n, k \in \mathbb{Z}_N$ is a DFT. The DFT can therefore be computed using the IDFT with the following steps: Compute the conjugate complex of the input sequence, transform the sequence with the IDFT algorithm, and compute the conjugate complex of the output sequence.

It is also possible to use the Kronecker product algorithms, i.e., the Winograd FFT, to compute the DFT directly. This leads to a slide-modified output index mapping, as the following example shows.

Example 6.20: A 12-point DFT can be computed using the following Kronecker product formulation:

$$\begin{bmatrix} X[0] \\ X[3] \\ X[6] \\ X[9] \\ X[4] \\ X[7] \\ X[10] \\ X[1] \\ X[8] \\ X[11] \\ X[2] \\ X[5] \end{bmatrix} = \begin{bmatrix} W_{12}^0 & W_{12}^0 & W_{12}^0 \\ W_{12}^0 & W_{12}^4 & W_{12}^8 \\ W_{12}^0 & W_{12}^8 & W_{12}^4 \end{bmatrix} \otimes \begin{bmatrix} W_{12}^0 & W_{12}^0 & W_{12}^0 & W_{12}^0 \\ W_{12}^0 & W_{12}^3 & W_{12}^6 & W_{12}^9 \\ W_{12}^0 & W_{12}^6 & W_{12}^0 & W_{12}^6 \\ W_{12}^0 & W_{12}^9 & W_{12}^6 & W_{12}^3 \end{bmatrix} \begin{bmatrix} x[0] \\ x[9] \\ x[6] \\ x[3] \\ x[4] \\ x[1] \\ x[10] \\ x[7] \\ x[8] \\ x[5] \\ x[2] \\ x[11] \end{bmatrix} \tag{6.54}$$

The input sequence $x[n]$ can be considered to be in the order used for Good-Thomas mapping, while in the (frequency) output index mapping for $X[k]$, each first and third element are exchanged, compared with the Good-Thomas mapping. 6.20

6.2.4 Comparison of DFT and FFT Algorithms

It should now be apparent that there are many ways to implement a DFT. The choice begins with the selection of a short DFT algorithm from among

Table 6.4. Number of real multiplications for a length-12 complex input FFT algorithm. (Twiddle-factor multiplications by W^0 are not counted). A complex multiplication is assumed to use four real multiplications.

DFT Method	Index Mapping: Good-Thomas Fig. 6.16 p. 241	Index Mapping: Cooley-Tukey Fig. 6.2 p. 228	Index Mapping: Winograd Example 6.16 p. 242
Direct	$4 \cdot 12^2 = 4 \cdot 144 = 576$		
RPFA	$4(3(4-1)^2 + 4(3-1)^2) = 172$	$4(43+6) = 196$	–
WFTA	$3 \cdot 0 \cdot 2 + 4 \cdot 2 \cdot 2 = 16$	$16 + 4 \cdot 6 = 40$	$2 \cdot 3 \cdot 4 = 24$

those shown in Fig. 6.1 (p.209). The short DFT can then be used to develop long DFTs, using the indexing schemes provided by Cooley-Tukey, Good-Thomas, or Winograd. A common objective in choosing an implementation is minimum multiplication complexity. This is a viable criterion in case that the implementation cost of multiplication is much higher compared with other operations, such as additions, data access, or index computation.

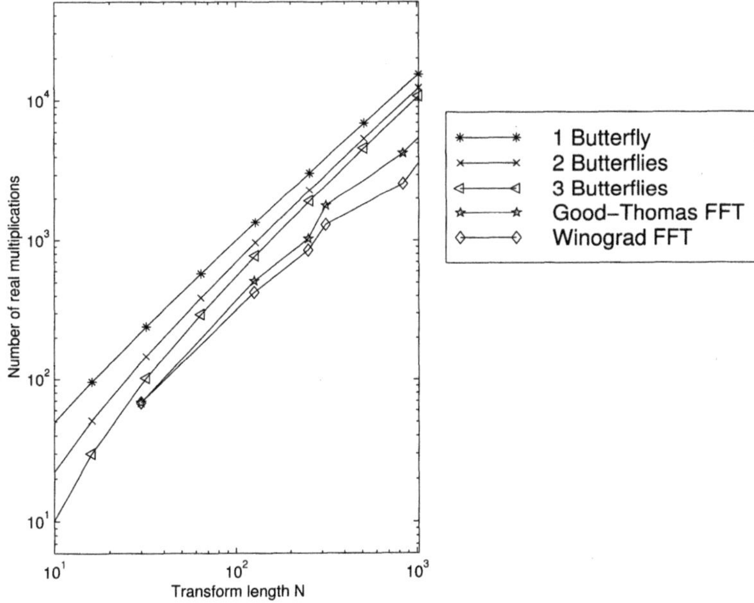

Fig. 6.17. Comparison of different FFT algorithm based on the number of necessary real multiplications.

Table 6.5. Important properties for FFT algorithms of length $N = \prod N_k$.

Property	Cooley-Tukey	Good-Thomas	Winograd
Any transform Length	yes	no $\gcd(N_k, N_l) = 1$	
Maximum order of W	N	$\max(N_k)$	
Twiddle factors needed	yes	no	no
# Multiplications	bad	fair	best
# Additions	fair	fair	fair
# Index computation effort	best	fair	bad
Data in-place	yes	yes	no
Implementation advantages	small Butterfly processor	can use RPFA, fast, simple FIR array	small size for full parallel, medium-size FFT (< 50)

Fig. 6.17 shows the number of multiplications required for various FFT lengths. It can be concluded that the Winograd FFT is most attractive, based purely on a multiply complexity criterion. In this chapter, the design of $N = 4 \times 3 = 12$-point FFTs has been presented in several forms. A comparison of a direct, Rader prime factor algorithms, and Winograd DFT algorithms used for the basic DFT blocks, and the three different index mappings called Good-Thomas, Cooley-Tukey and Winograd FFT, is presented in Table 6.4.

Besides the number of multiplications, other constraints must be considered, such as possible transform lengths, number of additions, index computation overhead, coefficient or data memory size, and run-time code length. In many cases, the Cooley-Tukey method provides the best overall solution, as suggested by Table 6.5.

With FPGAs reaching complexities of more than 1M gates today, full integration of an FFT on a single FPGA is viable. Because the design of such an FFT block is labor intensive, it most often makes sense to utilize commercially-available "intellectual property" (IP) blocks (sometimes also called "virtual components" VCs). See, for instance, the IP partnership programs at `www.xilinx.com` or `www.altera.com`. The majority of the commercially-available designs are based on radix-2 or radix-4.

Some of the published FPGA realizations are summarized in Table 6.6. The design by Goslin [118] is based on a radix-2 FFT, in which the butterflies have been realized using distributed arithmetic, discussed in Chapter 2. The design by Dandalis et al. [134], is based on an approximation of the DFT using the so-called arithmetic Fourier transform and will be discussed in Sect. 7.1. The ERNS FFT, from Meyer-Bäse et al. [135], uses the Rader

Table 6.6. Comparison of some FPGA FFT implementations [5].

Name	Data Type	FFT Type	N-point FFT time	Clock rate P	internal RAM/ ROM	Design aim/ source
Xilinx FPGA	8 Bit	Radix=2 FFT	$N = 256$ 102.4 μs	70 MHz 4.8 W @3.3V	no	573 CLBs [118]
Xilinx FPGA	16 Bit	AFT	$N = 256$ 82.48 μs	50 MHz 15.6 W @ 3.3 V	no	[134] 2602 CLBs
			42.08 μs	29.5 W		4922 CLBs
Xilinx FPGA ERNS-NTT	12.7 Bit	FFT using NTT	$N = 97$ 9.24 μs	26 MHz 3.5 W @3.3V	no	1178 CLBs [135]

algorithm in combination with the number theoretic transform, which will also be discussed in Chapter 7.

6.3 Fourier Related Transforms

The *discrete cosine transform* (DCT) and *discrete sine transform* (DST) are not DFTs, but they can be computed using a DFT. However, DCTs and DSTs can not be used directly to compute fast convolution, by multiplying the transformed spectra and an inverse transform, i.e., the convolution theorem does *not* hold. The applications for DCTs and DSTs are therefore not as broad as those for FFTs, but in some applications, like image compression, DCTs are (due to their close relationship to the Kahunen-Loevé transform) very popular. However, because DCTs and DSTs are defined by sine and cosine "kernels," they have a close relation to the DFT, and will be presented in this chapter. We will begin with the definition and properties of DCTs and DSTs, and will then present an FFT-like fast computation algorithm to implement the DCT. All DCTs obey the following transform pattern observed by Wang [136]:

$$X[k] = \sum_{n} x[n] C_N^{nk} \longleftrightarrow x[n] = \sum_{k} X[k] C_N^{n,k} \qquad (6.55)$$

The kernel functions C_N^{nk}, for four different DCT instances, are defined by

$$\begin{array}{lll}
\text{DCT-I:} & C_N^{n,k} = \sqrt{2/N}c[n]c[k]\cos\left(nk\frac{\pi}{N}\right) & n,k = 0,1,\ldots,N \\
\text{DCT-II:} & C_N^{n,k} = \sqrt{2/N}c[n]\cos\left(k(n+\frac{1}{2})\frac{\pi}{N}\right) & n,k = 0,1,\ldots,N-1 \\
\text{DCT-III:} & C_N^{n,k} = \sqrt{2/N}c[k]\cos\left(n(k+\frac{1}{2})\frac{\pi}{N}\right) & n,k = 0,1,\ldots,N-1 \\
\text{DCT-IV:} & C_N^{n,k} = \sqrt{2/N}\cos\left((k+\frac{1}{2})(n+\frac{1}{2})\frac{\pi}{N}\right) & n,k = 0,1,\ldots,N-1
\end{array}$$

where $c[p] = 1$ except $c[0] = 1/\sqrt{2}$. The DST has the same structure, but the cosine terms are replaced by sine terms. DCTs have the following properties:

1) DCTs implement functions using cosine bases.
2) All transforms are *orthogonal*, i.e., $\boldsymbol{C} \cdot \boldsymbol{C}^t = k[n]\boldsymbol{I}$.
3) A DCT is a real transform, unlike the DFT.
4) DCT-I is its own inverse.
5) DCT-II is the inverse of DCT-III, and vice versa.
6) DCT-IV is its own inverse. Type IV is symmetric, i.e., $\boldsymbol{C} = \boldsymbol{C}^t$.
7) The convolution property of the DCT is not the same as the convolution multiplication relationship in the DFT.
8) The DCT is an approximation of the Kahunen-Loevé transformation (KLT).

The two-dimensional 8×8 transform of the DCT-II is used most often in image compression, i.e., in the H.261, H.263, and MPEG standards for video and in the JPEG standard for still images. Because the two-dimensional transform is *separable* into two dimensions, we compute the two-dimensional DCT by row transforms followed by column transforms, or vice versa (Exercise 6.18, p. 254). We will therefore focus on the implementation of one-dimensional transforms.

6.3.1 Computing the DCT Using the DFT

Narasimha and Peterson have introduced a scheme describing how to compute the DCT with the help of the DFT [137] [138, p. 50]. The mapping of the DCT to the DFT is attractive because we can then use the wide variety of FFT-type algorithms. Because DCT-II is used most often, we will further develop the relationship of the DFT and DCT-II. To simplify the representation, we will skip the scaling operation, since it can be included at the end of the DFT or FFT computation. Assuming that the transform length is even, we can rewrite the DCT-II transform

$$X[k] = \sum_{n=0}^{N-1} x[n]\cos\left(k\left(n+\frac{1}{2}\right)\frac{\pi}{N}\right) \tag{6.56}$$

using the following permutation

$$y[n] = x[2n] \quad \text{and} \quad y[N-n-1] = x[2n+1]$$
$$\text{for} \quad n = 0,1,\ldots,N/2-1$$

It follows then that

$$X[k] = \sum_{n=0}^{N/2-1} y[n] \cos\left(k(2n + \frac{1}{2})\frac{\pi}{N}\right)$$
$$+ \sum_{n=0}^{N/2-1} y[N-n-1] \cos\left(k(2n + \frac{3}{2})\frac{\pi}{N}\right)$$
$$X[k] = \sum_{n} y[n] \cos\left(k(2n + \frac{1}{2})\frac{\pi}{N}\right) \tag{6.57}$$

If we now compute the DFT of $y[n]$ denoted with $Y[k]$, we find that

$$X[k] = \Re\left(W_{4N} Y[k]\right) \tag{6.58}$$
$$= \cos\left(\frac{\pi k}{2N}\right) \Re(Y[k]) - \sin\left(\frac{\pi k}{2N}\right) \Im(Y[k]) \tag{6.59}$$

This can be easily transformed in a C or MATLAB program (see Exercise 6.18, p. 254), and can be used to compute the DCT with the help of a DFT or FFT.

6.3.2 Fast Direct DCT Implementation

The symmetry properties of DCTs have been used by Byeong Lee [139] to construct an FFT-like DCT algorithm. Because of its similarities to a radix-2 Cooley-Tukey FFT, the resulting algorithm is sometimes referred to as the *Fast DCT* or simply FCT. Alternatively, a fast DCT algorithm can be developed using a matrix structure [140]. A DCT can be obtained by "transposing" an inverse DCT (IDCT) since the DCT is known to be an orthogonal transform. IDCT Type II was introduced in (6.55) and, noting that $\hat{X}[k] = c[k]X[k]$, it follows that

$$x[n] = \sum_{k=0}^{N-1} \hat{X}[k] C_N^{n,k}, \qquad n = 0, 1, \ldots, N-1. \tag{6.60}$$

Decomposing $x[n]$ into even and odd parts it can be shown that $x[n]$ can be reconstructed by two $N/2$ DCTs, namely

$$G[k] = \hat{X}[2k], \tag{6.61}$$
$$H[k] = \hat{X}[2k+1] + \hat{X}[2k-1], \qquad k = 0, 1, \ldots, N/2-1. \tag{6.62}$$

In the time domain, we get

$$g[n] = \sum_{k=0}^{N/2-1} G[k] C_{N/2}^{n,k}, \tag{6.63}$$
$$h[n] = \sum_{k=0}^{N/2-1} H[k] C_{N/2}^{n,k}, \qquad k = 0, 1, \ldots, N/2-1. \tag{6.64}$$

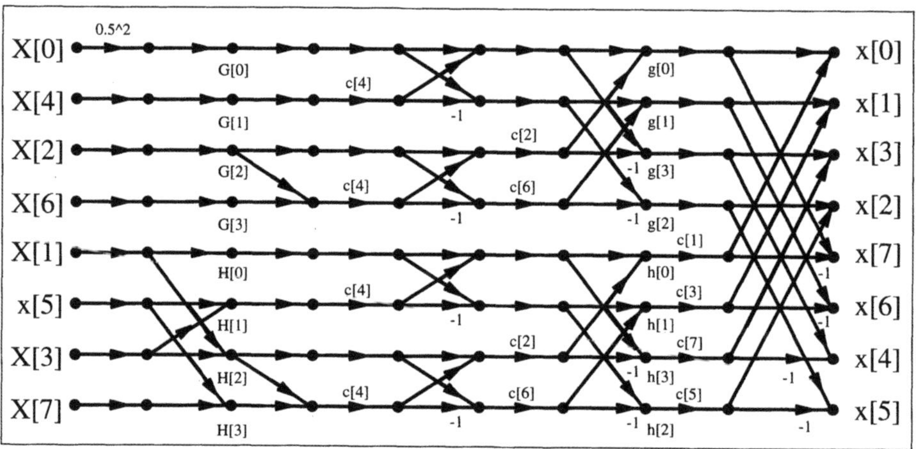

Fig. 6.18. 8-point fast DCT flow graph with the short hand notation $c[p] = 1/(2\cos(p\pi/16))$.

The reconstruction becomes

$$x[n] = g[n] + 1/(2C_N^{n,k})h[n], \tag{6.65}$$

$$x[N-1-n] = g[n] - 1/(2C_N^{n,k})h[n], \tag{6.66}$$

$$n = 0, 1, \ldots, N/2 - 1.$$

By repeating this process, we can decompose the DCT further. Comparing (6.63) with the radix-2 FFT twiddle-factor shown in Fig. 6.12 (p. 231) shows that a division seems to be necessary for the FCT. The twiddle-factors $1/(2C_N^{n,k})$ should therefore be precomputed and stored in a table. Such a table approach is also appropriate for the Cooley-Tukey FFT, because the "online" computation of the trigonometric function is, in general, too time consuming. We will to demonstrate the FCT with the following example.

Example 6.21: A 8-point FCT

For an 8-point FCT Equations (6.61) to (6.66) become

$$G[k] = \hat{X}[2k], \tag{6.67}$$

$$H[k] = \hat{X}[2k+1] + \hat{X}[2k-1], \qquad k = 0, 1, 2, 3. \tag{6.68}$$

and in the time domain we get

$$g[n] = \sum_{k=0}^{3} G[k]C_4^{n,k}, \tag{6.69}$$

$$h[n] = \sum_{k=0}^{3} H[k]C_4^{n,k}, \qquad n = 0, 1, 2, 3. \tag{6.70}$$

The reconstruction becomes

$$x[n] = g[n] + 1/(2C_8^{n,k})h[n], \tag{6.71}$$

$$x[N-1-n] = g[n] - 1/(2C_8^{n,k})h[n], \qquad n = 0, 1, 2, 3. \tag{6.72}$$

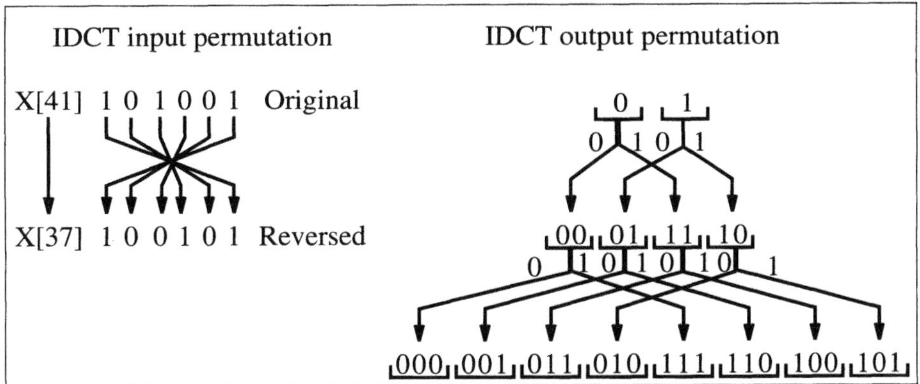

Fig. 6.19. Input and output permutation for the 8-point fast DCT.

The Equations (6.67) and (6.68) form the first stage in the flow graph in Fig. 6.18, and (6.71) and (6.72) build the last stage in the flow graph. 6.21

In Fig. 6.18, the input sequence $\hat{X}[k]$ is applied in bit-reversed order. The order of the output sequence $x[n]$ is generated in the following manner: starting with the set $(0, 1)$ we form the new set by adding a prefix 0 and 1. For the prefix 1, all bits of the previous pattern are inverted. For instance, from the sequence 10 we get the two babies 010 and $1\bar{1}\bar{0} = 101$. This scheme is graphically interpreted in Fig. 6.19.

Exercises

6.1: Compute the 3 dB bandwidth, first zero, maximum sidelobe, and decrease per octave, for a rectangular and triangular window using the Fourier transform.

6.2: **(a)** Compute the cyclic convolution of $x[n] = \{3, 1, -1\}$ and $f[n] = \{2, 1, 5\}$.
(b) Compute the DFT matrix $\boldsymbol{W}_3$ for $N = 3$.
(c) Compute the DFT of $x[n] = \{3, 1, -1\}$ and $f[n] = \{2, 1, 5\}$.
(d) Now compute $Y[k] = X[k]F[k]$, followed by $\boldsymbol{y} = \boldsymbol{W}_3^{-1}\boldsymbol{Y}$, for the signals from part (c).
(Note: use a C compiler or MATLAB for part (c) and (d).)

6.3: A single spectral component $X[k]$ in the DFT computation

$$X[k] = x[0] + x[1]W_N^k + x[2]W_N^{2k} + \ldots + x[N-1]W_N^{(N-1)k}.$$

can be rearranged by collecting all common factors W_N^k, such that we get

$$X[k] = x[0] + W_N^k(x[1] + W_N^k(x[2] + \ldots + W_N^k x[N-1]) \ldots)).$$

This results in a possibly recursive computation of $X[k]$. This is called the Goertzel algorithm and is graphically interpreted by Fig. 6.5 (p. 216). The Goertzel algorithm

can be attractive if only a few spectral components must be computed. For the whole DFT, the effort is of order N^2 and there is no advantage compared with the direct DFT computation.
(a) Construct the recursive signal flow graph, including input and output register, to compute a single $X[k]$ for $N = 5$.
(b) For $N = 5$ and $k = 1$, compute all registers contents for an impulse response $x[n] = \delta[n]$ for the design in (a) for a possible FPGA implementation.

6.4: Assume a given recursive system $Y(z) = X(z)/(1 + 2\cos(2\pi k/N)z^{-1} + z^{-2})$ of second-order, with fixed k and N.
(a) Draw the signal flow graph of the system.
(b) Let us assume $N = 6$ and $k = 1$. Compute $y[n]$ for $n = 0, 1, \ldots, N-1$, depending on $x[0]$ to $x[5]$.
(c) Determine the relationship between this second-order system and the Goertzel algorithm from Sect. 6.1.3 (p. 215).

6.5: The Bluestein chirp-z algorithm was defined in Sect. 6.1.4 (p. 216). This algorithm is graphically interpreted in Fig. 6.6 (p. 217).
(a) Determine the CZT algorithms for $N = 4$.
(b) Using C or MATLAB, determine the CZT for the triangular sequence $x[n] = \{0, 1, 2, 3\}$.
(c) Using C or MATLAB, extend the length to $N = 256$, and check the CZT results with an FFT of the same length. Use a triangular input sequence, $x[n] = n$.

6.6: **(a)** Design a direct implementation of the nonrecursive filter for the $N = 7$ Rader algorithm.
(b) Determine the coefficients that can be combined.
(c) Compare the realizations from (a) and (b) in terms of realization effort.

6.7: Design a length $N = 3$ Winograd DFT algorithm.

6.8: **(a)** Using the two dimensional index transform $n = 3n_1 + 2n_2 \bmod 6$, determine the mapping (6.18) on p. 227. Is this mapping bijective?
(b) Using the two dimensional index transform $n = 2n_1 + 2n_2 \bmod 6$, determine the mapping (6.18) on p. 227. Is this mapping bijective?
(c) For $\gcd(N_1, N_2) > 1$, Burrus [109] found the following conditions such that the mapping is bijective:

$$A = aN_2 \text{ and } B \neq bN_1 \text{ and } \gcd(a, N_1) = \gcd(B, N_2) = 1$$

$$or$$

$$A \neq aN_2 \text{ and } B = bN_1 \text{ and } \gcd(A, N_1) = \gcd(b, N_2) = 1$$

with $a, b \in \mathbb{Z}$. Suppose $N_1 = 9$ and $N_2 = 15$. For $A = 15$, compute all possible values for $B \in \mathbb{Z}_{20}$.

6.9: For $\gcd(N_1, N_2) = 1$, Burrus [109] found that in the following conditions the mapping is bijective:

$$A = aN_2 \text{ and/or } B = bN_1 \text{ and } \gcd(A, N_1) = \gcd(B, N_2) = 1 \quad (6.73)$$

with $a, b \in \mathbb{Z}$. Assume $N_1 = 5$ and $N_2 = 8$. Determine whether the following mappings are possibly bijective index mappings:
(a) $A = 8, B = 5$.
(b) $A = 8, B = 10$.
(c) $A = 24, B = 15$.

(d) For $A = 7$, compute all valid $B \in \mathbb{Z}_{40}$.
(e) For $A = 8$, compute all valid $B \in \mathbb{Z}_{40}$.

6.10: **(a)** Draw the signal flow graph for a radix-2 DIF algorithm where $N = 16$.
(b) Write a C or MATLAB program for the DIF radix-2 FFT.
(c) Test your FFT program with a triangular input $x[n] = n+jn$ with $n \in [0, N-1]$.

6.11: **(a)** Draw the signal flow graph for a radix-2 DIT algorithm where $N = 8$.
(b) Write a C or MATLAB program for the DIT radix-2 FFT.
(c) Test your FFT program with a triangular input $x[n] = n+jn$ with $n \in [0, N-1]$.

6.12: Compute the index mapping for an $N = 16$ radix-4 FFT.
Draw the signal flow graph for the $N = 16$ radix-4 FFT.

6.13: Draw the signal flow graph for an $N = 12$ Good-Thomas FFT, such that no crossings occur in the signal flow graph.
(Hint: Use a 3D representation of the row and column DFTs)

6.14: The index transform for FFTs by Burrus and Eschenbacher [141] is given by

$$n = N_2 n_1 + N_1 n_2 \bmod N \qquad \begin{cases} 0 \leq n_1 \leq N_1 - 1 \\ 0 \leq n_2 \leq N_2 - 1 \end{cases} \tag{6.74}$$

and

$$k = N_2 k_1 + N_1 k_2 \bmod N \qquad \begin{cases} 0 \leq k_1 \leq N_1 - 1 \\ 0 \leq k_2 \leq N_2 - 1 \end{cases} \tag{6.75}$$

(a) Compute the mapping for n and k with $N_1 = 3$ and $N_2 = 4$.
(b) Compute W^{nk}.
(c) Substitute W^{nk} from (b) in the DFT matrix.
(d) What type of FFT algorithm is this?
(e) Can the Rader algorithm be used to compute the DFTs of length N_1 or N_2?

6.15: **(a)** Compute the DFT matrices $\boldsymbol{W}_2$ and $\boldsymbol{W}_3$.
(b) Compute the Kronecker product $\boldsymbol{W}'_6 = \boldsymbol{W}_2 \otimes \boldsymbol{W}_3$.
(c) Compute the index for the vectors $\boldsymbol{X}$ and $\boldsymbol{x}$, such that $\boldsymbol{X} = \boldsymbol{W}'_6 \boldsymbol{x}$ is a DFT of length 6.
(d) Compute the index mapping for $x[n]$ and $X[k]$, with $\boldsymbol{x} = \boldsymbol{W}_2^* \otimes \boldsymbol{W}_3^* \boldsymbol{X}$ being the IDFT.

6.16: The discrete Hartley-Transformation (DHT) is a transform for real signals. A length N transform is defined by

$$H[n] = \sum_{k=0}^{N-1} \mathrm{cas}(2\pi nk/N)\, h[k] \tag{6.76}$$

with $\mathrm{cas}(x) = \sin(x) + \cos(x)$. The relation with the DFT ($f[k] \stackrel{\mathrm{DFT}}{\longleftrightarrow} F[n]$) is

$$H[n] = \Re\{F[n]\} - \Im\{F[n]\} \tag{6.77}$$

$$F[n] = E[n] - jO[n] \tag{6.78}$$

$$E[n] = \frac{1}{2}\left(H[n] + H[-n]\right) \tag{6.79}$$

$$O[n] = \frac{1}{2}\left(H[n] - H[-n]\right) \tag{6.80}$$

where $\Re$ is the real part, $\Im$ the imaginary part, $E[n]$ the even part of $H[n]$, and $O[n]$ the odd part of $H[n]$.
(a) Compute the equation for the inverse DHT.
(b) Compute (using the frequency convolution of the DFT) the steps to compute a convolution with the DHT.
(c) Show possible simplifications for the algorithms from (b), if the input sequence is even.

6.17: The DCT-II form is:

$$X[k] = c[k]\sqrt{\frac{2}{N}}\sum_{n=0}^{N-1} x[n]\cos\left(\frac{2\pi}{4N}(2n+1)k\right) \tag{6.81}$$

$$c[k] = \begin{cases} \sqrt{1/2} & k = 0 \\ 1 & \text{otherwise} \end{cases} \tag{6.82}$$

(a) Compute the equations for the inverse transform.
(b) Compute the DCT matrix for $N = 4$.
(c) Compute the transform of $x[n] = \{1, 2, 2, 1\}$ and $x[n] = \{1, 1, -1, -1\}$.
(d) What can you say about the DCT of even or odd symmetric sequences?

6.18: The following MATLAB code can be used to compute the DCT-II transform (assuming even length $N = 2^n$), with the help of a radix-2 FFT (see Exercise 6.10).

```
function X = DCTII(x)
  N = length(x);                        % get length
  y = [ x(1:2:N); x(N:-2:2) ];  % re-order elements
  Y = fft(y);                           % Compute the FFT
  w = 2*exp(-i*(0:N-1)'*pi/(2*N))/sqrt(2*N); % get weights
  w(1) = w(1) / sqrt(2);                          % make it unitary
  X = real(w .* Y);                     % compute pointwise product
```

(a) Compile the program with C or MatLab.
(b) Compute the transform of $x[n] = \{1, 2, 2, 1\}$ and $x[n] = \{1, 1, -1, -1\}$.

6.19: Like the DFT, the DCT is a separable transform and, we can therefore implement a 2D DCT using 1D DCTs. The 2D $N \times N$ transform is given by

$$X[n_1, n_2] = \frac{c[n_1]c[n_2]}{2}\sum_{k=0}^{N-1}\sum_{l=0}^{N-1} x[k,l]\cos\left(n_1(k+\frac{1}{2})\frac{\pi}{N}\right)\cos\left(n_2(l+\frac{1}{2})\frac{\pi}{N}\right) \tag{6.83}$$

where $c[0] = 1/\sqrt{2}$ and $c[m] = 1$ for $m \neq 0$.
Use the program introduced in Exercise 6.18 to compute an 8×8 DCT transform by
(a) First row followed by column transforms.
(b) First column followed by row transforms.
(c) Direct implementation of (6.83).
(d) Compare the results from (a) and (b) for the test data $x[k,l] = k + l$ with $k, l \in [0, 7]$

Exercises Using MaxPlusII

6.20: **(a)** Implement a first-order system according to Exercise 6.3, to compute the Goertzel algorithm for $N = 5$ and $n = 1$, and 8-bit coefficient and input data, using

MaxPlusII.
(b) Determine the number of LCs and the **Registered Performance**.
(c) Simulate the design with the three input sequences $\{20, 40, 60, 80, 100\}$, $\{j20, j40, j60, j80, j100\}$, and $\{20 + j20, 40 + j40, 60 + j60, 80 + j80, 100 + j100\}$.

6.21: **(a)** Design a **Component** to compute the (real input) 4-point Winograd DFT (from Example 6.16, p. 242) using MaxPlusII. The input and output precision should be 8 bits.
(b) Determine the number of LCs and the **Registered Performance**.
(c) Simulate the design with the input sequences $\{40+j100, 60+j80, 80+j60, 100+j40\}$.

6.22: **(a)** Design a **Component** to compute the (real input) 3-point complex Winograd DFT (from Example 6.16, p. 242) using MaxPlusII. The input and output precision should be 8 bits.
(b) Determine the number of LCs and the **Registered Performance**.
(c) Simulate the design with the input sequences $\{40 + j100, 60 + j80, 80 + j60\}$.

6.23: **(a)** Using the designed 3- and 4-point **Components** from Exercises 6.21 and 6.22, use component instantiation to design a fully parallel 12-point Good-Thomas FFT similar to that shown in Fig. 6.16 (p. 241), using MaxPlusII. The input and output precision should be 8 bits.
(b) Determine the number of LCs and the **Registered Performance**.
(c) Simulate the design with the input sequences $x[n] = 10n$ with $0 \leq n < 12$.

6.24: **(a)** Design a component **ccmulp** similar to the one shown in Example 6.11 (p. 233), to compute the twiddle-factor multiplication. Use three pipeline stages for the multiplier and one for the input subtraction $X - Y$, using MaxPlusII. The input and output precision should again be 8 bits.
(b) Conduct a simulation to ensure that the pipelined multiplier correctly computes $(70 + j50)(121 + j39)$.
(d) Determine the number of LCs and the **Registered Performance** of the twiddle-factor multiplier.
(e) Now implement the whole pipelined butterfly processor.
(f) Conduct a simulation, with the data from Example 6.12 (p. 236).
(g) Determine the number of LCs and the **Registered Performance** of the whole pipelined butterfly processor.

7. Advanced Topics

Several algorithms exist that enable FPGAs to outperform PDSPs by an order of magnitude, due to the fact that FPGAs can be built with bitwise implementations. Such applications are the focus of this chapter.

For Number Theoretic Transforms (NTTs), the essential advantage of FPGAs is that it's possible to implement modulo arithmetic in any desired bit width. NTTs are discussed in detail in Sect. 7.1.

For error control and cryptography, two basic building blocks are used: Galois field arithmetic and linear feedback shift registers (LFSR). Both can be efficiently implemented with FPGAs, and are discussed in Sect. 7.2. If, for instance, an N-bit LFSR is used as an M-multistep number generator, this will give an FPGA at least an MN speed advantage over a PDSPs or microprocessor.

Finally, in Sect. 7.3, communication systems designed with FPGAs will demonstrate low system costs, high throughput, and the possibility of fast prototyping. A comprehensive discussion of both coherent and incoherent receivers will close this chapter.

7.1 Rectangular and Number Theoretic Transforms (NTTs)

Fast implementation of convolution, and discrete Fourier transform (DFT) computations, are frequent problems in signal and image processing. In practice these operations are most often implemented using fast Fourier transform (FFT) algorithms. NTTs can, in some instances, outperform FFT-based systems. In addition, it is also possible to use a rectangular transform, like the Walsh/Hadamard or the arithmetic Fourier transform, to get an approximation of the DFT or convolution, as will be discussed at the end of Section 7.1.

In 1971, Pollard [142] defined the NTT, over a finite group, as the transform pair

$$x[n] = N^{-1} \sum_{k=0}^{N-1} X[k]\alpha^{-nk} \bmod M \leftrightarrow X[k] = \sum_{n=0}^{N-1} x[k]\alpha^{kn} \bmod M, \quad (7.1)$$

where $N \cdot N^{-1} \equiv 1$ exists, and $\alpha \in \mathbb{Z}_M$ ($\mathbb{Z}_M = \{0,1,2,\ldots,M-1\}$, and $\mathbb{Z}_M \cong \mathbb{Z}/M\mathbb{Z}$) is an element of order N, i.e., $\alpha^N \equiv 1$ and $\alpha^k \not\equiv 1$ for all $k \in \{1,2,3,\ldots,N-1\}$ in the finite group $(\mathbb{Z}_M, \cdot)$ (see Exercise 7.1, p. 329).

It is important to be able to ensure that, for a given tuple (α, M, N), such a transform pair exists. Clearly, α must be of order N modulo M. In order to ensure that the inverse NTT (INTT) exists, other requirements are:

1) The multiplicative inverse N^{-1} mod M must exist, i.e., the equation $x \cdot N \equiv 1$ mod M must have a solution $x \in \mathbb{Z}_M$.
2) The determinant of the transform matrix $|\boldsymbol{A}| = |[\alpha^{kn}]|$ must be nonzero so that the matrix is invertable, i.e., $\boldsymbol{A}^{-1}$ exists.

1) It can only be concluded that a multiplicative inverse exists if α and M do not share a common factor, or in short notation, $\gcd(\alpha, M) = 0$.

2) For the second condition, a well-known fact from algebra is used: The NTT matrix is a special case of the *Vandermonde matrix* (with $a[k] = \alpha_N^k$), and it follows for the determinant

$$\det(\boldsymbol{V}) = \begin{vmatrix} 1 & a[0] & a[0]^2 & \cdots & a[0]^{L-1} \\ 1 & a[1] & a[1]^2 & \cdots & a[1]^{L-1} \\ \vdots & \vdots & \cdots & \ddots & \vdots \\ 1 & a[L-1] & a[L-1]^2 & \cdots & a[L-1]^{L-1} \end{vmatrix} = \prod_{k>l} (a[k] - a[l]) \quad (7.2)$$

For $\det(\boldsymbol{V}) \neq 0$, it is required that $a[k] \neq a[l] \; \forall \; k \neq l$. Since the calculations are, in fact, modulo M, a second constraint arises. Specifically, there cannot be a zero multiplier in the determinant (i.e., $\gcd\left(\prod_{k>l} a_k - a_l, M\right) = 1$).

In conclusion, to check the existence of an NTT, it must be verified that:

Theorem 7.1: Existence of an NTT over $\mathbb{Z}_M$

An NTT of length N for α defined over $\mathbb{Z}_M$ exists, if:

1) $\gcd(\alpha, M) = 1$
2) α is of order N, i.e.

$$\alpha^n \bmod M \begin{cases} = 1 & n = N \\ \neq 1 & 1 \leq n < N \end{cases} \quad (7.3)$$

3) The inverse $\det(\boldsymbol{A})^{-1}$ exist, i.e., $\gcd(\alpha^l - 1, M) = 1$ for $l = 1,2,\ldots,N-1$.

For $\mathbb{Z}_p, p =$ prime, all the conditions shown above are automatically satisfied. In $\mathbb{Z}_p$ elements up to an order $p-1$ can be found. But transforms length $p-1$ are, in general, of limited practical interest, since in this case "general" multiplications and modulo reductions are necessary, and it is more appropriate to use a "normal" FFT in binary or QRNS arithmetic [143],[35, paper 5-6].

There are no useful transforms in the ring $M = 2^b$. But it is possible to use the next neighbors, $2^b \pm 1$. If primes are used, then conditions 1 and 3 are automatically satisfied. We therefore need to discuss what kind of primes $2^b \pm 1$ are known.

Mersenne and Fermat Numbers. Primes of the form $2^b - 1$ were first investigated by the French mathematician Marin Mersenne (1588-1648). Using the geometry series

$$\left(1 + 2^q + 2^{2q} + \ldots + 2^{qr-1}\right)(2^q - 1) = 2^{qr} - 1$$

it can be concluded that exponent b of a Mersenne prime must also be a prime. This is necessary, but not sufficient, as the example $2^{11} - 1 = 23 \cdot 89$ shows. The first Mersenne primes $2^b - 1$ have exponents

$$b = 2, 3, 5, 7, 13, 17, 31, 61, 89, 107, 127, 521, 607, 1279. \tag{7.4}$$

Primes of the type $2^b + 1$ are known from one of Fermat's old letters. Fermat conjectured that all numbers $2^{(2^t)} + 1$ are primes but, as for Mersenne primes, this is necessary but not sufficient. It is necessary because if b is odd, i.e., $b = q2^t$ then

$$2^{q2^t} = \left(2^{(2^t)} + 1\right)\left(2^{(q-1)2^t} - 2^{(q-2)2^t} + 2^{(q-3)2^t} - \cdots + 1\right)$$

is not prime, as in the case of $(2^4 + 1)|(2^{12} + 1)$, i.e., $17|4097$. There are five known Fermat primes

$$F_0 = 3 \quad F_1 = 5 \quad F_2 = 17 \quad F_3 = 257 \quad F_4 = 65537 \tag{7.5}$$

but Euler (1707-1783) showed that 641 divides $F_5 = 2^{32} + 1$. Up to F_{21} there are no Fermat primes, which reduce the possible prime Fermat primes for NTTs to the first five.

7.1.1 Arithmetic Modulo $2^b \pm 1$

In Chapter 2, the one's complement (1C) and diminished-by-one (D1) coding were reviewed. Consult Table 2.1 (p. 32) for C1 and D1 coding. It was claimed that C1 coding can efficiently represent arithmetic modulo $2^b - 1$. This is used to build Mersenne NTTs, as suggested by Rader [144]. D1 coding efficiently represents arithmetic modulo $2^b + 1$, and is therefore preferred for Fermat NTTs, as suggested by Leibowitz [145].

The following table illustrates again the 1C and D1 arithmetic for computing addition.

1C	D1
$s = a + b + c_N$	if$((a == 0)\&\&(b == 0))s = 0$ else $s = a + b + \overline{c_N}$

where a and b are the input operands, s is the sum and c_N the carry bit of the intermediate sum $a+b$ without modulo reduction. To implement the 1C addition, first form the intermediate B-bit sum. Then add the carry of the MSB c_N to the LSB. In D1 arithmetic, the carry must first be inverted before adding it to the LSB. The hardware requirement to add modulo $2^B \pm 1$ is therefore a total of two adders. The second adder may be built using half adders, because one operand, besides the carry in the LSB, is zero.

Example 7.2: As an example, compute $10+7 \bmod M$.

Decimal	1C $M=15$	D1 $M=17$
7 +10	0111 +1010	00110 +01001
17 Correction	10001 +1 = 0010	01111 +1 = 1.0000
Check:	$17_{10} \bmod 15 = 2$	$17_{10} \bmod 17 = 0$

7.2

Subtraction is defined in terms of an *additive inverse*. Specifically, $B = -A$ is said to be the additive inverse of A if $A+B=0$. How the additive inverse is built can easily be seen by consulting Table 2.1 (p. 32). Additive inverse production is

1C	D1
$\overline{a}$	if(zf$(a)! = 1)\overline{a}$

It can be seen that a bitwise complement must first be computed. That is sufficient in case of 1C, and for the nonzero elements in D1, coding. But for the zero in D1, the bitwise complement should be inhibited.

Example 7.3: The computation of the inverse of two is as follows

Decimal	1C $M=15$	D1 $M=17$
2 −2	0010 1101	0001 1110

which can be verified using the data provided in Table 2.1 (p. 32)

7.3

The simplest α for an NTT is 2. Depending on $M = 2^b \pm 1$, the arithmetic codings (C1 for Mersenne transforms and D1 for Fermat NTTs) is selected first. The only necessary multiplications are then those with $\alpha^k = 2^k$. These multiplications are implemented, as shown in Chapter 2, by a binary (left) rotation by k bit positions. The leftmost outgoing bit, i.e., carry c_N, is copied to the LSB. For the D1 coding (other than where $A = 0$) a complement of the carry bit must be computed, as the following table shows:

1C	D1
$\text{shl}(X, k, c_N)$	$\text{if}(X! = 0)\ \text{shl}(X, k, \overline{c_N})$

The following example illustrates the multiplications by $\alpha^k = 2^k$ used most frequently in NTTs.

Example 7.4: Multiplication by 2^k for 1C and D1 Coding

The following table shows the multiplication of ± 2 by 2, and finally a multiplication of 2 by $8 = 2^3$ to demonstrate the modulo operation for 1C and D1 coding.

Decimal	1C $M = 15$	D1 $M = 17$
$2 \cdot 2^1$	0010	0001
$= 4$	0100	0011
$-2 \cdot 2^1$	1101	1110
$= -4$	1011	1100
$2 \cdot 2^3$	0010	0001
$= 16$	0001	1111

which can be verified using the data found in Table 2.1 (p. 32). 7.4

7.1.2 Efficient Convolutions Using NTTs

In the last section we saw that with α being a power of two, multiplication was reduced to data shifts that can be built efficiently and fast with FPGAs, if the modulus is $M = 2^b \pm 1$. Obviously this can be extended to complex α's of the kind $2^u \pm j2^v$. Multiplication of complex $\alpha's$ can also be reduced to simple data shifts.

In order to avoid general multiplications and general modulo operations, the following constraints when building NTTs should be taken into account:

Theorem 7.5: Constraints for Practical Useful NTTs

A NTT is only of practical interest if

1) The arithmetic is modulo $M = 2^b \pm 1$.
2) All multiplications $x[k]\alpha^{kn}$ can be realized with a maximum of 2 modulo additions.

7.1.3 Fast Convolution Using NTTs

Fast cyclic convolution of two sequences $\boldsymbol{x}$ and $\boldsymbol{h}$ may be performed by multiplying two transformed sequences [54, 131, 144], as described by the following theorem.

Theorem 7.6: Convolution by NTT

Let $\boldsymbol{x}$ and $\boldsymbol{y}$ be sequences of length N defined modulus M, and $\boldsymbol{z} = \langle \boldsymbol{x} \circledast \boldsymbol{y} \rangle_M$ be the circular convolution of $\boldsymbol{x}$ and $\boldsymbol{y}$. Let $\boldsymbol{X} = \mathrm{NTT}(\boldsymbol{x})$, and $\boldsymbol{Y} = \mathrm{NTT}(\boldsymbol{y})$ be the length-N NTTs of $\boldsymbol{x}$ and $\boldsymbol{y}$ computed over M. Then

$$\boldsymbol{z} = \mathrm{NTT}^{-1}(\boldsymbol{X} \odot \boldsymbol{Y}). \tag{7.6}$$

To prove the theorem, it must first be known that the commutative, associative, and distributive laws hold in a ring modulo M. That these properties hold is obvious, since $\mathbb{Z}$ is an integral domain (a commutative ring with unity) [146, 147].

Specifically, the circular convolution outcome, $y[n]$, is given by

$$y[n] = \left\langle N^{-1} \sum_{l=0}^{N-1} \left(\sum_{m=0}^{N-1} x[m]\alpha^{ml} \right) \left(\sum_{k=0}^{N-1} h[k]\alpha^{kl} \right) \alpha^{-ln} \right\rangle_M . \tag{7.7}$$

Applying the properties of commutation, associatiation, and distribution, the sums and products can be rearranged, giving

$$y[n] = \left\langle \sum_{k=0}^{N-1} \sum_{m=0}^{N-1} x[m]h[k] \left(N^{-1} \sum_{l=0}^{N-1} \alpha^{(m+k-n)l} \right) \right\rangle_M . \tag{7.8}$$

Clearly for combinations of m, n, and k such that $\langle m + n - k \rangle \equiv 0 \bmod N$, the sum over l gives N ones and is therefore equal to N. However, for $\langle m + n - k \rangle_N \equiv r \not\equiv 0$, the sum is given by

$$\sum_{l=0}^{N-1} \alpha^{rl} = 1 + \alpha^r + \alpha^{2r} + \ldots + \alpha^{r(N-1)} = \frac{1 - \alpha^{rN}}{1 - \alpha^r} \equiv 0 \tag{7.9}$$

for $\alpha^r \not\equiv 1$. Because α is of order N, and $r < N$, it follows that $\alpha^r \not\equiv 1$. It follows that for the sum over l, Eq. (7.8) becomes

$$N^{-1} \sum_{l=0}^{N-1} \alpha^{(m+k-n)l} = \begin{cases} \langle NN^{-1} \equiv 1 \rangle_M & \text{for } m + l - n \equiv 0 \bmod N \\ 0 & \text{for } m + l - n \not\equiv 0 \bmod N \end{cases} .$$

It's now possible to eliminate either the sum over k, using $k \equiv \langle n - m \rangle$, or the sum over m, using $m \equiv \langle n - k \rangle$. The first case gives

$$y[n] = \left\langle \textstyle\sum_{m=0}^{N-1} x[m]h[\langle n - m \rangle_N] \right\rangle_M , \tag{7.10}$$

while the second case gives

$$y[n] = \left\langle \textstyle\sum_{k=0}^{N-1} h[k]x[\langle n - k \rangle_N] \right\rangle_M \qquad \square \tag{7.11}$$

The following example demonstrates the convolution.

Example 7.7: Fermat NTT of Length 4

Compute the cyclic convolution of length-4 time series $x[n] = \{1, 1, 0, 0\}$ and $h[n] = \{1, 0, 0, 1\}$, using a Fermat NTT modulo 257.

Solution: For the NTT of length 4 modulo $M = 257$, the element $\alpha = 16$ has order 4. In addition, using a symmetric range $[-128, \ldots, 128]$, we need $4^{-1} \equiv -64 \bmod 257$ and $16^{-1} \equiv -64 \bmod 257$. The transform and inverse transform matrices are given by

$$\boldsymbol{T} = \begin{bmatrix} 1 & 1 & 1 & 1 \\ 1 & 16 & -1 & -16 \\ 1 & -1 & 1 & -1 \\ 1 & -16 & -1 & 16 \end{bmatrix} \qquad \boldsymbol{T}^{-1} = \begin{bmatrix} 1 & 1 & 1 & 1 \\ 1 & -16 & -1 & 16 \\ 1 & -1 & 1 & -1 \\ 1 & 16 & -1 & -16 \end{bmatrix} \tag{7.12}$$

The transform of $x[n]$ and $h[n]$ is followed by the multiplication element by element, of the transformed sequence. The result for $y[n]$, using the INTT, is shown in the following

$$\begin{array}{rcl} n, k & = & \{0,\ 1,\ 2,\ 3\ \} \\ x[n] & = & \{1,\ 1,\ 0,\ 0\ \} \\ X[k] & = & \{2,\ 17,\ 0,\ -15\} \\ h[n] & = & \{1,\ 0,\ 0,\ 1\ \} \\ H[k] & = & \{2,\ -15,\ 0,\ 17\ \} \\ X[k] \cdot H[k] & = & \{4\ \ 2,\ \ 0\ \ 2\ \} \\ y[n] = x[n] \circledast h[n] & = & \{2\ \ 1\ \ 0\ \ 1\ \} \end{array}$$

7.7

Wordlength limitations for NTT. When using an NTT to perform convolution, remember that all elements of the output sequence $y[n]$ must be bounded by M. This is true (for simplicity, unsigned coding is assumed) if

$$x_{\max} h_{\max} L \leq M. \tag{7.13}$$

If the bit widths $B_x = \log_2(x_{\max})$, $B_h = \log_2(h_{\max})$, $B_L = \log_2(L)$, and $B_M = \log_2(M)$ are used, it follows that for $B_x = B_h$ the maximum bit width of the input is bounded by

$$\boxed{B_x = \frac{B_M - B_L}{2},} \tag{7.14}$$

with the additional constraint that $M = 2^b \pm 1$, and α is a power of two. It follows that very few prime M transforms exist. Table 7.1 displays the most useful choices of α's, and the attendant transform length (i.e., order of α's) of Mersenne and Fermat NTTs.

If complex transforms and non-prime Ms are also considered, then the number and length of the transform becomes larger, and the complexity also increases. In general, for nonprime modul, the conditions from Theorem 7.1 (p. 258) should be checked. It is still possible to utilize Mersenne or Fermat arithmetic, by using the following congruence

$$a \bmod u \equiv (a \bmod (u \cdot v)) \bmod u. \tag{7.15}$$

Table 7.1. Prime $M = 2^b \pm 1$ NTTs including complex transforms.

Mersenne $M = 2^b - 1$		Fermat $M = 2^b + 1$	
α	$\mathrm{ord}_M(\alpha)$	α	$\mathrm{ord}_M(\alpha)$
2	b	2	b
-2	$2b$	$\sqrt{2}$	$2b$
$\pm 2j$	$4b$	$1+j$	$4b$
$1 \pm j$	$8b$		

which states that everything is first computed modulo $M = u \cdot v = 2^b \pm 1$, and only the output sequence need be computed modulo u, which is the valid module regarding Theorem 7.1. Although using $M = u \cdot v = 2^b \pm 1$ increases the internal bit width, 1C or D1 arithmetic can be used. They have lower complexity than modulo arithmetic modulo u, and this will, in general, reduce the overall effort for the system.

Such nonprime NTTs are called *pseudo* transforms, i.e., *pseudo-Mersenne* transforms or *pseudo-Fermat* transforms. The following example demonstrates the construction for a pseudo-Fermat transform.

Example 7.8: A Fermat NTT of Length 50

Using the MATLAB utility `order.m` (see Exercise 7.1, p. 329), it can be determined that $\alpha = 2$ is of order 50 modulo $2^{25} + 1$. From Theorem 7.1, we know that $\gcd(\alpha^2 - 1, M) = 3$, and a length 50 transform does *not* exist modulo $2^{25} + 1$. It is therefore necessary to identify the "bad" factors in $M = (2^b \pm 1)$, those which do not have order 50, and exclude these factors by using the final modulo operation in (7.15).

Solution: Using the standard MATLAB function **`factor(2^25+1)`**, the prime-factors of M are:

$$2^{25} + 1 = 3 \cdot 11 \cdot 251 \cdot 4051. \tag{7.16}$$

The order of $\alpha = 2$ for the single factor can be computed with the algorithm given in Exercise 7.1 on p. 329. They are

$$\begin{array}{ll} \mathrm{ord}_3(2) = 2 & \mathrm{ord}_{11}(2) = 10 \\ \mathrm{ord}_{251}(2) = 50 & \mathrm{ord}_{4051}(2) = 50 \end{array} \tag{7.17}$$

In order to have an NTT of length 50, a final modulo reduction with $(2^{25} + 1)/33$ must be computed. 7.8

Comparing Fermat and Mersenne NTT implementations, consider that

- A Mersenne NTT of length b, with b primes, can be converted by the chirp-z transform (CZT), or the Rader prime factor theorem (PFT)[130], into a cyclic convolution, as shown in Fig. 7.1(a). In addition this allows a simplified bus structure if a multi FPGA implementation [135] is used.
- Fermat NTTs with $M = 2^{(2^t)} + 1$ have a power-of-two length $N = 2^t$, and can therefore be implemented with the usual Cooley-Tukey radix-2-type FFT algorithm which we discussed in Chapter 6.

7.1.4 Multidimensional Index Maps for NTTs and the Agarwal-Burrus NTT

For NTTs, in general the transform length N is proportional to the bit width b. This constraint makes it *impossible* to build long (one-dimensional) transforms, because the necessary bit width will be tremendous. It is possible to try the multidimensional index maps, called Good-Thomas and Cooley-Tukey, which we discussed in Chapter 6. If these methods are applied to NTTs, the following problems arise:

- In Cooley-Tukey algorithms of length $N = N_1 N_2$, an element of order N in the twiddle factors is needed. It follows that the transform length is not increased, compared with the one-dimensional case, and will result in large bit width. It is therefore not attractive.
- If Good-Thomas mapping is applied, there is *no* need for an element of length N, for a length $N = N_1 N_2$ transform. However, two co-prime length transforms N_1 and N_2 are needed for the same M. That's impossible for NTTs, if the transforms listed in Table 7.1 (p. 264) are used. The only way to make Good-Thomas NTTs work is to use different extension fields, as reported in [135], or to use them in combination with Winograd short-convolution algorithms, but this will also increase the complexity of the implementation.

An alternative method suggested by Agarwal and Burrus [148] seems to be more attractive. In the Agarwal-Burrus algorithm, a one-dimensional array is also first mapped into a two-dimensional array, but in contrast to the Good-Thomas methods, the lengths N_1 and N_2 must *not* be co-prime. The Agarwal-Burrus algorithm can be understood as a generalization of the overlap-save method, where periodic extensions of the signals are built. If an α of order $2L$ is used, a convolution of size

$$\boxed{N = 2L^2} \tag{7.18}$$

can be built. From Table 7.2, it can be seen that this two-dimensional method improves the maximum length of the transforms.

To compute the Agarwal-Burrus-NTT, the following five steps are used:

Algorithm 7.9: Agarwal-Burrus NTT

The cyclic convolution of $\boldsymbol{x} \circledast \boldsymbol{h}$ of length $N = 2L^2$, with an NTT of length L, is accomplished with the following steps:

1) Index transformation of the one-dimensional sequence into a two-dimensional array according to

$$\boldsymbol{x} = \begin{bmatrix} x[0] & x[L] & \cdots & x[N-L] \\ x[1] & x[L+1] & \cdots & x[N-L+1] \\ \vdots & \vdots & \ddots & \vdots \\ x[L-1] & x[2L-1] & \cdots & x[N-1] \\ 0 & 0 & \cdots & 0 \\ \vdots & \vdots & \ddots & \vdots \\ 0 & 0 & 0 & 0 \end{bmatrix} \tag{7.19}$$

$$\boldsymbol{h} = \begin{bmatrix} h[N-L+1] & h[L] & \cdots & h[N-2L+1] \\ \vdots & \vdots & \ddots & \vdots \\ h[N-1] & h[L-1] & \cdots & h[N-L-1] \\ h[0] & h[L] & \cdots & h[N-L] \\ h[1] & h[L+1] & \cdots & h[N-L+1] \\ \vdots & \vdots & \ddots & \vdots \\ h[L-1] & h[2L-1] & \cdots & h[N-1] \end{bmatrix}. \tag{7.20}$$

2) Computation of the row transforms $\begin{bmatrix} \rightarrow \\ \rightarrow \\ \vdots \end{bmatrix}$ followed by the column transforms $[\downarrow\downarrow\cdots]$.

3) Computation of the element-by-element matrix multiplication, $\boldsymbol{Y} = \boldsymbol{H} \odot \boldsymbol{X}$.

4) Inverse transforms of the columns $[\downarrow\downarrow\cdots]$ followed by the inverse row transforms $\begin{bmatrix} \rightarrow \\ \rightarrow \\ \vdots \end{bmatrix}$.

5) Reconstruction of the output sequence from the lower part of $\boldsymbol{y}$, according to

$$\boldsymbol{y} = \begin{bmatrix} \vdots & \vdots & \ddots & \vdots \\ y[0] & y[L] & \cdots & y[N-L] \\ y[1] & y[L+1] & \cdots & y[N-L+1] \\ \vdots & \vdots & \ddots & \vdots \\ y[L-1] & y[2L-1] & \cdots & y[N-1] \end{bmatrix}. \tag{7.21}$$

The Agarwal-Burrus NTT can be demonstrated with the following example:

Example 7.10: Length 8 Agarwal-Burrus NTT

An NTT modulo 257 of length 4 exists for $\alpha = 16$. Compute the convolution of $X(z) = 1 + z^{-1} + z^{-2} + z^{-3}$ with $F(z) = 1 + 2z^{-1} + 3z^{-2} + 4z^{-3}$ using a Fermat NTT modulo 257.

Solution: First, the index maps and transforms of $x[n]$ and $f[n]$ are computed. It follows that

$$\boldsymbol{x} = \begin{bmatrix} 1\,1\,0\,0 \\ 1\,1\,0\,0 \\ 0\,0\,0\,0 \\ 0\,0\,0\,0 \end{bmatrix} \longleftrightarrow \boldsymbol{X} = \begin{bmatrix} 4 & 34 & 0 & 227 \\ 34 & 32 & 0 & 2 \\ 0 & 0 & 0 & 0 \\ 227 & 2 & 0 & 225 \end{bmatrix} \tag{7.22}$$

$$\boldsymbol{f} = \begin{bmatrix} 0\,0\,0\,0 \\ 0\,2\,4\,0 \\ 1\,3\,0\,0 \\ 2\,4\,0\,0 \end{bmatrix} \longleftrightarrow \boldsymbol{F} = \begin{bmatrix} 16 & 143 & 255 & 112 \\ 253 & 114 & 66 & 206 \\ 249 & 212 & 255 & 51 \\ 253 & 45 & 195 & 145 \end{bmatrix} \tag{7.23}$$

Now, an element-by-element multiplication is computed, which results in

$$\boldsymbol{Y} = \begin{bmatrix} 64 & 236 & 0 & 238 \\ 121 & 50 & 0 & 155 \\ 0 & 0 & 0 & 0 \\ 120 & 90 & 0 & 243 \end{bmatrix} \longleftrightarrow \boldsymbol{y} = \begin{bmatrix} 2 & 6 & 4 & 0 \\ 0 & 2 & 6 & 4 \\ 1 & 6 & 9 & 4 \\ 3 & 10 & 7 & 0 \end{bmatrix} \tag{7.24}$$

From the lower half of $\boldsymbol{y}$, the element of $y[n] = \{1, 3, 6, 10, 9, 7, 4, 0\}$ can be seen. 7.10

With the Agarwal-Burrus NTT, a double-size intermediate memory is needed, but much longer transforms can be computed. The two-dimensional principle can easily be extended to three-dimensional index maps, but most often the transform length achieved with the two-dimensional method will be sufficient. For instance, for $\alpha = 2$ and $b = 32$, the transform length is increased from 64 in the one-dimensional case to $2^{11} = 2048$ in the two-dimensional case.

Table 7.2. Data for some Agarwal-Burrus NTTs, to compute cyclic convolution using real Fermat NTTs ($b = 2^t, t = 0$ to 4) or pseudo-Fermat NTTs $t = 5, 6$.

Module	α	1D	2D
$2^b + 1$	2	$2b$	$2b^2$
$2^b + 1$	$\sqrt{2}$	$4b$	$8b^2$

7.1.5 Computing the DFT Matrix with NTTs

Most often DFTs and NTTs are used to compute convolution, and it can be attractive to use NTTs to compute this convolution with FPGAs, because 1C and D1 can be efficiently implemented. But sometimes it is necessary to compute the DFT to estimate the Fourier spectrum. Then a question arises: Is

Table 7.3. Building blocks to compute DFT with Fermat NTT.

DFT length	Number ring	α	Number of real Mul.	Shift-Add.
3	$F_1, F_2, F_3, F_4, F_5, F_6$	$2^2, 2^4, 2^8, 2^{16}, 2^{32}, 2^{64}$	2	6
5	$F_1, F_2, F_3, F_4, F_5, F_6$	$2, 2^2, 2^4, 2^8, 2^{16}, 2^{32}$	4	20
17	F_3, F_4, F_5, F_6	$2, 2^2, 2^4, 2^8$	16	144
257	F_6	$\sqrt{2}$	256	4544
13	$F_1, F_2, F_3, F_4, F_5, F_6$	$2, 2^2, 2^4, 2^8, 2^{16}, 2^{32}$	16	104
97	F_4, F_5, F_6	$2, 2^2, 2^3, 2^4$	128	1408
193	F_5, F_6	$2, 2^2$	256	3200
769	F_6	$\sqrt{2}$	1024	16448

it possible to use the more efficient NTT to compute the DFT? This question has been addressed in detail by Siu and Constantinides [149].

The idea is as follows: For prime p-length DFTs, the Rader algorithm can be used, which converts the task into a length $p-1$ cyclic convolution. This cyclic convolution is then computed by an NTT of the original sequence and the DFT twiddle factors in the NTT domain, multiplication element-wise, and the back conversion. These processing steps are illustrated in Fig. 7.1(b). The principle is demonstrated in the following example.

Example 7.11: Rader algorithm for $N = 5$

For $N = 5$, a generator is $g = 2$, which gives the following index map, $\{2^0, 2^1, 2^2, 2^3\} \bmod 5 \equiv \{1, 2, 4, 3\}$. First, the DC component is computed with

$$X[0] = \sum_{n=0}^{4} x[n] = x[0] + x[1] + x[2] + x[3] + x[4]$$

and in the second step, $X[k] - x[0]$, the cyclic convolution

$$\{x[1], x[2], x[4], x[3]\} \circledast \{W_5^1, W_5^2, W_5^4, W_5^3\}.$$

Now the NTT is applied to the (reordered) sequences $x[n]$ and W_5^k, as shown in Example 7.7 (p. 263). The transformed sequences are then multiplied element by element, in the NTT domain, and finally the INTT is computed.

7.11

For Mersenne NTTs a problem arises, in that the NTT itself is of prime length, and therefore the length increased by one can *not* be of prime length. But for a Fermat NTT, the length is 2^t, since $M = 2^t + 1$, which is a prime. Siu, et al., found eight such short-length DFT building blocks to be useful. These basic building blocks are summarized in Table 7.3.

The first part of Table 7.3 shows blocks which do not need an index transform. In the second part are listed the building blocks which have two co-prime factors. They are $13 - 1 = 3 \cdot 4$, $97 - 1 = 3 \cdot 32$, $193 - 1 = 3 \cdot 64$, and $769 - 1 = 3 \cdot 256$. The disadvantage of the two-factor case is that, in a

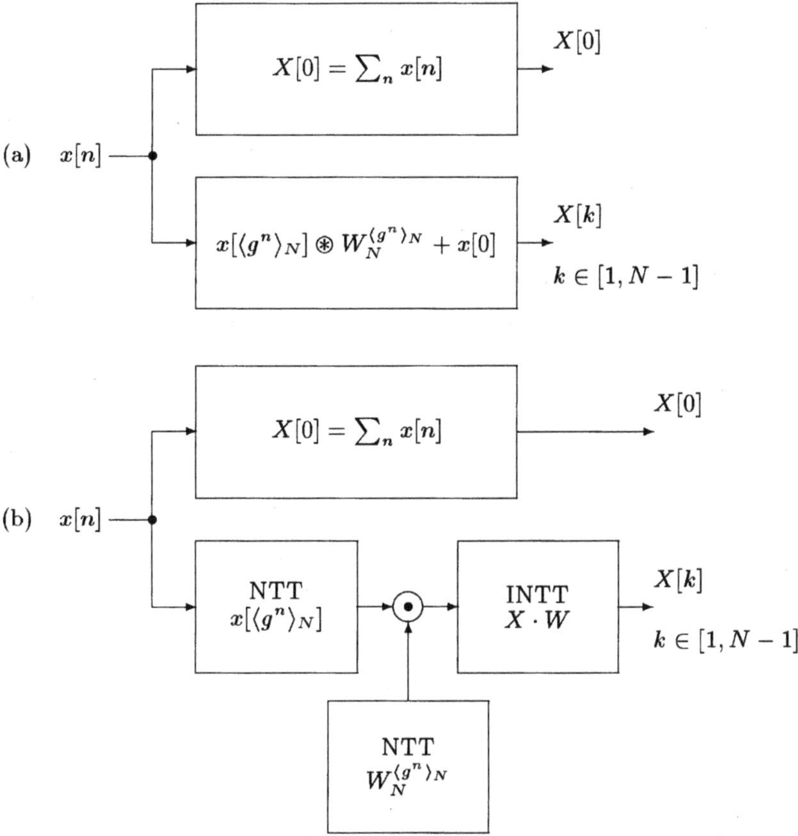

Fig. 7.1. The use of NTTs in Rader's prime-length algorithm for computing the DFT. **(a)** Rader's original algorithm. **(b)** Modification of the Rader prime algorithm using NTTs.

two-dimensional index map, for only one dimension every second transform of the twiddle factor becomes zero.

In the multidimensional map, it is also possible to implement a radix-2 FFT-like algorithm, or to combine Fermat NTTs with other NTT algorithms, such as the (pseudo-) Fermat NTT transform, (pseudo-) Mersenne transform, Lagrange interpolation, Eisenstein NTTs or a short convolution algorithm such as the Winograd algorithm [149, 54].

In the following, the techniques for the two-factor case using a length $13 - 1 = 3 \cdot 4$ multidimensional index map are reviewed. This is similar to the discussion in Chapter 6 for FFTs.

7.1.6 Index Maps for NTTs

To directly realize the NTT matrix is generally too expensive. This problem may be resolved by suitable multidimensional techniques. Burrus [109] gives a systematic overview of different common and prime factor maps, from one dimension to multiple dimensions. The mapping is explained for the two dimensional case. Higher-order mapping is equivalent. The mapping from the one-dimensional cyclic length-N convolution from (7.1), into a two-dimensional convolution with dimension $N = N_1 \cdot N_2$, can be written in linear form as follows:

$$n = M_1 n_1 + M_2 n_2 \bmod N \tag{7.25}$$

where $n_1 \in \{0, 1, 2, \ldots, N_1 - 1\}$ and $n_2 \in \{0, 1, 2, \ldots, N_2 - 1\}$. For $\gcd(N_1, N_2) \neq 1$, the well-known Cooley-Tukey FFT algorithm may be used. Burrus [109] shows that the map is cyclic in both dimensions if and only if N_1 and N_2 are relatively prime, i.e., $\gcd(N_1, N_2) = 1$. In order for this map to be one-to-one and onto (i.e., a bijection), the mapping constants M_1 and M_2 must satisfy certain conditions. For the relatively prime case, the conditions to make the mapping bijective are:

$$\begin{aligned} &[M_1 = \beta N_1 \; and/or \; M_2 = \gamma N_1] \; and \\ &\quad \gcd(M_1, N_1) = \gcd(M_2, M_2) = 1 \end{aligned} \tag{7.26}$$

As an example, consider $N_1 = 3$ and $N_2 = 4$, $N = 12$. From condition (7.26) we see that it is necessary to choose M_1 (a multiple of N_2), or M_2 (a multiple of N_1), or both. Make M_1 the simplest multiple of N_2, i.e., $M_1 = N_2 = 4$, which also satisfies $\gcd(M_1, N_1) = \gcd(4, 3) = 1$. Then, noting that $\gcd(M_2, N_2) = \gcd(M_2, 4) = 1$, the possible values for M_2 are $\{1, 3, 5, 7, 9, 11\}$. As a simple choice, select $M_2 = N_1 = 3$. The map becomes $n = \langle 4n_1 + 3n_2 \rangle_{12}$. Now let's apply the map to consider a 12-point convolution example. The transform of the one-dimensional cyclic array $x[n]$ into a 3×4 two-dimensional array $x[n_1, n_2]$, produces

$$[x[0] x[1] x[2] \ldots x[11]] \leftrightarrow \begin{bmatrix} x[0] & x[3] & x[6] & x[9] \\ x[4] & x[7] & x[10] & x[1] \\ x[8] & x[11] & x[2] & x[5] \end{bmatrix}. \tag{7.27}$$

To recover the sequence $X[k]$ from the $X[k_1, k_2]$, use the Chinese Remainder Theorem, as suggested by Good[132],

$$k = \left\langle (N_2^{-1} \bmod N_1) N_2 k_1 + (N_1^{-1} \bmod N_2) N_1 k_2 \right\rangle_N. \tag{7.28}$$

The α matrix can now be rewritten as

$$X[k_1, k_2] = \sum_{n_1=0}^{N_1-1} \left(\sum_{n_2=0}^{N_2-1} x[n_1, n_2] \alpha_{N_2}^{n_2 k_2} \right) \alpha_{N_1}^{n_1 k_1}, \tag{7.29}$$

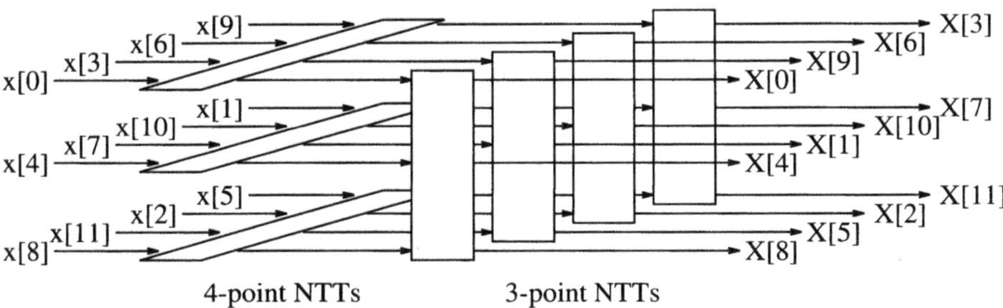

Fig. 7.2. Two-dimensional map. First stage: three 4-point NTTs. Second stage: four 3-point NTTs.

where α_{N_i} is an element of order N_i. Having mapped the original sequence $x[n]$ into the two-dimensional array $x[n_1, n_2]$, the desired matrix can be evaluated by the following two steps:

1) Perform an N_2-point NTT on each row of the matrix $x[n_1, n_2]$.
2) Perform an N_1-point NTT on each column of the resultant matrix, to yield $X[k_1, k_2]$

These processing steps are shown in Fig. 7.2. The input map is given by (7.27), while the output map can be computed with (7.28),

$$k = \left\langle \langle 4^{-1} \rangle_3 4k_1 + \langle 3^{-1} \rangle_4 3k_2 \right\rangle_{12} = \langle 4k_1 + 9k_2 \rangle_{12}. \tag{7.30}$$

The array $X[k_1, k_2]$ will therefore have the following arrangement:

$$[X[0]X[1]X[2]\ldots X[11]] \leftrightarrow \begin{bmatrix} X[0] & X[9] & X[6] & X[3] \\ X[4] & X[1] & X[10] & X[7] \\ X[8] & X[5] & X[2] & X[11] \end{bmatrix}. \tag{7.31}$$

Length 97 DFT case study. In recent years programmable digital signal processors (e.g., TMS320; Motorola 56K; AT&T 32C) have become the dominant vehicle to implement fast convolution via FFT algorithms. These PDSPs provide a fast (real) multiplier with typical cycle times of 10 to 50 ns. There are also some NTT implementations [150], but NTT implementations need modulo arithmetic, which is not supported by general purpose PDSPs. Dedicated accelerators, such as the FNT from J. McClellan [150], use 90 standard ECL 10K ICs. In recent years, field programmable gate arrays (FPGAs) have become dense enough and fast enough to implement typical high-speed DSP applications [4, 119]. It's possible to implement several arithmetic cores with only one FPGA, producing good packaging, speed, and power characteristics. FPGAs, with their fine granularity, can implement modulo arithmetic efficiently, without penalty, as in the PDSP case.

In NTT implementation of Fermat number arithmetic, the previously discussed speed and hardware advantages, compared with conventional FFT implementations, become an even bigger advantage for an FPGA implementation. By implementing the DFT algorithm with the Rader prime convolution strategy, the required I/O performance can be further reduced.

To clarify the NTT design paradigm, a length-97 DFT in the Fermat number system, F_4 and F_5, for real input data, will be shown. A Xilinx XC4K multi-FPGA board has been used to implement this design, as reported in [135].

For modulo Fermat number arithmetic (modulo 2^n+1) it is advantageous to use, instead of the usual two's complement arithmetic (2C), the "Diminished one" (D1) number system from Leibowitz [145]. Negative numbers are the same as in 2C, and positive numbers are diminished by one. The zero is encoded as a zero string and the MSB "ZERO-FLAG" is one. Therefore the diminished system consists of a ZERO-FLAG and integer bits x_k. For 2C, the MSB is the sign bit, while for D1 the second MSB is the sign bit. With this encoding the basic operations of 2C↔D1 conversion, negation, addition, and multiplication by 2^m can easily be determined, as shown in Sect. 7.1.1 (p. 259).

The rough processing steps of the 97-point transform are shown in Fig. 7.1(b). A direct length-96 implementation for a single NTT will cost at least 96×2 barrel shifters and 96×2 accumulators and, therefore, approximately $96(2 \times 32 + 2 \times 18) = 9600$ Xilinx combinatorial logic blocks (CLBs). Therefore it seemed reasonable to use a 32×3 index map, as described in the last section. The length-32 FFT now becomes a simpler length-32 Fermat NTT, and the length-3 transform has $\alpha^k = 1; \hat{\jmath}$ and $-1 - \hat{\jmath}$ with $\hat{\jmath}^2 = \hat{\jmath} + 1$. The 32-point FNT can be realized with the usual radix-2 FFT-type algorithm, while the length-3 transform can be implemented by a two-tap FIR filter. The following table gives CLB utilization estimates for Xilinx XC4000 FPGAs, for F_4:

Length-32 FNT	Length-3 FIR NTT	14 Multipliers 32-bit	Length-3 NTTS^{-1}	Two length-32 FNT^{-1}
104	108	462	288	216

The design consumes a total of 1178 CLBs. To get high throughput, the buffer memory between the blocks must be doubled. Two real buffers for the first FNT, and three complex buffers, are required. If the buffers are realized internally, an additional 748 CLBs are required, which will also minimize the I/O requirements. If 80% utilization is assumed, then about six XC4010s are needed for the design, including the buffer memory.

The time-critical path in the design is the length-32 FNT. To maximize the throughput, a three-stage pipeline is used inside the butterfly. For a 5 ns FPGA, the butterfly speed is 28 ns for F_4, and 38.5 ns for F_5. For three length-32 FNTs, five stages, each with 16 butterflies, must be computed. This gives a total transform time of 7.15 μs for F_4, and 9.24 μs for F_5, for the length-97 DFT. To set this result in perspective, the time for the butterfly computation gives a fair comparison. A TMS320C50 PDSP with a 50 ns cycle

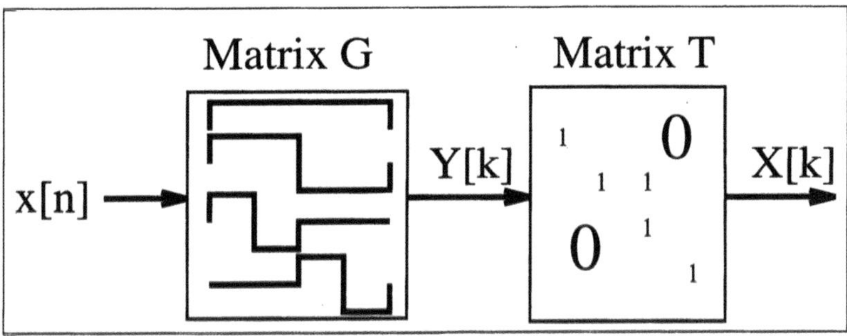

Fig. 7.3. DFT computation using rectangular transform and map matrix $\boldsymbol{T}$.

time needs 17 cycles for a butterfly [151], or 850 ns, assuming zero wait-state memory. Another "conventional" FPGA fixed-point arithmetic design [119] uses four serial/parallel multipliers (2 MHz), and therefore has a latency of 500 ns for the butterfly.

7.1.7 Using Rectangular Transforms to Compute the DFT

Rectangular transforms also map an input sequence in an image domain, but do not necessarily have the DFT structure, i.e., $\boldsymbol{A} = [\alpha^{nk}]$. Examples are Haar transforms [91], the Walsh/Hadamard transform [59], a ruff-quantized DFT [5], or the arithmetic Fourier transform [152, 153, 134]. These rectangular transforms in general do *not* support cyclic convolution, but they may be used to approximate the DFT spectrum [126]. The advantage of rectangular transforms is that the coefficients are from the set $\{-1, 0, 1\}$ and they do not need any multiplications.

How to compute the DFT is shown in Fig. 7.3. In order to have a useful system, it is assumed that the rectangular transform can be computed with low effort, and the second transform using the matrix $\boldsymbol{T}$, which maps the rectangular transform to the DFT vectors, has only a few nonzero elements.

Table 7.4. Comparison of different transforms to approximate the DFT [5].

Transform	Number of base	Algorithmic complexity	Zeros in 16×16 $\boldsymbol{T}$ Matrix
Walsh	N	$N \log_2(N)$	66
Hadamard	N	$N \log_2(N)$	66
Haar	N	$2N$	18
AFT	$N+1$	N^2	82
QDFT	$2N$	$(N/8)^2$+3N	86

Table 7.4 compares different implementations. The algorithmic complexity of the Walsh/Hadamard and Haar transforms is most interesting, but from the number of zeros in the second transform $\boldsymbol{T}$ it can be concluded that the arithmetic Fourier transform and the ruff-quantized DFT are more attractive for approximating the DFT.

7.2 Error Control and Cryptography

Modern communications systems, such as pagers, mobile phones or satellite transmission systems, use algorithms to correct transmission errors, since error-correction coding better utilize the band-limited channel capacity than special modulation schemes (see Fig. 7.4). In addition, most systems also use cryptography algorithms, not just to protect messages against unauthorized listeners, but also to protect messages against unauthorized changes.

In a typical transmission scheme, such as that shown in Fig. 7.5, the *encoder* (for error correction or cryptography) is placed between the data source and the actual modulation. On the receiver side, the *decoder* is located between demodulation and the data destination (sink). Often an encoder and decoder are combined in one circuit, referred to as a CODEC.

Typical error correction and cryptographic algorithms use finite field arithmetic and are therefore more suitable for FPGAs than they are for PDSPs [155]. Bitwise operations or linear feedback shift registers (LFSR) can be very efficiently realized with FPGAs. Some CODEC schemes use large tables, and one objective when selecting the appropriate algorithms for FPGAs is therefore to find out which algorithms are most suitable. The

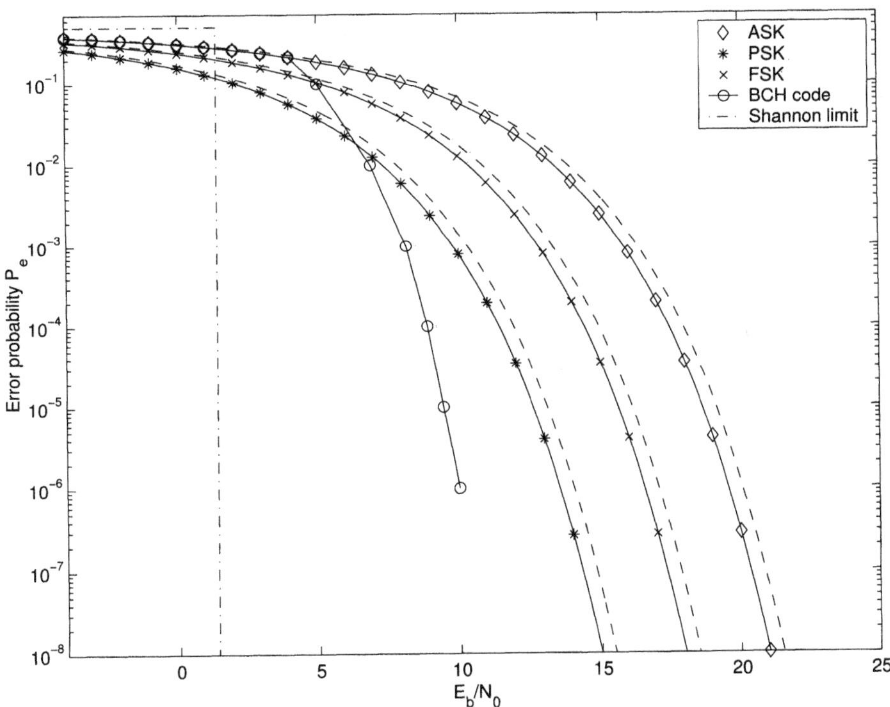

Fig. 7.4. Performance of modulation schemes [154].

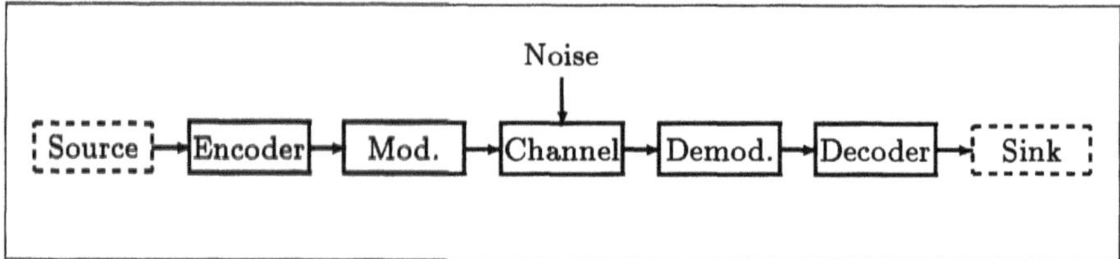

Fig. 7.5. Typical communications system configuration.

algorithms presented in this section are mainly based on previous publications [4] and have been used to develop a paging system for low frequencies [156, 157, 158, 159, 160], and an error-correction scheme for radio-controlled watches [161, 162].

It's impossible in a short section to present the whole theory of error correction and cryptography. We will present the basic ideas and suggest, for further investigation, one of the excellent text books in this area [124, 163, 92, 164, 165, 166, 167].

7.2.1 Basic Concepts from Coding Theory

The simplest way to protect a digital transmission against random errors is to repeat the message several times. This is called *repetition code.* For a repetition of 5, for instance, the message is sent five times, i.e.,

$$0 \Leftrightarrow 00000 \tag{7.32}$$

$$1 \Leftrightarrow 11111 \tag{7.33}$$

where the left side shows the k information bits and the right side the n-bit code words. The minimum distance between two codewords, also called the *Hamming distance* d^*, is also n and the repetition code is of the form $(n, k, d^*) = (5, 1, 5)$. With such a code it is possible to correct up to $\lfloor (n-1)/2 \rfloor$ random errors. But from the perspective of channel efficiency, this code is not very attractive. If our system is two-way then it is more efficient a to use a technique such as a parity check and an *automatic repeat request* (ARQ) for any detected parity error. Such parity checks are used, for instance, in PC memory.

Error correction using a Hamming code. If a few more parity check bits are added, it is possible to *correct* a word with a parity error.

If the parities $P_{1,0}, P_{1,1}, P_{1,2}$ and $P_{1,3}$ are computed using modulo 2 operations, i.e., XOR, according to

$$\begin{aligned}
P_{1,0} &= i_{21} \oplus i_{22} \oplus i_{23} \oplus i_{24}\ ;\oplus i_{25} \oplus i_{26} \oplus i_{27}\\
P_{1,1} &= i_{21} \oplus i_{23} \oplus i_{25} \oplus i_{27}\\
P_{1,2} &= i_{21} \oplus i_{22} \oplus i_{25} \oplus i_{26}\\
P_{1,3} &= i_{21} \oplus i_{22} \oplus i_{23} \oplus i_{24}
\end{aligned}$$

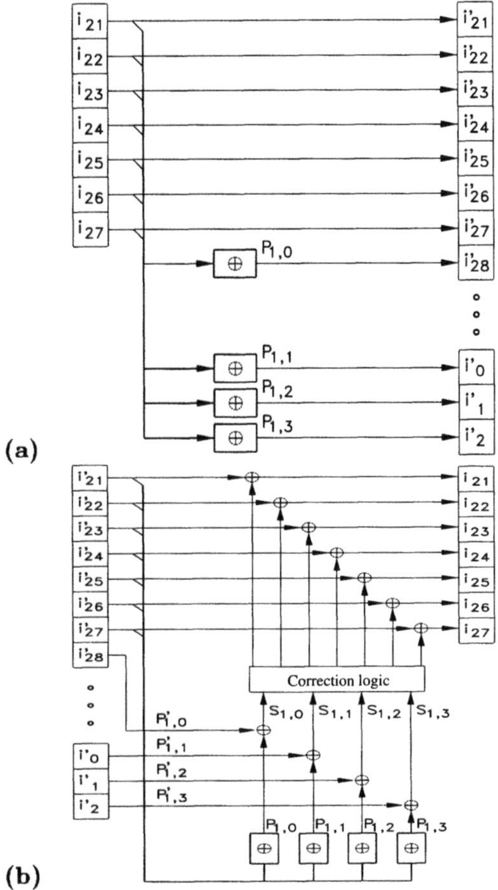

Fig. 7.6. **(a)** Coder and **(b)** Decoder for Hamming code.

then the parity detector is $i'_{28}(= P_{1,0})$ and three additional bits are necessary to locate the error position. Fig. 7.6(a) shows the encoder, and Fig. 7.6(b) the decoder including the correction logic. On the decoder side the incoming parities are XOR'd with the newly computed parities. This forms the so-called *Syndrome* $(S_{1,0} \cdots S_{1,3})$. The parities have been chosen in such a way that the syndrome pattern corresponds to the position of the bit in binary code, i.e., a $3 \rightarrow 7$ de-multiplexer can be used to decode the error location.

For a more compact representation of the decoder, the following parity check matrix $\boldsymbol{H}$ can be used

$$\boldsymbol{H} = \left[\boldsymbol{P}^T \vdots \boldsymbol{I}\right] = \begin{bmatrix} 1\,1\,1\,1\,1\,1\,1\,1\,0\,0\,0 \\ 1\,0\,1\,0\,1\,0\,1\,0\,1\,0\,0 \\ 1\,1\,0\,0\,1\,1\,0\,0\,0\,1\,0 \\ 1\,1\,1\,1\,0\,0\,0\,0\,0\,0\,1 \end{bmatrix} \tag{7.34}$$

Table 7.5. Estimated effort for error correction with Hamming code.

Block Hamming code	CLB effort for Minutes (11,7,3)	 Hours (10,6,3)	 Date (27,22,3)
Register	6	6	14
Syndrome computation	5	5	16
Correction logic	4	4	22
Output register	4	4	11
Sum	19	19	63
Total		101	

It is possible to describe the encoder using a generator matrix. $\boldsymbol{G} = [\boldsymbol{I} \vdots \boldsymbol{P}]$, i.e., the generator matrix consists of a systematic identity matrix $\boldsymbol{I}$ followed by the parity-bits matrix $\boldsymbol{P}$. A code word $\boldsymbol{v}$ is computed by multiplying (modulo 2) the information word $\boldsymbol{i}$ with the generator matrix $\boldsymbol{G}$:

$$\boldsymbol{v} = \boldsymbol{i} \cdot \boldsymbol{G} \tag{7.35}$$

The coders shown in Fig. 7.6 are those for a (11,7,3) Hamming code, and it is possible to detect *and* correct one error. In general it can be shown that for 4 parity bits, up to 15 information bits, can be used, i.e., a (15,11,3) Hamming code has been *shortened* to a (11,7,3) code.

A Hamming code with distance 3 generally has a $(2^m - 1, 2^m - m, 3)$ structure. The dates in radio-controlled watches, for instance, are coded with 22 bits, and a (31,26,3) Hamming code can be shortened to a (27,22,3) code to archieve a single-error correcting code. The parity check matrix becomes:

$$\boldsymbol{H} = \begin{bmatrix} 1\,0\,1\,0\,1\,0\,1\,0\,1\,0\,1\,0\,1\,0\,1\,0\,1\,0\,1\,0\,1\,0\,1\,0\,0\,0\,0 \\ 1\,1\,0\,0\,1\,1\,0\,0\,1\,1\,0\,0\,1\,1\,0\,0\,1\,1\,0\,0\,1\,1\,0\,1\,0\,0\,0 \\ 1\,1\,1\,1\,0\,0\,0\,0\,1\,1\,1\,1\,0\,0\,0\,0\,1\,1\,1\,1\,0\,0\,0\,0\,1\,0\,0 \\ 1\,1\,1\,1\,1\,1\,1\,1\,0\,0\,0\,0\,0\,0\,0\,0\,1\,1\,1\,1\,1\,1\,0\,0\,0\,1\,0 \\ 1\,1\,1\,1\,1\,1\,1\,1\,1\,1\,1\,1\,1\,1\,1\,1\,0\,0\,0\,0\,0\,0\,0\,0\,0\,0\,1 \end{bmatrix}$$

Again, the syndromes can be sorted in such a way that the correction logic is a simple $5 \rightarrow 22$ de-multiplexer.

Table 7.5 shows the estimated effort in CLBs using Xilinx XC3K FPGAs for an error-correction unit for radio-controlled watches that uses three separate data blocks for minutes, hours, and date.

In conclusion, with an additional 3+3+5=11 bits and the parity bits for the minutes using about 100 CLBs, it is possible to correct one error in each of the three blocks.

Survey of Error Correction Codes

After the introductory examples in the last section, commonly used codes and possible encoder and decoder implementations will be discussed next.

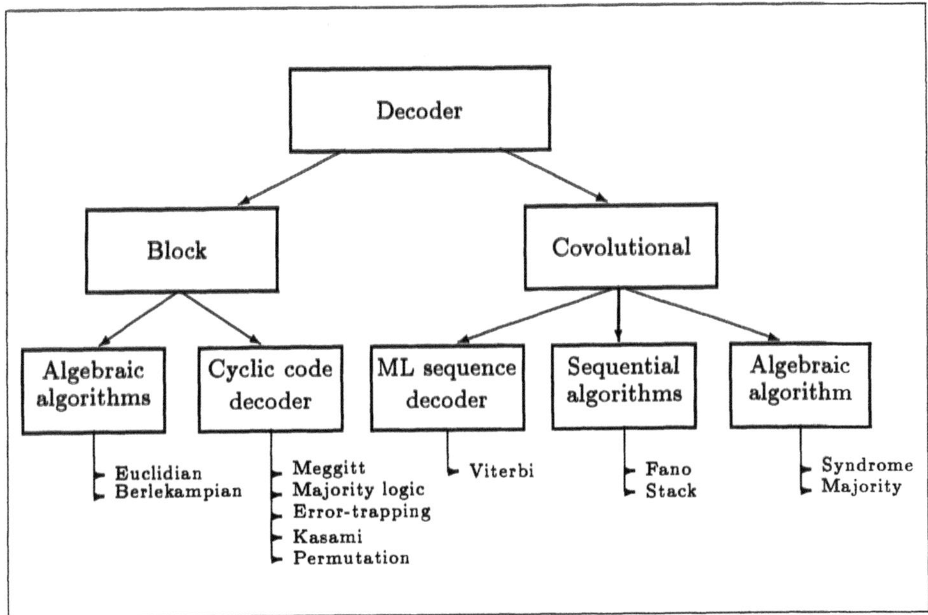

Fig. 7.7. Decoder for error correction.

Most often the effort for the decoder is of greater concern, since many communications systems like pager or radios use one sender and several receivers. Fig. 7.7 shows a diagram of possible decoders.

Some nearly optimal decoders use huge tables and are not included in Fig. 7.7. The difference between block and convolutional codes is based on whether "memory" is used in code generation. Both methods are characterized by the code rate R, which is the quotient of the information bits and the code length, i.e., $R = k/n$. For tree codes with memory, the actual output block, which is n bits long, depends not only on the present k information bits, but also on the previous m symbols, as shown in Fig. 7.8. Characteristics of convolution codes are the memory length $\nu = m \cdot k$, as well the distance profile, the free distance d_f, and the minimum distance d_m (see, for instance, [124]). Block codes can most often be constructed with algebraic methods using Galois fields, but tree codes are often only found in computer simulations.

Our discussion will be limited to *linear* codes, i.e., codes where the sum of two code words is again a code word, because this simplifies the decoder implementation. For linear codes, the Hamming distance can always be computed as the difference between a codeword and the zero word, which simplifies comparisons of the performance of the code. Linear tree codes are often called *convolutional codes*, because the codes can be built using an FIR-like structure. Convolutional codes may be *catastrophic* or *noncatastrophic*. In the case of catastrophic code, a single error will be propagated forever. It can

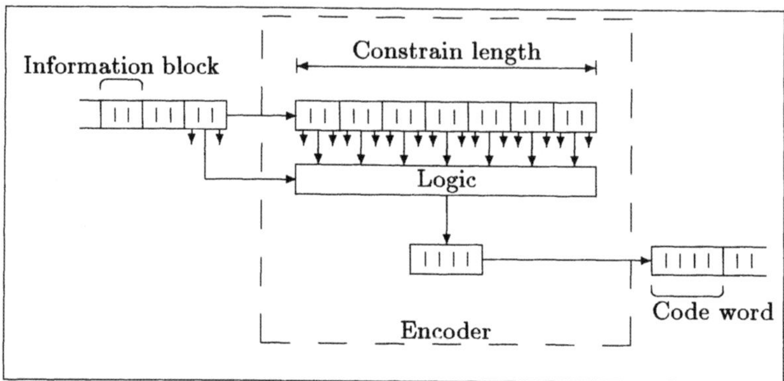

Fig. 7.8. Parameters of the convolutional encoders.

be shown that systematic convolutional codes are always noncatastrophic. It's also common to distinguish between *random error* correction codes and *burst error* correction codes. In burst error-correction, there may be a long burst of errors (or erasures). In random error-correction code, the capability to correct errors is not limited to consecutive bits – the error may have a random position in the received code word.

Coding bounds. With *coding bounds* we can compare different coding schemes. The bounds show the maximum error correction capability of the code. A decoder can never be better than the upper bound of the code, and sometimes to reduce the complexity of the decoder it's necessary to decode less than the theoretical bound.

A simple but still good, rough estimation is the Singleton bound or the Hamming bound. The *Singleton bound* states that the minimum Hamming distance d^* is upper bounded by the number of parity bits $(n-k)$. It's also known [124, p. 256] that the number of correctable errors t and the number of erasures e for a code is upper bounded by the Hamming distance. This gives the following bounds:

$$e + 2t + 1 \leq d^* \leq n - k + 1 \tag{7.36}$$

A code with $d^* = n - k + 1$ is called *maximum distance separable*, but besides the repetition code and the parity check code, there are no binary maximum distance separable codes [124, p. 431]. Following the example in the last section from Table 7.5, with 11 parity bits the upper bound can be used to correct up to five errors.

For a t-error-correcting binary code, the following *Hamming bound* provides a good estimation:

$$2^{n-k} \geq \sum_{m=0}^{t} \binom{n}{m} \tag{7.37}$$

Equation (7.37) says that the possible number of parity check patterns (2^{n-k}) must be greater than or equal to the number of error patterns. If the equal sign is valid in (7.37), such codes are called *perfect codes*. A perfect code is, for instance, the Hamming code discussed in the last section. If it is desired, for instance, to find a code to protect all 44 bits transmitted in one minute for radio-controlled watches, using the maximum-available 13 parity bits, then it follows that

$$2^{13} > \binom{44}{0} + \binom{44}{1} + \binom{44}{2} \quad \text{but} \tag{7.38}$$

$$2^{13} < \binom{44}{0} + \binom{44}{1} + \binom{44}{2} + \binom{44}{3} \tag{7.39}$$

i.e., it should be possible to find a code with the capability to correct two random errors but none with three errors. In the following sections we will review such block encoders and decoders, and then discuss convolutional encoders and decoders.

7.2.2 Block Codes

The linear cyclic binary BCH codes (from Bose, Chaudhuri, and Hocquenghem) and the subclass of Reed-Solomon codes, consist of a large class of block codes. BCH codes have various known efficient decoders in time and frequency domains. In the following, we will illustrate the shortening of a (63,50,6) to a (57,44,6) BCH code. The algorithm is discussed in details by Blahut [124, p. 162-6].

The code is based on a transformation of GF(2^6) to GF(2). To describe GF(2^6), a primitive polynomial of degree 6 is needed, such as $\mathrm{P}(x) = x^6+x+1$. To compute the generator polynomial, the least common multiple of the first $d-1 = 5$ minimal polynomials in GF(2^6) must be computed. If α denotes a primitive element in GF(2^6), it follows then that $\alpha^0 = 1$ and $m_{1(x)} = x-1$. The minimum polynomials of α, α^2 and α^4 are identical $m_{\alpha(x)} = x^6 + x + 1$, and the minimum polynomial to α^3 is $m_{\alpha^3(x)} = x^6 + x^4 + x^2 + x + 1$. It is now possible to build the generator polynomial, $g(x)$:

$$g(x) = m_{1(x)} \cdot m_{\alpha(x)} \cdot m_{\alpha^3(x)} \tag{7.40}$$

$$= x^{13} + x^{12} + x^{11} + x^{10} + x^9 + x^8 + x^6 + x^3 + x + 1 \tag{7.41}$$

Using this generator polynomial (to compute the parity bits), it is now a straight forward procedure to build the encoder and decoder.

Encoder. Since a systematic code is desired, the first code word bits are identical with the information bits. The parity bits $p(x)$ are computed by modulo reduction of the information bits $i(x)$ shifted in order to get a systematic code according to:

$$p(x) = i(x) \cdot x^{n-k} \bmod g(x) \tag{7.42}$$

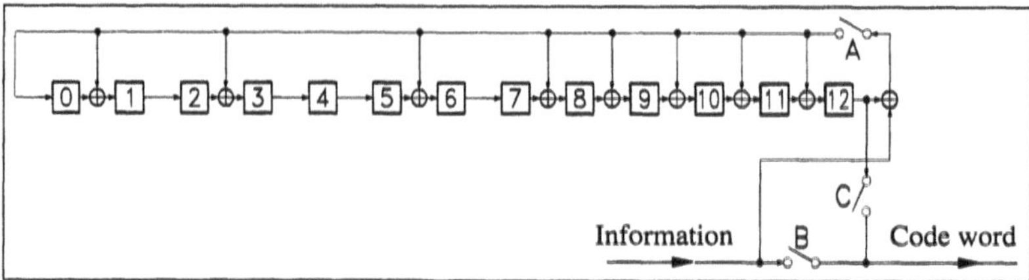

Fig. 7.9. Encoder for (57,44,6) BCH code.

Such a modulo reduction can be archieved with a recursive shift register as shown in Fig. 7.9.

The circuit works as follows: In the beginning, switch A and B are closed and C is open. Next, the information bits are applied (MSB first) and directly transferred to the codeword. At the same time, the recursive shift register computes the parity bits. After the information bits are all processed, switch A and B are opened and C is closed. The parity bits are now shifted into the code word.

Decoder. The decoder is usually more complex than the encoder. A Meggitt decoder can be used for decoding in the time domain, and frequency decoding is also possible, but it needs a detailed understanding of the algebraic properties of BCH codes ([124, p. 166-200], [92, p. 81-107], [163, p. 65-73]). Such frequency decoders for FPGAs are already available as Intellectual Property (IP) blocks, sometimes also called "virtual components," VC (see [168, 18, 19]).

The Meggitt decoder (shown in Fig. 7.10) is very efficient for codes with only a few errors to be corrected, since the decoder uses the cyclic properties of BCH codes. Only errors in the highest bit position are corrected and then a cyclic shift is computed, so that eventually all corrupted bits pass the MSB position and are corrected.

In order to use a shortened code and to regain the cyclic properties of the codes, a forward incoupling of the received data $a(x)$ must be computed. This condition can be gained for code shortened by b bits using the condition

$$s(x) = a(x)i(x) \bmod g(x) = x^{n-k+b} i(x) \bmod g(x) \tag{7.43}$$

For the shortened (57,44,6) BCH code this becomes

$$\begin{aligned} a(x) &= x^{63-50+6} \bmod g(x) \;=\; x^{19} \bmod g(x) \\ &= x^{19} \bmod (x^{13} + x^{12} + x^{11} + x^{10} + x^{9} + x^{8} + x^{6} + x^{3} + x + 1) \\ &= x^{10} + x^{7} + x^{6} + x^{5} + x^{3} + x + 1 \end{aligned}$$

The developed code has the ability to correct two errors. If only the error in the MSB need be corrected, a total of $1 + \binom{56}{1} = 1 + 56 = 57$ different error

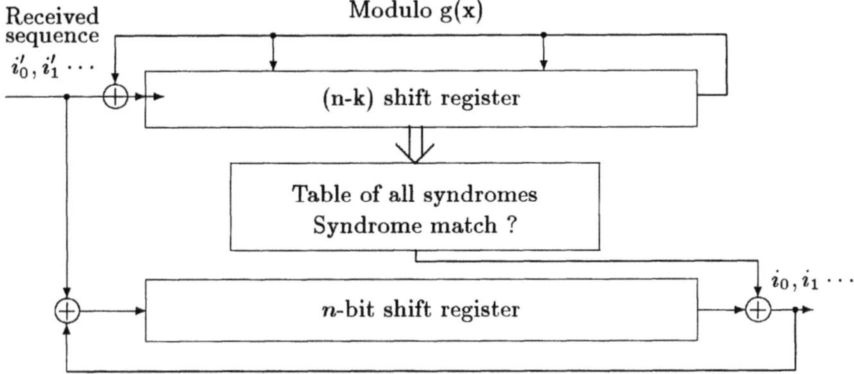

Fig. 7.10. Basic blocks of the Meggitt decoders.

patterns must be stored, as shown in Table 7.6. The 57 syndrome values can be computed through a simulation and are listed in [162, B.3].

Now all the building blocks are available for constructing the Meggitt decoder for the (57,44,6) BCH code. The decoder is shown in Fig. 7.11.

The Meggitt decoder has two stages. In the initialization phase, the syndrome is computed by processing the received bits modulo the generator polynomial $g(x)$. This takes 57 cycles. In the second phase, the actual error correction takes place. The content of the syndrome register is compared with the values of the syndrome table. If an entry is found, the table delivers a one, otherwise it delivers a zero. This hit bit is then XOR'd with the received bits in the shift register. In this way, the error is removed from the shift register. The hit bit is also wired to the syndrome register, to remove the error pattern from the syndrome register. Once again the syndrome and the shift register are clocked, and the next correction can be done. At the end, the shift register should include the corrected word, while the syndrome register

Table 7.6. Table of possible error patterns.

No.	Error pattern						
1	0	0	0	$\cdots$	0	0	1
2	0	0	0	$\cdots$	0	1	1
3	0	0	0	$\cdots$	1	0	1
$\vdots$	$\cdots$	$\cdots$	$\cdots$	$\cdots$	$\cdots$	$\cdots$	$\cdots$
56	0	1	0	$\cdots$	0	0	1
57	1	0	0	$\cdots$	0	0	1

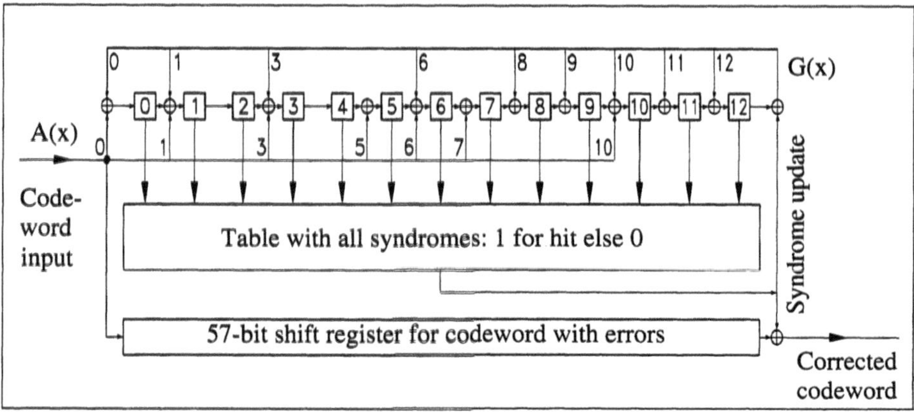

Fig. 7.11. Meggitt decoder for (57,44,6) BCH code.

should contain the all-zero word. If the syndrome is not zero, then more than two errors have occurred, and these can not be corrected with this BCH code.

Our only concern for an FPGA implementation of the Meggitt decoder is the large number (13) of inputs for the syndrome table, because the LUTs of FPGAs typically have 4 to 8 inputs. It's possible to use an external EPROM or (for Altera Flex 10K) four 2Kbit EABs to implement a table of size $2^{13} \times 1$. The syndrome is wired to the address lines, which deliver a hit (one) for the 57 syndromes, and otherwise a zero. It is also possible to use the logic synthesis tool to compute the table with internal logic blocks on the FPGA. The Xilinx XNFOPT (used in [162]) needs 132 LUTs, each with $2^4 \times 2$ bits. If modern binary decision diagrams (BBDs) synthesizer type [169, 170, 171] are used, this number can (at the cost of additional delays) be reduced to 58 LUTs with a size of $2^4 \times 2$ bits [172]. Table 7.7 shows the estimated effort, using Flex 10K, for the Meggitt decoder using the different kinds of syndrome tables.

Table 7.7. Estimated effort for Altera FLEX devices, for the three versions of the Meggitt decoder based on XC3K implementations [4]. (EABs are used as $2^{11} \times 1$ ROMs.)

	Syndrome table		
Function group	Using EABs	Only LCs	BDD [172]
Interface	36 LCs	36 LCs	36 LCs
Syndrome table	2 LCs, 4 EABs	264 LCs	116 LCs
64-bit FIFO	64 LCs	64 LCs	64 LCs
Meggitt decoder	12 LCs	12 LCs	12 LCs
State machine	21 LCs	21 LCs	21 LCs
Total	135 LCs, 4 EABs	397 LCs	249 LCs

7.2.3 Convolutional Codes

We also want to explore the kind of convolutional error-correcting decoders that are suitable for an FPGA realization. To simplify the discussion, the following constraints are defined, which are typical for communications systems:

- The code should minimize the complexity of the decoder. Encoder complexity is of less concern.
- The code is linear systematic.
- The code is convolutional.
- The code should allow random error correction.

A systematic code is stipulated to allow a power-down mode, in which only incoming bits are received without error correction [161]. A random error correction code is stipulated if the channel is slow-fading.

Fig. 7.12 shows a diagram of the possible tree codes, while Fig. 7.7 (p. 279) shows possible decoders. Fano and stack decoders are not very suitable for an FPGA implementation because of the complexity of organizing a stack [160]. A conventional μP/μC realization is much more suitable here. In the following sections, maximum-likelihood sequence decoders and algebraic algorithms are compared regarding hardware complexity, measured in CLBs usage for the Xilinx XC3K FPGA, and achievable error correction.

Viterbi maximum likelihood sequence decoder. The Viterbi decoder deals with an erroneous sequence by determing the corresponding sender sequence with the minimum Hamming distance. Put differently, the algorithm finds the optimal path through the trellis diagram, and is therefore an *optimal* memoryless noisy-sequence estimator (MLSE).

The advantage of the Viterbi decoder is its constant decoding time and MLSE optimality. The disadvantage lies in its high memory requirements and resulting limitation to codes with very short constraint length. Figs. 7.13 and 7.14 show an $R = k/n = 1/2$ encoder and the attendant trellis diagram. The constraint length $\nu = m \cdot k$ is 2, so the trellis has 2^{ν} nodes. Each node has $2^k = 2$ outgoing and at most $2^k = 2$ incoming edges. For a binary trellis ($k = 1$) like this, it is convenient to show a zero as an upward edge and a one as a downward edge.

For MLSE decoding it is sufficient to store only the 2^{ν} paths (and their metrics) passing through the nodes at a given level, because the MLSE path must pass through one of these nodes. Incoming paths with a smaller metric than the "survivor" with the highest metric need not be stored, because these paths will never be part of the MLSE path. Nevertheless, the maximum metric at any given time may not be part of the MLSE path if it is part of a short erroneous sequence. Voting down such a local error is analogous to demodulating a digital FM signal with memory [173]. Simulation results in [124, p. 381] and [92, p. 120-3] show that it's sufficient to construct a path

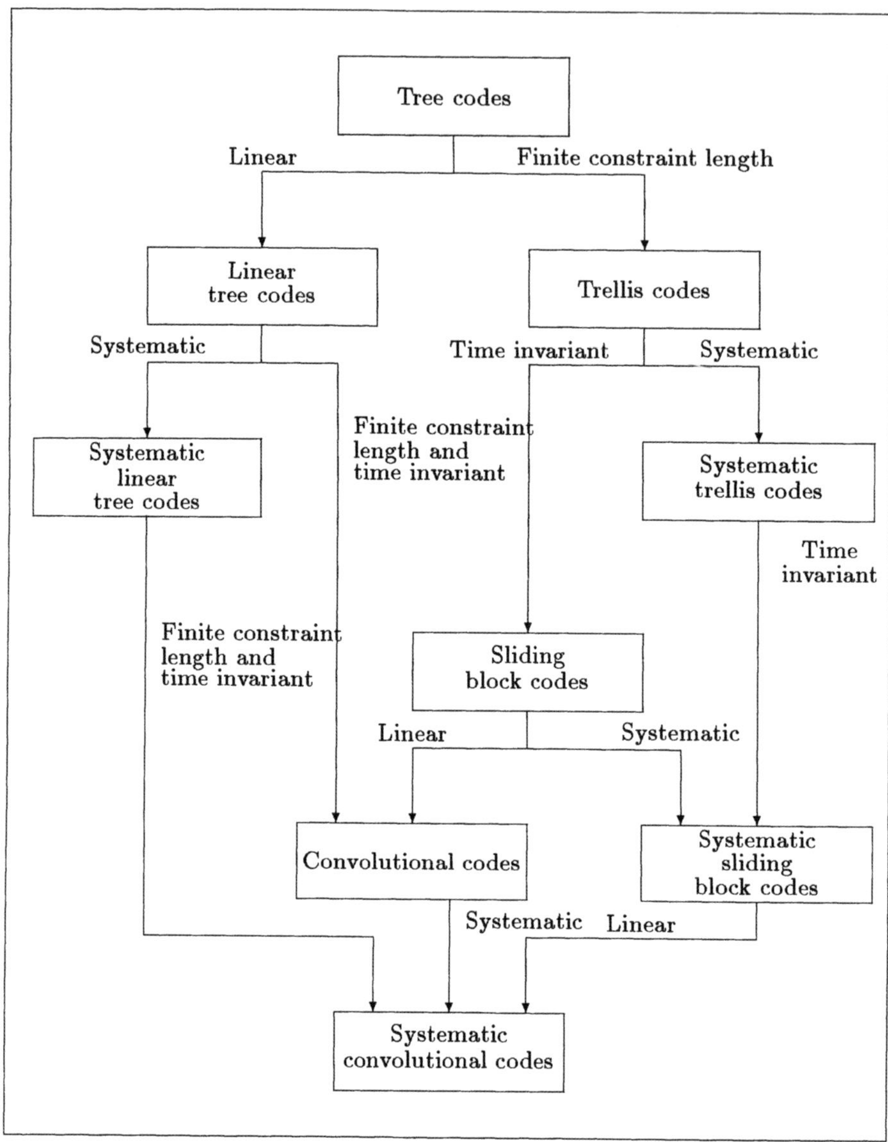

Fig. 7.12. Survey of tree codes [124].

memory of four to five times the constraint length. Infinite path memory yields no significant improvement.

The Viterbi decoder hardware consists of three main parts: path memory with output decoder (see Fig. 7.15), survivor computation, and maximum detection (see Fig. 7.16). The path memory is $4\nu 2^{\nu}$ bits, consuming $2\nu 2^{\nu}$ CLBs. The output decoder uses $(1 + 2 + \ldots + 2^{\nu-1})$ 2-to-1 multiplexers.

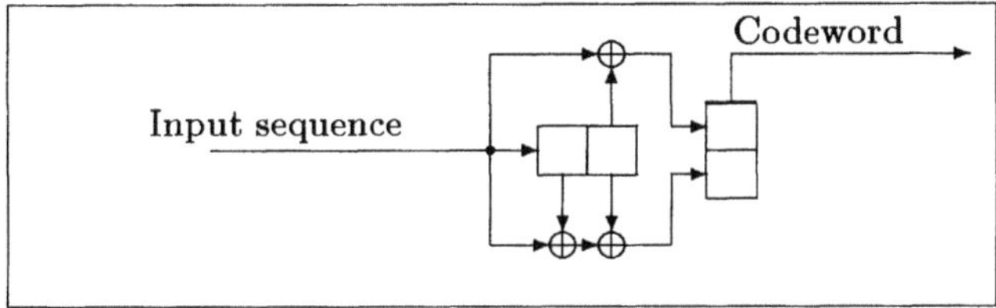

Fig. 7.13. Encoder for an $R = 1/2$ convolutional decoder.

Metric update adders, registers and comparisons are each $(\lceil \log_2(\nu * n) \rceil + 1)$ bits wide. For the maximum computation, additional comparisons, 2-to-1 multiplexers and a decoder are necessary.

The hardware for decoders with $k > 1$ seems too complex to implement with today's FPGAs. For $n > 2$ the information rate $R = 1/n$ is too low, so the most suitable code rate is $R = 1/2$. Table 7.8 lists the complexity in CLBs in a XC3K FPGA for constraint lengths $\nu = 2, 3, 4$, and the general case, for $R = 1/2$. It can be seen that complexity increases exponentially with constraint length ν, which should thus be as short as possible. Although very few errors can be corrected in the short window allowed by such a small constraint length, the MLSE algorithm guarantees acceptable performance.

Next it is necessary to choose an appropriate generating polynomial. It is shown in the literature ([174, p. 306-8], [175, p. 465], [164, p. 402-7], [124, p. 367]) that, for a given constraint length, *nonsystematic* codes have better performance than systematic codes, but using a nonsystematic code contradicts the demand for using the information bits without error correction. *Quick Look In* (QLI) codes are nonsystematic convolution codes with $R = 1/2$, providing *free distance* values as good as any known code for constraint lengths $\nu = 2$ to 4 [176]. The advantage of QLI codes is that only one XOR gate is necessary for the reconstruction of the information sequence. QLIs with $\nu = 2, 3$, and 4 have a free distance of $d_f = 5, 6$, and 7 respectively

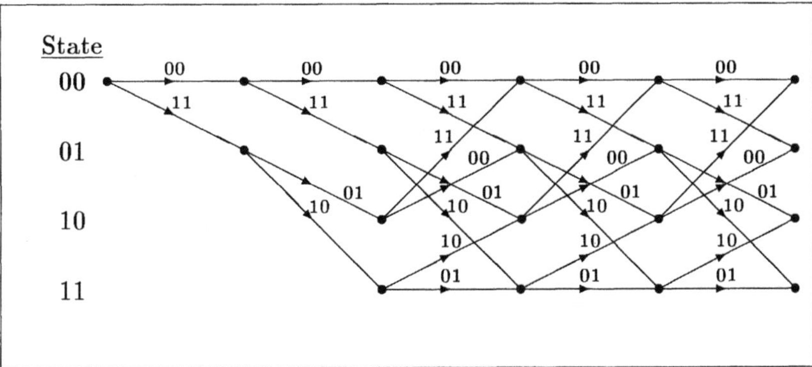

Fig. 7.14. Trellis for $R = 1/2$ convolutional decoder.

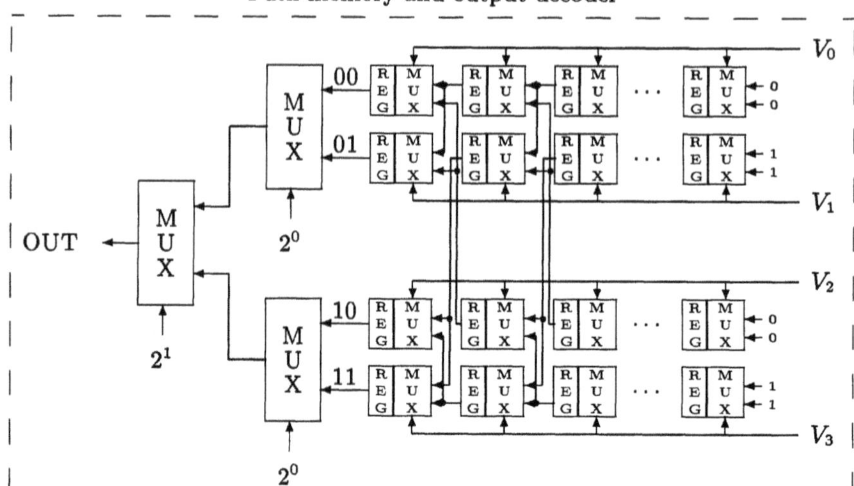

Fig. 7.15. Viterbi decoder with constraint length 4ν and $2^{\nu=2}$ nodes: path memory and output decoder.

[175, p. 465]. This seems to be a good compromise for low power consump-

Table 7.8. Hardware complexity in CLBs for an $R = 1/2$ Viterbi decoder for $\nu = 2, 3, 4$, and the general case.

Function	$\nu = 2$	$\nu = 3$	$\nu = 4$	$\nu \in \mathbb{N}$
Path memory	16	48	128	$4 \cdot \nu \cdot 2^{\nu-1}$
Output decoder	1,5	3,5	6,5	$1 + 2 + \ldots + 2^{\nu-2}$
Metric ΔM	4	4	4	4
Metric clear	1	2	4	$\lceil (2 + 4 + \ldots + 2^{\nu-1})/4 \rceil$
Metric adder	24	64	128	$(\lceil \log_2(n\nu) \rceil + 1) \cdot 2^{\nu+1}$
Survivor-MUX	6	24	48	$(\lceil \log_2(n\nu) \rceil + 1) \cdot 2^{\nu-1}$
Metric compare	6	24	48	$(\lceil \log_2(n\nu) \rceil + 1) \cdot 2^{\nu-1}$
Maximum compare	4,5	14	30	$(\lceil \log_2(n\nu) \rceil + 1) \cdot \frac{1}{2} \cdot (1 + 2 + \ldots 2^{\nu-1})$
MUX	3	12	28	$(2 + \ldots + 2^{\nu-1}) \cdot \frac{1}{2} \cdot (\lceil \log_2(n\nu) \rceil + 1)$
Decoder	1	2	4	$\lceil (2 + \ldots + 2^{\nu-1})/4 \rceil$
State Machine	4	4	4	
Sum:	67	197.5	428.5	

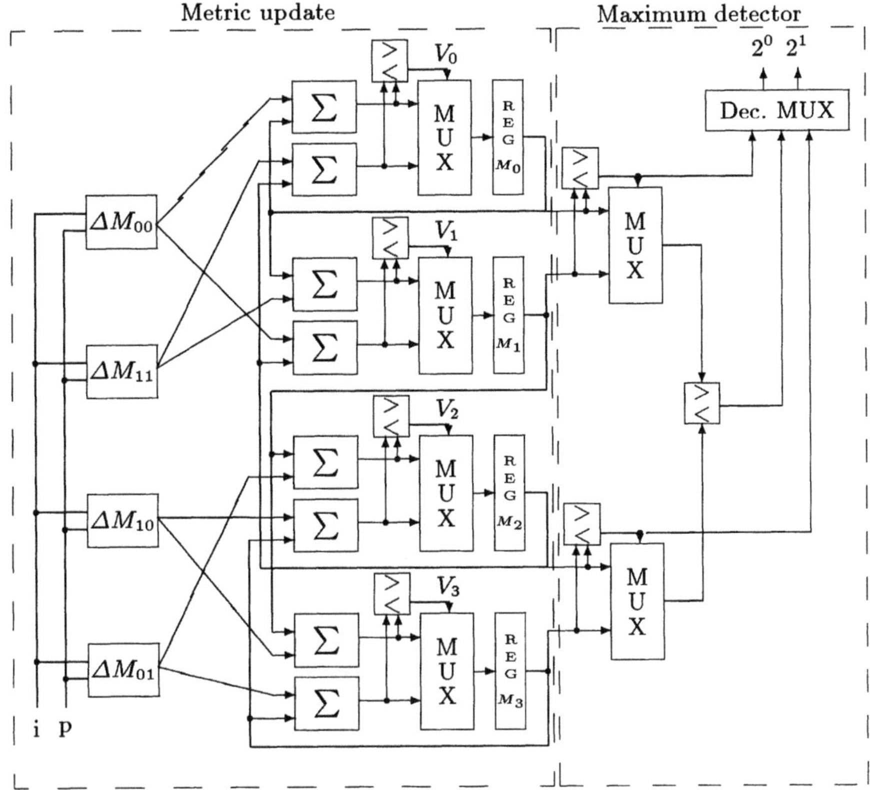

Fig. 7.16. Viterbi decoder with constraint length 4ν and $2^{\nu=2}$ nodes: metric calculation.

tion. The upper part of Table 7.9 shows the generating polynomials in octal notation.

Error-correction performance of the QLI decoder. To compute the error-correction performance of the QLI decoder, it is convenient to use the "union bound" method. Because QLI codes are linear, error sequences can be computed as a difference from the zero sequence. An MLSE decoder will make an incorrect decision if a sequence that starts at the null state, and differs from the null-word at j separate time steps, contains at least $j/2$ ones. The probability of this occurrence is

$$P_j = \begin{cases} \sum_{i=(j+1)/2}^{j} \binom{j}{i} p^i q^{j-i} & \text{for odd j} \\ \frac{1}{2}\binom{j}{j/2} p^{j/2} q^{j/2} + \sum_{i=j/2+1}^{j} \binom{j}{i} p^i q^{j-i} & \text{for even j.} \end{cases} \tag{7.44}$$

Now the only thing necessary for a bit-error probability formula is to compute the number w_j of paths with weight j for the code, which is an easily programmable task [160, C.4]. Because P_j decreases exponentially with

Table 7.9. Union-bound weights for a Viterbi decoder with $\nu = 2$ to 4 using QLI codes.

Code	O1 = 7 O2 = 5	O1 = 74 O2 = 54	O1 = 66 O2 = 46
Constraint length	$\nu = 2$	$\nu = 3$	$\nu = 4$
Distance		Weight w_j	
0-4	0	0	0
5	1	0	0
6	4	2	0
7	12	7	4
8	32	18	12
9	80	49	26
10	192	130	74
11	448	333	205
12	1024	836	530
13	2304	2069	1369
14	5120	5060	3476
15	11264	12255	8470
16	24576	29444	19772
17	53079	64183	43062
18	109396	126260	83346
19	103665	223980	147474
20	262144	351956	244458

increasing j, only the first few w_j must be computed. Table 7.9 shows the w_j for $j = 0$ to 20. The total error probability can now be computed with:

$$P_b < \frac{1}{k} \sum_{j=0}^{\infty} w_j P_j. \tag{7.45}$$

Syndrome algebraic decoder. The syndrome decoder (Fig. 7.17) and encoder (Fig. 7.18), like standard block decoders, computes a number of parity bits from the data sequence. The decoder's newly computed parity bits are XOR'd with the received parity bits to create the "syndrome" word, which will be nonzero if an error occurs in transmission. The error position and value are determined from the syndrome value. In contrast to block codes, where only one generator polynomial is used, convolutional codes at data rate $R = k/n$ have $k+1$ generating polynomials. The complete generator may be written in a compact $n \times k$ generator matrix. For the encoder of Fig. 7.18 the matrix is

$$\boldsymbol{G}(x) = \begin{bmatrix} 1 & \quad x^{21} + x^{20} + x^{19} + x^{17} + x^{16} + x^{13} + x^{11} + 1 \end{bmatrix} \tag{7.46}$$

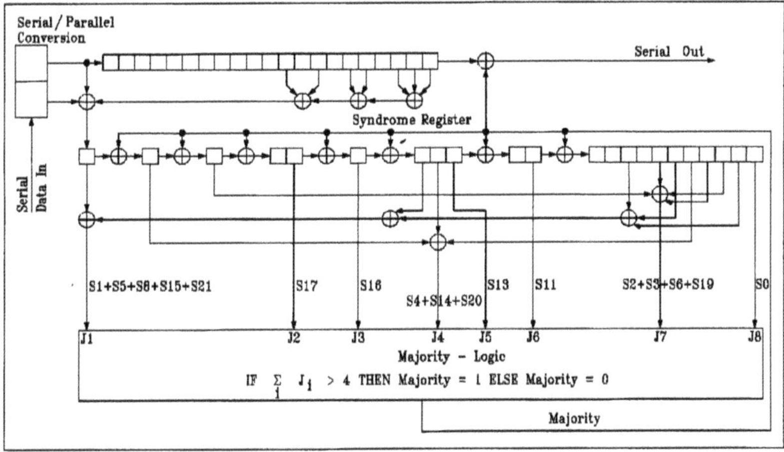

Fig. 7.17. Trial and error majority decoder with $J = 8$.

For a systematic code the matrix has the form $\boldsymbol{G}(x) = [\boldsymbol{I} \vdots \boldsymbol{P}(x)]$. The parity check matrix $\boldsymbol{H}(x) = [-\boldsymbol{P}(x)^T \vdots \boldsymbol{I}]$ is easily computed, given that $\boldsymbol{G} \cdot \boldsymbol{H}^T = \boldsymbol{0}$. The desired syndrome vector is thus $\boldsymbol{S} = \boldsymbol{v} \cdot \boldsymbol{H}^T$, where $\boldsymbol{v}$ is the received bit sequence.

The syndrome decoder now looks up the calculated syndrome in a table to find the correct sequence. To keep the table small, only sequences with an error at the first bit position are included. If the decoder needs to correct errors of more than one bit, we cannot clear the syndrome after the correction. Instead, the syndrome value must be subtracted from a syndrome register (see the "Majority" signal in Fig. 7.17).

A 22-bit table would be necessary for the standard convolutional decoder, but it is unfortunately difficult to implement a good FPGA look-up table with greater than 4 to 11 bit addresses [161]. Majority codes, a special class of syndrome-decodable codes, offer an advantage here. This type of canonical self-orthogonal code (CSOC) has exclusively ones in the first row of the $\{A_k\}$ parity check matrix (where the J columns are used as an orthogonal set to compute the syndrome) [164, p. 284]. Thus, every error in the first bit position

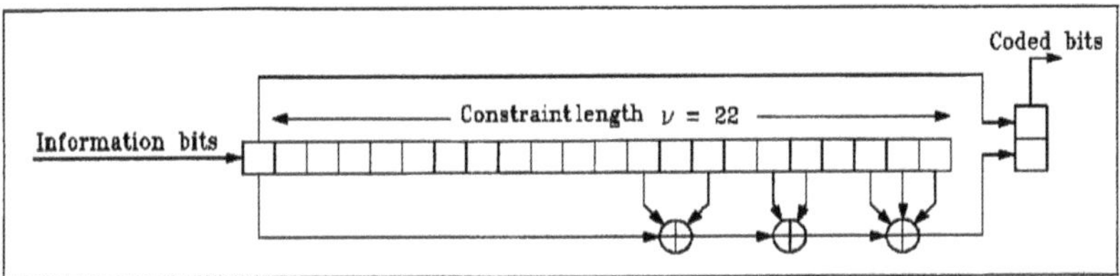

Fig. 7.18. Systematic (44, 22) encoder with rate $R = 1/2$ and constraint length $\nu = 22$.

Table 7.10. Some majority-decodable "trial and error" codes [164, p. 406].

J	t_{MD}	ν	Generating polynomial	Orthogonal equation
2	1	2	$1+x$	s_0, s_1
4	2	6	$1+x^3+x^4+x^5$	s_0, s_3, s_4, s_1+s_5
6	3	12	$1+x^6+x^7+x^9+x^{10}+x^{11}$	$s_0, s_6, s_7, s_9, s_1+s_3+s_{10}, s_4+s_8+s_{11}$
8	4	22	$1+x^{11}+x^{13}+x^{16}+x^{17}+x^{19}+x^{20}+x^{21}$	$s_0, s_{11}, s_{13}, s_{16}, s_{17}, s_2+s_3+s_6+s_{19}, s_4+s_{14}+s_{20}, s_1+s_5+s_8+s_{15}+s_{21}$
10	5	36	$1+x^{18}+x^{19}+x^{27}+x^{28}+x^{29}+x^{30}+x^{32}+x^{33}+x^{35}$	$s_0, s_{18}, s_{19}, s_{27}, s_1+s_9+s_{28}, s_{10}+s_{20}+s_{29}, s_{11}+s_{30}+s_{31}, s_{13}+s_{21}+s_{23}+s_{32}, s_{14}+s_{33}+s_{34}, s_2+s_3+s_{16}+s_{24}+s_{26}+s_{35}$

will cause at least $\lceil J/2 \rceil$ ones in the syndrome register. The decoding rule is therefore

$$e_0^i = \begin{cases} 1 & \text{for} \quad \sum_{k=1}^{J} A_k > \lceil J/2 \rceil \\ 0 & \text{otherwise} \end{cases} \tag{7.47}$$

Thus the name "majority code": instead of the expensive syndrome table only a majority vote is needed. Massey [164, p. 289] has designed a class of majority codes, called trial and error codes, which, instead of evaluating the syndrome vector directly, manipulate a combination of syndrome bits to get a vector orthogonal to e_0^i. This small additional hardware cost results in slightly better error correction performance than the conventional CSOC codes. Table 7.10 lists some trial and error codes with data rate $R = 1/2$. Fig. 7.17 shows a trial and error decoder with $J = 8$. Table 7.11 shows the complexity in CLBs of decoders with $J = 4$ to 10.

Error-correction capability of the trial and error decoder. To calculate the error-correction performance of trial and error codes, we must first

Table 7.11. Complexity in CLBs of a majority decoder with $J = 4$ to 10.

Function	$J=4$	$J=6$	$J=8$	$J=10$
Register	6	12	22	36
XOR-Gate	2	4	7	11
Majority-circuit	1	5	7	15
Sum	9	22	36	62

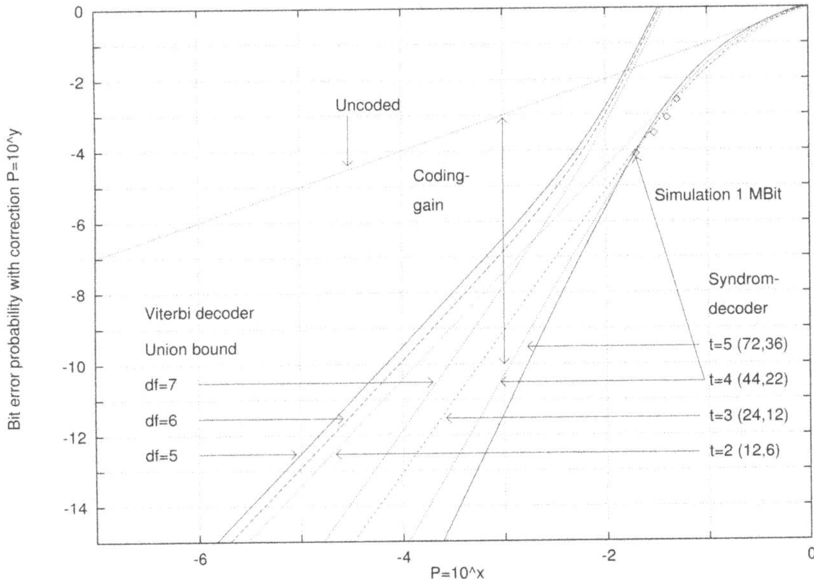

Fig. 7.19. Performance comparison of Viterbi and majority decoders.

note that in a window twice the constraint length, the codes allow up to $\lfloor J/2 \rfloor$-bit errors [124, p. 440]:

$$P(J) = \sum_{k=0}^{\lfloor J/2 \rfloor} \binom{2\nu}{k} p^k (1-p)^{2\nu-k} \tag{7.48}$$

A computer simulation of 10^6 bits, in Fig. 7.19, reveals good agreement with this equation. The equivalent single-error probability P_B of an (n, k) code can be computed with

$$P(J) = P(0) = (1 - P_B)^k \tag{7.49}$$

$$\rightarrow P_B = 1 - e^{\ln(P(J))/k} \tag{7.50}$$

Final comparison. Fig. 7.19 shows the error-correction performance of Viterbi and majority decoders. For a comparable hardware cost (Viterbi, $\nu = 2$, $d_f = 5$, 67 CLBs and trial and error, $t = 5$, 62 CLBs) the better performance of the majority decoder, due to the greater constraint length permitted, is immediately apparent. The optimal MLSE property of the Viterbi algorithm cannot compensate for its short constraint length.

7.2.4 Cryptography Algorithms for FPGAs

Many communication systems use data stream ciphers to protect relevant information, as shown in Fig. 7.20. The key sequence K is more or less a "pseudo-random sequence" (known to the sender and the receiver), and with the modulo 2 property of the XOR function, the plain text P can be reconstructed at receiver side, because

$$P \oplus K \oplus K = P \oplus 0 = P \tag{7.51}$$

In the following, we compare an algorithm based on a linear-feedback shift register (LFSR) and a "Data Encryption Standard" (DES) cryptographic algorithm. Neither algorithm requires large tables and both are suitable for an FPGA implementation.

Linear Feedback Shift Registers Algorithm

LFSRs with maximal sequence length are a good approach for an ideal security key, because they have good statistical properties (see, for instance [177], [178]). In other words, it is difficult to analyze the sequence in a cryptographic attack, an analysis called *cryptoanalysis*. Because bitwise designs are possible with FPGAs, such LFSRs are more efficiently realized with FPGAs than PDSPs. Two possible realizations of a LFSR of length 8 are shown in Fig. 7.21.

For the XOR LFSR there is always the possibility of the all-zero word, which should never be reached. If the cycle starts with any nonzero word, the cycle length is always $2^l - 1$. Sometimes, if the FPGA wakes up with an all-zero state, it's more convenient to use a "mirrored" or inverted LFSR circuit. If the all-zero word is a valid pattern and produces exactly the inverse sequence, it is necessary to substitute the XOR with a "not XOR" or XNOR gate. Such LFSRs can easily be designed using a `PROCESS` statement in VHDL, as the following example shows.

Example 7.12: Length 6 LFSR

The following VHDL code[2] implements a LFSR of length 6.

[2] The equivalent Verilog code `lfsr.v` for this example can be found in Appendix A on page 383.

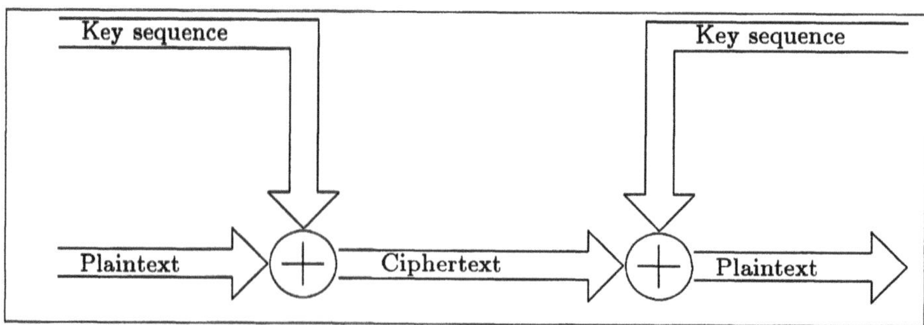

Fig. 7.20. The principle of a synchronous data stream cipher.

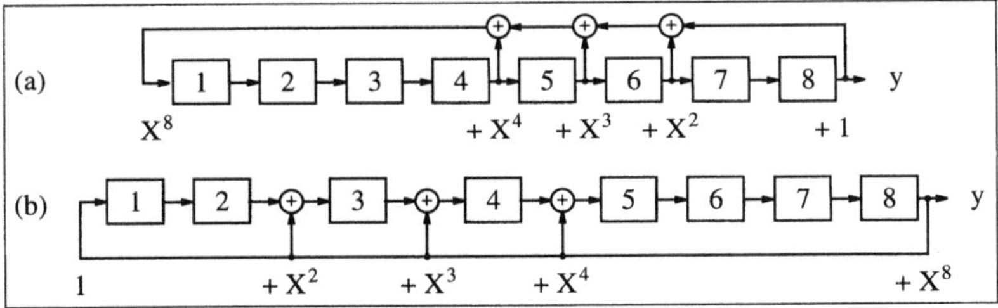

Fig. 7.21. Possible realizations of LFSRs.

```
LIBRARY ieee;
USE ieee.std_logic_1164.ALL;
USE ieee.std_logic_arith.ALL;

ENTITY lfsr IS                              ------> Interface
  PORT ( clk : IN  STD_LOGIC;
         y   : OUT STD_LOGIC_VECTOR(6 DOWNTO 1));
END lfsr;

ARCHITECTURE flex OF lfsr IS

  SIGNAL  ff  :   STD_LOGIC_VECTOR(6 DOWNTO 1);

BEGIN

  PROCESS           -- Implement length 6 LFSR with xnor
  BEGIN
    WAIT UNTIL clk = '1';
    ff(1) <= NOT (ff(5) XOR ff(6));
    FOR I IN 6 DOWNTO 2 LOOP      -- Tapped delay line:
      ff(I) <= ff(I-1);           -- shift one
    END LOOP;
  END PROCESS ;

  PROCESS (ff)
  BEGIN                  -- Connect to I/O cell
    FOR k IN 1 TO 6 LOOP
      y(k) <= ff(k);
    END LOOP;
  END PROCESS;

END flex;
```

From the simulation of the design in Fig. 7.22, it can be concluded that the LFSR goes through all possible bit patterns, which results in the maximum sequence length $2^6 - 1 = 63 \approx 630\text{ns}/10\text{ns}$. The clock cycle was 10 ns, and the `Registered Performance` is 100 MHz. 7.12

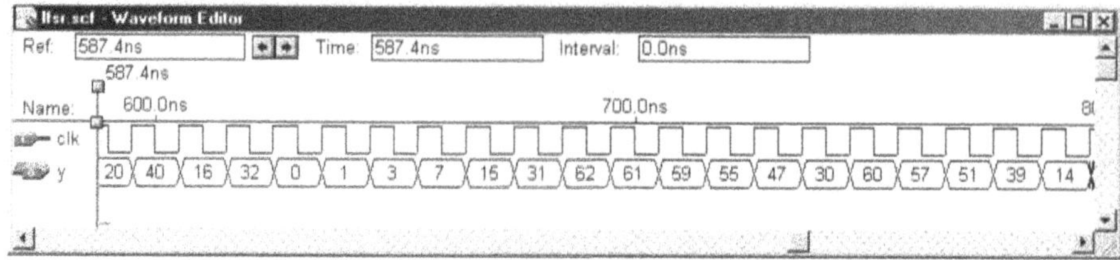

Fig. 7.22. LFSR simulation.

Note that a complete cycle of an LFSR sequence fulfils the three criteria for optimal length $2^l - 1$ pseudo-random sequences defined by Golomb [179, p. 188]:

1) The number of 1s and 0s in a cycle differs by no more than one.
2) Runs of length k (*e.g.*, $111\cdots$ sequence, $000\cdots$ sequence) have a total fractional part of all runs of $1/2^k$.
3) The autocorrelation function $C(\tau)$ is constant for $\tau \in [1, n-1]$.

LFSRs are usually constructed from primitive polynomials in GF(2) using the circuits shown in Fig. 7.21 . W. Stahnke [165] has compiled a list of such primitive polynomials up to order 168. This paper is available online at `http://www.jstor.org`. With today's available algebraic software packages like MAPLE, MUPAD, or MAGMA such a list can easily be extended. The following is a code example for MAPLE to compute the primitive polynomials of type $x^l + x^a + 1$ with the smallest a.

```
with(numtheory):
for l from 2 by 1 to 45 do
  for a from 1 by 1 to l-1 do
    if (Primitive(x^l+x^a+1) mod 2) then
    print(l,a);
    break;
  fi;
  od;
od;
```

Table 7.12 shows the necessary `XOR` list of the first 45 maximum length LFSRs according to Fig. 7.21(a). For instance, the entry for polynomial fourteen (14, 13, 11, 9) means the primitive polynomial is

$$\begin{aligned} p_{14}(x) &= x^{14} + x^{14-13} + x^{14-11} + x^{14-9} + 1 \\ &= x^{14} + x^5 + x^3 + x + 1. \end{aligned}$$

For $l > 2$ these primitive polynomials always have "twins," which are also primitive polynomials [180]. These are the "time" reversed versions $x^l + x^{l-a} + 1$.

Table 7.12. A list of the first 45 LFSR.

l	Exponents	l	Exponents	l	Exponents
1	1	16	16, 14, 13, 11	31	31, 28
2	2, 1	17	17, 14	32	32, 30, 29, 23
3	3, 2	18	18, 11	33	33, 20
4	4, 3	19	19, 18, 17, 14	34	34, 31, 30, 26
5	5, 3	20	20, 17	35	35, 33
6	6, 5	21	21, 19	36	36, 25
7	7, 6	22	22, 21	37	37, 36, 33, 31
8	8, 6, 5, 4	23	23, 18	38	37, 36, 33, 31
9	9, 5	24	24, 23, 21, 20	39	39, 35
10	10, 7	25	25, 22	40	40, 37, 36, 35
11	11, 9	26	26, 25, 24, 30	41	41, 38
12	12, 11, 8, 6	27	27, 26, 25, 22	42	42, 39, 38, 35
13	13, 12, 10, 9	28	28, 25	43	43, 41, 40, 36
14	14, 13, 11, 9	29	29, 27	44	44, 42, 41, 37
15	15, 14	30	30, 29, 26, 24	45	45, 44, 43, 41

Stahnke [165][18, XAPP52] has computed primitive polynomials of type $x^l + x^a + 1$. There are no primitive polynomials with four elements, i.e. $(x^l + x^b + x^a + 1)$ for $l < 45$. But it is possible to find polynomials of the type $x^l + x^{a+b} + x^b + x^a + 1$, which Stahnke used for those l where a polynomial of the type $x^l + x^a + 1$ ($l = 8, 12, 13$, etc.) does not exist.

The LFSRs with four elements in Table 7.12 were computed to have the maximum sum (i.e., $a + b$) for the tap exponents. We will see later that, for multistep LFSR implementations, this usually gives the minimum complexity.

If n random bits are used at once, it is possible to clock our LFSR n times. In general it is *not* a good idea to use just the lowest n bits of our LFSR, since this will lead to weak random properties, i.e., low cryptographic security. But it is possible to compute the equation for n-bit shifts, so that only one clock cycle is needed to generate n new random bits. The necessary equation can be computed more easily if a "state-space" description of the LFSR is used, as the following example shows.

Example 7.13: Three Steps-at-Once LFSR

Let's assume a primitive polynomial of length 6, e.g., $p = x^6 + x + 1$, is used to compute random sequences. The task now is to compute three "new" bits in one clock cycle. To obtain the required equation, the state-space description of our LFSR must first be computed, i.e., $\boldsymbol{x}(t+1) = \boldsymbol{A}\boldsymbol{x}(t)$

$$\begin{bmatrix} x_6(t+1) \\ x_5(t+1) \\ x_4(t+1) \\ x_3(t+1) \\ x_2(t+1) \\ x_1(t+1) \end{bmatrix} = \begin{bmatrix} 0\,1\,0\,0\,0\,0 \\ 0\,0\,1\,0\,0\,0 \\ 0\,0\,0\,1\,0\,0 \\ 0\,0\,0\,0\,1\,0 \\ 0\,0\,0\,0\,0\,1 \\ 1\,1\,0\,0\,0\,0 \end{bmatrix} \begin{bmatrix} x_6(t) \\ x_5(t) \\ x_4(t) \\ x_3(t) \\ x_2(t) \\ x_1(t) \end{bmatrix} \qquad (7.52)$$

With this state-space description, the actual values $\boldsymbol{x}(t)$ and the transition matrix $\boldsymbol{A}$ are used to compute the new values $\boldsymbol{x}(t+1)$. To compute the values for $\boldsymbol{x}(t+2)$, simply compute $\boldsymbol{x}(t+2) = \boldsymbol{A}\boldsymbol{x}(t+1) = \boldsymbol{A}^2\boldsymbol{x}(t)$. The next iteration gives $\boldsymbol{x}(t+3) = \boldsymbol{A}^3\boldsymbol{x}(t)$. The equations for an n-step-at-once LFSR can therefore be computed by evaluating $\boldsymbol{A}^n$ mod 2. For $n = 3$ it follows that

$$\boldsymbol{A}^3 \bmod 2 = \begin{bmatrix} 0&0&0&1&0&0 \\ 0&0&0&0&1&0 \\ 0&0&0&0&0&1 \\ 1&1&0&0&0&0 \\ 0&1&1&0&1&0 \\ 0&0&1&1&0&1 \end{bmatrix} \tag{7.53}$$

As expected, for the register x_6 to x_4 there is a shift of three positions, while the other three values x_1 to x_3 are computed using an EXOR operation. The following VHDL code[3] implements this three-step LFSR.

```
LIBRARY ieee;
USE ieee.std_logic_1164.ALL;
USE ieee.std_logic_arith.ALL;

ENTITY lfsr6s3 IS                          ------> Interface
  PORT ( clk : IN  STD_LOGIC;
         y   : OUT STD_LOGIC_VECTOR(6 DOWNTO 1));
END lfsr6s3;

ARCHITECTURE flex OF lfsr6s3 IS

  SIGNAL ff : STD_LOGIC_VECTOR(6 DOWNTO 1);

BEGIN

  PROCESS   -- Implement three-step length-6 LFSR with xnor
  BEGIN
    WAIT UNTIL clk = '1';
    ff(6) <= ff(3);
    ff(5) <= ff(2);
    ff(4) <= ff(1);
    ff(3) <= NOT (ff(5) XOR ff(6));
    ff(2) <= NOT (ff(4) XOR ff(5));
    ff(1) <= NOT (ff(3) XOR ff(4));
  END PROCESS ;

  PROCESS (ff)
  BEGIN                  -- Connect to I/O cell
    FOR k IN 1 TO 6 LOOP
      y(k) <= ff(k);
    END LOOP;
  END PROCESS;

END flex;
```

Fig. 7.23 shows a simulation for the three-step LFSR design. Comparing the simulation of this LFSR in Fig. 7.23 with the simulation of the single-step LFSR in Fig. 7.22, it can be concluded that now every third sequence value occurs. The cycle length is reduced from $2^6 - 1$ to $(2^6 - 1)/3 = 21$. 7.13

[3] The equivalent Verilog code **lfsr6s3.v** for this example can be found in Appendix A on page 384.

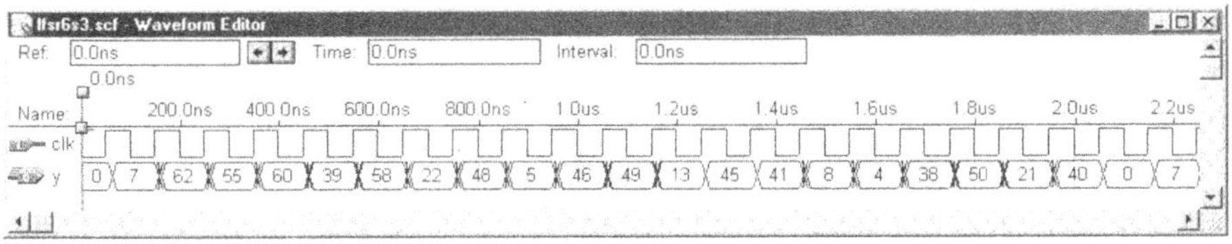

Fig. 7.23. Multistep LFSR simulation.

To implement such a multi-step LFSR, we want to select the primitive polynomial which results in the lowest circuit effort, which can be computed by counting the nonzero entries in the $\boldsymbol{A}^k$ mod 2 matrix, and/or the maximum fan-in for the register, which corresponds to the number of ones in each row. For a few shifts, the fan-in for the circuit from Fig. 7.21(a) may be advantageous. It can also be observed [4] that if the feedback signals are close in the $\boldsymbol{A}$ matrix, some entries in the $\boldsymbol{A}^k$ matrix may become zero, due to the modulo 2 operations. As mentioned earlier, the two and four-tap LFSR data in Table 7.12 were therefore computed to yield the maximum sum of all taps. For the same sum, the primitive polynomial that has the larger value for the smallest tap was selected, e.g., (11, 12) is better than (10, 13). This was chosen because tap l is mandatory for the maximum-length LFSR, and the other values should be close to this tap.

If, for instance, Stanke's $s_{14,a}(x) = x^{14} + x^{12} + x^{11} + x + 1$ primitive polynomial is used, this will result in 58 entries for an $n = 8$ multistep LFSR, while if the LFSR from Table 7.12, $p_{14} = x^{14} + x^5 + x^3 + x^1 + 1$ (i.e., taps 14,13,11,9) is used, the $\boldsymbol{A}^8$ mod 2 matrix has only 35 entries (Exercise 7.6, p. 331). Fig. 7.24 shows the total number of ones for the LFSR for the two polynomials with the two different implementations from Fig. 7.21, while Fig. 7.25 shows the maximum fan-in (i.e., the maximum needed input bit width for a LC) for this LFSR. It can be concluded from the two figures that a careful choice of the polynomial and LFSR structure can provide substantial savings. For the multistep LFSR synthesis, it can be seen from Fig. 7.25 that the LFSR of Fig. 7.21(b) has fewer fan-ins (i.e., smaller LC input bit width), but for longer multistep k, the effort seems similar for the primitive polynomials from Table 7.12.

Combining LFSR

An additional gain in performance in cryptographic security can be achieved if several LFSR registers are combined into one key generator. Several linear and nonlinear combinations exist [167],[166, p. 150-173]. Meaningful for

[4] It is obviously not applicable to select the LFSR with the smallest implementation effort, because there are $\phi(2^l - 1)/l$ primitive polynomials, where $\phi(x)$ is the Euler function which computes the number of coprimes to x. For instance, a 16-bit register has $\phi(2^{16} - 1)/16 = 2048$ different primitive polynomials [180]!

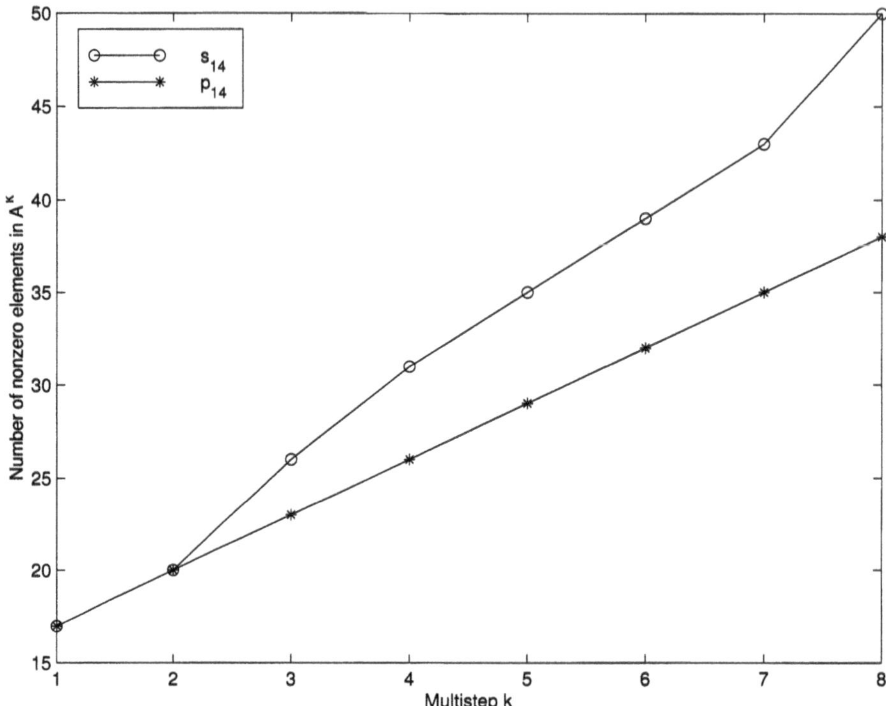

Fig. 7.24. Number of ones in $\boldsymbol{A}_{14}^{k}$.

implementation effort and security are nonlinear combinations with thresholds. For a combination of three different LFSRs with length L_1, L_2, and L_3 the *linear complexity*, which is the equivalent length of *one* LFSR (which may be synthesized with the Berlekamp-Massey algorithm, for instance [166, p. 141-9]), provides

$$L_{\mathrm{ges}} = L_1 \cdot L_2 + L_2 \cdot L_3 + L_1 \cdot L_3. \tag{7.54}$$

Fig. 7.26 shows a realization for such a scheme.

Since the key in the selected paging format has 50 bits, a total length of $2 \cdot 50 = 100$ registers was chosen, and the three feedback polynomials are:

$$p_{33}(x) = x^{33} + x^6 + x^4 + x + 1 \tag{7.55}$$
$$p_{29}(x) = x^{29} + x^2 + 1 \tag{7.56}$$
$$p_{38}(x) = x^{38} + x^6 + x^5 + x + 1 \tag{7.57}$$

All the polynomials are *primitive*, which guarantees that the length of all three shift-register sequences gives a maximum. For the linear complexity of the combination it follows that:

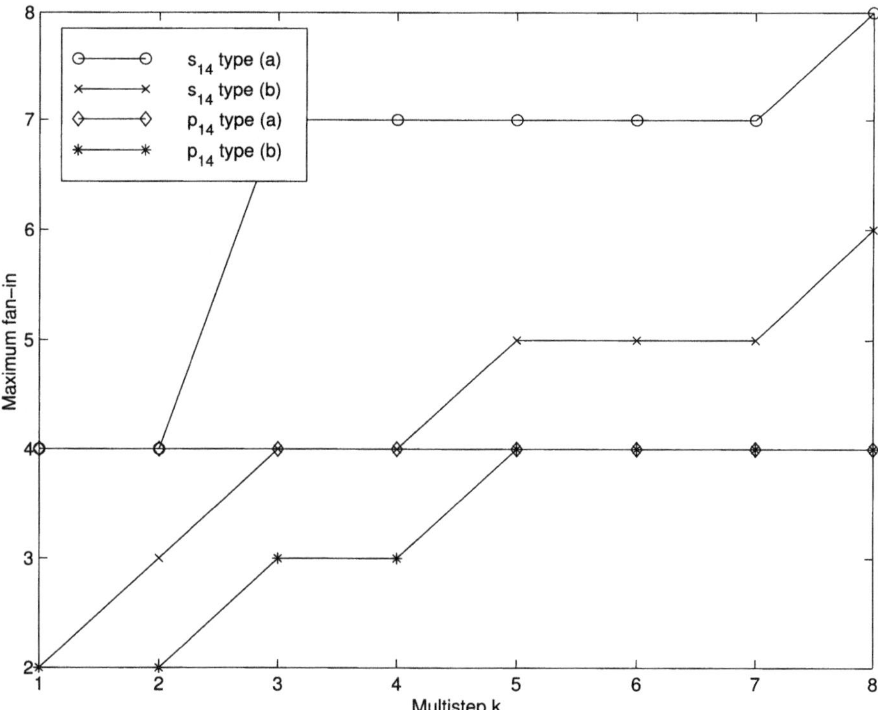

Fig. 7.25. Maximum fan-in for multistep length-14 LFSR.

$$L_1 = 33; \qquad L_2 = 29; \qquad L_3 = 38$$
$$L_{\text{total}} = 33 \cdot 29 + 33 \cdot 38 + 29 \cdot 38 = 3313$$

After each coding the key is lost, and an additional 50 registers are needed to store the key. The 50-bit key is used twice. Table 7.13 shows the hardware resources required with Xilinx FPGAs of the 3K family.

Table 7.13. Cost, measured in CLBs, of a 3K Xilinx FPGA.

Function group	CLBs
50-bit key register	25
100-bit shift register	50
Feedback	3
Threshold	0.5
XOR with message	0.5
Total	79

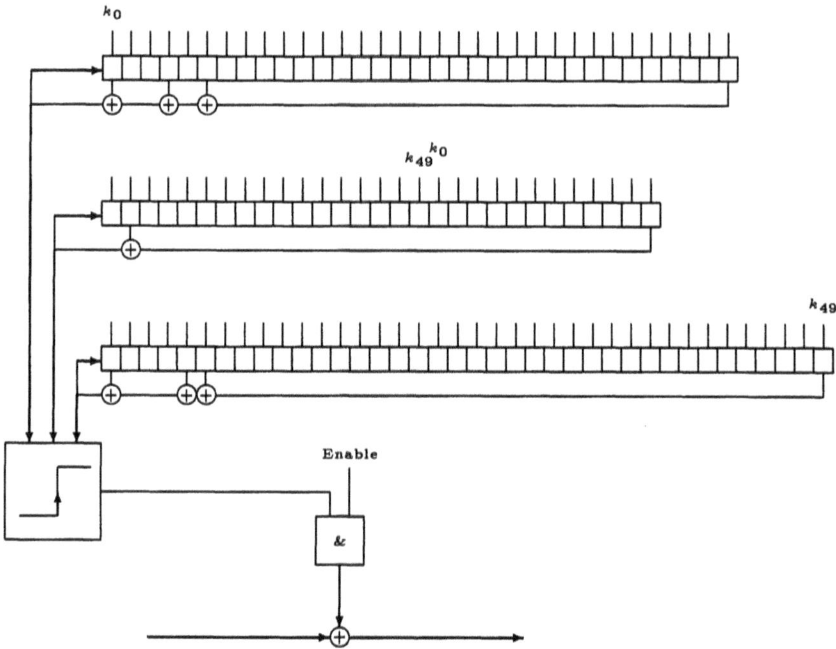

Fig. 7.26. Realization of the data stream cipher with 3 LFSR.

DES based algorithm. The Data Encryption Standard (DES), outlined in Fig. 7.27, is typically used in a block cipher. By selecting the "output feedback mode" (OFB) it is also possible to use the modified DES in a data stream cipher (see Fig. 7.28). The other modes (ECB, CBC or CFB) of the DES are, in general, not applicable for communication systems, due to the "avalanche effect": A single-bit error in the transmission will alter approximately 50% of all bits in a block.

We will review the principles of the DES algorithm and then discuss suitable modifications for FPGA implementations.

The DES comprises a finite-state machine translating plain text blocks into cipher text blocks. First the block to be substituted is loaded into the state register (32 bits). Next it is expanded (to 48 bits), combined with the key (also 48 bits) and substituted in eight 6→4 bit-width S-boxes. Finally, permutations of single bits are performed. This cycle may be (if desired, with a changing key) applied several times. In the DES, the key is usually shifted one or two bits so that after 16 rounds the key is back in the original position. Because the DES can therefore be seen as an iterative application of the Feistel cipher (shown in Fig. 7.29), the S-boxes must not be invertable.

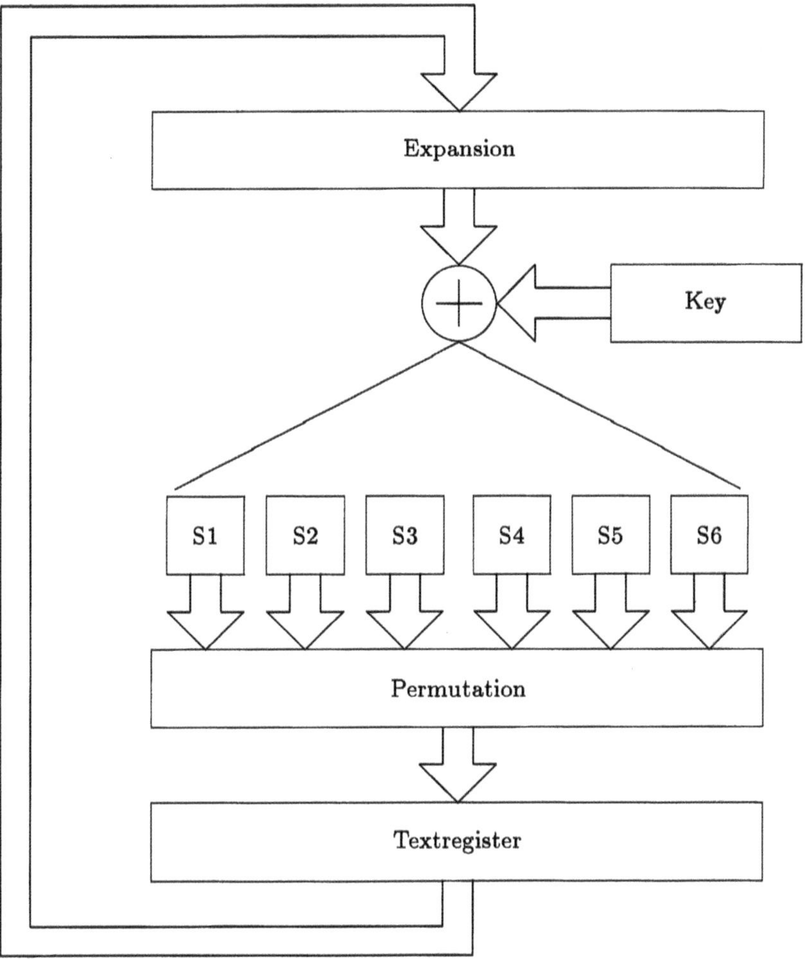

Fig. 7.27. State machine for a block encryption system (DES).

To simplify an FPGA realization some modifications are useful, such as a reduction of the length of the state register to 25 bits. No expansion is used. Use the final permutations as listed in Table 7.14.

Table 7.14. Table for permutation.

From bit no.	0	1	2	3	4	5	6	7	8	9	10	11	12
To bit no.	20	4	5	10	15	21	0	6	11	16	22	1	7
From bit no.	13	14	15	16	17	18	19	20	21	22	23	24	
To bit no.	12	17	23	2	8	13	18	24	3	9	14	19	

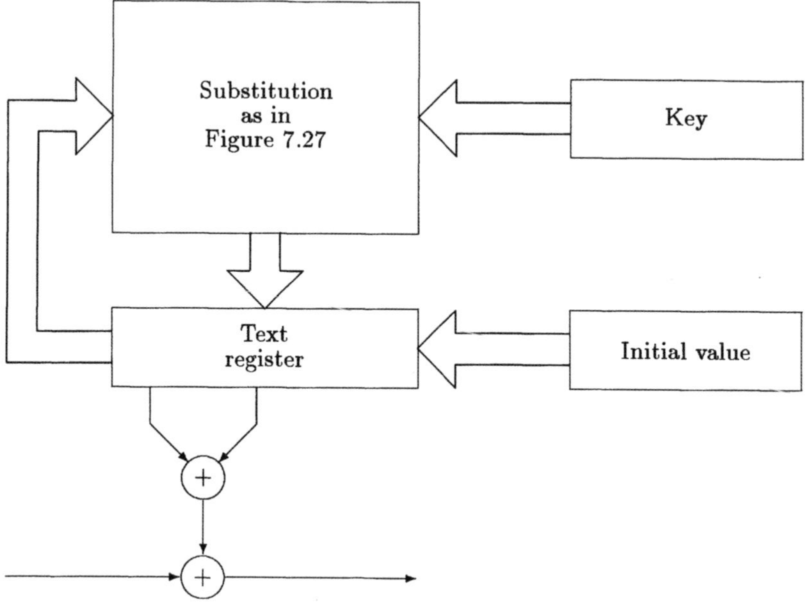

Fig. 7.28. Block cipher in the OFB-mode used as data stream cipher.

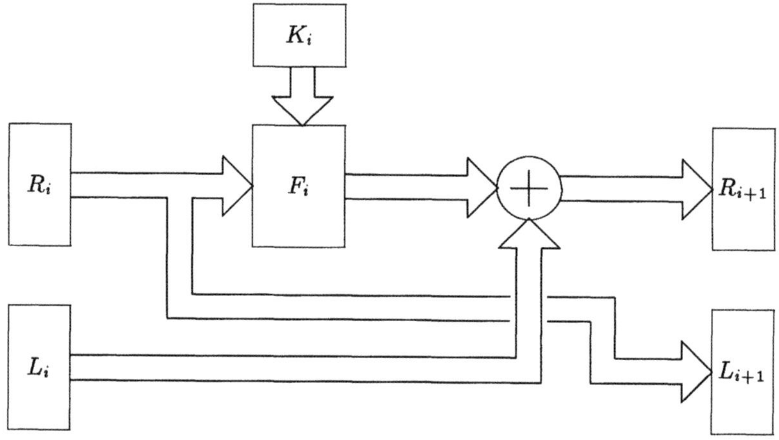

Fig. 7.29. Principle of the Feistel network.

Because most FPGAs only have four to five input look-up tables (LUTs), S-boxes with five inputs have been designed, as displayed in Table 7.15.

Although the intention was to use the OFB mode only, the S-boxes in Table 7.16 were generated in such a manner that they can be inverted. The

Table 7.15. The five new designed substitution boxes (S-boxes).

Input	Box 1	Box 2	Box 3	Box 4	Box 5
0	1E	F	14	19	6
1	13	1	1D	14	E
2	14	13	16	D	1A
3	1	1F	B	4	3
4	1A	19	5	1C	B
5	1B	1C	E	1A	1E
6	E	12	8	1E	0
7	B	11	F	1	2
8	D	8	4	C	1D
9	10	7	C	F	C
A	3	1B	1E	1B	18
B	0	0	13	1D	17
C	4	1A	10	5	1
D	6	C	1	15	15
E	A	1D	18	E	1B
F	17	2	17	13	9
10	19	B	1C	17	19
11	16	1E	A	9	A
12	7	18	1B	3	4
13	1C	D	3	10	14
14	1D	5	19	A	13
15	5	14	D	16	11
16	2	15	0	12	10
17	1F	9	2	1F	12
18	F	3	15	B	5
19	11	10	6	2	F
1A	C	6	7	6	8
1B	18	17	12	18	16
1C	9	4	1F	11	1C
1D	15	16	1A	8	7
1E	8	E	9	7	D
1F	12	A	11	0	1F

modified DES may therefore also be used as a normal block cipher (Electronic Code Book).

Table 7.16. Dependency matrix for the five substitution boxes (ideal value is 16).

Box 1

20	12	20	20	20
12	20	12	16	16
12	16	16	12	8
16	16	20	12	16
20	16	20	12	12

Box 2

20	16	20	12	20
20	20	20	16	16
12	20	20	16	8
16	24	12	16	12
16	20	16	20	20

Box 3

20	12	16	16	16
16	20	16	16	16
16	16	20	12	12
16	8	12	16	20
20	12	12	20	12

Box 4

20	16	20	20	16
12	16	12	16	20
20	16	16	20	16
20	16	16	20	24
16	12	28	20	16

Box 5

12	20	8	12	20
20	12	16	24	20
16	12	12	20	16
16	20	16	20	12
12	16	16	12	24

Table 7.17. Hardware effort of the modified DES based algorithm.

Function group	CLBs
25-bit key register	12.5
25-bit additions	12.5
25-bit state register	12.5
Five S-Boxes 5→5	25
Permutation	0
25-bit initialization vector	12.5
Multiplex: Initialization vector/S-box	12.5
XOR with message	1
Total	87.5

A reasonable test for S-boxes is the dependency matrix. This matrix shows, for every input/output combination, the probability that an output bit changes if an input bit is changed. With the avalanche effect the ideal probability is 1/2. Table 7.16 shows the dependency matrix for the new five S-boxes. Instead of the probability, the table shows the absolute number of occurrences. Since there are $2^5 = 32$ possible input vectors for each S-box, the ideal value is 16. A random generator was used to generate the S-boxes. The reason that some values differ much from the ideal 16 may lie in the desired inversion.

The hardware effort of the DES-based algorithm is summarized in Table 7.17.

Cryptographic performance comparison. We will next discuss the cryptographic performance analysis of the LFSR- and DES-based algorithms. Several security tests have been defined and the following comparison shows the two most interesting (the others do not show clear differences between the two schemes). For both tests, 100 random keys were generated.

1) Using different keys, the generated sequences were analyzed. In each random key, one bit was changed and the number of bit changes in the plain text was recorded. On average, about 50% of the bits should be inverted (avalanche effect).
2) Similar to Test 1, but this time the number of changes in the output sequence were analyzed, depending on the changed position of the key. Again, 50% of the bits should change in sign.

For both tests, plain text with 64-bit length were used (see again Fig. 7.20, p. 294). The plain text is arbitrary. For Test 1, all variations over each individual key position were accumulated. For Test 2, all changes depending on the position in the output sequence were accumulated. This test was performed for 100 random keys. For Test 1, the ideal value is $64 \cdot 0.5 \cdot 100 = 3200$ and for Test 2 the optimal value is $50 \cdot 0.5 \cdot 100 = 2500$. Figs. 7.30 and 7.31 display the

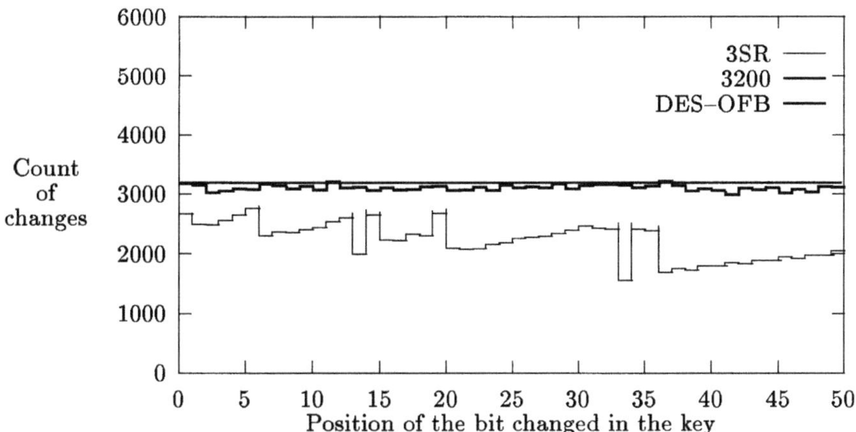

Fig. 7.30. Results of Test 1.

results. They clearly show that the DES-OFB scheme is much more sensitive to changes in the key than is the scheme with three LFSRs. The conclusion from Test 2 is that the SR scheme needs about 32 steps until a change in the key will affect the output sequence. For the DES-OFB scheme, only the first four samples differ considerably from the ideal value of 2500.

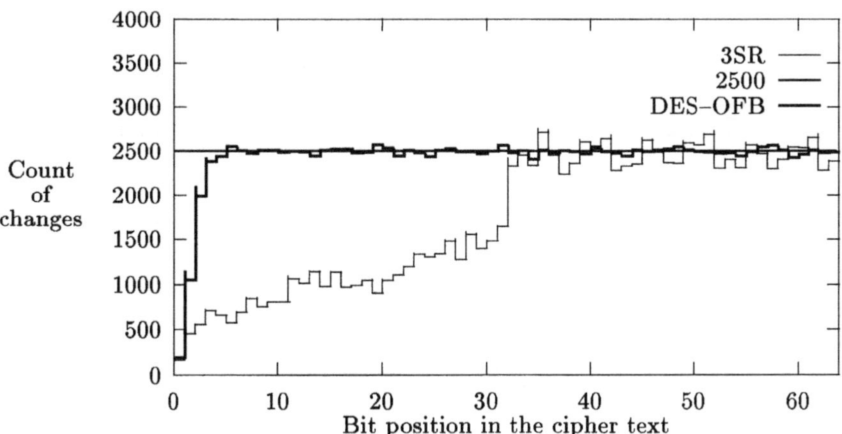

Fig. 7.31. Results of Test 2.

Due to the superior test results, the DES-OFB scheme may be preferred over the LFSR scheme.

A final note on encryption security. In general it is not easy to conclude that a encryption system is secure. Besides the fact that a key may be stolen, the fact that a fast crack algorithm is not *now* known does not *prove* that there are no such fast algorithms. There is also the problem of a "brute force attack" using more powerful computers and/or parallel attacks. A good example is the 56-bit key DES algorithm, which was the standard for many years but was finally declared insecure in 1997. The DES was first cracked by a network of volunteer computer owners on the Internet, which cracked the key in 39 days. Later, in July 1997, the Electronic Frontier Foundation (EFF) finished the design of a cracker machine. It has been documented in a book [181], including all schematics and software source code, which can be downloaded from `http://www.eff.org/`. This cracker machine performs an exhaustive key search and can crack any 56-bit key in less than five days. It was build out of custom chips, each of which has 24 cracker units. Each of the 29 boards used consists of 64 "Deep Crack" chips, i.e., a total of 1856 chips, or 44544 units, are in use. The system cost was $250,000. When DES was introduced in 1977 the system costs were estimated with $20 million, which corresponds to about $40 million today. This shows a good approximation to "Moore's law" which says that every 18 months the size or speed or price of microprocessors improves by a factor 2. From 1977 to 1998 the price of such a machine should drop to $40 \cdot 10^6/2^{22/1.5} \approx$ fifteen hundred dollars, i.e.,

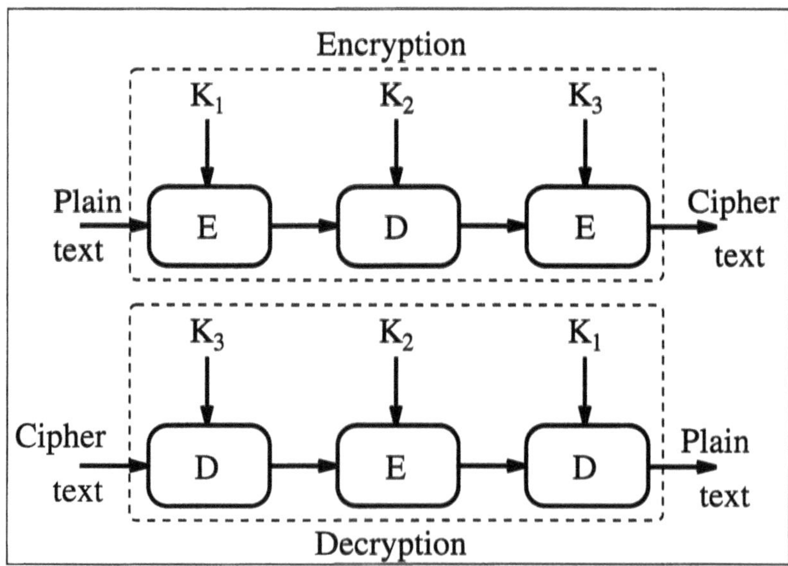

Fig. 7.32. Triple DES (K_l = keys; E=single Encryption; D=single Decryption).

Table 7.18. Encryption algorithms [182].

Algorithm	Key size (bits)	Mathematical operations/ principle	Symmetry	Developed by (year)
DES	56	XOR, fixed S-boxes	s	IBM (1977)
Triple DES	122 – 168	XOR, fixed S-boxes	s	
RSA	variable	Prime factors	a	Rivest/Shamir/ Adleman (1977)
IDEA	128	XOR, add., mult.	s	Massey/Lai (1991)
Blowfish	< 448	XOR, add. fixed S-boxes	s	Schneider (1993)
RC5	< 2048	XOR, add., rotation	s	Rivest (1994)
CAST-128	40 – 128	XOR, rotation, S-boxes	s	Adams/Tavares (1997)

it should be affordable to build a DES cracker today (as was proven by the EFF).

Therefore the 56-bit DES is not secure anymore, but it is now common to use triple DES, as displayed in Fig. 7.32, or other 128-bit key systems. Table 7.18 shows that these systems seemed to be secure for the next years. The EFF cracker, for instance, today will need about $5 \cdot 2^{112}$ days, or $7 \cdot 10^{31}$ years, to crack the triple DES.

The first column in Table 7.18 is the commonly used abbreviations for the algorithms. The second and third columns contain the typical parameters of the algorithm. Symmetric algorithms (designated in the fourth column with an "s") are usually based on Feistel's algorithm, while asymmetric algorithms can be used in a public/private key system. The last column displays the name of the developer and the year the algorithm was first published.

7.3 Modulation and Demodulation

For a long time the goal of communications system design was to realize fully digital receiver, consisting of only an antenna and a fully programmable circuit with digital filters, demodulators and/or decoders for error correction and cryptography on a single programmable chip. With today's FPGA gate count above one million gates this has become a reality. "FPGAs will clearly be a key technology for communication systems well into the 21st century" as predicted by W. Carter [183] . In this section, the design and implementation of a communication system is developed in the context of FPGAs.

7.3.1 Basic Modulation Concepts

A basic communication system transmits and receives information broadcast over a carrier frequency, say f_0. This carrier is *modulated* in amplitude, frequency or phase, proportional to the signal $x(t)$ being transmitted. Fig. 7.33 shows a modulated signal for a binary transmission. For binary transmission, the modulations are called amplitude shift keying (ASK), phase shift keying (PSK), and frequency shift keying (FSK).

In general, it is more efficient to describe a (real) modulated signal with a projection of a rotating arrow on the horizontal axis, according to

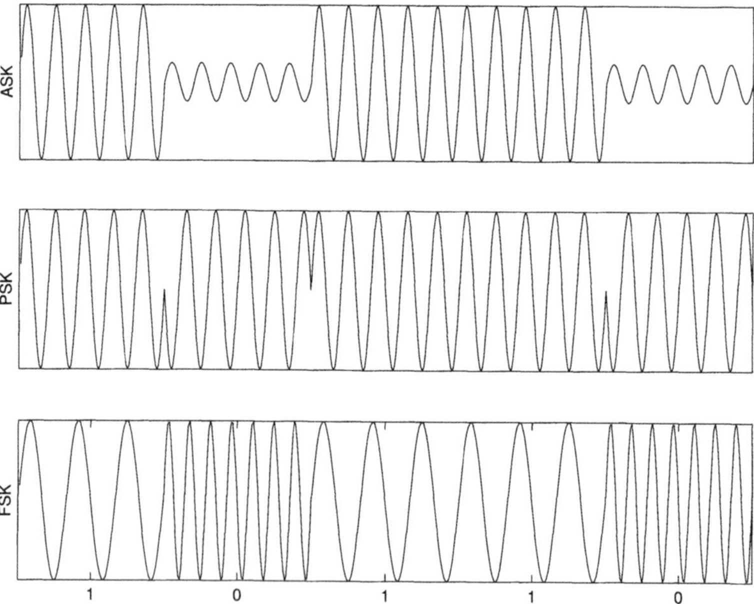

Fig. 7.33. ASK, PSK, and FSK modulation.

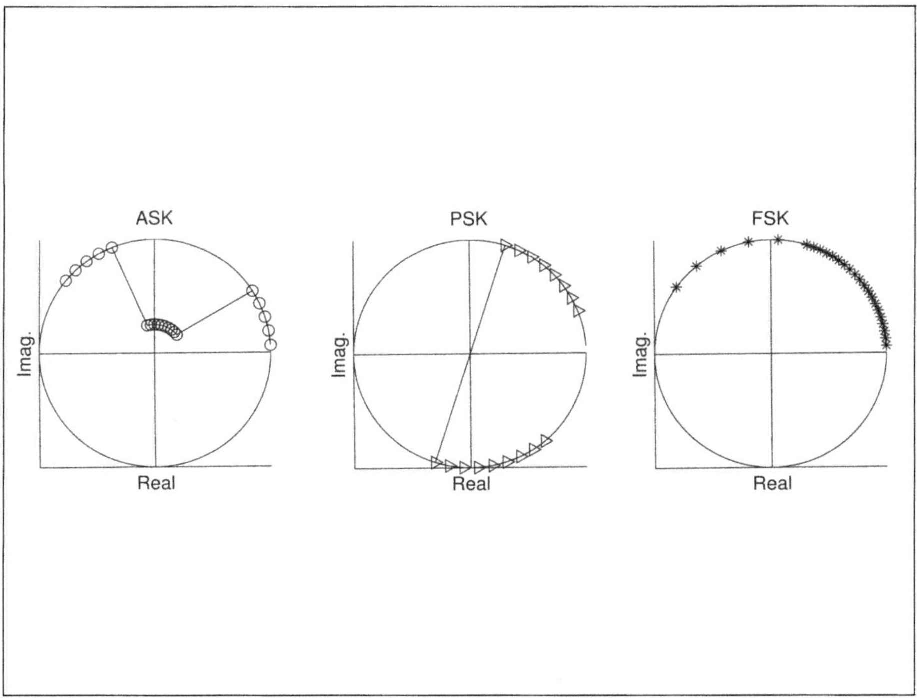

Fig. 7.34. Modulation in the complex plan.

$$\begin{aligned} s(t) &= \Re\left\{A(t)e^{j(2\pi f_0 t + \Delta\phi(t) + \phi_0)}\right\} \\ &= A(t)\cos(2\pi f_0 t + \Delta\phi(t) + \phi_0), \end{aligned} \tag{7.58}$$

where ϕ_0 is a (random) phase offset, $A(t)$ describes the part of the amplitude envelope, and $\Delta\phi(t)$ describes the frequency- or phase-modulated component, as shown in Fig. 7.34. As can be seen from (7.58), AM and PM/FM can be used separately to transmit different signals.

An efficient solution (that does not require large tables) for realizing universal modulator is the CORDIC algorithm discussed in Chapter 2 (p. 66). The CORDIC algorithm is used in the rotation mode, i.e., it is a coordinate converter from $(R, \theta) \rightarrow (X, Y)$. Fig. 7.35 shows the complete modulator for AM, PM, and FM.

To implement amplitude modulation, the signal $A(t)$ is directly connected with the radius R input of the CORDIC. In general, the CORDIC algorithm in rotation mode has an attendant linear increase in the radius. This corresponds to a change in the gain of an amplifier and need not be taken into consideration for the AM scheme. In case the linear increased radius (factor 1.6468, see Table 2.1, p. 32), is not desired, it is possible either to scale the input or the output by 1/1.6468 with a constant coefficient multiplier.

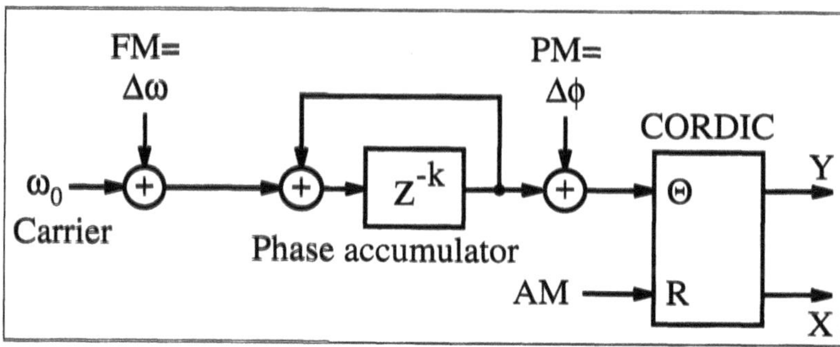

Fig. 7.35. Universal modulator using CORDIC.

The phase of the transmitted signal $\theta = 2\pi f_0 t + \Delta\phi(t)$ must also be computed. To generate the constant carrier frequency, a linear increasing phase signal according to $2\pi f_0 t$ must be generated, which can be done with an accumulator. If FM should be generated, it is possible to modify f_0 by Δf, or to use a second accumulator to compute $2\pi \Delta f t$, and to add the results of the two accumulators. For the PM signal, a constant offset (not increasing in time) is added to the phase of the signal. These phase signals are added and applied to the angle input z or θ of the CORDIC processor. The Y register is set to zero at the beginning of the iterations.

The following example demonstrates a fully pipelined version of the CORDIC modulator.

Example 7.14: Universal Modulator using CORDIC

A universal modulator for AM, PM, and FM according to Fig. 7.35, can be designed with the following VHDL code[5] of the CORDIC part.

```
PACKAGE nine_bit_int IS    -- User defined types
  SUBTYPE NINE_BIT IS INTEGER RANGE -256 TO 255;
  TYPE ARRAY_NINE_BIT IS ARRAY (0 TO 3) OF NINE_BIT;
END nine_bit_int;

LIBRARY work;
USE work.nine_bit_int.ALL;

LIBRARY ieee;
USE ieee.std_logic_1164.ALL;
USE ieee.std_logic_arith.ALL;

ENTITY ammod IS                         ------> Interface
       PORT (clk                  : IN  STD_LOGIC;
             r_in , phi_in        : IN  NINE_BIT;
             x_out, y_out, eps : OUT NINE_BIT);
END ammod;
```

[5] The equivalent Verilog code ammod.v for this example can be found in Appendix A on page 385.

```
ARCHITECTURE flex OF ammod IS

BEGIN

  PROCESS                           ------> Behavioral Style
    VARIABLE x, y, z : ARRAY_NINE_BIT; -- Tapped delay line
  BEGIN
  WAIT UNTIL clk = '1';        -- Compute last value first
    x_out <= x(3);             -- in sequential statements !!
    eps   <= z(3);
    y_out <= y(3);

    IF z(2) > 0 THEN                     -- Rotate 14 degrees
      x(3) := x(2) - y(2) /4;
      y(3) := y(2) + x(2) /4;
      z(3) := z(2) - 14;
    ELSE
      x(3) := x(2) + y(2) /4;
      y(3) := y(2) - x(2) /4;
      z(3) := z(2) + 14;
    END IF;

    IF z(1) > 0 THEN                     -- Rotate 26 degrees
      x(2) := x(1) - y(1) /2;
      y(2) := y(1) + x(1) /2;
      z(2) := z(1) - 26;
    ELSE
      x(2) := x(1) + y(1) /2;
      y(2) := y(1) - x(1) /2;
      z(2) := z(1) + 26;
    END IF;

    IF z(0) > 0 THEN                     -- Rotate  45 degrees
      x(1) := x(0) - y(0);
      y(1) := y(0) + x(0);
      z(1) := z(0) - 45;
    ELSE
      x(1) := x(0) + y(0);
      y(1) := y(0) - x(0);
      z(1) := z(0) + 45;
    END IF;

    IF phi_in > 90    THEN        -- Test for |phi_in| > 90
      x(0) := 0;                  -- Rotate 90 degrees
      y(0) := r_in;               -- Input in register 0
      z(0) := phi_in - 90;
    ELSIF phi_in < -90 THEN
      x(0) := 0;
      y(0) := - r_in;
      z(0) := phi_in + 90;
    ELSE
      x(0) := r_in;
      y(0) := 0;
```

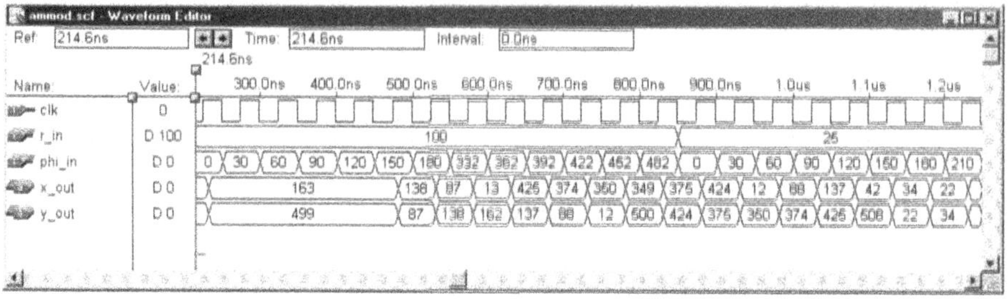

Fig. 7.36. Simulation of an AM modulator using the CORDIC algorithm.

```
        z(0) := phi_in;
      END IF;
    END PROCESS;

  END flex;
```

Fig. 7.36 reports the simulation of an AM signal. Note that the Altera simulation does not produce signed data, but rather unsigned binary data (where negative values have a 512 offset). A pipeline delay of four steps is seen and the value 100 is enlarged by a factor of 1.6. A switch in radius **r_in** from 100 to 25 results in the maximum value **x_out** dropping from 163 to 42. The CORDIC modulator runs at 40.65 MHz and uses 279 LCs. 7.14

Demodulation may be *coherent* or *incoherent*. A coherent receiver must recover the unknown carrier phase ϕ_0, while an incoherent does not need to do so. If the receiver uses an intermediate frequency (IF) band, this type of receiver is called a superhet or double superhet (two IF bands) receiver. IF receivers are also sometimes called *heterodyne* receivers. If no IF stages are employed, a *zero IF* or *homodyne* receiver results. Fig. 7.37 presents a systematic overview of the different types of receivers. Some of the receivers can only be used for one modulation scheme, while others can be used for multiple modes (e.g., AM, PM, and FM). The latter is called an universal receiver. We will first discuss the incoherent receiver, and then the coherent receiver.

All receivers use intensive filters (as discussed in Chapter 3 and 4), in order to select only the signal components of interest. In addition, for heterodyne receivers, filters are needed to suppress the mirror frequencies, which arise from the frequency shift

$$s(t) \cdot \cos(2\pi f_m t) \longleftrightarrow S(f + f_m) + S(f - f_m) \tag{7.59}$$

7.3.2 Incoherent Demodulation

In an incoherent demodulation scheme, it is assumed that the exact carrier frequency is known to the receiver, but the initial phase ϕ_0 is not.

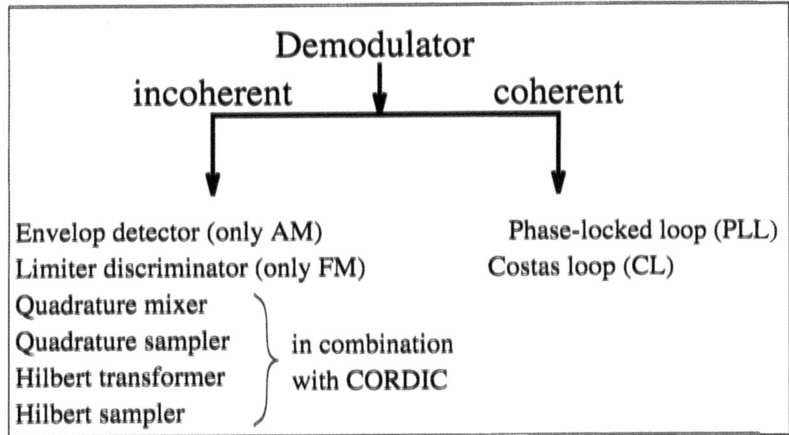

Fig. 7.37. Coherent and incoherent demodulation schemes.

If the signal component is successfully selected with digital or analog filtering, the question arises, whether only one demodulation mode (e.g., AM or FM) or universal demodulator is needed. An incoherent AM demodulator can be as simple as a full or halfwave rectifier and an additional lowpass filter. For FM or PM demodulation only, the *limiter/discriminator* type of demodulator is an efficient implementation. This demodulator builds a threshold of the input signal to limit the values to ± 1, and then basically "measures" the distance between the zero crossings. These receivers are easily implemented with FPGAs but sometimes produce 2π jumps in the phase signal (called "clicks" [184, 185]). There are other demodulators with better performance.

We will focus on universal receivers using in-phase and quadrature components. This type of receiver basically inverts the modulation scheme relative to (7.58) from p. 311. In a first step we have to compute, from the received cosines, components that are "in-phase" with the sender's sine components (which are in quadrature to the carrier, hence the name Q phase). These I and Q phases are used to reconstruct the arrow (rotating with the carrier frequency) in the complex plane. Now, the demodulation is just the inversion of the circuit from Fig. 7.35. It is possible to use the CORDIC algorithm in the vectoring mode, i.e., a coordinate conversion $X, Y \rightarrow R, \theta$ with $I = X$ and $Q = Y$ is used. Then the output R is directly proportional to the AM portion, and the PM/FM part can be reconstructed from the θ signal, i.e., the Z register.

A difficult part of demodulation is I/Q generation, and typically two methods are used: a quadrature scheme and a Hilbert transform.

In the quadrature scheme the input signal is multiplied with the two mixer signals, $2\cos(2\pi f_m t)$ and $-j2\sin(2\pi f_m t)$. If the signals in the IF band $f_{\mathrm{IF}} = f_0 - f_m$ are now selected with a filter, the complex sum of these signals is then a reconstruction of the complex rotating arrow. Fig. 7.38 shows this scheme

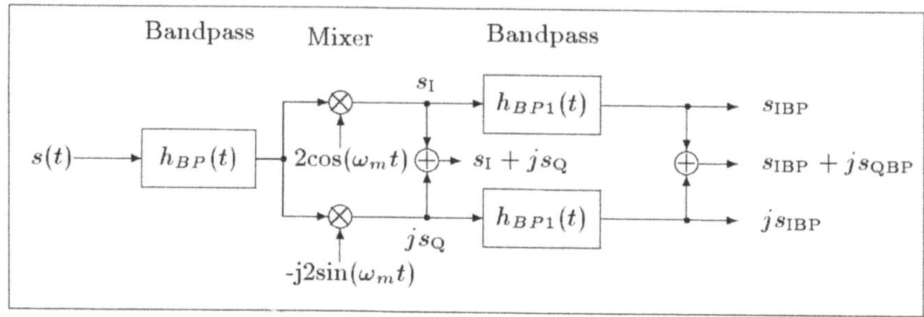

Fig. 7.38. Generation of I- and Q-phase using quadrature scheme.

while Fig. 7.39 displays an example of the I/Q generation. From Fig. 7.39 it can be seen that the final signal has no negative spectral components. This is typical for this type of incoherent receiver and these signals are called *analytic.*

To decrease the effort for the filters, it is desirable to have an IF frequency close to zero. In an analog scheme (especially for AM) this often introduces a new problem, that the amplifier drifts into saturation. But for a fully digital receiver, such a homodyne or zero IF receiver can be built. The bandpass filters then reduce to lowpass filters. Hogenauer's CIC filters (see

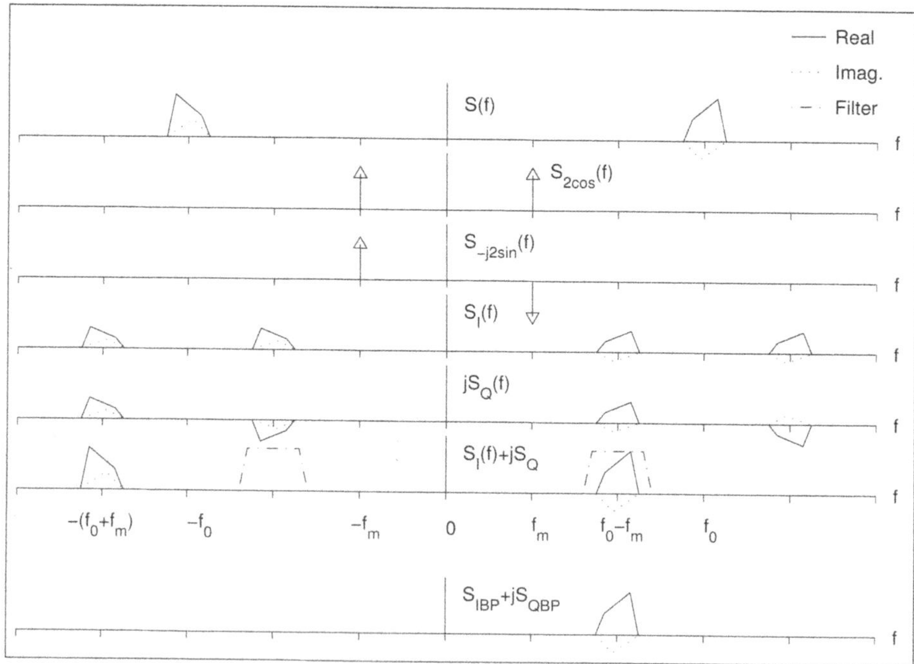

Fig. 7.39. Spectral example of the I/Q generation.

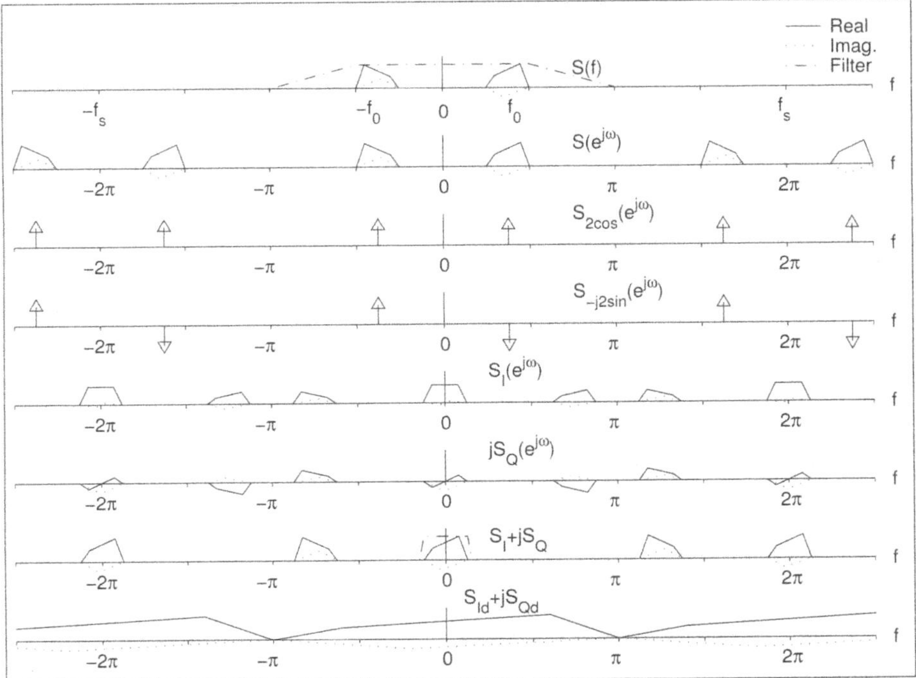

Fig. 7.40. Spectra for the zero IF receiver. Sampling frequency was 2π.

Chapter 5, p. 155) are efficient realizations of these high decimation filters. Fig. 7.40 shows the corresponding spectra. The real input signal is sampled at 2π. Then the signal is multiplied with a cosine signal $S_{2\cos}(e^{j\omega})$ and a sine signal $S_{-j2\sin}(e^{j\omega})$. This produces the in-phase component $S_{\mathrm{I}}(e^{j\omega})$ and the quadrature component $jS_{\mathrm{Q}}(e^{j\omega})$. These two signals are now combined into a complex analytic signal $S_{\mathrm{I}} + jS_{\mathrm{Q}}$. After the final lowpass filtering, a decimation in sampling rate can be applied.

Such a fully digital zero IF for LF has been built using FPGA technology [186].

Example 7.15: Zero IF Receiver

This receiver has an antenna, a programmable gain adjust (AGC), and a Cauer lowpass 7th order followed by an 8-bit video A/D converter. The receiver uses eight times oversampling (0.4 - 1.2 MHz) for the input range from 50 to 150 KHz. The quadrature multipliers are 8×8-bit array multipliers. Two-stage CIC filters were designed with 24- and 19-bit integrator precision, and 17- and 16-bits precision for the comb sections. The final sampling rate reduction was 64. The full design could fit on a single XC3090 Xilinx FPGA. The following table shows the effort for the single units:

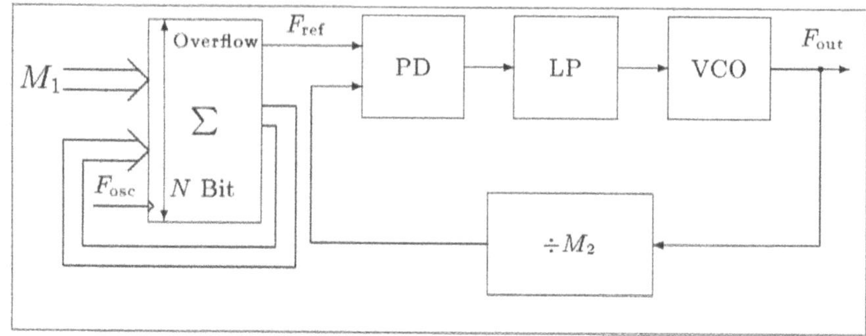

Fig. 7.41. PLL with accumulator as reference.

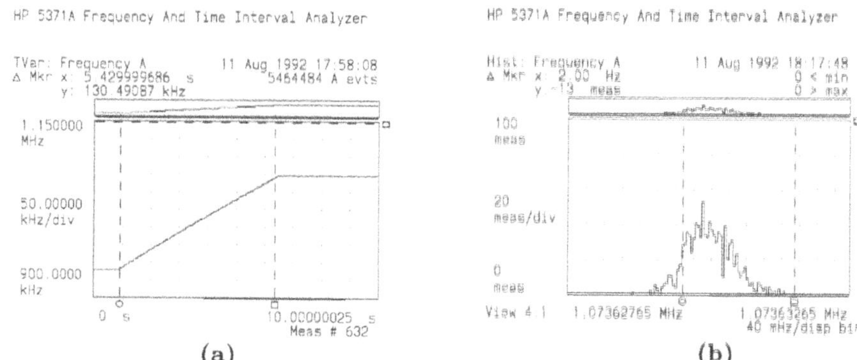

Fig. 7.42. PLL synthesizers with accumulator reference. **(a)** Behavior of the synthesizer for switching F_{out} from 900 KHz to 1.2 MHz. **(b)** Histogram of the frequency error, which is less than 2 Hz.

Design part	CLBs
Mixer with sin/cos tables	74
Two CIC filters	168
State machine and PDSP interface	18
Frequency synthesizer	32
Total	292

For the tunable frequency synthesizer an accumulator as reference for an analog phase-locked loop (PLL) was used [4]. Fig. 7.41 shows this type of frequency synthesizer and Fig. 7.42 displays the measured performance of the synthesizer. The accumulator synthesizer could be clocked very high due to the fact that only the overflow is needed. A bitwise carry save adder was therefore used. The accumulator was used as a reference for the PLL which produces $F_{\text{out}} = M_2 F'_{\text{in}} = M_1 M_2 F_{\text{in}}/2^N$. 7.15

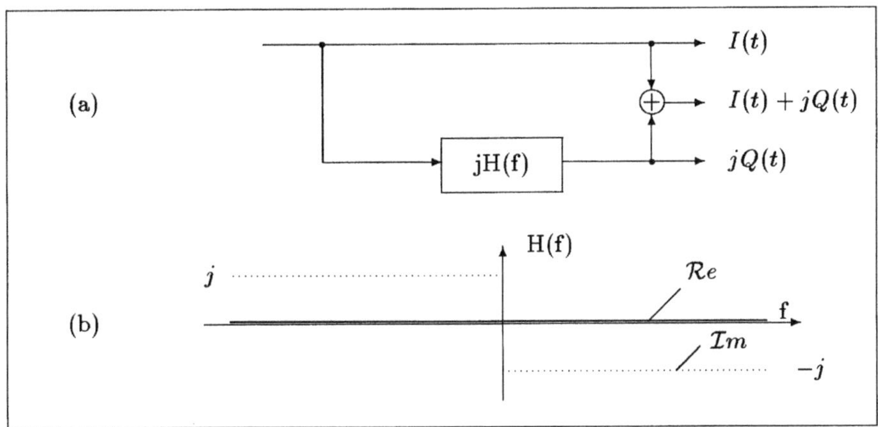

Fig. 7.43. Hilbert transformer. **(a)** Filter. **(b)** Spectrum of $H(f)$.

The *Hilbert transformer* scheme relies on the fact that a sine signal can be computed from the cosine signal by a phase delay of 90°. If a filter is used to produce this Hilbert transformer, the amplitude of the filter must be one and the phase must be 90° for all frequencies. Impulse response and transfer function can be found using the definition of the Fourier transform, i.e.,

$$h(t) = \frac{1}{\pi t} \longleftrightarrow \quad H(j\omega) = -j\gamma(\omega) = \begin{cases} j & -\infty < \omega < 0 \\ -j & 0 < \omega < \infty \end{cases} \tag{7.60}$$

with $\gamma(\omega) = -1 \; \forall \, \omega < 0$ and $\gamma(\omega) = 1 \; \forall \, \omega \geq 0$ as the sign function. A Hilbert filter can only be approximated by an FIR filter and resulting coefficients have been reported (see for instance [187], [188], [120, p. 168-174], or [65, p. 681]).

Simplification for narrow-band receivers. If the input signals are narrow-band signals, i.e., the transmitted bit rate is much smaller than the carrier frequency, some simplifications in the demodulation scheme are possible. In the *input sampling scheme* it is then possible to sample at the carrier rate, or at a multiple of the period $T_0 = 1/f_0$ of the carrier, in order to ensure that the sampled signals are already free of the carrier component.

The *quadrature scheme* becomes trivial if the zero IF receiver samples at $4f_0$. In this case, the sine and cosine components are elements of $0, 1$ or -1, and the carrier phase is $0, 90°, 180° \ldots$. This is sometimes referred to as "complex sampling" in the literature [189, 190]. It possible to use *under-sampling,* i.e., only every second or third carrier period is evaluated by the sampler. Then the sampled signal will still be free from the carrier frequency.

The *Hilbert transformer* can also be simplified if the signal is sampled at $T_0/4$. A Hilbert sampler of first order with $Q_{-1} = 1$ or a second-order type using the symmetric coefficients $Q_1 = -0.5; Q_{-1} = 0.5$ or asymmetric $Q_{-1} = 1.5; Q_{-3} = 0.5$ coefficients can be used [191].

Table 7.19. Coefficients of the Hilbert sampler.

Type	Coefficients	Bit	$\Delta f/f_0$
Zero order	$Q_{-1} = 1,0$	8	0.005069
	$Q_{-1} = 1,0$	12	0.000320
	$Q_{-1} = 1,0$	16	0.000020
First order asymmetric	$Q_{-1} = 1,5;Q_{-3} = 0.5$	8	0.032805
	$Q_{-1} = 1,5;Q_{-3} = 0.5$	12	0.008238
	$Q_{-1} = 1,5;Q_{-3} = 0.5$	16	0.002069
First order symmetric	$Q_1 = -0.5;Q_{-1} = 0.5$	8	0.056825
	$Q_1 = -0.5;Q_{-1} = 0.5$	12	0.014269
	$Q_1 = -0.5;Q_{-1} = 0.5$	16	0.003584

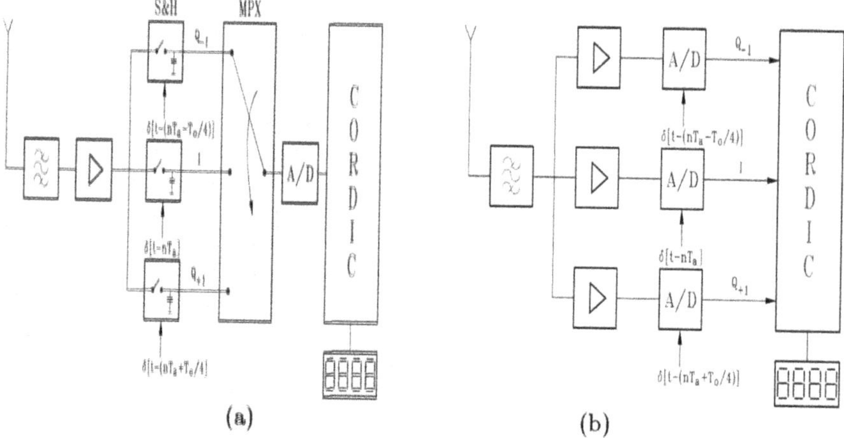

Fig. 7.44. (a) Two versions of the Hilbert sampler of first order.

Table 7.19 reports the three short-term Hilbert transformer coefficients and the maximum allowed frequency offset Δf of the modulation, for the Hilbert filter providing a specified accuracy.

Fig. 7.44 shows two possible realizations for the Hilbert that which have been used to demodulate radio control watch signals [192, 193]. The first method uses three Sample & Hold circuits and the second method uses three A/D converters to build a symmetric Hilbert sampler of first order.

Fig. 7.45 shows the spectral behavior of the Hilbert sampler with a direct undersampling by two.

7.3.3 Coherent Demodulation

If the phase ϕ_0 of the receiver is known, then demodulation can be accomplished by multiplication and lowpass filtering. For AM, the received signal

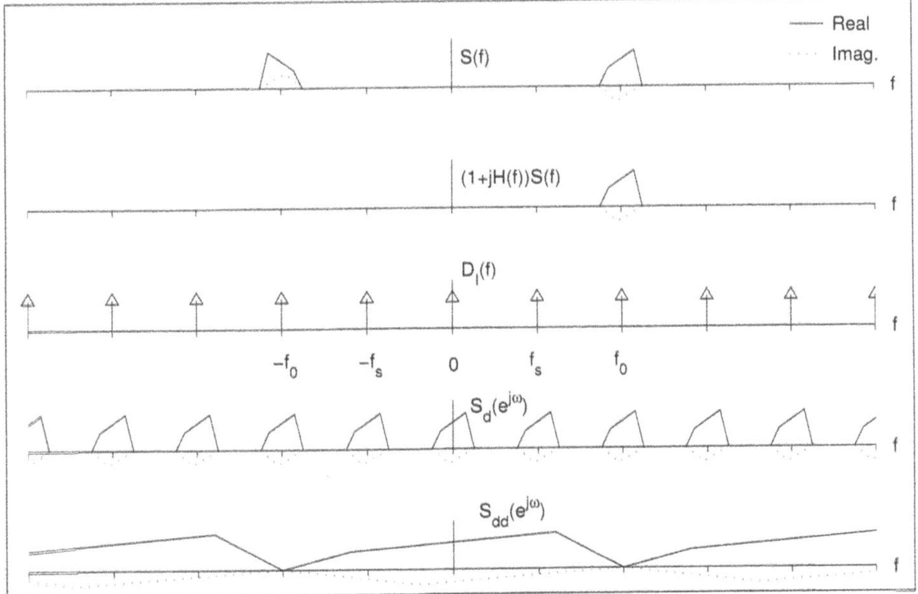

Fig. 7.45. Spectra for the Hilbert sampler with undersampling.

$s(t)$ is multiplied with $2\cos(\omega_0 t+\phi_0)$ and for PM or FM by $-2\sin(\omega_0 t+\phi_0)$. It follows

AM:

$$A(t)\cos(2\pi f_0 t + \phi_0)\cdot 2\cos(2\pi f_0 t + \phi_0)$$
$$= \underbrace{A(t)}_{\text{Lowpass component}} + A(t)\cos(4\pi f_0 t + 2\phi_0) \tag{7.61}$$

$$s_{\mathrm{AM}}(t) = A(t) - A_0 \tag{7.62}$$

PM:

$$-2\sin(2\pi f_0 t + \phi_0)\cdot\cos(2\pi f_0 t + \phi_0 + \Delta\phi(t))$$
$$= \underbrace{\sin(\Delta\phi(t))}_{\text{Lowpass component}} + \cos(4\pi f_0 t + 2\phi_0 + \Delta\phi(t)) \tag{7.63}$$

$$\sin(\Delta\phi(t)) \approx \Delta\phi(t) \tag{7.64}$$

$$s_{\mathrm{PM}}(t) = \frac{1}{\eta}\Delta\phi(t) \tag{7.65}$$

FM:

$$s_{\mathrm{FM}}(t) = \frac{1}{\eta}\frac{d\ \ \Delta\phi(t)}{d\,t} \tag{7.66}$$

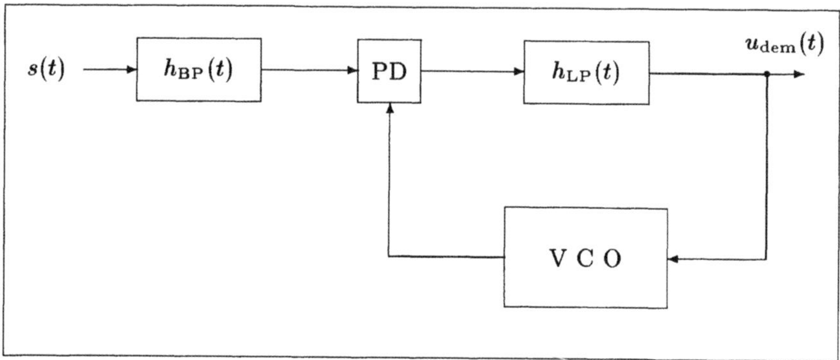

Fig. 7.46. Phase-locked loop (PLL) with (necessary) bandpass ($h_{\mathrm{BP}}(t)$), phase detector (PD), lowpass ($h_{\mathrm{LP}}(t)$), and voltage-controlled oscillator (VCO).

η is the so-called modulation index.

In the following we will discuss the types of coherent receivers which are suitable for an FPGA implementation. Typically, coherent receivers provide a 1 dB better signal-to-noise ratio than an incoherent receiver (see Fig. 7.4, p. 275). A synchronous or coherent FM receiver tracks the carrier phase of the incoming signal with a voltage-controlled oscillator (VCO) in a loop. The DC part of this voltage is directly proportional to the FM signal. PM signal demodulation requires integration of the VCO control signal, and AM demodulation requires the addition of a second mixer and a $\pi/2$ phase shifter in sequence with a lowpass filter. The risk with coherent demodulation is that for a low signal-to-noise channel, the loops may be out-of-lock, and performance will decrease tremendously.

There are two common types of coherent receiver loops: the phase-locked loop (PLL) and the Costas loop (CL). Figs. 7.46 and 7.47 are block diagrams of a PLL and CL, respectively, showing the nearly doubled complexity of the CL. Each loop may be realized as an analog (linear PLL/CL) or all-digital (ADPLL, ADCL) circuit (see [194], [195], [196], [197]). The stability analysis of these loops is beyond the scope of the book and is well covered in the literature ([198], [199], [200], [201], [202]). We will discuss efficient realizations of PLLs and CLs [203, 204]. The first PLL is a direct translation of an analog PLL to FPGA technology.

Linear phase-locked loop. The difference between linear and digital loops lies in the type of input signal to be processed. A linear PLL or CL uses a fast multiplier as a phase detector, providing a possibly multilevel input signal to the loop. A digital PLL or CL can process only binary input signals. (*Digital* refers to the quality of the input signal here, not to the hardware realization!)

As shown in Fig. 7.46, the linear PLL has three main blocks:

- Multiplier as phase detector
- Loop filter

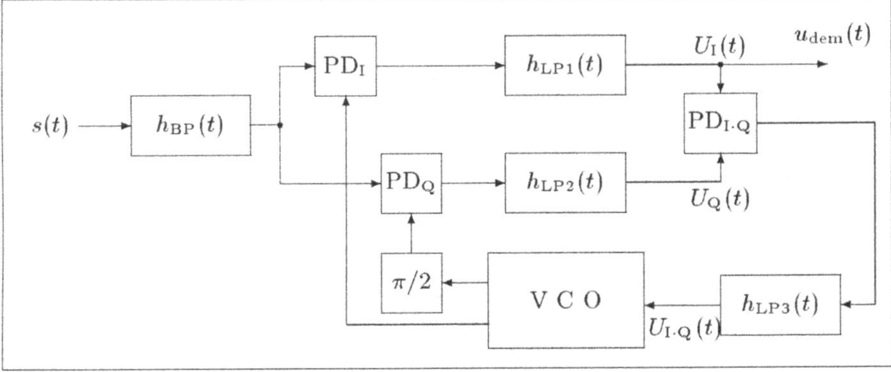

Fig. 7.47. Costas loop with (necessary) bandpass ($h_{\mathrm{BP}}(t)$), three phase detectors (PD), three lowpass filters ($h_{\mathrm{LP}}(t)$), and a voltage-controlled oscillator (VCO) with $\pi/2$ phase shifter.

- VCO

To keep the loop in lock, a loop output signal-to-noise ratio larger than 4 = 6 dB is required [197, p. 35]. Since the selection of typical antennas is not narrow enough to achieve this, an additional narrow bandpass filter has been added to Figs. 7.46 and 7.47 as a necessary addition to the demodulator [204]. The "cascaded bandpass comb" filter (see Table 5.4, p. 176) is an efficient example. However, the filter design is much easier if a fixed IF is used, as in the case of a superhet or double superhet receiver.

The VCO (or digitally controlled oscillator (DCO) for ADPLLs) oscillates with a frequency $\omega_2 = \omega_0 + K_0 \cdot U_f(t)$, where ω_0 is the resting point and K_0 the gain of the VCO/DCO. For sinusoidal input signals we have the signal

$$u_{\mathrm{dem}}(t) = K_d \sin\left(\Delta\phi(t)\right) \tag{7.67}$$

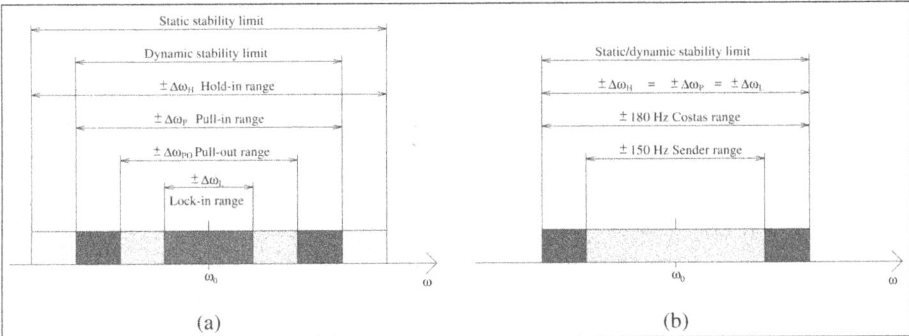

Fig. 7.48. (a) Operation area PLL/CL. **(b)** Operation area of the CL of Fig. 7.51.

Table 7.20. Cost in CLBs of a linear PLL universal demodulator in a Xilinx XC3000 FPGA.

Function group	FM only	FM, AM, and PM
Phase detector (8 × 8-bit multiplier)	65	72
Loop filter (two-stage CIC)	84	168
Frequency synthesizer	34	34
DCO ($N/N+K$ divider, sin/cos table)	16	16+2
PDSP Interface	15	15
Total	214	307

at the output of the lowpass, where $\Delta\phi(t)$ is the phase difference between the DCO output and the bandpass-filtered input signal. For small differences, the sine can be approximated by its argument, giving $u_{\text{dem}}(t)$ proportional to $\Delta\phi(t)$ (the loop stays in lock). If the input signal has a very sudden phase discontinuity, the loop will go out of lock. Fig. 7.48 shows the different operation areas of the loop. The hold-in range $\omega_0 \pm \Delta\omega_H$ is the static operation limit (useful only with a frequency synthesizer). The lock-in range is the area where the PLL will lock in within a single period of the frequency difference $\omega_1 - \omega_2$. Within the pull-in range, the loop will lock in within the capture time T_L, which may last more than one period of $\omega_1 - \omega_2$. The pull-out range is the maximum frequency jump the loop can sustain without going out of lock. $\omega_0 \pm \Delta\omega_{PO}$ is the dynamic operation limit used in demodulation. There is much literature optimizing PD, loop filter and the VCO gain; see [198], [199], [200], [201], [202].

The major advantage of the linear loop over the digital PLL and CL is its noise reduction capability, but the fast multiplier used for the PD in a linear PLL has a particularly high hardware cost, and this PD has an unstable rest point at $\pi/2$ phase shift ($-\pi/2$ is a stable point), which impedes lock-in. Table 7.20 estimates the hardware cost in CLBs of a linear PLL, using the same functional blocks as the 8-bit incoherent receiver (see Example 7.15, p. 317). Using a hardware multiplier in multiplex for AM and PM demodulation, the right-hand column reduces the circuit's cost by 58 CLBs, allowing it to fit into a 320 CLB Xilinx XC3090 device.

A comparison of these costs to those of an incoherent receiver, consuming 292 CLBs without CORDIC demodulation and 367.5 CLBs with an additional CORDIC processor [57], shows a slight improvement in the linear PLL realization. If only FM and PM demodulation are required, a digital PLL or CL, described in the next two sections, can reduce complexity dramatically.

These designs were developed to demodulate WeatherFAX pictures, which were transmitted in central Europe by the low-frequency radio stations DCF37 and DCF54 (Carrier 117.4 KHz and 134.2 KHz; frequency modulation F1C: ± 150 Hz).

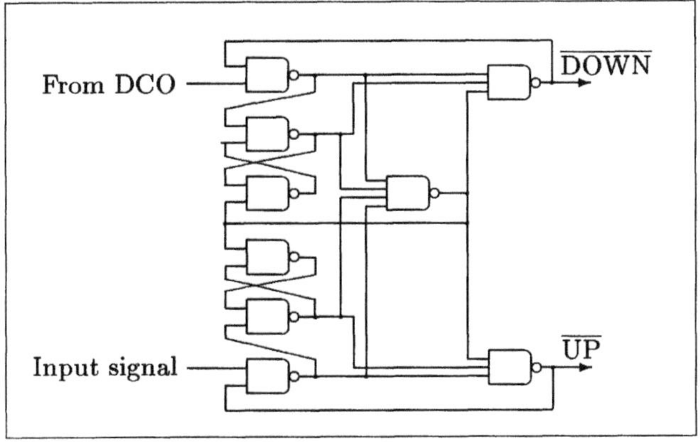

Fig. 7.49. Phase detector [18, p. 8-127].

Digital PLLs. As explained in the last section, a digital PLL works with binary input signals. Phase detectors for a digital PLL are simpler than the fast multipliers used for linear PLLs; usual choices are XOR gates, edge-triggered JK flip-flops, or paired RS flip-flops with some additional gates [197, p. 60-65]. The phase detector shown in Fig. 7.49 is the most complex, but it provides phase and frequency sensitivity and a quasi-infinite hold-in range.

Modified counters are used as loop filters for DPLLs. These may be $N/(N + K)$ counters or multistage counters, such as an N-by-M divider, where separate UP and DOWN counters are used, and a third counter measures the UP/DOWN difference. It is then possible to break off further signal processing if a certain threshold is not met. For the DCO, any typical all-digital frequency synthesizer, such as an accumulator, divider, or multiplicative generator may be used. The most frequently used synthesizer is the tunable divider, popular because of its low phase error. The low resolution of this synthesizer can be improved by using a low receiver IF [205].

One DPLL realization with very low complexity is the 74LS297 circuit, which utilizes a "pulse-stealing" design. This scheme may be improved with the phase- and frequency-sensitive J-K flip-flop, as displayed in Fig. 7.50. The PLL works as follows: the "detect flip-flop" runs with rest frequency

$$F_{\text{comp}} = \frac{F_{\text{in}}}{N} = \frac{F_{\text{osc}}}{KM} \tag{7.68}$$

To allow tracking of incoming frequencies higher than the rest frequency, the oscillator frequency $F_{\text{osc}+}$ is set slightly higher:

$$T_{\text{comp}} - T_{\text{comp}+} = \frac{1}{2} T_{\text{osc}}, \tag{7.69}$$

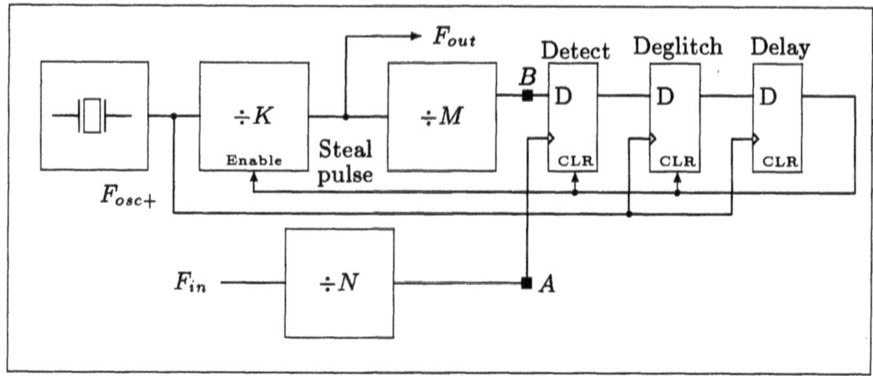

Fig. 7.50. "Pulse-stealing" PLL [206].

Table 7.21. Hardware complexity of a pulse-stealing DPLL [205].

Function group	CLBs
DCO	16
Phase detector	5
Loop filter	6
Averaging	11
PC–Interface	10
Frequency synthesizer	26
Stage machine	10
Total	84

such that the signal at point B oscillates half a period faster than the F_{osc} signal. After approximately two periods at the rest frequency, a one will be latched in the detector flip-flop. This signal runs through the de-glitch and delay flip-flops, and then inhibits one pulse of the $\div K$ divider (thus the name "pulse-stealing"). This delays the signal at B such that the phase of signal A runs after B, and the cycle repeats. The lock-in range of the PLL has a lower bound of $F_{\text{in}}|_{min}$=0 Hz. The upper bound depends on the maximum output frequency $F_{\text{osc+}}/K$, so the lock-in range becomes

$$\pm\Delta\omega_L = \pm N \cdot F_{\text{osc+}}/(K \cdot M) \tag{7.70}$$

A receiver can be simplified by leaving out the counters N and M. In a WeatherFAX image-decoding application the second IF of the double-superhet receiver is set in such a way that the frequency modulation of 300 Hz (δf = 0 Hz → white; δf = 300 Hz → black) corresponds to exactly 32 steal pulses, so that the steal pulses correspond directly to the grayscale level. We set a pixel rate of 1,920 Baud, and a IF of 16.6 KHz. For each pixel, four "steal values" (number of steal pulses in an interval) are determined, so a

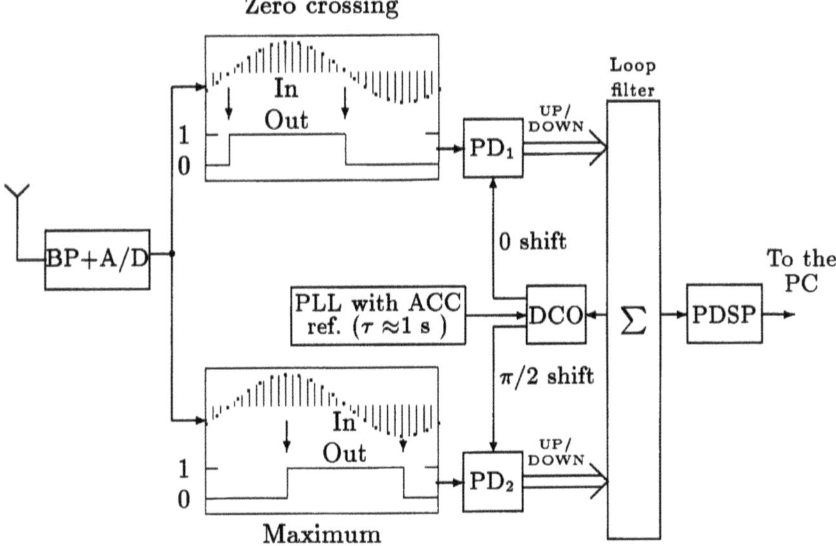

Fig. 7.51. Structure of the Costas loop.

total of $\log_2(2 \cdot 4) = 3$ bit shifts are used to compute the 16 gray-level values. Table 7.21 shows the hardware complexity of this PLL type.

Costas loop. This extended type of coherent loop was first proposed by John P. Costas in 1956, who used the loop for carrier recovery. As shown in Fig. 7.47, the CL has an *in-phase* and a *quadrature* path (subscripted I and Q there). With the $\pi/2$ phase shifter and the third PD and lowpass, the CL is approximately twice as complex as the PLL, but locks on to a signal twice as fast. Costas loops are very sensitive to small differences between in-phase and quadrature gain, and should therefore always be realized as all-digital circuits. The FPGA seems to be an ideal realization vehicle ([207], [208]).

For a signal $U(t) = A(t)\sin(\omega_0 t + \Delta\phi(t))$ we get, after the mixer and lowpass filters,

$$U_I(t) = K_d A(t) \cos\left(\Delta\phi(t)\right) \tag{7.71}$$

$$U_Q(t) = K_d A(t) \sin\left(\Delta\phi(t)\right) \tag{7.72}$$

where $2K_d$ is the gain of the PD. $U_\text{I}(t)$ and $U_\text{Q}(t)$ are then multiplied together in a third PD, and are lowpass filtered to get the DCO control signal:

$$U_{I \cdot Q}(t) \sim K_d \sin\left(2\Delta\phi(t)\right) \tag{7.73}$$

A comparison of (7.67) and (7.73) shows that, for small modulations of $\Delta\phi(t)$, the slope of the control signal $U_{I \cdot Q}(t)$ is twice the PLL's. As in the PLL, if only FM or PM demodulation is needed, the PDs may be all digital.

Table 7.22. Loop filter output and DCO correction values at 32-times oversampling.

Accumulator					
Under-flow	Over-flow	Sum	DCO-IN	gray value	$f_{\text{carrier}} \pm \delta f$
Yes	No	$s < -(2^{13}-1)$	3	0	+180 Hz
No	No	$-(2^{13}-1) \leq s < -2048$	2	0	+120 Hz
No	No	$-2048 \leq s < -512$	1	4	+60 Hz
No	No	$-512 \leq s < 512$	0	8	+0 Hz
No	No	$512 \leq s < 2048$	-1	12	−60 Hz
No	No	$2048 \leq s < 2^{13}-1$	-2	15	−120 Hz
No	Yes	$s \geq 2^{13}-1$	-3	15	−180 Hz

Fig. 7.51 shows a block diagram of a CL. The antenna signal is first filtered and amplified by a fourth-order Butterworth bandpass, then digitized by an 8-bit converter at a sampling rate 32 or 64 times the carrier base frequency. The resulting signal is split, and fed into a zero-crossing detector and a minimum/maximum detector. Two phase detectors compare the signals with a reference signal, and its $\pi/2$-shifted counterpart, synthesized by a high time-constant PLL with a reference accumulator [4, section 2]. Each phase detector has two edge detectors, which should generate a total of 4 UP and 4 DOWN signals. If more UP signals than DOWN are generated by the PDs, then the reference frequency is too low, and if more DOWN signals are generated, it is too high. The differences $\sum$ UP $-\sum$ DOWN are accumulated for one pixel-duration in a 13-bit accumulator acting as a loop filter. The loop filter data are passed to a pixel converter, which gives "correction values" to the DCO as shown in Table 7.22. The accumulated sums are also used as grayscale values for the pixel, and passed on to a PC to store and display the WeatherFAX pictures.

The smallest detectable phase offset for a 2 KBaud pixel rate is

$$f_{\text{carrier}+1} = \frac{1}{1/f_{\text{carrier}} - t_{ph37} \cdot 2\ kBaud/f_{\text{carrier}}} = 117.46 \text{ KHz} \quad (7.74)$$

where $t_{ph37} = 1/(32 \cdot 117 \text{ KHz}) = 266$ ns is the sampling period at 32-times oversampling. The frequency resolution is 117.46 KHz $-(f_{\text{carrier}} = 117.4$ KHz$) = 60$ Hz. With a frequency modulation of 300 Hz, five grayscale values can be distinguished. Higher sampling rates for the accumulator are not possible with the 3164-4ns FPGA and the limited A/D converter used. With a fast A/D converter and an Altera Flex or Xilinx XC4K FPGA, 128- and 256-times oversampling are possible.

For a maximum phase offset of π, the loop will require a maximum lock-in time T_L of $\lceil 16/3 \rceil = 6$ samples, or about 1.5 μs. Table 7.23 shows the complexity of the CL for 32 and 64 times oversampling.

Table 7.23. Complexity of a fully digital Costas loop [204, p. 60].

Function group	CLBs with oversampling 32 times	64 times
Frequency synthesizer	33	36
Zero detection	42	42
Maximum detection	16	16
Four phase detectors	8	8
Loop filter	51	51
DCO	12	15
TMS interface	11	11
Sum	173	179

Exercises

7.1: The following MATLAB code can be used to compute the order of an element.

```
function N = order(x,M)
% Compute the order of x modulo M
p = x; l=1;
while p ~= 1
  l = l+1; p = p * x;
  re =real(p); im = imag(p);
  p = mod(re,M) + i * mod(im,M);
end;
N=l;
```

If, for instance, the function is called with **order(2,2^25+1)** the result is 50. To compute the single factors of $2^{25}+1$, the standard MATLAB function **factor(2^25+1)** can be used.
For
(a) $\alpha = 2$ and $M = 2^{41}+1$
(b) $\alpha = -2$ and $M = 2^{29}-1$
(c) $\alpha = 1+j$ and $M = 2^{29}+1$
(d) $\alpha = 1+j$ and $M = 2^{26}-1$
compute the transform length, the "bad" factors ν (i.e., order not equal $\text{order}(\alpha, 2^B \pm 1)$), all "good" prime factors M/ν, and the available input bit width $B_x = (\log_2(M/\nu) - \log_2(L))/2$.

7.2: To compute the inverse $x^{-1} \bmod M$ for $\gcd(x, M) = 1$ of the value x, we can use the fact that the following diophantic equation holds:

$$\gcd(x, M) = u \cdot x + v \cdot M \qquad \text{with} \quad u, v \in \mathbb{Z} \tag{7.75}$$

(a) Explain how to use the MATLAB function **[g u v]=gcd(x,M)** to compute the multiplication inverse.
Compute the following multiplicative inverses if possible:
(b) $3^{-1} \bmod 73$;
(c) $64^{-1} \bmod 2^{32}+1$;
(d) $31^{-1} \bmod 2^{31}-1$;

(e) $89^{-1} \bmod 2^{11} - 1$;
(f) $641^{-1} \bmod 2^{32} + 1$.

7.3: The following MATLAB code can be used to compute Fermat NTTs for length 2, 4, 8, and 16 modulo 257.

```
function Y = ntt(x)
% Compute Fermat NTT of length 2,4,8 and 16 modulo 257
l = length(x);
switch (l)
  case 2, alpha=-1;
  case 4, alpha=16;
  case 8, alpha=4;
  case 16, alpha=2;
  otherwise, disp('NTT length not supported')
end
A=ones(l,l); A(2,2)=alpha;
%*********Computing second column
for m=3:l
 A(m,2)=mod(A(m-1,2)* alpha, 257);
end
%*********Computing rest of matrix
for m=2:l
  for n=2:l-1
    A(m,n+1)=mod(A(m,n)*A(m,2),257);
  end
end
%*********Computing NTT A*x
for k = 1:l
  C1 = 0;
  for j = 1:l
    C1 = C1 + A(k,j) * x(j);
  end
  X(k) = mod(C1, 257);
end
Y=X;
```

(a) Compute the NTT $\boldsymbol{X}$ of $\boldsymbol{x} = \{1,1,1,1,0,0,0,0\}$.
(b) Write the code for the appropriate INTT. Compute the INTT of $\boldsymbol{X}$ from part (a).
(c) Compute the element-by-element product $\boldsymbol{Y} = \boldsymbol{X} \odot \boldsymbol{X}$ and $\mathrm{INTT}(\boldsymbol{Y}) = \boldsymbol{y}$.
(d) Extend the code for a complex Fermat NTT and INTT for $\alpha = 1 + j$. Test your program with the identity $\boldsymbol{x} = \mathrm{INTT}(\mathrm{NTT}(\boldsymbol{x}))$.

7.4: The Walsh transform for $N = 4$ is given by:

$$\boldsymbol{W}_4 = \begin{bmatrix} 1 & 1 & 1 & 1 \\ 1 & 1 & -1 & -1 \\ 1 & -1 & -1 & 1 \\ 1 & -1 & 1 & -1 \end{bmatrix}$$

(a) Compute the scalar product of the row vector. What property does the matrix

have?
(b) Use the results from (a) to compute the inverse $\boldsymbol{W}_4^{-1}$.
(c) Compute the 8×8 Walsh matrix $\boldsymbol{W}_8$, by scaling the original row vector by two (i.e., $h[n/2]$) and computing an additional two "children" $h[n] + h[n-4]$ and $h[n] - h[n-4]$ from row 3 and 4. There should be no zero in the resulting $\boldsymbol{W}_8$ matrix.
(d) Draw a function tree to construct Walsh matrices of higher order.

7.5: The Hadamard matrix can be computed using the following iteration

$$\boldsymbol{H}_{2^{l+1}} = \begin{bmatrix} \boldsymbol{H}_{2^l} & \boldsymbol{H}_{2^l} \\ \boldsymbol{H}_{2^l} & -\boldsymbol{H}_{2^l} \end{bmatrix} \tag{7.76}$$

with $\boldsymbol{H}_1 = [1]$.
(a) Compute $\boldsymbol{H}_2, \boldsymbol{H}_4$ and $\boldsymbol{H}_8$.
(b) Find the appropriate index for the rows in $\boldsymbol{H}_4$ and $\boldsymbol{H}_8$, compared with Walsh matrix $\boldsymbol{W}_4$ and $\boldsymbol{W}_8$ from Exercise 7.4.
(c) Determine the general rule to map a Walsh matrix into a Hadamard matrix.
Hint: First compute the index in binary notation.

7.6: The following MATLAB code can be used to compute the state-space description for $p_{14} = x^{14} + x^5 + x^3 + x^1 + 1$, the nonzero elements using **nnz**, and the maximum fan-in.

```
p= input('Please define power of matrix = ')
A=zeros(14,14);
for m=1:13
  A(m,m+1)=1;
end
A(14,14)=1;
A(14,13)=1;
A(14,11)=1;
A(14,9)=1;
Ap=mod(A^p,2);
nnz(Ap)
max(sum(Ap,2))
```

(a) Compute the number of nonzero elements and fan-in for $p = 2$ to 8.
(b) Modify the code to compute the twin $p_{14} = x^{14} + x^{13} + x^{11} + x^9 + 1$. Compute the number of nonzero elements for the modified polynomial for $p = 2$ to 8.
(c) Modify the original code to compute the alternative LFSR implementation (see Fig. 7.21, p. 295) for (a) and (b) and compute the nonzero elements for $p = 2$ to 8.

Exercises Using MaxPlusII

7.7: **(a)** Compile the code for the length-6 LFSR **lfsr.vhd** from Example 7.12 (p. 294) using MaxPlusII.
(b) For the line

```
ff(1) <= NOT (ff(5) XOR ff(6));
```

substitute

```
ff(1) <= ff(5) XNOR ff(6);
```

and compile with MaxPlusII.
(c) Now change the **Compiler** settings **Interfaces** → **VHDL Netlist Reader Settings** from **VHDL 1987** to **VHDL 1993** and compile again. Explain the results.

References

1. B. Dipert: "EDN's First Annual PLD Directory," *EDN* pp. 54–84 (2000) Http://www.ednmag.com/ednmag/reg/2000/08172000/17cs.htm
2. S. Brown, Z. Vranesic: *Fundamentals of Digital Logic with VHDL Design* (Prentice Hall, Englewood Cliffs, New Jersey, 1999)
3. D. Smith: *HDL Chip Design* (Doone Publications, Madison, Alabama, USA, 1996)
4. U. Meyer-Bäse: *The Use of Complex Algorithm in the Realization of Universal Sampling Receiver Using FPGAs* (in German) (VDI/Springer, Düsseldorf, 1995), Vol. 10, No. 404, 215 pages
5. U. Meyer-Bäse: *Fast Digital Signal Processing* (in German) (Springer, Heidelberg, 1999), 370 pages
6. P. Lapsley, J. Bier, A. Shoham, E. Lee: *DSP Processor Fundamentals* (IEEE Press, New York, 1997)
7. D. Shear: "EDN's DSP Benchmarks," *EDN* **33**, 126–148 (1988)
8. GEC Plessey Semiconductors: "Data Sheet," ERA60100 (1990)
9. J. Greene, E. Hamdy, S. Beal: "Antifuse Field-Programmable Gate Arrays," *Proceedings of the IEEE* pp. 1042–56 (1993)
10. J. Rose, A. Gamal, A. Sangiovanni-Vincentelli: "Architecture of Field-Programmable Gate Arrays," *Proceedings of the IEEE* pp. 1013–29 (1993)
11. Xilinx: "PREP Benchmark Observations," in *Xilinx-Seminar* San Jose (1993)
12. Altera Corporation: "PREP Benchmarks Reveal FLEX 8000 is Biggest, MAX 7000 is Fastest," in *Altera Corporation News & Views* San Jose (1993)
13. Actel: "PREP Benchmarks Confirm Cost Effectiveness of Field-Programmable Gate Arrays," in *Actel-Seminar* (1993)
14. E. Lee: "Programmable DSP Architectures: Part I," *IEEE Transactions on Acoustics, Speech and Signal Processing Magazine* pp. 4–19 (1988)
15. E. Lee: "Programmable DSP Architectures: Part II," *IEEE Transactions on Acoustics, Speech and Signal Processing Magazine* pp. 4–14 (1989)
16. R. Petersen, B. Hutchings: "An Assessment of the Suitability of FPGA-Based Systems for Use in Digital Signal Processing," *Lecture Notes in Computer Science* **975**, 293–302 (Springer, Heidelberg, 1995)
17. J. Villasenor, B. Hutchings: "The Flexibility of Configurable Computing," *IEEE Signal Processing Magazine* pp. 67–84 (1998)
18. Xilinx: "Data Book," XC2000, XC3000 and XC4000 (1993)
19. Altera Corporation: "Data Sheet," FLEX 10K CPLD Family (1996)
20. Altera Corporation: "Manual," Max+PlusII Getting Started (1995)
21. O. Spaniol: *Computer Arithmetic: Logic and Design* (John Wiley & Sons, New York, 1981)
22. I. Koren: *Computer Arithmetic Algorithms* (Prentice Hall, Englewood Cliffs, New Jersey, 1993)

23. E.E. Swartzlander: *Computer Arithmetic, Vol. I* (Dowden, Hutchingon and Ross, Inc., Stroudsburg, Pennsylvania, 1980), also reprinted by IEEE Computer Society Press (1990)
24. E. Swartzlander: *Computer Arithmetic, Vol. II* (IEEE Computer Society Press, Stroudsburg, Pennsylvania, 1990)
25. K. Hwang: *Computer Arithmetic: Principles, Architecture and Design* (John Wiley & Sons, New York, 1979)
26. N. Takagi, H. Yasuura, S. Yajima: "High-Speed VLSI Multiplication Algorithm with a Redundant Binary Addition Tree," *IEEE Transactions on Computers* **34**(2) (1985)
27. D. Bull, D. Horrocks: "Reduced-Complexity Digital Filtering Structures Using Primitive Operations," *Electronics Letters* pp. 769–771 (1987)
28. D. Bull, D. Horrocks: "Primitive Operator Digital Filters," *IEE Proceedings-G* **138**, 401–411 (1991)
29. A. Dempster, M. Macleod: "Use of Minimum-Adder Multiplier Blocks in FIR Digital Filters," *IEEE Transactions on Circuits and Systems II* **42**, 569–577 (1995)
30. A. Dempster, M. Macleod: "Comments on 'Minimum Number of Adders for Implementing a Multiplier and Its Application to the Design of Multiplierless Digital Filters'," *IEEE Transactions on Circuits and Systems II* **45**, 242–243 (1998)
31. F. Taylor, R. Gill, J. Joseph, J. Radke: "A 20 Bit Logarithmic Number System Processor," *IEEE Transactions on Computers* **37**(2) (1988)
32. P. Lee: "An FPGA Prototype for a Multiplierless FIR Filter Built Using the Logarithmic Number System," *Lecture Notes in Computer Science* **975**, 303–310 (Springer, Heidelberg, 1995)
33. J. Mitchell: "Computer Multiplication and Division Using Binary Logarithms," *IRE Transactions on Electronic Computers* **EC-11**, 512–517 (1962)
34. N. Szabo, R. Tanaka: *Residue Arithmetic and Its Applications to Computer Technology* (McGraw-Hill, New York, 1967)
35. M. Soderstrand, W. Jenkins, G. Jullien, F. Taylor: *Residue Number System Arithmetic: Modern Applications in Digital Signal Processing*, IEEE Press Reprint Series (IEEE Press, New York, 1986)
36. U. Meyer-Bäse, A. Meyer-Bäse, J. Mellott, F. Taylor: "A Fast Modified CORDIC-Implementation of Radial Basis Neural Networks," *Journal of VLSI Signal Processing* pp. 211–218 (1998)
37. V. Hamann, M. Sprachmann: "Fast Residual Arithmetics with FPGAs," in *Proceedings of the Workshop on Design Methodologies for Microelectronics* Smolenice Castle, Slovakia (1995), pp. 253–255
38. G. Jullien: "Residue Number Scaling and Other Operations Using ROM Arrays," *IEEE Transactions on Communications* **27**, 325–336 (1978)
39. M. Griffin, M. Sousa, F. Taylor: "Efficient Scaling in the Residue Number System," in *IEEE International Conference on Acoustics, Speech, and Signal Processing* (1989), pp. 1075–1078
40. G. Zelniker, F. Taylor: "A Reduced-Complexity Finite Field ALU," *IEEE Transactions on Circuits and Systems* **38**(12) (1991)
41. IEEE Task P754: "A Proposed Standard for Binary Floating-Point Arithmetic," *IEEE Transactions on Computers* **14**(12) (1981)
42. N. Shirazi, P. Athanas, A. Abbott: "Implementation of a 2-D Fast Fourier Transform on an FPGA-Based Custom Computing Machine," *Lecture Notes in Computer Science* **975**, 282–292 (Springer, Heidelberg, 1995)
43. M. Bayoumi, G. Jullien, W. Miller: "A VLSI Implementation of Residue Adders," *IEEE Transactions on Circuits and Systems* **34**(3), 284–288 (1987)

44. A. Garcia, U. Meyer-Bäse, F. Taylor: "Pipelined Hogenauer CIC Filters Using Field-Programmable Logic and Residue Number System," in *IEEE International Conference on Acoustics, Speech, and Signal Processing* Vol. 5 (1998), pp. 3085–3088
45. L. Turner, P. Graumann, S. Gibb: "Bit-Serial FIR Filters with CSD Coefficients for FPGAs," *Lecture Notes in Computer Science* **975**, 311–320 (Springer, Heidelberg, 1995)
46. A. Croisier, D. Esteban, M. Levilion, V. Rizo: (1973), "Digital Filter for PCM Encoded Signals," US Patent No. 3777130
47. A. Peled, B. Liu: "A New Realization of Digital Filters," *IEEE Transactions on Acoustics, Speech and Signal Processing* **22**(6), 456–462 (1974)
48. K. Yiu: "On Sign-Bit Assignment for a Vector Multiplier," *Proceedings of the IEEE* **64**, 372–373 (1976)
49. K. Kammeyer: "Quantization Error on the Distributed Arithmetic," *IEEE Transactions on Circuits and Systems* **24**(12), 681–689 (1981)
50. F. Taylor: "An Analysis of the Distributed-Arithmetic Digital Filter," *IEEE Transactions on Acoustics, Speech and Signal Processing* **35**(5), 1165–1170 (1986)
51. S. White: "Applications of Distributed Arithmetic to Digital Signal Processing: A Tutorial Review," *IEEE Transactions on Acoustics, Speech and Signal Processing Magazine* pp. 4–19 (1989)
52. K. Kammeyer: "Digital Filter Realization in Distributed Arithmetic," in *Proc. European Conf. on Circuit Theory and Design* (1976), Genoa, Italy
53. F. Taylor: *Digital Filter Design Handbook* (Marcel Dekker, New York, 1983)
54. H. Nussbaumer: *Fast Fourier Transform and Convolution Algorithms* (Springer, Heidelberg, 1990)
55. H. Schmid: *Decimal Computation* (John Wiley & Sons, New York, 1974)
56. Y. Hu: "CORDIC-Based VLSI Architectures for Digital Signal Processing," *IEEE Signal Processing Magazine* pp. 16–35 (1992)
57. U. Meyer-Bäse, A. Meyer-Bäse, W. Hilberg: "**CO**ordinate **R**otation **DI**gital **C**omputer (CORDIC) Synthesis for FPGA," *Lecture Notes in Computer Science* **849**, 397–408 (Springer, Heidelberg, 1994)
58. J.E. Volder: "The CORDIC Trigonometric Computing Technique," *IRE Transactions on Electronics Computers* **8**(3), 330–4 (1959)
59. J. Walther: "A Unified Algorithm for Elementary Functions," *Spring Joint Computer Conference* pp. 379–385 (1971)
60. X. Hu, R. Huber, S. Bass: "Expanding the Range of Convergence of the CORDIC Algorithm," *IEEE Transactions on Computers* **40**(1), 13–21 (1991)
61. D. Timmermann (1990): "CORDIC-Algorithmen, Architekturen und monolithische Realisierungen mit Anwendungen in der Bildverarbeitung," Ph.D. thesis, VDI/Springer, Düsseldorf, Vol. 10, No. 152
62. H. Hahn (1991): "Untersuchung und Integration von Berechnungsverfahren elementarer Funktionen auf CORDIC-Basis mit Anwendungen in der adaptiven Signalverarbeitung," Ph.D. thesis, VDI/Springer, Düsseldorf, Vol. 9, No. 125
63. G. Ma (1989): "A Systolic Distributed Arithmetic Computing Machine for Digital Signal Processing and Linear Algebra Applications," Ph.D. thesis, University of Florida, Gainesville
64. Y.H. Hu: "The Quantization Effects of the CORDIC Algorithm," *IEEE Transactions on Signal Processing* pp. 834–844 (1992)
65. A.V. Oppenheim, R.W. Schafer: *Discrete-Time Signal Processing* (Prentice Hall, Englewood Cliffs, New Jersey, 1992)

66. D.J. Goodman, M.J. Carey: "Nine Digital Filters for Decimation and Interpolation," *IEEE Transactions on Acoustics, Speech and Signal Processing* **25**(2) 121–126 (1977)
67. R. Hartley: "Subexpression Sharing in Filters Using Canonic Signed Digital Multiplier," *IEEE Transactions on Circuits and Systems II* **30**(10), 677–688 (1996)
68. R. Saal: *Handbook of filter design* (AEG-Telefunken, Frankfurt, Germany, 1979)
69. C. Barnes, A. Fam: "Minimum Norm Recursive Digital Filters that Are Free of Overflow Limit Cycles," *IEEE Transactions on Circuits and Systems* pp. 569–574 (1977)
70. A. Fettweis: "Wave Digital Filters: Theorie and Practice," *Proceedings of the IEEE* pp. 270–327 (1986)
71. R. Crochiere, A. Oppenheim: "Analysis of Linear Digital Networks," *Proceedings of the IEEE* **63**(4), 581–595 (1995)
72. A. Dempster, M. Macleod: "Multiplier Blocks and Complexity of IIR Structures," *Electronics Letters* **30**(22), 1841–1842 (1994)
73. A. Dempster, M. Macleod: "IIR Digital Filter Design Using Minimum Adder Multiplier Blocks," *IEEE Transactions on Circuits and Systems II* **45**, 761–763 (1998)
74. A. Dempster, M. Macleod: "Constant Integer Multiplication Using Minimum Adders," *IEE Proceedings - Circuits, Devices & Systems* **141**, 407–413 (1994)
75. K. Parhi, D. Messerschmidt: "Pipeline Interleaving and Parallelism in Recursive Digital Filters - Part I: Pipelining Using Scattered Look-Ahead and Decomposition," *IEEE Transactions on Acoustics, Speech and Signal Processing* **37**(7), 1099–1117 (1989)
76. H. Loomis, B. Sinha: "High-Speed Recursive Digital Filter Realization," *Circuits, Systems, Signal Processing* **3**(3), 267–294 (1984)
77. M. Soderstrand, A. de la Serna, H. Loomis: "New Approach to Clustered Look-Ahead Pipelined IIR Digital Filters," *IEEE Transactions on Circuits and Systems II* **42**(4), 269–274 (1995)
78. J. Living, B. Al-Hashimi: "Mixed Arithmetic Architecture: A Solution to the Iteration Bound for Resource-Efficient FPGA and CPLD Recursive Digital Filters," in *IEEE International Symposium on Circuits and Systems* Vol. I (1999), pp. 478–481
79. H. Martinez, T. Parks: "A Class of Infinite-Duration Impulse Response Digital Filters for Sampling Rate Reduction," *IEEE Transactions on Acoustics, Speech and Signal Processing* **26**(4), 154–162 (1979)
80. K. Parhi, D. Messerschmidt: "Pipeline Interleaving and Parallelism in Recursive Digital Filters - Part II: Pipelined Incremental Block Filtering," *IEEE Transactions on Acoustics, Speech and Signal Processing* **37**(7), 1118–1134 (1989)
81. M. Shajaan, J. Sorensen: "Time-Area Efficient Multiplier-Free Recursive Filter Architectures for FPGA Implementation," in *IEEE International Conference on Acoustics, Speech, and Signal Processing* (1996), pp. 3269–3272
82. P. Vaidyanathan: *Multirate Systems and Filter Banks* (Prentice Hall, Englewood Cliffs, New Jersey, 1993)
83. S. Winograd: "On Computing the Discrete Fourier Transform," *Mathematics of Computation* **32**, 175–199 (1978)
84. Z. Mou, P. Duhamel: "Short-Length FIR Filters and Their Use in Fast Nonrecursive Filtering," *IEEE Transactions on Signal Processing* **39**, 1322–1332 (1991)

85. P. Balla, A. Antoniou, S. Morgera: "Higher Radix Aperiodic-Convolution Algorithms," *IEEE Transactions on Acoustics, Speech and Signal Processing* **34**(1), 60–68 (1986)
86. E.B. Hogenauer: "An Economical Class of Digital Filters for Decimation and Interpolation," *IEEE Transactions on Acoustics, Speech and Signal Processing* **29**(2), 155–162 (1981)
87. Harris Semiconductor: "Data Sheet," HSP43220 Decimating Digital Filter (1992)
88. Motorola Inc.: "Pilkingtons Patent," *Elektronik Praxis* pp. 40–43 (1993)
89. O. Six (1996): "Design and Implementation of a Xilinx Universal XC-4000 FPGAs board," Master's thesis, Institute for Data Technics, Darmstadt University of Technology
90. S. Dworak (1996): "Design and Realization of a New Class of Frequency Sampling Filters for Speech Processing Using FPGAs," Master's thesis, Institute for Data Technics, Darmstadt University of Technology
91. A. Haar: "Zur Theorie der orthogonalen Funktionensysteme," *Mathematische Annalen* **69**, 331–371 (1910). Dissertation Göttingen 1909
92. P. Sweeney: *Error Control Coding* (Prentice Hall, New York, 1991)
93. C. Herley, M. Vetterli: "Wavelets and Recursive Filter Banks," *IEEE Transactions on Signal Processing* **41**, 2536–2556 (1993)
94. I. Daubechies: *Ten Lectures on Wavelets* (Society for Industrial and Applied Mathematics (SIAM), Philadelphia, 1992)
95. I. Daubechies, W. Sweldens: "Factoring Wavelet Transforms into Lifting Steps," *The Journal of Fourier Analysis and Applications* **4**, 365–374 (1998)
96. G. Strang, T. Nguyen: *Wavelets and Filter Banks* (Wellesley-Cambridge Press, Wellesley MA, 1996)
97. D. Esteban, C. Galand: "Applications of Quadrature Mirror Filters to Split Band Voice Coding Schemes," in *IEEE International Conference on Acoustics, Speech, and Signal Processing* (1977), pp. 191–195
98. M. Smith, T. Barnwell: "Exact Reconstruction Techniques for Tree-Structured Subband Coders," *IEEE Transactions on Acoustics, Speech and Signal Processing* pp. 434–441 (1986)
99. M. Vetterli, J. Kovacevic: *Wavelets and Subband Coding* (Prentice Hall, Englewood Cliffs, New Jersey, 1995)
100. R. Crochiere, L. Rabiner: *Multirate Digital Signal Processing* (Prentice Hall, Englewood Cliffs, New Jersey, 1983)
101. M. Acheroy, J.M. Mangen, Y. Buhler.: "Progressive Wavelet Algorithm Versus JPEG for the Compression of METEOSAT Data," in *SPIE*, San Diego (1995)
102. T. Ebrahimi, M. Kunt: "Image Compression by Gabor Expansion," *Optical Engineering* **30**, 873–880 (1991)
103. D. Gabor: "Theory of Communication," *J. Inst. Elect. Eng* (London)**93**, 429–457 (1946)
104. A. Grossmann, J. Morlet: "Decomposition of Hardy Functions into Square Integrable Wavelets of Constant Shape," *SIAM J. Math. Anal.* **15**, 723–736 (1984)
105. U. Meyer-Bäse: "High-Speed Implementation of Gabor and Morlet Wavelet Filterbanks Using RNS Frequency Sampling Filters," in *Aerosense 98 *SPIE**, Orlando (1998), pp. 522–533
106. U. Meyer-Bäse: "Die Hutlets – eine biorthogonale Wavelet-Familie: Effiziente Realisierung durch multipliziererfreie, perfekt rekonstruierende Quadratur Mirror Filter," *Frequenz* pp. 39–49 (1997)

107. U. Meyer-Bäse, F. Taylor: "The Hutlets - a Biorthogonal Wavelet Family and Their High-Speed Implementation with RNS, Multiplier-Free, Perfect Reconstruction QMF," in *Aerosense 97 *SPIE**, Orlando (1997), pp. 670–681
108. M. Heideman, D. Johnson, C. Burrus: "Gauss and the History of the Fast Fourier Transform," *IEEE Transactions on Acoustics, Speech and Signal Processing Magazine* **34**, 265–267 (1985)
109. C. Burrus: "Index Mappings for Multidimensional Formulation of the DFT and Convolution," *IEEE Transactions on Acoustics, Speech and Signal Processing* **25**, 239–242 (1977)
110. B. Baas: (1998), "SPIFFEE an Energy-Efficient Single-Chip 1024-Point FFT Processor," http://www.stanford.edu/ bbaas/spiffee1.html
111. G. Sunada, J. Jin, M. Berzins, T. Chen: "COBRA: An 1.2 Million Transistor Exandable Column FFT Chip," in *Proceedings of the International Conference on Computer Design: VLSI in Computers and Processors* (IEEE Computer Society Press, Los Alamitos, California, USA, 1994), pp. 546–550
112. Texas Memory Systems: "TM-66 swiFFT Chip," http://www.texmemsys.com (1996)
113. Sharp Microeletronics: "BDSP9124 Digital Signal Processor," http://www.butterflydsp.com (1997)
114. J. Mellott (1997): "Long Instruction Word Computer," Ph.D. thesis, University of Florida, Gainesville
115. P. Lavoie: "A High-Speed CMOS Implementation of the Winograd Fourier Transform Algorithm," *IEEE Transactions on Signal Processing* **44**(8), 2121–2126 (1996). Http://www.dreo.dnd.ca/pages/electdev/ewd011.htm
116. G. Panneerselvam, P. Graumann, L. Turner: "Implementation of Fast Fourier Transforms and Discrete Cosine Transforms in FPGAs," in *Lecture Notes in Computer Science* **1142**, 272–281 (Springer, Heidelberg, 1996),
117. Altera Corporation: "Fast Fourier Transform," in *Solution Brief 12, Altera Corparation* (1997)
118. G. Goslin: "Using Xilinx FPGAs to Design Custom Digital Signal Processing Devices," in *Proceedings of the DSP^X* (1995), pp. 595–604
119. C. Dick: "Computing 2-D DFTs Using FPGAs," *Lecture Notes in Computer Science: Field-Programmable Logic* **1142**, pp. 96–105 (Springer, Heidelberg, 1996)
120. S.D. Stearns, D.R. Hush: *Digital Signal Analysis* (Prentice Hall, Englewood Cliffs, New Jersey, 1990)
121. K. Kammeyer, K. Kroschel: *Digitale Signalverarbeitung* (Teubner Studienbücher, Stuttgart, 1989)
122. E. Brigham: *FFT*, 3rd edn. (Oldenbourg Verlag, München Wien, 1987)
123. R. Ramirez: *The FFT: Fundamentals and Concepts* (Prentice Hall, Englewood Cliffs, New Jersey, 1985)
124. R.E. Blahut: *Theory and Practice of Error Control Codes* (Addison-Wesley, Melo Park, California, 1984)
125. C. Burrus, T. Parks: *DFT/FFT and Convolution Algorithms* (John Wiley & Sons, New York, 1985)
126. D. Elliott, K. Rao: *Fast Transforms: Algorithms, Analyses, Applications* (Academic Press, New York, 1982)
127. A. Nuttall: "Some Windows with Very Good Sidelobe Behavior," *IEEE Transactions on Acoustics, Speech and Signal Processing* **ASSP-29**(1), 84–91 (1981)
128. U. Meyer-Bäse, K. Damm (1988): "Fast Fourier Transform Using Signal Processor," Master's thesis, Department of Information Science, Darmstadt University of Technology

129. M. Narasimha, K. Shenoi, A. Peterson: "Quadratic Residues: Application to Chirp Filters and Discrete Fourier Transforms," in *IEEE International Conference on Acoustics, Speech, and Signal Processing* (1976), pp. 12–14
130. C. Rader: "Discrete Fourier Transform When the Number of Data Samples is Prime," *Proceedings of the IEEE* **56**, 1107–8 (1968)
131. J. McClellan, C. Rader: *Number Theory in Digital Signal Processing* (Prentice Hall, Englewood Cliffs, New Jersey, 1979)
132. I. Good: "The Relationship between Two Fast Fourier Transforms," *IEEE Transactions on Computers* **20**, 310–317 (1971)
133. L. Thomas: "Using a Computer to Solve Problems in Physics," in *Applications of Digital Computers* (Ginn, Dordrecht, 1963)
134. A. Dandalis, V. Prasanna: "Fast Parallel Implementation of DFT Using Configurable Devices," *Lecture Notes in Computer Science* **1304**, 314–323 (Springer, Heidelberg, 1997)
135. U. Meyer-Bäse, S. Wolf, J. Mellott, F. Taylor: "High Performance Implementation of Convolution on a Multi FPGA Board Using NTT's Defined Over the Eisenstein Residuen Number System," in *Aerosense 97 *SPIE**, Orlando (1997), pp. 431–442
136. Z. Wang: "Fast Algorithms for the Discrete W Transfrom and for the discrete Fourier Transform," *IEEE Transactions on Acoustics, Speech and Signal Processing* pp. 803–816 (1984)
137. M. Narasimha, A. Peterson: "On the Computation of the Discrete Cosine Transform," *IEEE Transaction on Communications* **26**(6), 934–936 (1978)
138. K. Rao, P. Yip: *Discrete Cosine Transform* (Academic Press, San Diego, California, 1990)
139. B. Lee: "A New Algorithm to Compute the Discrete Cosine Transform," *IEEE Transactions on Acoustics, Speech and Signal Processing* **32**(6), 1243–1245 (1984)
140. S. Ramachandran, S. Srinivasan, R. Chen: "EPLD-Based Architecture of Real Time 2D Discrete Cosine Transform and Quantization for Image Compression," in *IEEE International Symposium on Circuits and Systems* Vol. III (1999), pp. 375 – 378
141. C. Burrus, P. Eschenbacher: "An In-Place, In-Order Prime Factor FFT Algorithm," *IEEE Transactions on Acoustics, Speech and Signal Processing* **29**(4), 806–817 (1981)
142. J. Pollard: "The Fast Fourier Transform in a Finite Field," *Mathematics of Computation* **25**, 365–374 (1971)
143. F. Taylor: "An RNS Discrete Fourier Transform Implementation," *IEEE Transactions on Acoustics, Speech and Signal Processing* **38**, 1386–1394 (1990)
144. C. Rader: "Discrete Convolutions via Mersenne Transforms," *IEEE Transactions on Computers* **C-21**, 1269–1273 (1972)
145. Leibowitz: "A Simplified Binary Arithmetic for the Fermat Number Transform," *IEEE Transactions on Acoustics, Speech and Signal Processing* **24**, 356–359 (1976)
146. N. Bloch: *Abstract Algebra with Applications* (Prentice Hall, Englewood Cliffs, New Jersey, 1987)
147. J. Lipson: *Elements of Algebra and Algebraic Computing* (Addison-Wesley, London, 1981)
148. R. Agrawal, C. Burrus: "Fast Convolution Using Fermat Number Transforms with Applications to Digital Filtering," *IEEE Transactions on Acoustics, Speech and Signal Processing* **22**, 87–97 (1974)
149. W. Siu, A. Constantinides: "On the Computation of Discrete Fourier Transform Using Fermat Number Transform," *IEE Proceedings F* **131**, 7–14 (1984)

150. J. McClellan: "Hardware Realization of the Fermat Number Transform," *IEEE Transactions on Acoustics, Speech and Signal Processing* **24**(3), 216–225 (1976)
151. Texas Instruments: "User's Guide," TMS320C50, p. 7-51 (1993)
152. I. Reed, D. Tufts, X. Yu, T. Truong, M.T. Shih, X. Yin: "Fourier Analysis and Signal Processing by Use of the Möbius Inversion Formula," *IEEE Transactions on Acoustics, Speech and Signal Processing* **38**(3), 458–470 (1990)
153. H. Park, V. Prasanna: "Modular VLSI Arichitectures for Computing the Arithmetic Fourier Transform," *IEEE Transactions on Signal Processing* **41**(6), 2236–2246 (1993)
154. H. Lüke: *Signalübertragung* (Springer, Heidelberg, 1988)
155. D. Herold, R. Huthmann (1990): "Decoder for the Radio Data System (RDS) Using Signal Processor TMS320C25," Master's thesis, Institute for Data Technics, Darmstadt University of Technology
156. U. Meyer-Bäse, R. Watzel: "A Comparison of DES and LFSR-based FPGA Implementable Cryptography Algorithms," in *3rd International Symposium on Communication Theory & Applications* (1995), pp. 291–298
157. U. Meyer-Bäse, R. Watzel: "An Optimized Format for Long Frequency Paging Systems," in *3rd International Symposium on Communication Theory & Applications* (1995), pp. 78–79
158. U. Meyer-Bäse: "Convolutional Error Decoding with FPGAs," *Lecture Notes in Computer Science* **1142**, 376–175 (Springer, Heidelberg, 1996)
159. R. Watzel (1993): "Design of Paging Scheme and Implementation of the Suitable Cryto-Controller Using FPGAs," Master's thesis, Institute for Data Technics, Darmstadt University of Technology
160. J. Maier, T. Schubert (1993): "Design of Convolutional Decoders Using FPGAs for Error Correction in a Paging System," Master's thesis, Institute for Data Technics, Darmstadt University of Technology
161. Y. Gao, D. Herold, U. Meyer-Bäse: "Zum bestehenden Übertragungsprotokoll kompatible Fehlerkorrektur," in *Funkuhren Zeitsignale Normalfrequenzen* (1993), pp. 99–112
162. D. Herold (1991): "Investigation of Error Corrections Steps for DCF77 Signals Using Programmable Gate Arrays," Master's thesis, Institute for Data Technics, Darmstadt University of Technology
163. D. Wiggert: *Error-Control Coding and Applications* (Artech House, Dedham, Mass., 1988)
164. G. Clark, J. Cain: *Error-Correction Coding for Digital Communications* (Plenum Press, New York, 1988)
165. W. Stahnke: "Primitive Binary Polynomials," *Mathematics of Computation* pp. 977–980 (1973)
166. W. Fumy, H. Riess: *Kryptographie* (R. Oldenbourg Verlag, München, 1988)
167. B. Schneier: *Applied Cryptography* (John Wiley & Sons, New York, 1996)
168. M. Langhammer: "Reed-Solomon Codec Design in Programmable Logic," *Communication System Design* (www.csdmag.com) pp. 31–37 (1998)
169. B. Akers: "Binary Decusion Diagrams," *IEEE Transactions on Computers* pp. 509–516 (1978)
170. R. Bryant: "Graph-Based Algorithms for Boolean Function Manipulation," *IEEE Transactions on Computers* pp. 677–691 (1986)
171. A. Sangiovanni-Vincentelli, A. Gamal, J. Rose: "Synthesis Methods for Field Programmable Gate Arrays," *Proceedings of the IEEE* pp. 1057–83 (1993)
172. R. del Rio (1993): "Synthesis of boolean Functions for Field Programmable Gate Arrays," Master's thesis, Univerity of Frankfurt, FB Informatik

173. U. Meyer-Bäse: "Optimal Strategies for Incoherent Demodulation of Narrow Band FM Signals," in *3rd International Symposium on Communication Theory & Applications* (1995), pp. 30–31
174. J. Proakis: *Digital Communications* (McGraw-Hill, New York, 1983)
175. R. Johannesson: "Robustly Optimal One-Half Binary Convolutional Codes," *IEEE Transactions on Information Theory* pp. 464–8 (1975)
176. J. Massey, D. Costello: "Nonsystematic Convolutional Codes for Sequential Decoding in Space Applications," *IEEE Transactions on Communications* pp. 806–813 (1971)
177. F. MacWilliams, J. Sloane: "Pseudo-Random Sequences and Arrays," *Proceedings of the IEEE* pp. 1715–29 (1976)
178. T. Lewis, W. Payne: "Generalized Feedback Shift Register Pseudorandom Number Algorithm," *Journal of the Association for Computing Machinery* pp. 456–458 (1973)
179. P. Bratley, B. Fox, L. Schrage: *A Guide to Simulation* (Springer-Lehrbuch, Heidelberg, 1983), pp. 186–190
180. M. Schroeder: *Number Theory in Science and Communication* (Springer, Heidelberg, 1990)
181. Electronic Frontier Foundation: *Cracking DES* (O'Reilly & Associates, Sebastopol, 1998)
182. W. Stallings: "Encryption Choices Beyond DES," *Communication System Design (www.csdmag.com)* pp. 37–43 (1998)
183. W. Carter: "FPGAs: Go Reconfigure," *Communication System Design (www.csdmag.com)* p. 56 (1998)
184. J. Anderson, T. Aulin, C.E. Sundberg: *Digital Phase Modulation* (Plenum Press, New York, 1986)
185. U. Meyer-Bäse (1989): "Investigation of Threshold Improving Limiter/Discriminator Demodulator for FM Signals Through Computer Simulations," Master's thesis, Department of Information Science, Darmstadt University of Technology
186. E. Allmann, T. Wolf (1991): "Design and Implementation of a Fully Digital Zero IF Receiver Using Programmable Gate-Arrays and Floating-Point DSPs," Master's thesis, Institute for Data Technics, Darmstadt University of Technology
187. O. Herrmann: "Quadraturfilter mit rationalem Übertragungsfaktor," *Archiv der elektrischen Übertragung* pp. 77–84 (1969)
188. O. Herrmann: "Transversalfilter zur Hilbert-Transformation," *Archiv der elektrischen Übertragung* pp. 581–587 (1969)
189. V. Considine: "Digital Complex Sampling," *Electronics Letters* pp. 608–609 (1983)
190. T.E. Thiel, G.J. Saulnier: "Simplified Complex Digital Sampling Demodulator," *Electronics Letters* pp. 419–421 (1990)
191. U. Meyer-Bäse, W. Hilberg: (1992), "Schmalbandempfänger für Digitalsignale," German Patent No. 4219417.2-31
192. B. Schlanske (1992): "Design and Implementation of a Universal Hilbert Sampling Receiver with CORDIC Demodulation for LF Fax Signals Using digital Signal Processor," Master's thesis, Institute for Data Technics, Darmstadt University of Technology
193. A. Dietrich (1992): "Realisation of a Hilbert Sampling Receiver with CORDIC Demodulation for DCF77 Signals Using Floating-Point Signal Processors," Master's thesis, Institute for Data Technics, Darmstadt University of Technology

194. A. Viterbi: *Principles of Coherent Communication* (McGraw-Hill, New York, 1966)
195. F. Gardner: *Phaselock Techniques* (John Wiley & Sons, New York, 1979)
196. H. Geschwinde: *Einführung in die PLL-Technik* (Vieweg, Braunschweig, 1984)
197. R. Best: *Theorie und Anwendung des Phase-locked Loops* (AT Press, Schwitzerland, 1987)
198. W. Lindsey, C. Chie: "A Survey of Digital Phase-Locked Loops," *Proceedings of the IEEE* pp. 410–431 (1981)
199. R. Sanneman, J. Rowbotham: "Unlock Characteristics of the Optimum Type II Phase-Locked Loop," *IEEE Transactions on Aerospace and Navigational Electronics* pp. 15–24 (1964)
200. J. Stensby: "False Lock in Costas Loops," *Proceedings of the 20th Southeastern Symposium on System Theory* pp. 75–79 (1988)
201. A. Mararios, T. Tozer: "False-Lock Performance Improvement in Costas Loops," *IEEE Transactions on Communications* pp. 2285–88 (1982)
202. A. Makarios, T. Tozer: "False-Look Avoidance Scheme for Costas Loops," *Electronics Letters* pp. 490–2 (1981)
203. U. Meyer-Bäse: "Coherent Demodulation with FPGAs," *Lecture Notes in Computer Science* **1142**, 166–175 (Springer, Heidelberg, 1996)
204. J. Guyot, H. Schmitt (1993): "Design of a Fully Digital Costas Loop Using Programmable Gate Arrays for Coherent Demodulation of Low Frequency Signals," Master's thesis, Institute for Data Technics, Darmstadt University of Technology
205. R. Resch, P. Schreiner (1993): "Design of Fully Digital Phase-Locked Loops Using programmable Gate-Arrys for a Low-Frequency Reciever," Master's thesis, Institute for Data Technics, Darmstadt University of Technology
206. D. McCarty: "Digital PLL suits FPGAs," *Elektronic Design* p. 81 (1992)
207. J. Holmes: "Tracking-Loop Bias Due to Costas Loop Arm Filter Imbalance," *IEEE Transactions on Communications* pp. 2271–3 (1982)
208. H. Choi: "Effect of Gain and Phase Imbalance on the Performance of Lock Detector of Costas Loop," *IEEE International Conference on Communications*, Seattle pp. 218–222 (1987)

A. Verilog Source Code

```
//*********************************************************
// IEEE STD 1364-1995 Verilog file: example.v
// Author-EMAIL: Uwe.Meyer-Baese@ieee.org
//*********************************************************
//`include "220model.v"   // Using predefined components

module example (clk, a, b, op1, sum, d);  //----> Interface

  parameter WIDTH =8;    // Bit width
  input  clk;
  input  [WIDTH-1:0] a, b, op1;
  output [WIDTH-1:0] sum, d;

  wire [WIDTH-1:0]  c;          // Auxiliary variables
  reg  [WIDTH-1:0]  s;          // Infer FF with always
  wire [WIDTH-1:0] op2, op3;

  wire  clkena, ADD, ena, aset, sclr, sset, aload, sload,
               aclr, ovf1, cin1; // Auxiliary lpm signals

// Default for add:
  assign cin1=0; assign aclr=0; assign ADD=1;

  assign ena=1; assign aclr=0; assign aset=0;
  assign sclr=0; assign sset=0; assign aload=0;
  assign sload=0; assign clkena=0; // Default for FF

  assign op2 = b;        // Only one vector type in Verilog;
          // no conversion int -> logic vector necessary

// Note when using 220model.v ALL component's signals
// must be defined, default values can only be used for
// the parameters.
```

```
  lpm_add_sub add1          //----> Component instantiation
  ( .result(op3), .dataa(op1), .datab(op2)); // Used ports
//  .cin(cin1),.cout(cr1), .add_sub(ADD), .clken(clkena),
//  .clock(clk), .overflow(ovl1), .aclr(aclr)); // Unused
    defparam add1.lpm_width = WIDTH;
    defparam add1.lpm_representation = "SIGNED";

  lpm_ff reg1
  ( .data(op3), .q(sum), .clock(clk));  // Used ports
//  .enable(ena), .aclr(aclr), .aset(aset), .sclr(sclr),
//  .sset(sset), .aload(aload), .sload(sload)); // Unused
    defparam reg1.lpm_width = WIDTH;

  assign c = a + b; //----> Continuous assignment statement

  always @(posedge clk)  //----> Behavioral style
  begin : p1             // Infer register
    s = c + s;           // Signal assignment statement
  end
  assign d = s;

endmodule

//*********************************************************
// IEEE STD 1364-1995 Verilog file: fun_text.v
// Author-EMAIL: Uwe.Meyer-Baese@ieee.org
//*********************************************************
//  A 32 bit function generator using accumulator and ROM
//`include "220model.v"

module fun_text (M, sin, acc, clk); //----> Interface

  parameter WIDTH = 32;   // Bit width
  input  [WIDTH-1:0]  M;
  output [7:0]  sin, acc;
  input         clk;

  wire [WIDTH-1:0] s, acc32;
  wire [7:0]    msbs;                 // Auxiliary vectors
  wire  ADD, ena, aset, sclr, sset; // Auxiliary signals
  wire aload, sload, aclr, ovf1, cin1, clkena;

  // Default for add:
  assign clkena=0; assign cin1=0; assign ADD=1;
  //default for FF:
```

```
  assign ena=1; assign aclr=0; assign aset=0; assign sclr=0;
  assign sset=0; assign aload=0; assign sload=0;

  lpm_add_sub add_1                             // Add M to acc32
  ( .result(s), .dataa(acc32), .datab(M)); // Used ports
//  .cout(cr1), .add_sub(ADD), .overflow(ovl1),  // Unused
//  .clock(clk),.cin(cin1), .clken(clkena), .aclr(aclr));
//
    defparam add_1.lpm_width = WIDTH;
    defparam add_1.lpm_representation = "UNSIGNED";


  lpm_ff reg_1                                      // Save accu
  ( .data(s), .q(acc32), .clock(clk));        // Used ports
//  .enable(ena), .aclr(aclr), .aset(aset), // Unused ports
//  .sset(sset), .aload(aload), .sload(sload),.sclr(sclr));
    defparam reg_1.lpm_width = WIDTH;

  assign msbs = acc32[WIDTH-1:WIDTH-8];
  assign acc  = msbs;

  lpm_rom rom1
  ( .q(sin), .inclock(clk), .outclock(clk),
                               .address(msbs)); // Used ports
//                         .memenab(ena) ) ;      // Unused port
    defparam rom1.lpm_width = 8;
    defparam rom1.lpm_widthad = 8;
    defparam rom1.lpm_file = "sine.mif";

endmodule

//*********************************************************
// IEEE STD 1364-1995 Verilog file: add_1p.v
// Author-EMAIL: Uwe.Meyer-Baese@ieee.org
//*********************************************************
//`include "220model.v"

module add_1p (x, y, sum, clk);
  parameter WIDTH   = 15, // Total bit width
            WIDTH1  = 7,  // Bit width of LSBs
            WIDTH2  = 8;  // Bit width of MSBs

  input  [WIDTH-1:0] x,y;  // Inputs
  output [WIDTH-1:0] sum;  // Result
  input              clk;  // Clock
```

```
  reg  [WIDTH1-1:0] l1, l2; // LSBs of inputs
  wire [WIDTH1-1:0] q1, r1; // LSBs of inputs
  reg  [WIDTH2-1:0] l3, l4; // MSBs of input
  wire [WIDTH2-1:0] r2, q2, u2; // MSBs of input
  reg  [WIDTH-1:0]  s;    // Output register
  wire cr1,cq1; // LSBs carry signal
  wire [WIDTH2-1:0] h2; // Auxiliary MSBs of input

  wire  clkena, ADD, ena, aset, sclr;  // Auxiliary signals
  wire sset, aload, sload, aclr, ovf1, cin1;

// Default for add:
  assign cin1=0; assign aclr=0; assign ADD=1;

  assign ena=1; assign aclr=0;             // Default for FF
  assign sclr=0; assign sset=0; assign aload=0;
  assign sload=0; assign clkena=0; assign aset=0;

  // Split in MSBs and LSBs and store in registers
  always @(posedge clk) begin
    // Split LSBs from input x,y
    l1[WIDTH1-1:0] <= x[WIDTH1-1:0];
    l2[WIDTH1-1:0] <= y[WIDTH1-1:0];
    // Split MSBs from input x,y
    l3[WIDTH2-1:0] <= x[WIDTH2-1+WIDTH1:WIDTH1];
    l4[WIDTH2-1:0] <= y[WIDTH2-1+WIDTH1:WIDTH1];
  end
/************* First stage of the adder  *****************/
  lpm_add_sub add_1                     // Add LSBs of x and y
  ( .result(r1), .dataa(l1), .datab(l2), .cout(cr1));
                                                 // Used ports
//  .overflow(ovl1), .clken(clkena), .add_sub(ADD),
//  .cin(cin1), .clock(clk), .aclr(aclr)); // Unused ports
    defparam add_1.lpm_width = WIDTH1;
    defparam add_1.lpm_direction = "add";

  lpm_ff reg_1                            // Save LSBs of x+y
  ( .data(r1), .q(q1), .clock(clk));             // Used ports
//  .enable(ena), .aclr(aclr), .aset(aset), .sclr(sclr),
//  .sset(sset), .aload(aload), .sload(sload)); // Unused
    defparam reg_1.lpm_width = WIDTH1;

  lpm_ff reg_2       // Save LSBs carry
```

```
  ( .data(cr1), .q(cq1), .clock(clk));        // Used ports
//  .enable(ena), .aclr(aclr), .aset(aset), .sclr(sclr),
//  .sset(sset), .aload(aload), .sload(sload)); // Unused
    defparam reg_2.lpm_width = 1;

  lpm_add_sub add_2                          // Add MSBs of x and y
  ( .dataa(l3), .datab(l4), .result(r2)     ); // Used ports
//  .add_sub(ADD), .cout(cout1), .cin(cin1), .clken(clkena),
//  .overflow(ovl1),  .clock(clk), .aclr(aclr)); // Unused
   defparam add_2.lpm_width = WIDTH2;
   defparam add_2.lpm_direction = "add";

 lpm_ff reg_3                                 // Save MSBs of x+y
 ( .data(r2), .q(q2), .clock(clk));           // Used ports
// .enable(ena), .aclr(aclr), .aset(aset), .sclr(sclr),
// .sset(sset), .aload(aload), .sload(sload)); // Unused
    defparam reg_3.lpm_width = WIDTH2;


/************** Second stage of the adder ****************/
// One operand is zero
  assign h2 = {WIDTH2{1'b0}};

  lpm_add_sub add_3   // Add MSBs (x+y) and carry from LSBs
  ( .cin(cq1), .dataa(q2), .datab(h2), .result(u2));
                                              // Used ports
//  .cout(cout1), .overflow(ovl1), .clken(clkena),// Unused
//  .add_sub(ADD), .clock(clk), .aclr(aclr));     // ports
    defparam add_3.lpm_width = WIDTH2;
    defparam add_3.lpm_direction = "add";

  always @(posedge clk) begin  // Build a single registered
    s = {u2[WIDTH2-1:0],q1[WIDTH1-1:0]};      // output word
  end                               // of WIDTH=WIDTH1+WIDHT2

  assign sum = s ;     // Connect s to output pins

endmodule

//*********************************************************
// IEEE STD 1364-1995 Verilog file: add_2p.v
// Author-EMAIL: Uwe.Meyer-Baese@ieee.org
//*********************************************************
// 22-bit adder with two pipeline stages
// uses four components: csa7.v; csa7cin.v;
```

```
//                    add_ff8.v; add_ff8cin.v

//`include "220model.v"
//`include "csa7.v"
//`include "csa7cin.v"
//`include "add_ff8.v"
//`include "add_ff8cin.v"

module add_2p (x, y, sum, clk);
  parameter WIDTH   = 22, // Total bit width
            WIDTH1  = 7,  // Bit width of LSBs
            WIDTH2  = 7,  // Bit width of middle s
            WIDTH12 = 14,  // Sum WIDTH1+WIDTH2
            WIDTH3  =  8;  // Bit width of MSBs

  input  [WIDTH-1:0] x,y;   // Inputs
  output [WIDTH-1:0] sum;   // Result
  input              clk;   // Clock

  reg  [WIDTH1-1:0] l1, l2; // LSBs of inputs
  wire [WIDTH1-1:0] q1, v1, s1; // LSBs of inputs
  reg  [WIDTH2-1:0] l3, l4; // Middle bits
  wire [WIDTH2-1:0] q2, h2, v2, s2; // Middle bits
  reg  [WIDTH3-1:0] l5, l6; // MSBs of input
  wire [WIDTH3-1:0] q3, h3, v3, s3; // MSBs of input
  wire [WIDTH-1:0]  s;    // Output register
  wire cq1, cq2, cv2; // Carry signals
  wire ena, aset, sclr, sset, aload, sload, aclr;
                                    // Auxiliary FF signals
  assign ena=1; assign aclr=0; assign aset=0;
  assign sclr=0; assign sset=0; assign aload=0;
  assign sload=0;                          // Default for FF

  // Split in MSBs and LSBs and store in registers
  always @(posedge clk) begin
    // Split LSBs from input x,y
    l1[WIDTH1-1:0]  <= x[WIDTH1-1:0];
    l2[WIDTH1-1:0]  <= y[WIDTH1-1:0];
    // Split middle bits from input x,y
    l3[WIDTH2-1:0]  <= x[WIDTH2-1+WIDTH1:WIDTH1];
    l4[WIDTH2-1:0]  <= y[WIDTH2-1+WIDTH1:WIDTH1];
    // Split MSBs from input x,y
    l5[WIDTH3-1:0]  <= x[WIDTH3-1+WIDTH12:WIDTH12];
    l6[WIDTH3-1:0]  <= y[WIDTH3-1+WIDTH12:WIDTH12];
```

```
  end

//************** First stage of the adder  ****************
  csa7 add_1        // Add LSBs of x and y
  ( .a(l1), .b(l2), .clock(clk), .s(q1), .c(cq1));

  csa7 add_2        // Add LSBs of x and y
  ( .a(l3), .b(l4), .clock(clk), .s(q2), .c(cq2) );

        add_ff8 add_3             // Add MSBs of x and y
        ( .a(l5), .b(l6), .clock(clk), .s(q3));

//************* Second stage of the adder *****************
  // Two operands are zero
  assign h2 = {WIDTH2{1'b0}};
  assign h3 = {WIDTH3{1'b0}};

  lpm_ff reg_1                  // Save q1
  ( .data(q1), .q(v1), .clock(clk));  // Used ports
//    .enable(ena), .aclr(aclr), .aset(aset), .sclr(sclr),
//    .sset(sset), .aload(aload), .sload(sload)); // Unused
  defparam reg_1.lpm_width = WIDTH1;

// Add result of middle bits (x+y) and carry from LSBs
  csa7cin add_4
  (.a(q2), .b(h2), .cin(cq1), .clock(clk), .s(v2),.c(cv2));

// Add result of MSBs bits (x+y) and carry from middle
  add_ff8cin add_5
  ( .a(q3), .b(h3), .cin(cq2), .clock(clk), .s(v3));

//************* Third stage of the adder *****************
  lpm_ff reg_2                  // Save v1
  ( .data(q1), .q(s1), .clock(clk));  // Used ports
//  .enable(ena), .aclr(aclr), .aset(aset), .sclr(sclr),
//  .sset(sset), .aload(aload), .sload(sload)); // Unused
    defparam reg_2.lpm_width = WIDTH1;

  lpm_ff reg_3                  // Save v2
  ( .data(v2), .q(s2), .clock(clk));  // Used ports
//  .enable(ena), .aclr(aclr), .aset(aset), .sclr(sclr),
//  .sset(sset), .aload(aload), .sload(sload)); // Unused
    defparam reg_3.lpm_width = WIDTH1;
```

```
// Add result of MSBs bits (x+y) and 2. carry from middle
  add_ff8cin add_6
  ( .a(v3), .b(h3), .cin(cv2), .clock(clk), .s(s3));

// Build a single output word of WIDTH=WIDTH1+WIDHT2+WIDTH3
  assign s ={s3[WIDTH3-1:0],s2[WIDTH2-1:0],s1[WIDTH1-1:0]};

  assign sum = s;    // Connect s to output pins

endmodule

//*********************************************************
// IEEE STD 1364-1995 Verilog file: add_3p.v
// Author-EMAIL: Uwe.Meyer-Baese@ieee.org
//*********************************************************
// 29-bit adder with three pipeline stages
// uses four components: csa7.v; csa7cin.v;
//                  add_ff8.v; add_ff8cin.v

//`include "220model.v"
//`include "csa7.v"
//`include "csa7cin.v"
//`include "add_ff8.v"
//`include "add_ff8cin.v"

module add_3p (x, y, sum, clk);
  parameter WIDTH    = 29, // Total bit width
            WIDTH0   = 7,  // Bit width of LSBs
            WIDTH1   = 7,  // Bit width of 2. LSBs
            WIDTH01  = 14, // Sum WIDTH0+WIDTH1
            WIDTH2   = 7,  // Bit width of 2. MSBs
            WIDTH012 = 21,  // Sum WIDTH0+WIDTH1+WIDTH2
            WIDTH3   =  8;  // Bit width of MSBs

  input  [WIDTH-1:0] x,y;  // Inputs
  output [WIDTH-1:0] sum;  // Result
  input              clk;  // Clock

  reg  [WIDTH0-1:0] l0, l1; // LSBs of inputs
  wire [WIDTH0-1:0] q0, v0, r0, s0; // LSBs of inputs
  reg  [WIDTH1-1:0] l2, l3; // 2. LSBs of input
  wire [WIDTH1-1:0] q1, v1, r1, s1; // 2. LSBs of input
  reg  [WIDTH2-1:0] l4, l5; // 2. MSBs bits
  wire [WIDTH2-1:0] q2, v2, r2, s2, h7; // 2. MSBs bits
  reg  [WIDTH3-1:0] l6, l7; // MSBs of input
```

```
  wire [WIDTH3-1:0] q3, v3, r3, s3, h8; // MSBs of input
  wire [WIDTH-1:0]  s;    // Output register
  wire cq0, cq1, cq2, cv1, cv2, cr2; // Carry signals
  wire ena, aset, sclr, sset, aload, sload, aclr;
                                      // Auxiliary FF signals

  assign ena=1; assign aclr=0; assign aset=0;
  assign sclr=0; assign sset=0; assign aload=0;
  assign sload=0;                             // Default for FF

// Split in MSBs and LSBs and store in registers
always @(posedge clk) begin
  // Split LSBs from input x,y
  l0[WIDTH0-1:0] <= x[WIDTH0-1:0];
  l1[WIDTH0-1:0] <= y[WIDTH0-1:0];
  // Split 2. LSBs from input x,y
  l2[WIDTH1-1:0] <= x[WIDTH1-1+WIDTH0:WIDTH0];
  l3[WIDTH1-1:0] <= y[WIDTH1-1+WIDTH0:WIDTH0];
  // Split 2. MSBs from input x,y
  l4[WIDTH2-1:0] <= x[WIDTH2-1+WIDTH01:WIDTH01];
  l5[WIDTH2-1:0] <= y[WIDTH2-1+WIDTH01:WIDTH01];
  // Split MSBs from input x,y
  l6[WIDTH3-1:0] <= x[WIDTH3-1+WIDTH012:WIDTH012];
  l7[WIDTH3-1:0] <= y[WIDTH3-1+WIDTH012:WIDTH012];
end

//************* First stage of the adder  ****************
  csa7 add_0                    // Add LSBs of x and y
  ( .a(l0), .b(l1), .clock(clk), .s(q0), .c(cq0));
  csa7 add_1                    // Add 2. LSBs of x and y
  ( .a(l2), .b(l3), .clock(clk), .s(q1), .c(cq1) );
  csa7 add_2                    // Add 2. MSBs of x and y
  ( .a(l4), .b(l5), .clock(clk), .s(q2), .c(cq2) );
  add_ff8 add_3                // Add MSBs of x and y
  ( .a(l6), .b(l7), .clock(clk), .s(q3) );

//************* Second stage of the adder ****************
  // Two operands are zero
  assign h7 = {WIDTH2{1'b0}};
  assign h8 = {WIDTH3{1'b0}};

  lpm_ff reg_1                  // Save q0
  ( .data(q0), .q(v0), .clock(clk)); // Used ports
//  .enable(ena), .aclr(aclr), .aset(aset), .sclr(sclr),
```

```
//   .sset(sset), .aload(aload), .sload(sload)); // Unused
     defparam reg_1.lpm_width = WIDTH0;

// Add result of 2. LSBs (x+y) and carry from LSBs
   csa7cin add_4
   (.a(q1), .b(h7), .cin(cq0), .clock(clk), .s(v1),.c(cv1));

// Add result of 2. MSBs (x+y) and carry from 2. LSBs
   csa7cin add_5
  (.a(q2), .b(h7), .cin(cq1), .clock(clk), .s(v2), .c(cv2));

// Add result of MSBs (x+y) and carry from 2. MSBs
   add_ff8cin add_6
   ( .a(q3), .b(h8), .cin(cq2), .clock(clk), .s(v3));

//************** Third stage of the adder ****************
lpm_ff reg_2                        // Save v0
( .data(v0), .q(r0), .clock(clk));  // Used ports
//   .enable(ena), .aclr(aclr), .aset(aset), .sclr(sclr),
//   .sset(sset), .aload(aload), .sload(sload)); // Unused
     defparam reg_2.lpm_width = WIDTH0;

   lpm_ff reg_3                        // Save v1
   ( .data(v1), .q(r1), .clock(clk));  // Used ports
//   .enable(ena), .aclr(aclr), .aset(aset), .sclr(sclr),
//   .sset(sset), .aload(aload), .sload(sload)); // Unused
     defparam reg_3.lpm_width = WIDTH1;

// Add result of 2. MSBs (x+y) and carry from 2. LSBs
csa7cin add_7
( .a(v2), .b(h7), .cin(cv1), .clock(clk), .s(r2), .c(cr2));

// Add result of MSBs (x+y) and carry from 2. MSBs
add_ff8cin add_8
( .a(v3), .b(h8), .cin(cv2), .clock(clk), .s(r3) );

//************ Fourth stage of the adder *****************
   lpm_ff reg_4                        // Save r0
   ( .data(r0), .q(s0), .clock(clk)); // Used ports
//   .enable(ena), .aclr(aclr), .aset(aset), .sclr(sclr),
//   .sset(sset), .aload(aload), .sload(sload)); //Unused
     defparam reg_4.lpm_width = WIDTH0;

   lpm_ff reg_5                        // Save r1
```

```
  ( .data(r1), .q(s1), .clock(clk));  // Used ports
//  .enable(ena), .aclr(aclr), .aset(aset), .sclr(sclr),
//  .sset(sset), .aload(aload), .sload(sload)); //Unused
    defparam reg_5.lpm_width = WIDTH1;

  lpm_ff reg_6                      // Save r2
  ( .data(r2), .q(s2), .clock(clk));  // Used ports
//  .enable(ena), .aclr(aclr), .aset(aset), .sclr(sclr),
//  .sset(sset), .aload(aload), .sload(sload)); //Unused
    defparam reg_6.lpm_width = WIDTH2;

// Add result of MSBs (x+y) and carry from 2. MSBs
add_ff8cin add_9
( .a(r3), .b(h8), .cin(cr2), .clock(clk), .s(s3));

// Build a single output word of
// WIDTH = WIDTH0 + WIDTH1 + WIDHT2 + WIDTH3
assign s = {s3[WIDTH3-1:0], s2[WIDTH2-1:0],
                            s1[WIDTH1-1:0],s0[WIDTH0-1:0]};

assign sum = s ;    // Connect s to output pins

endmodule

//*********************************************************
// IEEE STD 1364-1995 Verilog file: mul_ser.v
// Author-EMAIL: Uwe.Meyer-Baese@ieee.org
//*********************************************************
module mul_ser (clk, x, a, y);  //----> Interface

  input          clk;
  input  [7:0]   x, a;
  output [15:0]  y;
  reg    [15:0]  y;


  always @(posedge clk) //-> Multiplier in behavioral style
  begin : States
    parameter s0=0, s1=1, s2=2;
    reg [2:0] count;
    reg [1:0] state;
    reg  [15:0] p, t;        // Double bit width
    reg  [7:0] a_reg;
    case (state)
      s0 : begin          // Initialization step
```

```
        a_reg <= a;
        state <= s1;
        count = 0;
        p <= 0;              // Product register reset
        t <= {{8{x[7]}},x}; // Set temporary shift register
      end                                            // to x
      s1 : begin           // Processing step
        if (count == 7)    // Multiplication ready
          state <= s2;
        else          // Note that MaxPlusII does not does
          begin       // not allow variable bit selects,
          if (a_reg[0] == 1) // see (LRM Sec. 4.2.1)
            p <= p + t;      // Add 2^k
          a_reg <= a_reg >> 1;// Use LSB for the bit select
          t <= t << 1;
          count = count + 1;
          state <= s1;
        end
      end
      s2 : begin        // Output of result to y and
        y <= p;         // start next multiplication
        state <= s0;
      end
    endcase
  end

endmodule

//*********************************************************
// IEEE STD 1364-1995 Verilog file: cordic.v
// Author-EMAIL: Uwe.Meyer-Baese@ieee.org
//*********************************************************
module cordic (clk, x_in , y_in, r, phi, eps);

  parameter W = 7;  // Bit width - 1
  input         clk;
  input  [W:0] x_in, y_in;
  output [W:0] r, phi, eps;
  reg    [W:0] r, phi, eps;

// There is no bit access in 2D array types
// in Verilog, therefore use single vectors
  reg [W:0] x0, y0, z0;
  reg [W:0] x1, y1, z1;
  reg [W:0] x2, y2, z2;
```

```
reg [W:0] x3, y3, z3;

always @(posedge clk) begin //----> Infer register
  if (x_in > 0)              // Test for x_in < 0 rotate
    begin                    // 0, +90, or -90 degrees
    x0 <= x_in; // Input in register 0
    y0 <= y_in;
    z0 <= 0;
    end
  else if (y_in > 0)
    begin
    x0 <= y_in;
    y0 <= - x_in;
    z0 <= 90;
    end
  else
    begin
    x0 <= - y_in;
    y0 <= x_in;
    z0 <= -90;
    end

  if (y0 > 0)                   // Rotate 45 degrees
    begin
    x1 <= x0 + y0;
    y1 <= y0 - x0;
    z1 <= z0 + 45;
    end
  else
    begin
    x1 <= x0 - y0;
    y1 <= y0 + x0;
    z1 <= z0 - 45;
    end

  if (y1 > 0)                  // Rotate 26 degrees
    begin
    x2 <= x1 + {y1[W],y1[W:1]}; // i.e. x1 + y1 /2
    y2 <= y1 - {x1[W],x1[W:1]}; // i.e. y1 - x1 /2
    z2 <= z1 + 26;
    end
  else
    begin
    x2 <= x1 - {y1[W],y1[W:1]}; // i.e. x1 - y1 /2
```

```
      y2 <= y1 + {x1[W],x1[W:1]}; // i.e. y1 + x1 /2
      z2 <= z1 - 26;
      end

    if (y2 > 0)                          // Rotate 14 degrees
      begin
        x3 <= x2 + {y2[W],y2[W],y2[W:2]}; // i.e. x2 + y2/4
        y3 <= y2 - {x2[W],x2[W],x2[W:2]}; // i.e. y2 - x2/4
        z3 <= z2 + 14;
      end
    else
      begin
        x3 <= x2 - {y2[W],y2[W],y2[W:2]}; // i.e. x2 - y2/4
        y3 <= y2 + {x2[W],x2[W],x2[W:2]}; // i.e. y2 + x2/4
        z3 <= z2 - 14;
      end

    r   <= x3;
    phi <= z3;
    eps <= y3;
  end

endmodule

//*********************************************************
// IEEE STD 1364-1995 Verilog file: fir_gen.v
// Author-EMAIL: Uwe.Meyer-Baese@ieee.org
//*********************************************************
// This is a generic FIR filter generator
// It uses W1 bit data/coefficients bits
module fir_gen (clk, Load_x, x_in, c_in, y_out);

  parameter W1 = 9,   // Input bit width
            W2 = 18,  // Multiplier bit width 2*W1
            W3 = 19,  // Adder width = W2+log2(L)-1
            W4 = 11,  // Output bit width
            L  = 4,   // Filter length
            Mpipe = 3; // Pipeline steps of multiplier
  input clk, Load_x;  // std_logic
  input [W1-1:0] x_in, c_in;  // Inputs
  output [W3-1:0] y_out;  // Results

  reg [W1-1:0]  x;
  wire [W3-1:0]  y;
// 2D array types i.e. memories not supported by MaxPlusII
```

```
// in Verilog, use therefore single vectors
  reg  [W1-1:0] c0, c1, c2, c3; // Coefficient array
  wire [W2-1:0] p0, p1, p2, p3; // Product array
  reg  [W3-1:0] a0, a1, a2, a3; // Adder array

  wire  [W2-1:0] sum;  // Auxilary signals
  wire  clken, aclr;

  assign sum=0; assign aclr=0; // Default for mult
  assign clken=0;

//----> Load Data or Coefficient
  always @(posedge clk)
    begin: Load
    if (! Load_x) begin
      c3 <= c_in; // Store coefficient in register
      c2 <= c3;   // Coefficients shift one
      c1 <= c2;
      c0 <= c1;
      end
    else begin
      x <= x_in; // Get one data sample at a time
    end
  end

//----> Compute sum-of-products
  always @(posedge clk)
    begin: SOP
  // Compute the transposed filter additions
    a0 <= {p0[W2-1], p0} + a1;
    a1 <= {p1[W2-1], p1} + a2;
    a2 <= {p2[W2-1], p2} + a3;
    a3 <= {p3[W2-1], p3}; // First TAP has only a register
  end
  assign y = a0;

// Instantiate L pipelined multiplier
  lpm_mult mul_0            // Multiply  x*c0 = p0
    (.clock(clk), .dataa(x), .datab(c0), .result(p0));
//   .sum(sum), .clken(clken), .aclr(aclr)); // Unused ports
    defparam mul_0.lpm_widtha = W1;
    defparam mul_0.lpm_widthb = W1;
    defparam mul_0.lpm_widthp = W2;
    defparam mul_0.lpm_widths = W2;
```

```
    defparam mul_0.lpm_pipeline = Mpipe;
    defparam mul_0.lpm_representation = "SIGNED";

  lpm_mult mul_1               // Multiply  x*c1 = p1
    (.clock(clk), .dataa(x), .datab(c1), .result(p1));
//   .sum(sum), .clken(clken), .aclr(aclr)); // Unused ports
    defparam mul_1.lpm_widtha = W1;
    defparam mul_1.lpm_widthb = W1;
    defparam mul_1.lpm_widthp = W2;
    defparam mul_1.lpm_widths = W2;
    defparam mul_1.lpm_pipeline = Mpipe;
    defparam mul_1.lpm_representation = "SIGNED";

  lpm_mult mul_2               // Multiply  x*c2 = p2
    (.clock(clk), .dataa(x), .datab(c2), .result(p2));
//   .sum(sum), .clken(clken), .aclr(aclr)); // Unused ports
    defparam mul_2.lpm_widtha = W1;
    defparam mul_2.lpm_widthb = W1;
    defparam mul_2.lpm_widthp = W2;
    defparam mul_2.lpm_widths = W2;
    defparam mul_2.lpm_pipeline = Mpipe;
    defparam mul_2.lpm_representation = "SIGNED";

  lpm_mult mul_3               // Multiply  x*c3 = p3
    (.clock(clk), .dataa(x), .datab(c3), .result(p3));
//   .sum(sum), .clken(clken), .aclr(aclr)); // Unused ports
    defparam mul_3.lpm_widtha = W1;
    defparam mul_3.lpm_widthb = W1;
    defparam mul_3.lpm_widthp = W2;
    defparam mul_3.lpm_widths = W2;
    defparam mul_3.lpm_pipeline = Mpipe;
    defparam mul_3.lpm_representation = "SIGNED";

  assign y_out = y[W3-1:W3-W4];

endmodule

//*********************************************************
// IEEE STD 1364-1995 Verilog file: fir_srg.v
// Author-EMAIL: Uwe.Meyer-Baese@ieee.org
//*********************************************************
module fir_srg (clk, x, y);  //----> Interface

  input        clk;
  input  [7:0] x;
```

```
  output [7:0] y;
  reg    [7:0] y;
// Tapped delay line array of bytes
  reg    [7:0] tap0, tap1, tap2, tap3;
// For bit access use single vectors in Verilog

  always @(posedge clk)  //----> Behavioral Style
  begin : p1
   // Compute output y with the filter coefficients weight.
   // The coefficients are [-1  3.75  3.75  -1].
   // Multiplication and division for Altera MaxPlusII can
   // be done in Verilog with sign extensions and shifts!
    y <= (tap1<<1) + tap1 + {tap1[7],tap1[7:1]}
         + {tap1[7],tap1[7],tap1[7:2]} + (tap2<<1) + tap2
         + {tap2[7],tap2[7:1]}
         + {tap2[7],tap2[7],tap2[7:2]} - tap3 - tap0;

    tap3 <= tap2;  // Tapped delay line: shift one
    tap2 <= tap1;
    tap1 <= tap0;
    tap0 <= x;    // Input in register 0
  end

endmodule

//*********************************************************
// IEEE STD 1364-1995 Verilog file: dafsm.v
// Author-EMAIL: Uwe.Meyer-Baese@ieee.org
//*********************************************************
`include "case3.v" // User defined component

module dafsm (clk, x_in0, x_in1, x_in2, y); //--> Interface

  input         clk;
  input  [2:0]  x_in0, x_in1, x_in2;
  output [5:0]  y;
  reg    [5:0]  y;
  reg    [2:0]  x0, x1, x2;
  wire   [2:0]  table_in, table_out;

  reg [5:0] p;  // temporary register

  assign table_in[0] = x0[0];
  assign table_in[1] = x1[0];
  assign table_in[2] = x2[0];
```

```
  always @(posedge clk)  //----> DA in behavioral style
  begin : DA
    parameter s0=0, s1=1;
    reg [0:0] state;
    reg [1:0] count;   // Counts the shifts
    case (state)
      s0 : begin       // Initialization
        state <= s1;
        count = 0;
        p  <= {6{1'b0}};
        x0 <= x_in0;
        x1 <= x_in1;
        x2 <= x_in2;
      end
      s1 : begin                  // Processing step
        if (count == 3) begin     // Is sum of product done?
          y <= p;                // Output of result to y and
          state <= s0;           // start next sum of product
        end
        else begin
          p <= {p[5],p[5:1]} + {1'b0,table_out,2'b00};
          x0[0] <= x0[1];
          x0[1] <= x0[2];
          x1[0] <= x1[1];
          x1[1] <= x1[2];
          x2[0] <= x2[1];
          x2[1] <= x2[2];
          count = count + 1;
          state <= s1;
        end
      end
    endcase
  end

  case3 LC_Table0
  ( .table_in(table_in), .table_out(table_out));

endmodule

//*********************************************************
// IEEE STD 1364-1995 Verilog file: case3.v
// Author-EMAIL: Uwe.Meyer-Baese@ieee.org
//*********************************************************
module case3 (table_in, table_out);
```

```
  input  [2:0] table_in;  // Three bit
  output [2:0] table_out; // Range 0 to 6

  reg [2:0] table_out;

// This is the DA CASE table for
// the 3 coefficients: 2, 3, 1

  always @(table_in)
  begin
    case (table_in)
      0 :     table_out =  0;
      1 :     table_out =  2;
      2 :     table_out =  3;
      3 :     table_out =  5;
      4 :     table_out =  1;
      5 :     table_out =  3;
      6 :     table_out =  4;
      7 :     table_out =  6;
      default : ;
    endcase
  end

endmodule

//*********************************************************
// IEEE STD 1364-1995 Verilog file: case5p.v
// Author-EMAIL: Uwe.Meyer-Baese@ieee.org
//*********************************************************
module case5p (clk, table_in, table_out);

  input        clk;
  input  [4:0] table_in;
  output [4:0] table_out;     // range 0 to 25

  reg [4:0] table_out;
  reg [3:0] lsbs;
  reg [1:0] msbs0;
  reg [4:0] table0out00, table0out01;

// These are the distributed arithmetic CASE tables for
// the 5 coefficients: 1, 3, 5, 7, 9

  always @(posedge clk) begin
```

```
      lsbs[0] = table_in[0];
      lsbs[1] = table_in[1];
      lsbs[2] = table_in[2];
      lsbs[3] = table_in[3];
      msbs0[0] = table_in[4];
      msbs0[1] = msbs0[0];
    end

  // This is the final DA MPX stage.
    always @(posedge clk) begin
      case (msbs0[1])
        0 : table_out <= table0out00;
        1 : table_out <= table0out01;
        default : ;
      endcase
    end

  // This is the DA CASE table 00 out of 1.
    always @(posedge clk) begin
      case (lsbs)
        0  : table0out00 = 0;
        1  : table0out00 = 1;
        2  : table0out00 = 3;
        3  : table0out00 = 4;
        4  : table0out00 = 5;
        5  : table0out00 = 6;
        6  : table0out00 = 8;
        7  : table0out00 = 9;
        8  : table0out00 = 7;
        9  : table0out00 = 8;
        10 : table0out00 = 10;
        11 : table0out00 = 11;
        12 : table0out00 = 12;
        13 : table0out00 = 13;
        14 : table0out00 = 15;
        15 : table0out00 = 16;
        default ;
      endcase
    end

  // This is the DA CASE table 01 out of 1.
    always @(posedge clk) begin
      case (lsbs)
        0  : table0out01 = 9;
```

```
      1  : table0out01 = 10;
      2  : table0out01 = 12;
      3  : table0out01 = 13;
      4  : table0out01 = 14;
      5  : table0out01 = 15;
      6  : table0out01 = 17;
      7  : table0out01 = 18;
      8  : table0out01 = 16;
      9  : table0out01 = 17;
      10 : table0out01 = 19;
      11 : table0out01 = 20;
      12 : table0out01 = 21;
      13 : table0out01 = 22;
      14 : table0out01 = 24;
      15 : table0out01 = 25;
      default ;
    endcase
  end

endmodule

//*********************************************************
// IEEE STD 1364-1995 Verilog file: darom.v
// Author-EMAIL: Uwe.Meyer-Baese@ieee.org
//*********************************************************
//`include "220model.v"

module darom (clk, x_in0, x_in1, x_in2, y); //--> Interface

  input          clk;
  input   [2:0]  x_in0, x_in1, x_in2;
  output  [5:0]  y;
  reg     [5:0]  y;
  reg     [2:0]  x0, x1, x2;
  wire    [2:0]  table_in, table_out;

  reg [5:0] p;  // Temporary register
  wire ena;

  assign ena=1;

  assign table_in[0] = x0[0];
  assign table_in[1] = x1[0];
  assign table_in[2] = x2[0];
```

```
  always @(posedge clk)  //----> DA in behavioral style
  begin : DA
    parameter s0=0, s1=1;
    reg [0:0] state;
    reg [1:0] count;   // Counts the shifts
    case (state)
      s0 : begin  // Initialization
        state <= s1;
        count = 0;
        p  <= 0;
        x0 <= x_in0;
        x1 <= x_in1;
        x2 <= x_in2;
      end
      s1 : begin                    // Processing step
        if (count == 3) begin   // Is sum of product done?
          y <= p;                   // Output of result to y and
          state <= s0;              // start next sum of product
        end
        else begin
          p <= {p[5],p[5:1]} + {1'b0,table_out,2'b00};
          x0[0] <= x0[1];
          x0[1] <= x0[2];
          x1[0] <= x1[1];
          x1[1] <= x1[2];
          x2[0] <= x2[1];
          x2[1] <= x2[2];
          count = count + 1;
          state <= s1;
        end
      end
      default : ;
    endcase
  end

  lpm_rom rom_1
  (.address(table_in), .q(table_out));        // Used ports
// .inclock(clk), .outclock(clk), .memenab(ena)); // Unused
    defparam rom_1.lpm_width = 3;
    defparam rom_1.lpm_widthad = 3;
    defparam rom_1.lpm_outdata = "UNREGISTERED";
    defparam rom_1.lpm_address_control = "UNREGISTERED";
    defparam rom_1.lpm_file = "darom3.mif";
```

```
endmodule

//*********************************************************
// IEEE STD 1364-1995 Verilog file: dasign.v
// Author-EMAIL: Uwe.Meyer-Baese@ieee.org
//*********************************************************
`include "case3s.v" // User defined component

module dasign (clk, x_in0, x_in1, x_in2, y); //-> Interface

  input         clk;
  input  [3:0]  x_in0, x_in1, x_in2;
  output [6:0]  y;
  reg    [6:0]  y;
  reg    [3:0]  x0, x1, x2;
  wire   [2:0]  table_in;
  wire   [3:0]  table_out;

  reg [6:0] p;  // Temporary register

  assign table_in[0] = x0[0];
  assign table_in[1] = x1[0];
  assign table_in[2] = x2[0];

  always @(posedge clk)  //----> DA in behavioral style
  begin : DA
    parameter s0=0, s1=1;
    integer k;
    reg [0:0] state;
    reg [2:0] count;            // Counts the shifts
    case (state)
      s0 : begin                // Initialization step
        state <= s1;
        count = 0;
        p  <= 0;
        x0 <= x_in0;
        x1 <= x_in1;
        x2 <= x_in2;
      end
      s1 : begin                // Processing step
        if (count == 4) begin   // Is sum of product done?
          y <= p;               // Output of result to y and
          state <= s0;          // start next sum of product
        end
        else begin    // Subtract for last accumulator step
```

```
          if (count ==3)    // i.e. p/2 +/- table_out * 8
            p <= {p[6],p[6:1]} - (table_out << 3);
          else              // Accumulation for all other steps
            p <= {p[6],p[6:1]} + (table_out << 3);
          for (k=0; k<=2; k= k+1) begin     // Shift bits
            x0[k] <= x0[k+1];
            x1[k] <= x1[k+1];
            x2[k] <= x2[k+1];
          end
          count = count + 1;
          state <= s1;
        end
      end
    endcase
  end

  case3s LC_Table0
  ( .table_in(table_in), .table_out(table_out));

endmodule

//*********************************************************
// IEEE STD 1364-1995 Verilog file: case3s.v
// Author-EMAIL: Uwe.Meyer-Baese@ieee.org
//*********************************************************
module case3s (table_in, table_out);

  input  [2:0] table_in;  // Three bit
  output [3:0] table_out; // Range -2 to 4 -> 3 + sign bit

  reg [3:0] table_out;

// This is the DA CASE table for
// the 3 coefficients: -2, 3, 1

  always @(table_in)
  begin
    case (table_in)
      0 :     table_out =  0;
      1 :     table_out =  -2;
      2 :     table_out =  3;
      3 :     table_out =  1;
      4 :     table_out =  1;
      5 :     table_out =  -1;
      6 :     table_out =  4;
```

```
      7 :     table_out =  2;
      default : ;
    endcase
  end

endmodule

//*********************************************************
// IEEE STD 1364-1995 Verilog file: dapara.v
// Author-EMAIL: Uwe.Meyer-Baese@ieee.org
//*********************************************************
`include "case3s.v" // User defined component

module dapara (clk, x_in, y); //----> Interface

  input        clk;
  input  [3:0] x_in;
  output [6:0] y;
  reg    [6:0] y;

  reg  [2:0] x0, x1, x2, x3;
  wire [3:0] y0, y1, y2, y3;
  reg  [4:0] s0, s1;
  reg  [3:0] t0, t1, t2, t3;

  always @(posedge clk)  //----> DA in behavioral style
  begin : DA
    integer k;
    for (k=0; k<=1; k=k+1) begin     // Shift all four bits
       x0[k] <= x0[k+1];
       x1[k] <= x1[k+1];
       x2[k] <= x2[k+1];
       x3[k] <= x3[k+1];
    end
    x0[2] <= x_in[0];     // Load x_in in the
    x1[2] <= x_in[1];     // MSBs of register 2
    x2[2] <= x_in[2];
    x3[2] <= x_in[3];
    y <= {{3{y0[3]}},y0} + {{2{y1[3]}},y1,1'b0}
         + {y2[3],y2,2'b00} - (y3 << 3);
// Sign extensions, pipeline register, and adder tree:
//  t0 <= y0; t1 <= y1; t2 <= y2; t3 <= y3;
//  s0 <= {t0[3],t0} + (t1 << 1);
//  s1 <= {t2[3],t2} - (t3 <<1);
//  y  <= {{2{s0[4]}},s0} + (s1 << 2);
```

```
  end

  case3s LC_Table0 ( .table_in(x0), .table_out(y0));
  case3s LC_Table1 ( .table_in(x1), .table_out(y1));
  case3s LC_Table2 ( .table_in(x2), .table_out(y2));
  case3s LC_Table3 ( .table_in(x3), .table_out(y3));

endmodule

//*********************************************************
// IEEE STD 1364-1995 Verilog file: iir.v
// Author-EMAIL: Uwe.Meyer-Baese@ieee.org
//*********************************************************
module iir ( x_in,       // Input
             y_out,      // Result
             clk);
  parameter W = 14;     // Bit width - 1
  input  [W:0] x_in;
  output [W:0] y_out;
  input         clk;

  reg [W:0] x, y;

// initial begin
//  y=0;
//  x=0;
// end

// Use FFs for input and recursive part
always @(posedge clk) begin    // Note: there is no signed
  x  <= x_in;                  // integer in Verilog
  y  <= x + {y[W],y[W:1]} + {{2{y[W]}},y[W:2]};
                                 // i.e. x + y / 2 + y / 4;
end

assign  y_out = y;            // Connect y to output pins

endmodule

//*********************************************************
// IEEE STD 1364-1995 Verilog file: iir_pipe.v
// Author-EMAIL: Uwe.Meyer-Baese@ieee.org
//*********************************************************
module iir_pipe (x_in, y_out, clk); //----> Interface
```

```
  parameter W = 14; // Bit width - 1
  input          clk;
  input  [W:0]  x_in;    // Input
  output [W:0]  y_out;   // Result

  reg [W:0] x, x3, sx;
  reg [W:0] y, y9;

  always @(posedge clk)  // Infer FFs for input, output and
  begin                  // pipeline stages;
    x   <= x_in;         // use non-blocking FF assignments
    x3  <= {x[W],x[W:1]} + {x[W],x[W],x[W:2]};
                              // i.e. x / 2 + x / 4 = x*3/4
    sx  <= x + x3; // Sum of x element i.e. output FIR part
    y9  <= {y[W],y[W:1]} + {{4{y[W]}},y[W:4]};
                             // i.e. y / 2 + y / 16 = y*9/16
    y   <= sx + y9;                      // Compute output
  end

  assign y_out = y ;   // Connect register y to output pins

endmodule

//*********************************************************
// IEEE STD 1364-1995 Verilog file: iir_par.v
// Author-EMAIL: Uwe.Meyer-Baese@ieee.org
//*********************************************************
module iir_par (clk, x_in, clk2, y_out); //----> Interface

  parameter W = 14; // bit width - 1
  input          clk;
  input  [W:0]  x_in;
  output [W:0]  y_out;
  output         clk2;

  reg [W:0] x_even, x_odd, xd_odd, x_wait;
  reg [W:0] y_even, y_odd, y_wait, y;
  reg [W:0] x_e, x_o, y_e, y_o;
  reg [W:0] sum_x_even, sum_x_odd;
  reg        clk_div2;

  always @(posedge clk)        // Split x into even
  begin : Multiplex            // and odd samples;
    parameter even=0, odd=1;   // recombine y at clk rate
```

```
    reg [0:0] state;
    case (state)
      even : begin
          x_even <= x_in;
          x_odd <= x_wait;
          clk_div2 <= 1;
          y <= y_wait;
          state <= odd;
      end
      odd : begin
         x_wait <= x_in;
         y <= y_odd;
         y_wait <= y_even;
         clk_div2 <= 0;
         state <= even;
      end
    endcase
  end

  assign y_out = y;
  assign clk2  = clk_div2;

  always @(negedge clk_div2)
  begin: Arithmetic
    sum_x_even <= x_odd + {x_even[W],x_even[W:1]}
                        + {x_even[W],x_even[W],x_even[W:2]};
                      // i.e. x_odd + x_even / 2 + x_even /4
    y_even <= sum_x_even + {y_even[W],y_even[W:1]}
                             + {{4{y_even[W]}},y_even[W:4]};
                // i.e. sum_x_even + y_even / 2 + y_even /16
    xd_odd <= x_odd;
    sum_x_odd <= x_even + {xd_odd[W],xd_odd[W:1]}
                        + {xd_odd[W],xd_odd[W],xd_odd[W:2]};
                     // i.e. x_even + xd_odd / 2 + xd_odd /4
    y_odd  <= sum_x_odd + {y_odd[W],y_odd[W:1]}
                               + {{4{y_odd[W]}},y_odd[W:4]};
                  // i.e. sum_x_odd + y_odd / 2 + y_odd / 16
  end

endmodule

//*********************************************************
// IEEE STD 1364-1995 Verilog file: cic3r32.v
// Author-EMAIL: Uwe.Meyer-Baese@ieee.org
//*********************************************************
```

```
module cic3r32 (clk, x_in, y_out); //----> Interface

  input        clk;
  input  [7:0] x_in;
  output [8:0] y_out;
  reg    [7:0] y;

  parameter hold=0, sample=1;
  reg [1:0] state;
  reg [4:0]  count;
  reg  clk2;
  reg  [7:0]  x;      // Registered input
  wire [25:0] sxtx;   // Sign extended input
  reg  [25:0] i0, i1 , i2;         // I section  0, 1, and 2
  reg  [25:0] i2d1, i2d2, i2d3, i2d4, c1, c0; // I + COMB 0
  reg  [25:0] c1d1, c1d2, c1d3, c1d4, c2; // COMB section 1
  reg  [25:0] c2d1, c2d2, c2d3, c2d4, c3; // COMB section 2

  always @(negedge clk)
  begin : FSM
    case (state)
      hold : begin
        if (count < 31)
           state <= hold;
        else
           state <= sample;
        end
      default :
        state <= hold;
    endcase
  end

  assign sxtx = {{18{x[7]}},x};

  always @(posedge clk)
  begin : Int
      x     <= x_in;
      i0    <= i0 + sxtx;
      i1    <= i1 + i0 ;
      i2    <= i2 + i1 ;
    case (state)
      sample : begin
        c0    <= i2;
        count <= 0;
```

```
      end
      default :
        count <= count + 1;
    endcase
    if ((count > 8) && (count < 16))
      clk2  <= 1;
    else
      clk2  <= 0;
  end

  always @(posedge clk2)
  begin : Comb
      i2d1 <= c0;
      i2d2 <= i2d1;
      c1   <= c0 - i2d2;
      c1d1 <= c1;
      c1d2 <= c1d1;
      c2   <= c1  - c1d2;
      c2d1 <= c2;
      c2d2 <= c2d1;
      c3   <= c2  - c2d2;
  end

  assign y_out = c3[25:17];

endmodule

//*********************************************************
// IEEE STD 1364-1995 Verilog file: cic3s32.v
// Author-EMAIL: Uwe.Meyer-Baese@ieee.org
//*********************************************************
module cic3s32 (clk, x_in, y_out); //----> Interface

  input         clk;
  input  [7:0]  x_in;
  output [8:0]  y_out;
  reg    [7:0]  y;

  parameter hold=0, sample=1;
  reg [1:0] state;
  reg [4:0]  count;
  reg  clk2;
  reg  [7:0]  x;                         // Registered input
  wire [25:0] sxtx;                      // Sign extended input
  reg  [25:0] i0;                        // I section 0
```

```
reg  [20:0] i1;                     // I section 1
reg  [15:0] i2;                     // I section 2
reg  [13:0] i2d1, i2d2, i2d3, i2d4, c1, c0; // I + COMB 0
reg  [12:0] c1d1, c1d2, c1d3, c1d4, c2; // COMB section 1
reg  [11:0] c2d1, c2d2, c2d3, c2d4, c3; // COMB section 2

always @(negedge clk)
begin : FSM
  case (state)
    hold : begin
      if (count < 31)
         state <= hold;
      else
         state <= sample;
      end
    default :
      state <= hold;
  endcase
end

assign sxtx = {{18{x[7]}},x};

always @(posedge clk)
begin : Int
    x   <= x_in;
    i0  <= i0 + sxtx;
    i1  <= i1 + i0[25:5];
    i2  <= i2 + i1[20:5];
  case (state)
    sample : begin
      c0    <= i2[15:2];
      count <= 0;
    end
    default :
      count <= count + 1;
  endcase
  if ((count > 8) && (count < 16))
    clk2  <= 1;
  else
    clk2  <= 0;
end

always @(posedge clk2)
begin : Comb
```

```
      i2d1 <= c0;
      i2d2 <= i2d1;
      c1   <= c0 - i2d2;
      c1d1 <= c1[13:1];
      c1d2 <= c1d1;
      c2   <= c1[13:1] - c1d2;
      c2d1 <= c2[12:1];
      c2d2 <= c2d1;
      c3   <= c2[12:1] - c2d2;
  end

  assign y_out = c3[11:3];

endmodule

//*********************************************************
// IEEE STD 1364-1995 Verilog file: db4poly.v
// Author-EMAIL: Uwe.Meyer-Baese@ieee.org
//*********************************************************
module db4poly (clk, x_in, clk2, x_e, x_o, g0, g1, y_out);

  input           clk;
  output          clk2;
  input  [7:0]    x_in;
  output [16:0]   x_e, x_o, g0, g1; // Test signals
  output [8:0]    y_out;

  reg   [7:0] x_odd, x_even, x_wait;
  wire [16:0] x_odd_sxt, x_even_sxt;
  reg   clk_div2;

// Register for multiplier, coefficients, and taps
  reg   [16:0] m0, m1, m2, m3, r0, r1, r2, r3;
  reg   [16:0] x33, x99, x107;
  reg   [16:0] y;

  always @(posedge clk)  // Split into even and odd
  begin : Multiplex      // samples at clk rate
    parameter even=0, odd=1;
    reg [0:0] state;
    case (state)
      even : begin
        x_even <= x_in;
        x_odd  <= x_wait;
        clk_div2 = 1;
```

```
        state <= odd;
      end
      odd : begin
        x_wait <= x_in;
        clk_div2 = 0;
        state <= even;
      end
    endcase
  end

  assign x_odd_sxt = {{9{x_odd[7]}},x_odd};
  assign x_even_sxt = {{9{x_even[7]}},x_even};

  always @(x_odd_sxt or x_even_sxt)
  begin : RAG
// Compute auxiliary multiplications of the filter
    x33  = (x_odd_sxt << 5) + x_odd_sxt;
    x99  = (x33 << 1) + x33;
    x107 = x99 + (x_odd_sxt << 3);
// Compute all coefficients for the transposed filter
    m0 = (x_even_sxt << 7) - (x_even_sxt << 2); // m0 = 124
    m1 = x107 << 1;                              // m1 = 214
    m2 = (x_even_sxt << 6) - (x_even_sxt << 3)
                              + x_even_sxt;  // m2 =  57
    m3 = x33;                                    // m3 = -33
  end

  always @(negedge clk_div2) // Infer registers;
  begin : AddPolyphase       // use non-blocking assignments
//---------- Compute filter G0
    r0 <=  r2 + m0;          // g0 = 128
    r2 <=  m2;               // g2 = 57
//---------- Compute filter G1
    r1 <=  -r3 + m1;         // g1 = 214
    r3 <=  m3;               // g3 = -33
// Add the polyphase components
    y <= r0 + r1;
  end

// Provide some test signal as outputs
  assign x_e = x_even;
  assign x_o = x_odd;
  assign clk2 = clk_div2;
  assign g0 = r0;
```

```
  assign g1 = r1;

  assign y_out = y[16:8]; // Connect y / 256 to output

endmodule

//*********************************************************
// IEEE STD 1364-1995 Verilog file: db4latti.v
// Author-EMAIL: Uwe.Meyer-Baese@ieee.org
//*********************************************************
module db4latti (clk, x_in, clk2, x_e, x_o, g, h);

  input           clk;
  output          clk2;
  input  [7:0]   x_in;
  output [16:0]   x_e, x_o;
  output [8:0]    g, h;
  reg    [8:0]    g, h;

  reg  [7:0]  x_wait;
  wire [16:0] x_wait_sxt, x_in_sxt;
  reg  [16:0] sx_up, sx_low;
  wire [24:0] sx_up_sxt, sx_low_sxt;
  reg  clk_div2;
  wire [16:0] sxa0_up, sxa0_low;
  wire [16:0] up0, up1, low1;
  reg  [16:0] low0;
  wire [24:0] up0_sxt, low0_sxt;

  assign x_in_sxt = {{9{x_in[7]}},x_in};
  assign x_wait_sxt = {{9{x_wait[7]}},x_wait};

  always @(posedge clk)     // Split into even and odd
  begin : Multiplex         // samples at clk rate
    parameter even=0, odd=1;
    reg [0:0] state;
    case (state)
      even : begin
        // Multiply with 256*s=124
        sx_up   <= (x_in_sxt << 7) - (x_in_sxt << 2);
        sx_low  <= (x_wait_sxt << 7) - (x_wait_sxt << 2);
        clk_div2 <= 1;
        state <= odd;
      end
      odd : begin
```

```
        x_wait <= x_in;
        clk_div2 <= 0;
        state <= even;
      end
    endcase
  end

//******** Multipy a[0] = 1.7321
  assign sx_up_sxt = {{8{sx_up[16]}},sx_up};
  assign sx_low_sxt = {{8{sx_low[16]}},sx_low};
                                    // i.e. sign extensions
// Compute: (2*sx_up  - sx_up /4)-(sx_up /64 + sx_up /256)
  assign sxa0_up  = ((sx_up_sxt << 1)  - (sx_up_sxt >> 2))
                  - ((sx_up_sxt >> 6) + (sx_up_sxt >> 8));
// Compute: (2*sx_low - sx_low/4)-(sx_low/64 + sx_low/256)
  assign sxa0_low = ((sx_low_sxt << 1) - (sx_low_sxt >> 2))
                 - ((sx_low_sxt >> 6) + (sx_low_sxt >> 8));
//******** First stage -- FF in lower tree
  assign up0 = sxa0_low + sx_up;
  always @(negedge clk_div2)
  begin: LowerTreeFF
      low0 <= sx_low - sxa0_up;
  end

//******** Second stage: a[1]=0.2679
// Compute:   (up0 - low0/4) - (low0/64 + low0/256);
  assign up0_sxt = {{8{up0[16]}},up0};
  assign low0_sxt = {{8{low0[16]}},low0};
  assign up1  = (up0_sxt - (low0_sxt >> 2))
                 - ((low0_sxt >> 6) + (low0_sxt >> 8));
// Compute: (low0 + up0/4) + (up0/64  +  up0/256)
  assign low1 = (low0_sxt + (up0_sxt >> 2))
                      + ((up0_sxt >> 6) + (up0_sxt >> 8));

  assign x_e  = sx_up;        // Provide some extra
  assign x_o  = sx_low;       // test signals
  assign clk2 = clk_div2;

  always @(negedge clk_div2)
  begin: OutputScale
    g <= up1[16:8];       // i.e. up1 / 256
    h <= low1[16:8];      // i.e. low1 / 256;
  end
```

```
endmodule

//*********************************************************
// IEEE STD 1364-1995 Verilog file: rader7.v
// Author-EMAIL: Uwe.Meyer-Baese@ieee.org
//*********************************************************
module rader7 (clk, x_in, y_real, y_imag); //---> Interface

  input         clk;
  input  [7:0]  x_in;
  output [10:0] y_real, y_imag;
  reg    [10:0] y_real, y_imag;

  reg [10:0]  accu;                           // Signal for X[0]
  // Note: No direct bit access of 2D vector in Verilog
  // use auxiliary signal for this purpose
  reg [18:0]  imag0, imag1, imag2, imag3, imag4, imag5,
              real0, real1, real2, real3, real4, real5;
                                  // Tapped delay line array
  reg [18:0]  x57, x111, x160, x200, x231, x250 ;
                                  // The filter coefficients
  reg [18:0]  x5, x25, x110, x125, x256;
                              // Auxiliary filter coefficients
  reg [7:0]   x, x_0;                      // Signals for x[0]
  wire [18:0] x_sxt, x_0_sxt;

  assign x_sxt = {{9{x[7]}},x}; // Sign extension of input
  assign x_0_sxt = {{9{x_0[7]}},x_0};            // and x[0]

  always @(posedge clk)  // State machine for RADER filter
  begin : States
    parameter Start=0, Load=1, Run=2;
    reg [1:0] state;
    reg [4:0] count;
    case (state)
      Start : begin          // Initialization step
        state <= Load;
        count <= 1;
        x_0 <= x_in;         // Save x[0]
        accu <= 0 ;          // Reset accumulator for X[0]
        y_real  <= 0;
        y_imag  <= 0;
      end
      Load : begin   // Apply x[5],x[4],x[6],x[2],x[3],x[1]
        if (count == 8)      // Load phase done ?
```

```
        state <= Run;
      else begin
        state <= Load;
        accu <= accu + x_sxt;
      end
      count <= count + 1;
    end
    Run : begin   // Apply again x[5],x[4],x[6],x[2],x[3]
      if (count == 15) begin // Run phase done ?
        y_real  <= accu;        // X[0]
        y_imag  <= 0;  // Only re inputs i.e. Im(X[0])=0
        state <= Start;       // Output of result
      end                     // and start again
      else begin
        y_real  <= (real0 >> 8) + x_0_sxt;
                                  // i.e. real[0]/256+x[0]
        y_imag  <= (imag0 >> 8);      // i.e. imag[0]/256
        state <= Run;
      end
      count <= count + 1;
    end
  endcase
end

always @(posedge clk)     // Structure of the two FIR
begin : Structure         // filters in transposed form
  x <= x_in;
  // Real part of FIR filter in transposed form
  real0 <= real1 + x160  ;    // W^1
  real1 <= real2 - x231  ;    // W^3
  real2 <= real3 - x57   ;    // W^2
  real3 <= real4 + x160  ;    // W^6
  real4 <= real5 - x231  ;    // W^4
  real5 <= -x57;              // W^5

  // Imaginary part of FIR filter in transposed form
  imag0 <= imag1 - x200  ;    // W^1
  imag1 <= imag2 - x111  ;    // W^3
  imag2 <= imag3 - x250  ;    // W^2
  imag3 <= imag4 + x200  ;    // W^6
  imag4 <= imag5 + x111  ;    // W^4
  imag5 <= x250;              // W^5
end
```

```
  always @(posedge clk)               // Note that all signals
  begin : Coeffs                      // are globally defined
  // Compute the filter coefficients and use FFs
    x160   <= x5 << 5;          // i.e. 160 = 5 * 32;
    x200   <= x25 << 3;         // i.e. 200 = 25 * 8;
    x250   <= x125 << 1;        // i.e. 250 = 125 * 2;
    x57    <= x25 + (x << 5);   // i.e. 57 = 25 + 32;
    x111   <= x110 + x;         // i.e. 111 = 110 + 1;
    x231   <= x256 - x25;       // i.e. 231 = 256 - 25;
  end

  always  @(x_sxt or x5 or x25)       // Note that all signals
  begin : Factors                     // are globally defined
  // Compute the auxiliary factor for RAG without an FF
    x5     = (x_sxt << 2) + x_sxt;  // i.e. 5 = 4 + 1;
    x25    = (x5 << 2) + x5;         // i.e. 25 = 5*4 + 5;
    x110   = (x25 << 2) + (x5 << 2);// i.e. 110 = 25*4+5*4;
    x125   = (x25 << 2) + x25;       // i.e. 125 = 25*4+25;
    x256   = x_sxt << 8;             // i.e. 256 = 2 ** 8;
  end

endmodule

//*********************************************************
// IEEE STD 1364-1995 Verilog file: ccmul.v
// Author-EMAIL: Uwe.Meyer-Baese@ieee.org
//*********************************************************
//`include "220model.v"

module ccmul (clk, x_in, y_in, c_in,
              cps_in, cms_in, r_out, i_out);

  parameter W2 = 17,    // Multiplier bit width
            W1 = 9,     // Bit width c+s sum
            W  = 8;     // Input bit width
  input clk;  // Clock for the output register
  input [W-1:0] x_in, y_in, c_in;  // Inputs
  input [W1-1:0]  cps_in, cms_in;  // Inputs
  output [W-1:0]     r_out, i_out;  // Results
  reg    [W-1:0]     r_out, i_out;  // Results


  wire [W-1:0] x, y, c ;                     // Inputs and outputs
  wire [W2-1:0] r, i, cmsy, cpsx, xmyc, sum;     // Products
  wire [W1-1:0] xmy, cps, cms, sxtx, sxty;   // x-y etc.
```

```
  wire  clken, cr1, ovl1, cin1, aclr, ADD, SUB;
                                              // Auxiliary signals
  assign cin1=0; assign aclr=0; assign ADD=1; assign SUB=0;
  assign cr1=0; assign sum=0; assign clken=0;
                                                // Default for add
  assign x   = x_in;   // x
  assign y   = y_in;   // j * y
  assign c   = c_in;   // cos
  assign cps = cps_in; // cos + sin
  assign cms = cms_in; // cos - sin

  always @(posedge clk) begin
    r_out <= r[W2-2:W];        // Scaling and FF for output
    i_out <= i[W2-2:W];
  end

//********* ccmul with 3 mul. and 3 add/sub  **************
  assign sxtx  = {x[W-1],x};     // Possible growth for
  assign sxty  = {y[W-1],y};     // sub_1 -> sign extension

  lpm_add_sub sub_1                        // Sub:  x - y
  ( .result(xmy), .dataa(sxtx), .datab(sxty));// Used ports
//  .add_sub(SUB), .cout(cr1), .overflow(ovl1), .cin(cin1),
//   .clken(clken), .clock(clk), .aclr(aclr));  // Unused
    defparam sub_1.lpm_width = W1;
    defparam sub_1.lpm_representation = "SIGNED";
    defparam sub_1.lpm_direction = "sub";

  lpm_mult mul_1                      // Multiply  (x-y)*c = xmyc
  ( .dataa(xmy), .datab(c), .result(xmyc)); // Used ports
//  .sum(sum), .clock(clk), .clken(clken), .aclr(aclr));
                                                 // Unused ports
    defparam mul_1.lpm_widtha = W1;
    defparam mul_1.lpm_widthb = W;
    defparam mul_1.lpm_widthp = W2;
    defparam mul_1.lpm_widths = W2;
    defparam mul_1.lpm_representation = "SIGNED";

  lpm_mult mul_2                       // Multiply (c-s)*y = cmsy
  ( .dataa(cms), .datab(y), .result(cmsy)); // Used ports
//  .sum(sum), .clock(clk), .clken(clken), .aclr(aclr));
                                                 // Unused ports
```

```
    defparam mul_2.lpm_widtha = W1;
    defparam mul_2.lpm_widthb = W;
    defparam mul_2.lpm_widthp = W2;
    defparam mul_2.lpm_widths = W2;
    defparam mul_2.lpm_representation = "SIGNED";

  lpm_mult mul_3                     // Multiply (c+s)*x = cpsx
  ( .dataa(cps), .datab(x), .result(cpsx)); // Used ports
//  .sum(sum), .clock(clk), .clken(clken), .aclr(aclr));
                                               // Unused ports
    defparam mul_3.lpm_widtha= W1;
    defparam mul_3.lpm_widthb = W;
    defparam mul_3.lpm_widthp = W2;
    defparam mul_3.lpm_widths = W2;
    defparam mul_3.lpm_representation = "SIGNED";

  lpm_add_sub add_1         // Add:  r <= (x-y)*c + (c-s)*y
  ( .dataa(cmsy), .datab(xmyc), .result(r));  // Used ports
//  .add_sub(ADD), .cout(cr1), .overflow(ovl1), .cin(cin1),
//  .clken(clken), .clock(clk), .aclr(aclr));  // Unused
    defparam add_1.lpm_width = W2;
    defparam add_1.lpm_representation = "SIGNED";
    defparam add_1.lpm_direction = "add";

  lpm_add_sub sub_2           // Sub: i <= (c+s)*x - (x-y)*c
  ( .dataa(cpsx), .datab(xmyc), .result(i));  // Used ports
// .add_sub(SUB), .cout(cr1), .overflow(ovl1), .clock(clk),
//  .cin(cin1),  .clken(clken), .aclr(aclr));  // Unused
    defparam sub_2.lpm_width = W2;
    defparam sub_2.lpm_representation = "SIGNED";
    defparam sub_2.lpm_direction = "sub";

endmodule

//*********************************************************
// IEEE STD 1364-1995 Verilog file: bfproc.v
// Author-EMAIL: Uwe.Meyer-Baese@ieee.org
//*********************************************************
//`include "220model.v"
//`include "ccmul.v"

module bfproc (clk, Are_in, Aim_in, Bre_in, Bim_in, c_in,
       cps_in, cms_in, Dre_out, Dim_out, Ere_out, Eim_out);

  parameter W2 = 17,   // Multiplier bit width
```

```
          W1 = 9,     // Bit width c+s sum
          W  = 8;     // Input bit width
input clk;  // Clock for the output register
input [W-1:0] Are_in, Aim_in;        // 8-bit inputs
input [W-1:0] Bre_in, Bim_in, c_in;  // 8-bit inputs
input [W1-1:0]  cps_in, cms_in;  // 9-bit coefficients
output [W-1:0]  Dre_out, Dim_out, Ere_out, Eim_out;
reg    [W-1:0]  Dre_out, Dim_out;  // 8-bit registered
                                   // results
reg  [W-1:0] dif_re, dif_im;      // Bf out
reg  [W-1:0] Are, Aim, Bre, Bim;  // Inputs as integers
reg  [W-1:0] c;                   // Input
reg  [W1-1:0] cps, cms;           // Coefficient in

always @(posedge clk)   // Compute the additions of the
begin                   // butterfly using integers
  Are     <= Are_in;    // and store inputs
  Aim     <= Aim_in;    // in flip-flops
  Bre     <= Bre_in;
  Bim     <= Bim_in;
  c       <= c_in;            // Load from memory cos
  cps     <= cps_in;          // Load from memory cos+sin
  cms     <= cms_in;          // Load from memory cos-sin
  Dre_out <= ({Are[W-1],Are} + {Bre[W-1],Bre}) >> 1;
                                    // i.e. Are/2 + Bre/2
  Dim_out <= ({Aim[W-1],Aim} + {Bim[W-1],Bim}) >> 1;
end                                 // i.e. Aim/2 + Bim/2

   // No FF because butterfly difference "diff" is not an
always @(Are or Bre or Aim or Bim)         // output port
begin
  dif_re = ({Are[W-1],Are} - {Bre[W-1],Bre}) >> 1;
                                    // i.e. Are/2 - Bre/2
  dif_im = ({Aim[W-1],Aim} - {Bim[W-1],Bim}) >> 1;
end                                 // i.e. Aim/2 - Bim/2

//*** Instantiate the complex twiddle factor multiplier
ccmul ccmul_1                     // Multiply (x+jy)(c+js)
( .clk(clk), .x_in(dif_re), .y_in(dif_im),  .c_in(c),
  .cps_in(cps), .cms_in(cms), .r_out(Ere_out),
                                        .i_out(Eim_out));

endmodule
```

```
//*********************************************************
// IEEE STD 1364-1995 Verilog file: lfsr.v
// Author-EMAIL: Uwe.Meyer-Baese@ieee.org
//*********************************************************
module lfsr (clk, y); //----> Interface

  input         clk;
  output [6:1]  y;  // Result

  reg [6:1] ff; // Note that reg is keyword in Verilog and
                             // can not be variable name
  integer i;

  always @(posedge clk) begin // Length 6 LFSR with xnor
    ff[1] <= ff[5] ~^ ff[6]; // Use non-blocking assignment
    for (i=6; i>=2 ; i=i-1)  // Tapped delay line: shift one
      ff[i] <= ff[i-1];
  end

  assign   y = ff;         // Connect to I/O cell

endmodule

//*********************************************************
// IEEE STD 1364-1995 Verilog file: lfsr6s3.v
// Author-EMAIL: Uwe.Meyer-Baese@ieee.org
//*********************************************************
module lfsr6s3 (clk, y); //----> Interface

  input         clk;
  output [6:1]  y;  // Result

  reg [6:1] ff; // Note that reg is keyword in Verilog and
                             // can not be variable name

  always @(posedge clk) begin // Implement 3 step length
    ff[6] <= ff[3];           // 6 LFSR with xnor; use
    ff[5] <= ff[2];           // non-blocking assignments
    ff[4] <= ff[1];
    ff[3] <= ff[5] ~^ ff[6];
    ff[2] <= ff[4] ~^ ff[5];
    ff[1] <= ff[3] ~^ ff[4];
  end

  assign  y = ff;
```

```
endmodule

//*********************************************************
// IEEE STD 1364-1995 Verilog file: ammod.v
// Author-EMAIL: Uwe.Meyer-Baese@ieee.org
//*********************************************************
module ammod (clk, r_in , phi_in,
                   x_out, y_out, eps); //----> Interface

  parameter W = 8;  // Bit width - 1
  input         clk;
  input  [W:0] r_in, phi_in;
  output [W:0] x_out, y_out, eps;
  reg    [W:0] x_out, y_out, eps;
  reg    [W:0] r, phi;

  reg [W:0] x0, y0, z0; // There is no bit access in 2D
  reg [W:0] x1, y1, z1; // array types in Verilog,
  reg [W:0] x2, y2, z2; // therefore use single vectors
  reg [W:0] x3, y3, z3;

  always @(posedge clk) begin //----> Infer register
    if (phi_in > 90)            // Test for |phi_in| > 90
      begin                     // Rotate 90 degrees
      x0 <= 0;
      y0 <= r_in;               // Input in register 0
      z0 <= phi_in -'d90;
      end
    else if ((phi_in > 331) && (phi_in < 423))
      begin
      x0 <= 0;
      y0 <= - r_in;
      z0 <= phi_in + 'd90;
      end
    else
      begin
      x0 <= r_in;
      y0 <= 0;
      z0 <= phi_in;
     end

    if (z0 > 0)                     // Rotate 45 degrees
      begin
      x1 <= x0 - y0;
```

```
      y1 <= y0 + x0;
      z1 <= z0 - 'd45;
      end
    else
      begin
      x1 <= x0 + y0;
      y1 <= y0 - x0;
      z1 <= z0 + 'd45;
      end

    if (z1 > 0)                      // Rotate 26 degrees
      begin
      x2 <= x1 - {y1[W],y1[W:1]}; // i.e. x1 - y1 /2
      y2 <= y1 + {x1[W],x1[W:1]}; // i.e. y1 + x1 /2
      z2 <= z1 - 'd26;
      end
    else
      begin
      x2 <= x1 + {y1[W],y1[W:1]}; // i.e. x1 + y1 /2
      y2 <= y1 - {x1[W],x1[W:1]}; // i.e. y1 - x1 /2
      z2 <= z1 + 'd26;
      end

    if (z2 > 0)                             // Rotate 14 degrees
      begin
        x3 <= x2 - {y2[W],y2[W],y2[W:2]}; // i.e. x2 - y2/4
        y3 <= y2 + {x2[W],x2[W],x2[W:2]}; // i.e. y2 + x2/4
        z3 <= z2 - 'd14;
      end
    else
      begin
        x3 <= x2 + {y2[W],y2[W],y2[W:2]}; // i.e. x2 + y2/4
        y3 <= y2 - {x2[W],x2[W],x2[W:2]}; // i.e. y2 - x2/4
        z3 <= z2 + 'd14;
      end

    x_out <= x3;
    eps   <= z3;
    y_out <= y3;
  end

endmodule
```

B. VHDL and Verilog Coding

Unfortunately, today we find *two* HDL languages to be popular. The U.S. west coast and Asia prefer Verilog, while the U.S. east coast and Europe more frequently use VHDL. For digital signal processing with FPGAs, both languages seem to be well suited, but some VHDL examples are a little easier to read because of the supported signed arithmetic and multiply/divide operations in the IEEE VHDL 1076-1987 and 1076-1993 standards. This gap will disappear when the new Verilog IEEE standard 1364-1999 is approved, as it will also include signed arithmetic. Other constraints may include personal preferences, EDA library and tool availability, data types, readability, capability, and language extensions using PLIs, as well as commercial, business and marketing issues, to name just a few. A detailed comparison can be found in the book by Smith [3]. Tool providers acknowledge today that both languages need to be supported.

It is therefore a good idea to use an HDL code style that can easily be translated into either language. An important rule is to avoid any "keyword" in *both* languages in the HDL code when naming variables, labels, constants, user types, etc. The IEEE standard VHDL 1076-1987 uses 77 keywords and an extra 19 keywords are used in VHDL 1076-1993 (see VHDL 1076-1993 Language Reference Manual (LRM) on p. 179). New in VHDL 1076-1993 are:

```
GROUP, IMPURE, INERTIAL, LITERAL, POSTPONED, PURE, REJECT
 ROL, ROR, SHARED, SLA, SLL, SRA, SRL, UNAFFECTED, XNOR,
```

which are unfortunately *not* highlighted in the MaxPlusII editor. The IEEE standard Verilog 1364-1995, on the other hand, has 102 keywords (see LRM, p. 604). Together, both HDL languages have 182 keywords, including 17 in common. The Table B.1 shows VHDL 1076-1993 keywords in capital letters, Verilog 1364-1995 keywords in small letters, and the common keywords with a capital first letter.

Table B.1. VHDL 1076-1993 and Verilog 1364-1995 keywords.

ABS	event	OF	OF
ACCESS	EXIT	ON	SLA
AFTER	FILE	OPEN	SLL
ALIAS	For	Or	small
ALL	force	OTHERS	specify
always	forever	OUT	specparam
And	fork	output	SRA
ARCHITECTURE	Function	PACKAGE	SRL
ARRAY	GENERATE	parameter	strong0
ASSERT	GENERIC	pmos	strong1
assign	GROUP	PORT	SUBTYPE
ATTRIBUTE	GUARDED	posedge	supply0
Begin	highz0	POSTPONED	supply1
BLOCK	highz1	primitive	table
BODY	If	PROCEDURE	task
buf	ifnone	PROCESS	THEN
BUFFER	IMPURE	pull0	time
bufif0	IN	pull1	TO
bufif1	INERTIAL	pulldown	tran
BUS	initial	pullup	tranif0
Case	Inout	PURE	tranif1
casex	input	RANGE	TRANSPORT
casez	integer	rcmos	tri
cmos	IS	real	tri0
COMPONENT	join	realtime	tri1
CONFIGURATION	LABEL	RECORD	triand
CONSTANT	large	reg	trior
deassign	LIBRARY	REGISTER	trireg
default	LINKAGE	REJECT	TYPE
defparam	LITERAL	release	UNAFFECTED
disable	LOOP	REM	UNITS
DISCONNECT	macromodule	repeat	UNTIL
DOWNTO	MAP	REPORT	USE
edge	medium	RETURN	VARIABLE
Else	MOD	rnmos	vectored
ELSIF	module	ROL	Wait
End	Nand	ROR	wand
endcase	negedge	rpmos	weak0
endfunction	NEW	rtran	weak1
endmodule	NEXT	rtranif0	WHEN
endprimitive	nmos	rtranif1	While
endspecify	Nor	scalared	wire
endtable	Not	SELECT	WITH
endtask	notif0	SEVERITY	wor
ENTITY	notif1	SHARED	Xnor
	NULL	SIGNAL	Xor

B.1 List of Examples

The following table displays the results for all VHDL and Verilog examples given in this book.

Design	VHDL LCs	VHDL MHz	VHDL Page	Verilog LCs	Verilog MHz	Verilog Page
add_1p	26	97.08	50	26	97.08	345
add_2p	58	94.33	49	51	94.33	347
add_3p	105	91.74	49	105	91.74	350
ammod	279	40.65	312	277	42.01	385
bfproc	533	18.38	236	542	18.65	382
ccmul	493	–	233	504	–	380
cic3r32	332	38.46	159	332	38.75	370
cic3s32	199	41.66	167	200	41.32	372
cordic	256	43.47	72	253	45.04	354
dafsm	37	61.72	98	37	58.13	359
dapara	39	42.19	110	39	42.19	367
darom	34	32.25	105	34	32.25	363
dasign	65	42.19	107	65	39.84	365
db4latti	334	60.60	193	324	63.69	376
db4poly	208	90.90	148	191	99.00	374
example	25	125.00	13	25	125.00	343
fir_gen	890	46.72	81	882	48.54	356
fir_srg	97	19.53	93	97	19.76	358
fun_text	32	59.17	21	32	59.17	344
iir	31	55.86	117	31	55.86	368
iir_par	213	58.13	137	212	60.24	369
iir_pipe	64	73.52	131	64	73.52	368
lfsr	6	100.00	294	6	100.00	383
lfsr6s3	6	95.23	297	6	95.23	384
mul_ser	111	45.87	54	137	45.87	353
rader7	494	29.94	221	495	28.81	378

The following option of MaxPlusII version 9.23 were used:

- `Global Project Synthesis Style` to `FAST`.
- `Assign`→ `Global Project Logic Synthesis` option to `Optimize 10` (Speed).
- `Assign`→ `Global Project Logic Synthesis`→`Automatic Fast I/O`.
- `Assign`→ `Device` for `Device Family`. option `FLEX10K`.
- `Devices` → `EPF10K20RC240-4`.

The data in MHz are the `Registered Performance` (from the "timing analyzer output" files, i.e., `*.tao`) for the designs. The table is structured as follows: the first column shows the "entity" or module name of the design. Columns 2 to 4 are data for the VHDL designs: the number of LCs shown in the report file (`*.rpt`); `Registered Performance`; and the page with the source code. The same data are provided for the Verilog design examples, shown in column 5 to 8. Comparing the VHDL and Verilog synthesis results, we see that some data are not exactly identical. These results can be easily reproduced by using the scripts `maxplus.bat` in the VHDL or Verilog directories of the CD-ROM, and an "grep" through the report file (`*.rpt`) and the timing analyzer output files (`*.tao`).

B.2 Library of Parameterized Modules (LPM)

Throughout the book we use four different LPM megafunctions (see Fig. B.1), namely:

- `lpm_ff`, the flip-flop megafunction
- `lpm_rom`, the ROM megafunction.
- `lpm_add_sub`, the adder/subtractor megafunction, and
- `lpm_mult`, the multiplier megafunction.

These magafunctions are explained in the following, along with their port definitions, parameters, and resource usage. This information is also available using the MaxPlusII help under `VHDL` → `Megafunctions/LPM`, or in the "LPM Quick Reference Guide," located on the Altera digital library CD-ROM as PDF document `cat/lpm.pdf`

B.2.1 The Parameterized Flip-flop Megafunction (lpm_ff)

The `lpm_ff` function is useful if features are needed that are not available in the `DFF`, `DFFE`, `TFF`, and `TFFE` primitives, such as synchronous or asynchronous set, clear, and load inputs. We have used this megafunction for the following designs: `example`, p. 13, `fun_text`, p. 21, `add_1p`, p. 50, `add_2p`, p. 49, `add_3p`, p. 49 as well as for the components `csa7` and `csa7cin`.

Altera recommends instantiating this function as described in "Creating a Custom Megafunction Variation with the MegaWizard Plug-In Manager." The port names and order for Verilog HDL prototypes are:

```
module lpm_ff ( q,
                data, clock, enable,
                aclr, aset,
                sclr, sset,
                aload, sload) ;
```

The VHDL component declaration is shown below:

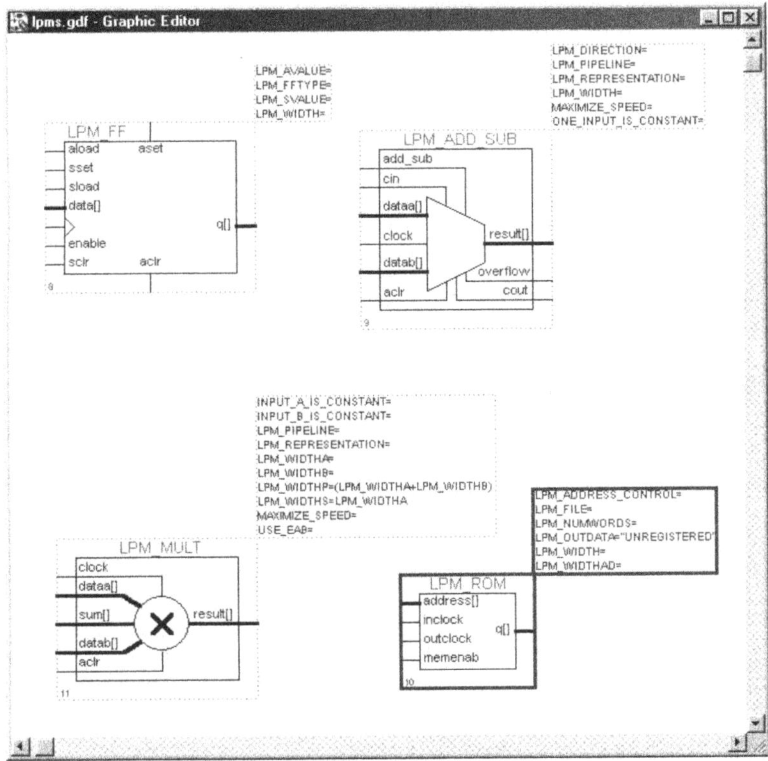

Fig. B.1. Four LPM megafunction used.

```
COMPONENT lpm_ff
   GENERIC (LPM_WIDTH: POSITIVE;
            LPM_AVALUE: STRING := "UNUSED";
            LPM_FFTYPE: STRING := "FFTYPE_DFF";
            LPM_TYPE: STRING := "L_FF";
            LPM_SVALUE: STRING := "UNUSED";
            LPM_HINT: STRING := "UNUSED");
   PORT (data: IN STD_LOGIC_VECTOR(LPM_WIDTH-1 DOWNTO 0);
        clock: IN STD_LOGIC;
       enable: IN STD_LOGIC := '1';
        sload: IN STD_LOGIC := '0';
         sclr: IN STD_LOGIC := '0';
         sset: IN STD_LOGIC := '0';
        aload: IN STD_LOGIC := '0';
         aclr: IN STD_LOGIC := '0';
         aset: IN STD_LOGIC := '0';
            q: OUT STD_LOGIC_VECTOR(LPM_WIDTH-1 DOWNTO 0));
END COMPONENT;
```

Ports

The following table displays all INPUT ports of `lpm_ff`:

Port name	Required	Description	Comments
data	No	T-type flip-flop: Toggle enable DD-type flip-flop: Data input	Input port LPM_WIDTH wide. If the **data** input is not used, at least one of the **aset**, **aclr**, **sset**, or **sclr** ports must be used. Unused data inputs default to GND.
clock	Yes	Positive-edge-triggered clock	
enable	No	Clock Enable input	Default = 1.
sclr	No	Synchronous Clear input	If both **sset** and **sclr** are used and both are asserted, **sclr** is dominant. The **sclr** signal affects the output **q** values before polarity is applied to the ports.
sset	No	Synchronous set input	Sets **q** outputs to the value specified by LPM_SVALUE, if that value is present, or sets the **q** outputs to all 1's. If both **sset** and **sclr** are used and both are asserted, **sclr** is dominant. The **sset** signal affects the output **q** values before polarity is applied to the ports.
sload	No	Synchronous load input. Loads the flip-flop with the value on the **data** input on the next active clock edge.	Default = 0. If **sload** is used, **data** must be used. For load operation, **sload** must be high (1) and enable must be high (1) or unconnected. The sload port is ignored when the LPM_FFTYPE parameter is set to "DFF."
aclr	No	Asynchronous Clear input	If both **aset** and **aclr** are used and both are asserted, **aclr** is dominant. The **aclr** signal affects the output **q** values before polarity is applied to the ports.
aset	No	Asynchronous set input	Sets **q** outputs to the value specified by LPM_AVALUE, if that value is present, or sets the **q** outputs to all 1's.
aload	No	Asynchronous load input. Asynchronously loads the flip-flop with the value on the data input.	Default = 0. If **aload** is used, **data** must be used.

The following table displays all OUTPUT ports of `lpm_ff`:

Port Name	Re-quired	Description	Comments
q	Yes	Data output from D or T flip-flops	Output port LPM_WIDTH wide

Parameters

The following table shows the parameters of the lpm_ff component:

Parameter	Type	Re-quired	Description
LPM_WIDTH	Integer	Yes	Width of the data and q ports
LPM_AVALUE	Integer	No	Constant value that is loaded when aset is high. If omitted, defaults to all 1's. The LPM_AVALUE parameter is limited to a maximum of 32 bits.
LPM_SVALUE	Integer	No	Constant value that is loaded on the rising edge of clock when sset is high. If omitted, defaults to all 1's.
LPM_FFTYPE	String	No	Values are "DFF," "TFF," and "UNUSED." Type of flip-flop. If omitted, the default is "DFF." When the LPM_FFTYPE parameter is set to "DFF," the sload port is ignored.
LPM_HINT	String	No	Allows you to specify Altera-specific parameters in VHDL design files. The default is "UNUSED."
LPM_TYPE	String	No	Identifies the LPM entity name in VHDL design files.

Note that for Verilog LPM 220 synthesizable code (i.e., 220model.v) the following parameter ordering applies: lpm_type, lpm_width, lpm_avalue, lpm_svalue, lpm_pvalue, lpm_fftype, lpm_hint.

Function

The following table is an example of the T-type flip-flop behavior in lpm_ff:

Inputs							Outputs
aclr	aset	enable	clock	sclr	sset	sload	Q[LPM_WIDTH-1..0]
1	X	X	X	X	X	X	000...
0	1	X	X	X	X	X	111... or LPM_AVALUE
0	0	0	X	X	X	X	q[LPM_WIDTH-1..0]
0	0	1	ʃ	1	X	X	000...
0	0	1	ʃ	0	1	X	111... or LPM_SVALUE
0	0	1	ʃ	0	0	1	data[LPM_WIDTH-1..0]
0	0	1	ʃ	0	0	0	q[LPM_WIDTH-1..0] xor data[LPM_WIDTH-1..0]

Resource Usage

The megafunction `lpm_ff` uses one logic cell per bit.

B.2.2 The Parameterized Adder/Subtractor Megafunction (lpm_add_sub)

Altera recommends using the `lpm_add_sub` function to replace all other types of adder/subtractor functions, including old-style adder/subtractor macrofunctions. We have used this megafunction for the following designs: `example`, p. 13, `fun_text`, p. 21, `add_1p`, p. 50, `ccmul`, p. 233, as well as for the components `add_ff8`, `add_ff8cin`, `csa7`, and `csa7cin`.

Altera recommends instantiating this function as described in "Creating a Custom Megafunction Variation with the MegaWizard Plug-In Manager." The port names and order for Verilog HDL prototypes are:

```
module lpm_add_sub ( cin,
                     dataa, datab,
                     add_sub, clock, aclr,
                     result, cout, overflow);
```

The VHDL component declaration is shown below:

```
COMPONENT lpm_add_sub
  GENERIC (LPM_WIDTH: POSITIVE;
           LPM_REPRESENTATION: STRING := "SIGNED";
           LPM_DIRECTION: STRING := "UNUSED";
           LPM_HINT: STRING := "UNUSED";
           LPM_PIPELINE: INTEGER := 0;
           LPM_TYPE: STRING := "L_ADD_SUB");
  PORT (dataa, datab
              : IN STD_LOGIC_VECTOR(LPM_WIDTH-1 DOWNTO 0);
       aclr, clken, clock, cin : IN STD_LOGIC := '0';
       add_sub                 : IN STD_LOGIC := '1';
```

```
        result : OUT STD_LOGIC_VECTOR(LPM_WIDTH-1 DOWNTO 0);
        cout, overflow          : OUT STD_LOGIC);
END COMPONENT;
```

Ports

The following table displays all INPUT ports of **lpm_add_sub**:

Port name	Required	Description	Comments
cin	No	Carry-in to the low-order bit. If the operation is "**ADD**," low = 0 and high = +1. If the operation is "**SUB**," low = -1 and high = 0.	If omitted, the default is 0 (i.e., low if the operation is "**ADD**" and high if the operation is "**SUB**").
dataa	Yes	Augend/Minuend	Input port **LPM_WIDTH** wide
datab	Yes	Addend/Subtrahend	Input port **LPM_WIDTH** wide
add_sub	No	If the signal is high, the operation = **dataa** + **datab**. If the signal is low, the operation = **dataa** – **datab**.	If the **LPM_DIRECTION** parameter is used, **add_sub** cannot be used. If omitted, the default is "**ADD**." Altera recommends that you use the **LPM_DIRECTION** parameter to specify the operation of the **lpm_add_sub** function, rather than assigning a constant to the **add_sub** port.
clock	No	Clock for pipelined usage	The clock port provides pipelined operation for the **lpm_add_sub** function. For **LPM_PIPELINE** values other than 0 (default value), the **clock** port must be connected.
clken	No	Clock enable for pipelined usage	Available for VHDL only
aclr	No	Asynchronous Clear for pipelined usage	The pipeline initializes to an undefined (X) logic level. The **aclr** port can be used at any time to reset the pipeline to all 0's, asynchronously to the **clock** signal.

The following table displays all OUTPUT ports of lpm_add_sub:

Port Name	Required	Description	Comments
result	Yes	dataa + or − datab + or − cin	Output port LPM_WIDTH wide
cout	No	Carry-out (borrow-in) of the MSB	If overflow is used, cout cannot be used. The cout port has a physical interpretation as the carry-out (borrow-in) of the MSB. cout is most meaningful for detecting overflow in "UNSIGNED" operations.
overflow	No	Result exceeds available precision.	If overflow is used, cout cannot be used. The overflow port has a physical interpretation as the XOR of the carry-in to the MSB with the carry-out of the MSB. overflow is meaningful only when the LPM_REPRESENTATION parameter value is "SIGNED."

Parameters

The following table shows the parameters of the lpm_add_sub component:

Parameter	Type	Required	Description
LPM_WIDTH	Integer	Yes	Width of the dataa, datab, and result ports.
LPM_DIRECTION	String	No	Values are "ADD," "SUB," and "UNUSED." If omitted, the default is "DEFAULT," which directs the parameter to take its value from the add_sub port. The add_sub port cannot be used if LPM_DIRECTION is used. Altera recommends that you use the LPM_DIRECTION parameter to specify the operation of the lpm_add_sub function, rather than assigning a constant to the add_sub port.
LPM_REPRESENTATION	String	No	Type of addition performed: "SIGNED," "UNSIGNED," or "UNUSED." If omitted, the default is "SIGNED."
LPM_PIPELINE	Integer	No	Specifies the number of clock cycles of latency associated with the result output. A value of zero (0) indicates that no latency exists, and that a purely combinatorial function will be instantiated. If omitted, the default is 0 (non-pipelined).
LPM_HINT	String	No	Allows you to specify Altera-specific parameters in VHDL design files. The default is "UNUSED."
LPM_TYPE	String	No	Identifies the LPM entity name in VHDL design files.
ONE_INPUT_IS_CONSTANT	String	No	Altera-specific parameter. Values are "YES," "NO," and "UNUSED." Provides greater optimization, if one input is constant. If omitted, the default is "NO."
MAXIMIZE_SPEED	Integer	No	Altera-specific parameter. You can specify a value between 0 and 10. If used, MaxPlusII attempts to optimize a specific instance of the lpm_add_sub function for speed rather than area, and overrides the setting of the Optimize option in the Global Project Logic Synthesis dialog box (Assign menu). If MAXIMIZE_SPEED is unused, the value of the Optimize option is used instead. If the setting for MAXIMIZE_SPEED is 6 or higher, the compiler will optimize lpm_add_sub megafunctions for higher speed; if the setting is 5 or less, the compiler will optimize for smaller area.

Note that for Verilog LPM 220 synthesizable code (i.e., `220model.v`) the following parameter ordering applies: `lpm_type`, `lpm_width`, `lpm_direction`, `lpm_representation`, `lpm_pipeline`, `lpm_hint`.

Function

The following table is an example of the UNSIGNED behavior in `lpm_add_sub`:

Inputs			Outputs	
add_sub	dataa	datab	cout,result	overflow
1	a	b	a + b + cin	cout
0	a	b	a - b - cin	!cout

The following table is an example of the SIGNED behavior in `lpm_add_sub`:

Inputs			Outputs	
add_sub	dataa	datab	cout,sum	overflow
1	a	b	$a+b$+cin	$a \geq 0$ and $b \geq 0$ and $sum < 0$ or $a < 0$ and $b < 0$ and $sum \geq 0$
0	a	b	$a-b$−cin	$a >= 0$ and $b < 0$ and $sum < 0$ or $a < 0$ and $b \geq 0$ and $sum \geq 0$

Resource Usage

The following table summarizes the resource usage for an `lpm_add_sub` megafunction used to implement a 16-bit unsigned adder with a carry-in input and a carry-out output. Logic cell usage scales linearly in proportion to adder width.

Design goals		Design results		
Device family	Optimization	LCs	Speed (ns)	Notes
FLEX 6K, 8K, and 10K	Routability Speed	45 18	53 17	Speed for EPF8282A-2
MAX 5K, 7K, and 9K	Routability	28 (22)	23	Speed for EPM7128E-7

Numbers of shared expanders used are shown in parentheses.

B.2.3 The Parameterized Multiplier Megafunction (lpm_mult)

Altera recommends that you use lpm_mult to replace all other types of multiplier functions, including old-style multiplier macrofunctions. We have used this megafunction for the designs ccmul, p. 233, and fir_gen, p. 81.

Altera recommends instantiating this function as described in "Creating a Custom Megafunction Variation with the MegaWizard Plug-In Manager." The port names and order for Verilog HDL prototype are:

```
module lpm_mult ( dataa, datab, sum, aclr, clock,
                  result);
```

The VHDL component declaration is shown below:

```
COMPONENT lpm_mult
  GENERIC (LPM_WIDTHA: POSITIVE;
           LPM_WIDTHB: POSITIVE;
           LPM_WIDTHS: POSITIVE;
           LPM_WIDTHP: POSITIVE;
           LPM_REPRESENTATION: STRING := "UNSIGNED";
           LPM_PIPELINE: INTEGER := 0;
           LPM_TYPE: STRING := "L_MULT";
           LPM_HINT : STRING := "UNUSED");
  PORT (dataa : IN STD_LOGIC_VECTOR(LPM_WIDTHA-1 DOWNTO 0);
        datab : IN STD_LOGIC_VECTOR(LPM_WIDTHB-1 DOWNTO 0);
        aclr, clken, clock : IN STD_LOGIC := '0';
        sum   : IN STD_LOGIC_VECTOR(LPM_WIDTHS-1 DOWNTO 0)
                                        := (OTHERS => '0');
        result: OUT STD_LOGIC_VECTOR(LPM_WIDTHP-1 DOWNTO 0)
        );
END COMPONENT;
```

Ports

The following table displays all INPUT ports of lpm_mult:

Port name	Required	Description	Comments
dataa	Yes	Multiplicand	Input port LPM_WIDTHA wide
datab	Yes	Multiplier	Input port LPM_WIDTHB wide
sum	No	Partial sum	Input port LPM_WIDTHS wide
clock	No	Clock for pipelined usage	The clock port provides pipelined operation for the lpm_mult function. For LPM_PIPELINE values other than 0 (default value), the clock port must be connected.
clken	No	Clock enable for pipelined usage	Available for VHDL only.
aclr	No	Asynchronous Clear for pipelined usage	The pipeline initializes to an undefined (X) logic level. The aclr port can be used at any time to reset the pipeline to all 0's, asynchronously to the clock signal.

The following table displays all OUTPUT ports of lpm_mult:

Port Name	Required	Description	Comments
result	Yes	result = dataa * datab + sum. The product LSB is aligned with the sum LSB.	Output port LPM_WIDTHP wide. If LPM_WIDTHP < max (LPM_WIDTHA + LPM_WIDTHB, LPM_WIDTHS) or (LPM_WIDTHA + LPM_WIDTHS), only the LPM_WIDTHP MSBs are present.

Parameters

The following table shows the parameters of the lpm_mult component:

Parameter	Type	Re-quired	Description
LPM_WIDTHA	Integer	Yes	Width of the **dataa** port
LPM_WIDTHB	Integer	Yes	Width of the **datab** port
LPM_WIDTHP	Integer	Yes	Width of the **result** port
LPM_WIDTHS	Integer	Yes	Width of the **sum** port. Required even if the **sum** port is not used.
LPM_REPRESENTATION	String	No	Type of multiplication performed: "SIGNED," "UNSIGNED," or "UNUSED." If omitted, the default is "UNSIGNED."
LPM_PIPELINE	Integer	No	Specifies the number of clock cycles of latency associated with the **result** output. A value of zero (0) indicates that no latency exists, and that a purely combinatorial function will be instantiated. If omitted, the default is 0 (non-pipelined).
LPM_HINT	String	No	Allows you to assign Altera-specific parameters in VHDL Design Files. The default is "UNUSED."
LPM_TYPE	String	No	Identifies the LPM entity name in VHDL Design Files.
INPUT_A_IS_CONSTANT	String	No	Altera-specific parameter. Values are "YES," "NO," and "UNUSED." If **dataa** is connected to a constant value, setting INPUT_A_IS_CONSTANT to "YES" optimizes the multiplier for resource usage and speed. If omitted, the default is "NO."
INPUT_B_IS_CONSTANT	String	No	Altera-specific parameter. Values are "YES," "NO," and "UNUSED." If **datab** is connected to a constant value, setting INPUT_B_IS_CONSTANT to "YES" optimizes the multiplier for resource usage and speed. The default is "NO."

Parameter	Type	Required	Description
USE_EAB	String	No	Altera-specific parameter. Values are "ON," "OFF," and "UNUSED." Setting the USE_EAB parameter to "ON" allows MaxPlusII to use EABs to implement 4 × 4 or (8 × constant value) building blocks in FLEX 10K devices. Altera recommends that you set USE_EAB to "ON" only when LCELLS are in short supply. If you wish to use this parameter, when you instantiate the function in a GDF, you must specify it by entering the parameter name and value manually with the Edit Ports/Parameters dialog box (Symbol menu). You can also use this parameter name in a TDF or a Verilog design file. You must use the LPM_HINT parameter to specify the USE_EAB parameter in VHDL design files.
LATENCY	Integer	No	Altera-specific parameter. Same as LPM_PIPELINE. (This parameter is provided only for backward compatibility with MaxPlusII pre-version 7.0 designs. For all new designs, you should use the LPM_PIPELINE parameter instead.)
MAXIMIZE_SPEED	Integer	No	Altera-specific parameter. You can specify a value between 0 and 10. If used, MaxPlusII attempts to optimize a specific instance of the lpm_mult function for speed rather than area, and overrides the setting of the Optimize option in the Global Project Logic Synthesis dialog box (Assign menu). If MAXIMIZE_SPEED is unused, the value of the Optimize option is used instead. If the setting for MAXIMIZE_SPEED is 6 or higher, the compiler will optimize lpm_mult megafunctions for higher speed; if the setting is 5 or less, the compiler will optimize for smaller area.
LPM_HINT	String	No	Allows you to specify Altera-specific parameters in VHDL Design Files. The default is "UNUSED."

Note that specifying a value for MAXIMIZE_SPEED has an effect only if LPM_REPRESENTATION is set to "SIGNED."

Note that for Verilog LPM 220 synthesizable code (i.e., 220model.v) the following parameter ordering applies: lpm_type, lpm_widtha, lpm_widthb, lpm_widths, lpm_widthp, lpm_representation, lpm_pipeline, lpm_hint.

Function

The following table is an example of the UNSIGNED behavior in lpm_mult:

Inputs			Outputs
dataa	datab	sum	product
a	b	s	LPM_WIDTHP most significant bits of $a*b+s$

Resource Usage

The following table summarizes the resource usage for an `lpm_mult` function used to implement 4-bit and 8-bit multipliers with `LPM_PIPELINE` $=0$ and without the optional `sum` input. Logic cell usage scales linearly in proportion to the square of the input width.

Design goals		Design results			
Device family	Optimization	Width	LCs	Speed (ns)	Notes
FLEX 6K, 8K, and 10K	Routability Speed	8 8	121 163	80 52	Speed for EPF8282A-2
FLEX 6K, 8K, and 10K	Routability Speed	4 4	29 41	34 27	Speed for EPF8282A-2
MAX 5K, 7K, and 9K	Routability Speed	4 4	26 (11) 27 (4)	23 19	Speed for EPM7128E-7

Numbers of shared expanders used are shown in parentheses. In the FLEX 10K device family, the 4-bit by 4-bit multiplier example shown above can be implemented in a single EAB.

B.2.4 The Parameterized ROM Megafunction (lpm_rom)

Altera recommends that you use the `lpm_rom` function to implement all ROM functions. The `lpm_rom` function is available only for FLEX 10K devices. We have used this megafunction for the designs `fun_text`, p. 21 and `darom`, p. 105. The MaxPlusII compiler automatically implements suitable portions of this function in EABs in FLEX 10K devices. Therefore, it is not necessary to use the `''Implement in EAB''` logic option for this function, and doing so may cause warning messages to appear.

Altera recommends instantiating this function as described in "Creating a Custom Megafunction Variation with the MegaWizard Plug-In Manager." You can use the `genmem.exe` utility to create a simulation model for this function for use in third-party simulators. Type `genmem` -h at a DOS prompt for information on how to use this utility.

The port names and order for Verilog HDL prototype are:

```
module lpm_rom ( address, inclock, outclock, memenab,
                 q);
```

The VHDL component declaration is shown below:

```
COMPONENT lpm_rom
  GENERIC (LPM_WIDTH    : POSITIVE;
           LPM_TYPE     : STRING := "L_ROM";
           LPM_WIDTHAD  : POSITIVE;
           LPM_NUMWORDS : POSITIVE;
           LPM_FILE     : STRING;
           LPM_ADDRESS_CONTROL : STRING := "REGISTERED";
           LPM_OUTDATA  : STRING := "REGISTERED";
           LPM_HINT     : STRING := "UNUSED");
   PORT(address : IN STD_LOGIC_VECTOR(LPM_WIDTHAD-1 DOWNTO 0);
        inclock : IN STD_LOGIC := '1';
        outclock : IN STD_LOGIC := '1';
        memenab  : IN STD_LOGIC := '1';
        q        : OUT STD_LOGIC_VECTOR(LPM_WIDTH-1 DOWNTO 0)
        );
END COMPONENT;
```

Ports

The following table displays all INPUT ports of `lpm_rom`:

Port name	Re-quired	Description	Comments
`address`	Yes	Address input to the memory	Input port `LPM_WIDTHAD` wide
`inclock`	No	Clock for input registers	The `address` port is synchronous (registered) when the `inclock` port is connected, and is asynchronous (registered) when the `inclock` port is not connected.
`outclock`	No	Clock for output registers	The addressed memory content-to-`q` response is synchronous when the `outclock` port is connected, and is asynchronous when it is not connected.
`memenab`	No	Memory enable input	High = data output on `q`, Low = high-impedance outputs

The following table displays all OUTPUT ports of `lpm_rom`:

Port Name	Re-quired	Description	Comments
`q`	Yes	Output of memory	Output port `LPM_WIDTH` wide

Parameters

The following table shows the parameters of the `lpm_rom` component:

Parameter	Type	Required	Description
LPM_WIDTH	Integer	Yes	Width of the q port.
LPM_WIDTHAD	Integer	Yes	Width of the address port. LPM_WIDTHAD should be (but is not required to be) equal to $\log_2$(LPM_NUMWORDS). If LPM_WIDTHAD is too small, some memory locations will not be addressable. If it is too large, the addresses that are too high will return undefined logic levels.
LPM_NUMWORDS	Integer	Yes	Number of words stored in memory. In general, this value should be (but is not required to be) $2^{\text{LPM\_WIDTHAD}} - 1 <$ LPM_NUMWORDS $\leq 2^{\text{LPM\_WIDTHAD}}$. If omitted, the default is $2^{\text{LPM\_WIDTHAD}}$
LPM_FILE	String	No	Name of the **Memory Initialization File** (*.mif) or **Hexadecimal** (Intel-Format) File (*.hex) containing ROM initialization data ("<filename>"), or "UNUSED."
LPM_ADDRESS_CONTROL	String	No	Values are "REGISTERED," "UNREGISTERED," and "UNUSED." Indicates whether the address port is registered. If omitted, the default is "REGISTERED."
LPM_OUTDATA	String	No	Values are "REGISTERED," "UNREGISTERED," and "UNUSED." Indicates whether the q and eq ports are registered. If omitted, the default is "REGISTERED."
LPM_HINT	String	No	Allows you to specify Altera-specific parameters in VHDL Design Files. The default is "UNUSED."
LPM_TYPE	String	No	Identifies the LPM entity name in VHDL Design Files.

Note that for Verilog LPM 220 synthesizable code (i.e., `220model.v`) the following parameter ordering applies: `lpm_type`, `lpm_width`, `lpm_widthad`, `lpm_numwords`, `lpm_address_control`, `lpm_outdata`, `lpm_file`, `lpm_hint`.

Function

The following table shows the *synchronous* read from memory behavior of `lpm_rom`:

OUTCLOCK	MEMENAB	Function
X	L	q output is high impedance (memory not enabled)
l	H	No change in output
ʃ	H	The output register is loaded with the contents of the memory location pointed to by **address**. **q** outputs the contents of the output register.

Totally asynchronous memory operations occur when neither **inclock** nor **outclock** is connected. The output **q** is asynchronous and reflects the data in the memory to which **address** points. The following table shows the *asynchronous* memory operations behavior of **lpm_rom**:

MEMENAB	Function
L	**q** output is high-impedance (memory not enabled)
H	The memory location pointed to by **address** is read

Resource Usage

The Megafunction **lpm_rom** uses one embedded cell per memory bit.

C. Glossary

ACC	Accumulator
ACT	Actel FPGA family
ADCL	All-Digital CL
ADPLL	All-Digital PLL
ADC	Analog-to-Digital Converter
ADSP	Analog Devices Digital Signal Processor family
AFT	Arithmetic Fourier Transform
AHDL	Altera HDL
AM	Amplitude Modulation
ALU	Arithmetic Logic Unit
AMD	Advanced Micro Devices, Inc.
ASCII	American Standard Code for Information Interchange
ASIC	Application Specific IC
BDD	Binary Decision Diagram
BP	Bandpass
BRS	Base Removal Scaling
BS	Barrel Shifter
CAE	Computer-Aided Engineering
CAST	Carlisle Adams and Stafford Tavares
CBC	Cipher Block Chaining
CBIC	Cell-Based IC
CD	Compact Disc
CFA	Common Factor Algorithm
CFB	Cipher FeedBack
CIC	Cascaded Integrator Comb
CL	Costas Loop
CLB	Configurable Logic Block
CMOS	Complementary Metal Oxide Semiconductor
CODEC	Coder/Decoder
CORDIC	COordinate Rotation DIgital Computer
COTS	Commercial Off-The-Shelf technology
CPLD	Complex PLD
CPU	Central Processing Unit
CQF	Conjugate Quadrature Filter
CRNS	Complex RNS
CRT	Chinese Remainder Theorem
CSOC	Canonical Self-Orthogonal Code
CSD	Canonical Signed Digit

CWT	Continuous Wavelet Transform
CZT	Chirp-z Transform
DA	Distributed Arithmetic
DAC	Digital-to-Analog Converter
DB	DauBechies filter
DC	Direct Current
DCT	Discrete Cosine Transform
DCO	Digital Controlled Oscillator
DES	Data Encryption Standard
DFT	Discrete Fourier Transform
DHT	Discrete Hartley Transform
DIF	Decimation In Frequency
DIT	Decimation In Time
DMT	Discrete Morlet Transform
DPLL	Digital PLL
DSP	Digital Signal Processing
DWT	Discrete Wavelet Transform
EAB	Embedded Array Block
ECB	Electronic Code Book
ECL	Emitter Coupled Logic
EDIF	Electronic Design Interchange Format
EFF	Electronic Frontier Foundation
EPF	Altera FPGA family
EPROM	Electrically Programmable ROM
ERA	Plessey FPGA family
ERNS	Eisenstein RNS
ESA	European Space Agency
FCT	Fast Cosine Transform
FC2	FPGA Compiler II
FF	Flip-Flop
FFT	Fast Fourier Transform
FIR	Finite Impulse Response
FIFO	First-In First-Out
FLEX	Altera FPGA family
FM	Frequency Modulation
FNT	Fermat NTT
FPGA	Field-Programmable Gate Array
FPL	Field-Programmable Logic (combines CPLD and FPGA)
FSF	Frequency Sampling Filter
FSK	Frequency Shift Keying
FSM	Finite State Machine
GAL	Generic Array Logic
GF	Galois Field
HB	HalfBand filter
HI	HIgh frequency
HDL	Hardware Description Language
HSP	Harris Semiconductor DSP ICs

IBM	International Business Machines (corporation)
IC	Integrated Circuit
IDCT	Inverse DCT
IDEA	International Data Encryption Algorithm
IDFT	Inverse Discrete Fourier Transform
IEEE	Institute of Electrical and Electronics Engineers
IF	Inter Frequency
IFFT	Inverse Fast Fourier Transform
IIR	Infinite Impulse Response
INTT	Inverse NTT
JPEG	Joint Photographic Experts Group
LAB	Logic Array Block
LAN	Local Area Network
LC	Logic Cell
LE	Logic Element
LF	Low Frequency
LFSR	Linear Feedback Shift Register
LNS	Logarithmic Number System
LO	LOw frequency
LP	Low Pass
LPM	Library of Parameterized Modules
LRS	serial Left Right Shifter
LSB	Least Significant Bit
LSI	Large Scale Integration
LTI	Linear Time-Invariant
LUT	Look-Up Table
MAC	Multiplication and ACcumulate
MACH	AMD/Vantis FPGA family
MAG	Multiplier Adder Graph
MAX	Altera CPLD family
MIF	Memory Initialization File
MLSE	Maximum Likelihood Sequence Estimator
MNT	Mersenne NTT
MPEG	Moving Picture Experts Group
MPX	Multiplexer
MSPS	Millions of Sample Per Second
MRC	Mixed Radix Conversion
MSB	Most Significant Bit
MUL	Multiplication
NP	Non Polynomial complex problem
NRE	Non Reoccurring Engineering costs
NTT	Number Theoretic Transform
OFB	Open FeedBack (mode)
PC	Personal Computer
PD	Phase Detector
PFA	Prime Factor Algorithm
PLA	Programmable Logic Array

PLD	Programmable Logic Device
PLL	Phase-locked loop
PM	Phase Modulation
PREP	PRogrammable Electronic Performance cooperation
PRNS	Polynomial RNS
PROM	Programmable ROM
PSK	Phase Shift Keying
QDFT	Quantized DFT
QLI	Quick Look-In
QFFT	Quantized FFT
QMF	Quadrature Mirror Filter
QRNS	Quadratic RNS
RAM	Random Access Memory
RC	Resistor/Capacity
RF	Radio Frequency
RISC	Reduced Instruction Set Computer
RNS	Residue Number System
ROM	Read Only Memory
RPFA	Rader Prime Factor Algorithm
RS	serial Right Shifter
RSA	Rivest, Shamir, and Adelman
SD	Signed Digit
SM	Signed Magnitude
SPLD	Simple PLD
SPT	Signed Power of Two
SR	Shift Register
SRAM	Static Random Access Memory
STFT	Short Time Fourier Transform
TLU	Table Look-Up
TMS	Texas Instruments DSP family
TI	Texas Instruments
TTL	Transistor Transistor Logic
UART	Universal Asynchronous Receiver/Transmitter
VCO	Voltage Control Oscillator
VHDL	VHSIC Hardware Description Language
VHSIC	Very High Speed Integrated Circuit
VLSI	Very Large Integrated ICs
WFTA	Winograd Fourier Transform Algorithm
XC	Xilinx FPGA family
XNOR	eXclusive NOR gate

D. CD-ROM File: "1readme.ps"

The accompanying CD-ROM includes

- A full version of the MaxPlusII software
- Altera's digital library 2000 (Version 4) with many application notes, data sheets and manuals
- All VHDL/Verilog design examples and utility programs and files.

For instance, you find under **Ug/Maxiigs.pdf** the full "Getting Started" manual, which includes detailed tutorials on MaxPlusII, AHDL, and the graphical design entry. Click on **altera.pdf** to get an overview of the full set of Altera documentation. To install the MaxPlusII9.23 student edition software, first read the files **Se_read.txt** and **Readme.txt** on the CD-ROM. You have to install both the MaxPlusII package (**Mp2_92se.exe**) and the patch, i.e., copy **Vhd.ddl** in the install directory or use the self-extractor **student923_vhd.exe**. Note that often the system files, with ***.dll** extensions are hidden. In Windows Explorer, **View** → **Folder Options** then under the **View** tab select **Show all files**. Without the patch, the conversion function **CONV_INTEGER** (for instance) will not work. After the installation user must register the software through Altera's web page at **www.altera.com** in order to get a license key.

The design examples for the book are located in the directories **book/vhdl** and **book/verilog** for VHDL and Verilog examples, respectively. These directories contain, for each example, the following three files:

- The VHDL or Verilog source code (***.vhd** and ***.v**)
- The Assignment and Configuration File (***.acf**)
- The Simulator Channel File (***.scf**)

For the design **fun_graf**, the "Graphic Design File" (***.gdf**) is included in **book/vhdl**. For the examples which utilize EABs (i.e., **fun_text** and **darom**), the "Memory Initialization File" (***.mif**) and the same file in Intel hex format (***.hex**) can be found on the CD-ROM. To simplify the compilation and postprocessing, the source code directories include some additional (***.bat**) files shown below:

File	Comment
maxplus.bat	Script to compile all design examples. Note that the installation path of MaxPlusII is assumed to be **C:max9\maxplus2.exe**. Change the path globally (if necessary) with an editor before running the script.
clean.bat	Cleans all temporary compiler files, but not the report files (***.rep**) and timing analyzer output files (***.tao**)
veryclean.bat	Cleans all temporary compiler files, *including* the report files (***.rep**) and timing analyzer output files (***.tao**)

Use **maxplus.bat** to compile all design examples and then **clean.bat** to remove the unnecessary files. The results for all examples are summarized in the following files under **book/util**.

File	Description
Lcs.vhd	Pins, memory usage and LC count for all VHDL examples as shown in the report files *.rpt.
Mhz.vhd	"Worst case" path name, delay, and **Registered Performance** in MHz for all VHDL examples, as shown in the timing analyzer output files, *.tao.
Lcs.v	Pins, memory usage and LC count for all Verilog examples, as shown in the report files, *.rpt.
Mhz.v	"Worst case" path name, delay, and **Registered Performance** in MHz for all Verilog examples, as shown in the timing analyzer output files, *.tao.

Using Compilers Other Than MaxPlusII

FPGA_CompilerII

The main advantage of using the FPGA_CompilerII (FC2) from Synopsys is that it is now possible to synthesize examples for other devices like Xilinx, Vantis, Actel, or QuickLogic. The TCL scripts **vhdl.fc2**, and **verilog.fc2**, respectively, provide the necessary commands for the shell mode of FC2, i.e., **fc2_shell**.

Using FPGA_CompilerII and VHDL. The **FPGA_CompilerII** script **vhdl.fc2** to compile the VHDL examples is shown in the following:

```
#------------------------------------------------------------
# Synopsys FPGA Compiler II VHDL simulation script vhdl.fc2
# for the book: Digital Signal Processing with FPGAs
# Author-EMAIL: Uwe.Meyer-Baese@ieee.org
#------------------------------------------------------------
# Usage: fc2_shell -f vhdl.fc2
#------------------------------------------------------------
create_project -dir . fc2
#*** Chapter 1 entitys:
set ch1 "example fun_text"
#*** Chapter 2 components:
set ch2 "csa7 csa7cin add_ff8 add_ff8cin"
#*** Chapter 2 entitys:
set ch2e "add_1p add_2p add_3p mul_ser cordic"
#*** Chapter 3 components:
set ch3 "case3 case5p case3s"
#*** Chapter 3 entitys:
set ch3e "fir_gen fir_srg dafsm darom dasign dapara"
#*** Chapter 4 entitys:
set ch4 "iir iir_pipe iir_par"
#*** Chapter 5 entitys:
set ch5 "cic3r32 cic3s32 db4poly db4latti"
```

```
#*** Chapter 6 entitys:
set ch6 "rader7 ccmul bfproc"
#*** Chapter 7 entitys:
set ch7 "lfsr lfsr6s3 ammod"

#*** Make a single list of all VHDL files
set files "$ch1 $ch2 $ch2e $ch3 $ch3e $ch4 $ch5 $ch6 $ch7"

#*** Compile all files in the list and report results
foreach current $files {
  add_file $current.vhd
  analyze_file
  create_chip -progress -name $current  \
     -target FLEX10K -device EPF10K20RC240 -speed -4 \
     -frequency 30 $current
  current_chip $current
  optimize_chip -name $current-opt
  report_chip > $current.rpt
  export_chip -dir fc2
}

#------------------------------------------------------------
#*** You may select similar target devices from other
#*** vendors. Use the following options for the
#*** "create_chip" command:
# Xilinx:   -target XC4000E -device 4013EPG223 -speed -4
# Vantis    -target VF1 -device VF1020AMYTC -speed 1
# Actel:    -target A1400 -device  A14100BPRQ208 -speed STD
# QuickLogic:
#           -target QLOGIC -device  QL3040-PQ240 -speed -4
#------------------------------------------------------------

quit
```

The scripts produce a report file which shows the device utilization and estimated **Registered Performance**. Unfortunately, these speed data are not very exact and it's better to import the three files produced by **FC2**, namely

- The EDIF file (***.edf**)
- The Assignment and Configuration File (***.acf**)
- The Library Mapping File (***.lmf**) from Synopsys

into MaxPlusII and compile the EDIF file in order to get exact results.

Using FPGA_CompilerII and Verilog. Using the Verilog interface in combination with FPGA_CompilerII requires some additional effort, because **FC2** does not support the definition of the parameter of a component using **defparam**. There are, in general, two ways to alter parameter values: the *defparam statement,* which allows assignment to parameters using their hierarchical names (supported by MaxPlusII, but not by **FC2**), and the *module instance parameter value assignment*, which allows values to be assigned inline during module instantiation (supported by **FC2**, but not by MaxPlusII).

In MaxPlusII we have used the *defparam statement* in, for instance, the design **fun_text.v**, p. 344:

```
lpm_rom rom1
( .q(sin), .inclock(clk), .outclock(clk),
                            .address(msbs)); // Used ports
//                     .memenab(ena) ) ;      // Unused port
defparam rom1.lpm_width = 8;
defparam rom1.lpm_widthad = 8;
defparam rom1.lpm_file = "sine.mif";
```

For FC2 this has to be coded as

```
lpm_rom
  #(8,8,256,"UNREGISTERED","UNREGISTERED","sine.hex") rom1
  ( .q(sin), .inclock(clk), .outclock(clk),
                            .address(msbs),  // Used ports
                     .memenab(ena) ) ;      // Unused port
```

The disadvantage with this coding is that we must be very careful with the order of the parameters. Note also that the first parameter for FC2 is *not* lpm_type, as defined in the Verilog LPM 220 synthesizable code, i.e., 220model.v (see p. 394).

The modified design files can be found in the directory book/verilog/fc2 along with the TCL script verilog.fc2 to be used with fc2_shell.

Model Technology

By using the synthesizable public domain models provided by the EDIF organization (at www.edif.org), it is also possible to use other VHDL/Verilog simulators than MaxPlusII.

Using MTI and VHDL. For VHDL, the two files 220pack.vhd and 220model.vhd must first be compiled. For the ModelSim simulator vsim from Model Technology, the script mti_vhdl.csh can be used for a device-independent compilation and simulation of the design examples. The script is shown below:

```
#!/bin/csh -f
#------------------------------------------------------------
# Model Technology VHDL compiler script for the book
# Digital Signal Processing with FPGAs
# Author-EMAIL: Uwe.Meyer-Baese@ieee.org
#------------------------------------------------------------

if ( ! -d work ) then
  vlib work
  echo "Library directory work created."
else
  echo "Library directory work found."
endif

if ( ! -d lpm ) then
  vlib lpm
  echo "Library directory lpm created."
else
  echo "Library directory lpm found."
endif

echo Compile lpm package.
```

```
vcom -work lpm -explicit -quiet 220pack.vhd 220model.vhd

echo Compile chapter 1 entitys.
vcom -work lpm -quiet example.vhd fun_text.vhd

echo Compile chapter 2 components.
vcom -work lpm -explicit -quiet csa7.vhd csa7cin.vhd
vcom -work lpm -explicit -quiet add_ff8.vhd add_ff8cin.vhd
echo Compile chapter 2 entitys.
vcom -work lpm -explicit -quiet add_1p.vhd add_2p.vhd
vcom -work lpm -explicit -quiet add_3p.vhd mul_ser.vhd
vcom -work lpm -explicit -quiet cordic.vhd

echo Compile chapter 3 components.
vcom -work lpm -explicit -quiet case3.vhd case5p.vhd
vcom -work lpm -explicit -quiet case3s.vhd
echo Compile chapter 3 entitys.
vcom -work lpm -explicit -quiet fir_gen.vhd fir_srg.vhd
vcom -work lpm -explicit -quiet dafsm.vhd darom.vhd
vcom -work lpm -explicit -quiet dasign.vhd dapara.vhd

echo Compile chapter 4 entitys.
vcom -work lpm -explicit -quiet iir.vhd iir_pipe.vhd
vcom -work lpm -explicit -quiet iir_par.vhd

echo Compile chapter 5 entitys.
vcom -work lpm -explicit -quiet cic3r32.vhd cic3s32.vhd
vcom -work lpm -explicit -quiet db4poly.vhd db4latti.vhd

echo Compile chapter 6 entitys.
vcom -work lpm -explicit -quiet rader7.vhd ccmul.vhd
vcom -work lpm -explicit -quiet bfproc.vhd

echo Compile chapter 7 entitys.
vcom -work lpm -explicit -quiet rader7.vhd ccmul.vhd
vcom -work lpm -explicit -quiet bfproc.vhd
```

Using MTI and Verilog. Using the Verilog interface with the `lpm` library from EDIF, i.e., `220model.v`, needs some additional effort. When using `220model.v` it is necessary to specify *all* ports in the Verilog `lpm` components. There is an extra directory `book/verilog/mti`, that provides the design examples with a full set of `lpm` port specification. The designs use

```
'\include "220model.v"
```

at the begin of each Verilog file to include the `lpm` components, if necessary. Use the script `mti_v.csh` to compile all Verilog design examples with Model Technology's `vcom` compiler.

In order to load the "Memory Initialization File" (`*.mif`), it is required to be familiar with the "Programming Language Interface" (PLI) of the Verilog 1364-1995 IEEE standard (see LRM Section 17, p. 228 ff). With this powerful PLI interface, conventional `C` programs can be dynamically loaded into the Verilog compiler. In order to generate a dynamically loaded object of the program `convert_hex2ver.c`, the path for the include files **veriuser.h** and **acc_user.h** must be specified. Use **-I**, when using the **gcc** or **cc** compiler under SUN Solaris. Using, for instance, the **gcc** compiler under SUN Solaris for the Model Technology Compiler, the following commands are used to produce the shared object:

```
gcc -c -I/<install_dir>/modeltech/include convert_hex2ver.c
ld -G -B symbolic -o convert_hex2ver.sl convert_hex2ver.o
```

By doing so, **ld** will generate a warning "Symbol referencing errors," because all symbols are first resolved within the shared library at link time, but these warnings can be ignored.

It is then possible to use these shared objects, for instance with Model Technology **vsim** in the first design example **fun_text.v**, with

```
vsim  -pli convert_hex2verl.sl  lpm.fun_text
```

To learn more about PLIs, check out Verilog IEEE standard 1364-1995, or the vendor's User's Manual of your Verilog compiler.

We can use the script **mti_v.csh** in order to compile all Verilog examples with MTI's **vlog**. But **vlog** does not perform a check of the correct component port instantiations or shared objects. A second script, **mti_v.do**, can be used for this purpose. Start the **vsim** simulator (without loading a design) and execute the "**DO**" file with

```
do mti_v.do
```

to perform the check for all designs.

Utility Programs and Files

A couple of extra utility programs are also included on the CD-ROM and can be found in the directory **book/util**. These are the following programs:

File	Description
sine.exe	Program to generate the MIF files for the function generator in Chapter 1.
csd.exe	Program to find the "Canonical Signed Digit" representation of integers or fractions as used in Chapter 2.
dagen.exe	Program to generate VHDL code for the distributed arithmetic files used in Chapter 3.
cic.exe	Program to compute the parameters for a CIC filter as used in Chapter 5.

The programs are compiled using the author's MS Visual C++ "Standard Edition" software (available for $50 - 100 at all major retailers) for DOS window applications and should therefore run on Windows 95, 98, and NT machines. The DOS script **Testall.bat** produces the examples used in the book.

Also under **book/util** we find the following utility files:

File	Description
`quickver.pdf`	Quick reference card for Verilog HDL from QUALIS
`quickvhd.pdf`	Quick reference card for VHDL from QUALIS
`quicklog.pdf`	Quick reference card for IEEE 1164 logic package from QUALIS
`93vhdl.vhd`	The IEEE VHDL 1076-1983 keywords
`95key.v`	The IEEE Verilog 1364-1995 keywords
`95direct.v`	The IEEE Verilog 1364-1995 compiler directives
`95key.v`	The IEEE Verilog 1364-1995 system tasks and functions

In addition, the CD-ROM includes a collection of useful Internet links (see file **`dsp4fpga.html`** under **`book/util`**), such as device vendors, software tools, VHDL and Verilog resources, and links to on-line available HDL introductions, e.g., the "Verilog Handbook" by Dr. D. Hyde and "The VHDL Handbook Cookbook" by Dr. P. Ashenden. Altera's VHDL, Verilog, and AHDL manuals are available through Altera's literature service (set of all 3 at $89), and may be later become available through Altera's Web page.

Index

www.ingramcontent.com/pod-product-compliance
Ingram Content Group UK Ltd.
Pitfield, Milton Keynes, MK11 3LW, UK
UKHW021859190726
13853UKWH00003B/1336
* 9 7 8 3 6 6 2 0 4 6 1 4 2 *